Just
Playing

MIT Press Series
on Economic Learning
and Social Evolution

General Editor
Ken Binmore, Director of the Centre for
Economic Learning and Social Evolution
University College London

Game Theory
and the
Social Contract II

Just
Playing

Ken Binmore

The MIT Press

Cambridge, Massachusetts
London, England

This book was set in Computer Modern by Ken Binmore. Printed and bound in the United States of America.

Library of Congress Cataloging-in-Publication Data

Binmore, K. G., 1940–
    Just playing: game theory and the social contract II / Ken Binmore.
    v. cm.
    Includes bibliographical references and index.
    ISBN 0-262-02444-6
    1. Game theory. 2. Social contract. 3. Political science-
-Philosophy. I. Title.
HB144.B56 1994   93-29610
519.3—dc20   CIP

Just Playing

is dedicated
to my daughter

Joanna

and her baby Robert,
who knows
no other way
to play.

Blake inv & s

The Eagle.

William Blake's allegorical sketches are justly famous.
Here a critic has his claws in a newly hatched idea.

Apology

"I couldn't afford to learn it," said the Mock Turtle with a sigh. " I only took the regular course."

"What was that?" inquired Alice.

"Reeling and writing, of course, to begin with," the Mock Turtle replied; "and then the different branches of Arithmetic—Ambition, Distraction, Uglification and Derision."

Lewis Carroll's Alice in Wonderland

Playing Fair, the first volume of *Game Theory and the Social Contract*, began with with an old-fashioned apology in which I apologized for many things: for being an outsider in moral and political philosophy; for being simultaneously frivolous and long-winded; for being too mathematical and for not being mathematical enough; and for much else besides. I have nothing new to apologize for in this second volume—except perhaps for having taken so long to write it, and for having misjudged the length of the chapters to which I had committed myself. However, under the pretence of apologizing again for some of my old failings, I plan to take the opportunity of responding to some of the methodological criticism that the first volume provoked.

A legend circulates in the economics profession telling of a referee who rejected a paper on the grounds that its results were wrong. The author responded with a definitive refutation of the referee's counterexample, but the referee rejected the paper again—this time on the grounds that the results were correct, but trivial. Such a referee would have a field day with this book, since he could justly argue that it is simultaneously both wrong and trivial.

The book is wrong as a description of the way society works because the stripped-down models it employs are far too simple to come close to capturing the richness and diversity of the full range of human experience. But who ever got anywhere by trying to study the world-as-it-is in all its complexity? To make progress, one has no alternative but to suppress all but the factors that seem to matter most. The model that results will be wrong to the extent that it is not a full representation of reality. But it is not very interesting to make the obvious point that the world is more complex than the models we use to study it. The interesting question is whether simple models can be found that are enough like the world to make their analysis relevant to real problems. If not, then the prospect for saying anything worthwhile at all is very bleak—for nobody ever manages to solve difficult problems without learning to solve simpler problems first.

In spite of occasional episodes of mathematical huffing and puffing, the book is trivial because the conclusions follow more or less immediately from the assumptions. If my warnings about what was to come were insufficiently emphatic, those who began the first volume hoping to be entertained by deep theorems or cunning arguments therefore genuinely deserve an apology. However, at this stage of its development, it seems to me that political philosophy needs deep theorems and cunning arguments like it needs a hole in the head. What is missing is a comprehensive and systematic framework that is broadly consistent with modern discoveries about human evolutionary history, so that we can keep our feet firmly on the ground while debating ethical problems.

My aim in writing this book is therefore entirely mundane. I want to demonstrate the possibility of thinking about morality and social reform in a coherent and scientifically respectable manner. But this involves swallowing some bitter pills, since some much-loved philosophical orthodoxies are not consistent with others. But after the necessary winnowing has been done, the way ahead often becomes obvious. For example, I think that Rawls' *Theory of Justice* goes astray in adopting a Kantian foundation for the notion of the original position. I believe that his idea is intuitively attractive for the quite different reason that it serves as a stylized description of a fairness norm that has actually evolved in human societies for sound biological reasons. If this view is accepted, one can no longer follow Rawls in arguing that citizens should regard themselves as being bound by the terms of the deal reached in the original position. Nor is it possible to acquiesce in his rejection of standard Bayesian decision theory when describing how decisions are made behind the veil of ignorance. But once the idea of the original position has been reformulated to take these and other points on board, it becomes almost trivial to vindicate Rawls' basic intuitions about the workings of fairness norms in large societies.

I can understand the disappointment of those who see this book as trivial or wrong, but I find it more difficult to empathize with critics who think that my approach is not so much trivial or wrong as naive or nonsensical. The charge of logical incoherence has actually been put to me in public by more than one political scientist. These occasions can be embarrassing, since debating logic with an untrained opponent is like dueling with an enemy determined to impale himself on one's sword. Let me therefore insist that I see no grounds at all for being apologetic about my efforts to ensure that the ideas presented in this book are free of contradiction. Professional philosophers do not need to be reminded how easy it is to gain popular acceptance by telling people what they want to hear and papering over the inconsistencies that this entails by various obscurantist devices. But I am not interested in playing this kind of game. I must therefore sometimes defend propositions that everybody, myself included, would prefer to be false. But the fact that one does not like a conclusion does not make the argument that leads to it a *reductio ad absurdum.*

It is easy to shrug off the claim that my conclusions are not so much wrong as nonsensical, but I find it harder to respond adequately when my approach is dismissed as naive. My unsophisticated background doubtless merits an apology insofar as it prevents me from packaging my ideas in a more palatable form, but I do not see that the kind of naïveté involved in seeking to rethink a theory from ground zero is anything but admirable. As for my lack of sophistication, it is true that I am a Grandma Moses with regard to many of the conventional concerns of moral and political philosophy. Nor does an innocent like myself always know which authorities can safely be flouted and which must continue to be mentioned only in hushed tones of respect. I therefore rashly treat such dignatories as David Hume or Immanuel Kant as though they occupied the office across the hall, and offer my own views on free will and the like as though I were their equal. In this, I am encouraged by Popper's [408] robust treatment of Plato, Hegel and Marx. I can still remember the relief on reading his *Open Society and its Enemies* to find someone else who had noticed that Plato was a fascist.[1]

If I had been brought up in a different tradition, I would perhaps have learned to be more respectful of authority—but I hope not. When I read the works of celebrated philosophers, I can sometimes hardly believe that such shoddy arguments have survived the test of time. To an outsider like myself, it often seems that they continue as part of the canon only because insiders have evolved an Orwellian Newspeak that makes it impossible to express the obvious objections without being categorized as naive or stupid.[2] In particular, there is a strong tendency to hang the *naive* label

---

[1]Popper's opening chapter begins by quoting Plato on the subject of leadership. According to Plato's *Republic,* everybody needs a leader. Nobody should even "get up, or move, or wash, or take his meals" without permission. But our children continue to be taught that Plato was the first and greatest of the prophets of liberalism.

[2]We are all familiar with the standard conditioning technique that puts off children who question a received orthodoxy by claiming that all will be made clear later. But when the same question is asked in a more advanced course, it is dismissed as childish.

on simple illustrative models when these fail to confirm conventional wisdom. But what is the point of adopting such a sophisticated attitude when complicated models are too intractable for us to analyze? In such circumstances, one has no choice but to work with the simple model that seems to distort the real world least. I have doubtless made errors of judgment in the simplifications I have introduced. Perhaps I have even simplified too much in an attempt to ensure that only trivial arguments are necessary to derive the conclusions. But none of these considerations merit dismissing an approach out of hand as naive. After all, Albert Einstein was no less naive when he constructed the theory of special relativity.

Finally, I have to admit to being somewhat chastened at recently finding a piece of my prose featured in *Pseuds Corner*.[3] But I hope that my more sophisticated readers will appreciate that the purpler passages with which I try to brighten up the arguments are sometimes actually meant to raise a wry smile. I know the risks that accompany such departures from the deadpan style of respectable intellectual inquiry. Why else do Mill and Kant have a reputation for intellectual rigor denied to Bentham and Hume? But my experience is that minds which take their scholarship very seriously are unreceptive to challenges to traditional wisdom in whatever form they may be expressed.

I see that my attempt at writing an apology has turned out to be more polemical than apologetic. However, as in the first volume, I am only too aware that my critics are right to accuse me of writing on a subject that lies beyond my competence. Joseph Harrington, Larry Samuelson, Joe Swierzbinski, and John Weymark have set me straight on some issues, and I surely must also have gone astray on matters that lie outside their ken. Anyone willing to help constructively by pointing out further errors of fact and judgment can be sure of a grateful reception.

---

[3] A British satirical magazine of some notoriety called *Private Eye* devotes this column to poking fun at flakes, phonies and poseurs.

Contents

Series Foreword

The MIT Press series on Economic Learning and Social Evolution reflects the widespread renewal of interest in the dynamics of human interaction. This issue has provided a broad community of economists, psychologists, biologists, anthropologists, and others with a sense of common purpose so strong that traditional interdisciplinary boundaries have begun to melt away.

Some of the books in the series will be works of theory. Others will be philosophical or conceptual in scope. Some will have an experimental or empirical focus. Some will be collections of papers with a common theme and a linking commentary. Others will have an expository character. Yet others will be monographs in which new ideas meet the light of day for the first time. But all will have two unifying themes. The first will be the rejection of the outmoded notion that what happens away from equilibrium can safely be ignored. The second will be a recognition that it is no longer enough to speak in vague terms of bounded rationality and spontaneous order. As in all movements, the time comes to put the beef on the table—and the time for us is now.

Authors who share this ethos and would like to be part of the series are cordially invited to submit outlines of their proposed books for consideration. Within our frame of reference, we hope that a thousand flowers will bloom.

Ken Binmore,
Director of the Centre for
Economic Learning and Social Evolution
University College London
Gower Street
London WC1E 6BT, UK

The descent of Man into the Vale of Death.

'Tis here all meet'

Before committing yourself to a long and weary journey, it is best to be sure that you want to go where you are being led.

Reading Guide

Another damn thick square book. Always scribble,
scribble, scribble! Eh, Mr Gibbon?

William, Duke of Gloucester

How much to read? As the Reading Guide to Volume I explains, the strait and narrow path followed by those who read every word of this book is long and sometimes tortuous. Not only are there long detours through material that many readers will think superfluous, but I have made a point of repeating the really important ideas again and again, because I am weary of being misunderstood, and see no way to insist that I literally mean what I say other than by straightforward repetition. Often the recycling of an idea may be unobtrusive because of the change in context, but I cannot guarantee that this will always be the case. In short, although I hope that everything I have written will be of interest to someone, there will be much that any but the most serious readers will wish to skip.

Some heartless souls will skip pretty much everything unless it impinges directly on their own interests. Readers of this persuasion will find the core of my theory in Sections 4.6–4.8. Economic theorists probably won't need to follow up the references to earlier material to make sense of the argument, but some preliminary reading of Sections 2.5–2.7 will certainly be helpful. Evolutionary biologists seem to spend a good deal of their time reinventing wheels that game theorists have been rolling around for decades. I would particularly like to draw their attention to the discussions of equilibrium selection, kinship, and repeated games in Sections 2.4–2.5 and 3.3–3.4. Evolutionary psychologists will similarly find that Sections 3.4–3.5 and Sections 4.4–4.5 are relevant to their discipline. Any feedback on the anthopological speculations of Section 4.5 would be much appreciated. Political philosophers might care to test the water by reading Section 4.10, in which I reveal both my radical aspirations and my awareness of the inadequacy of the available theory. Moral philosophers seeking ammunition to use in pouring scorn on naturalistic theories of ethics are encouraged to begin with Sections 2.4–2.5, 3.2, and 3.8.

To help more dedicated readers find their way about the book, I have used the three symbols shown below to help distinguish one passage from another. The Reading Guide to Volume I explains their meaning at length. By way of summary, let me observe that first-level material provides the core of the book. Technical and expository material is offered at the second-level. Third-level material mostly addresses philosophical issues that are peripheral to the main thrust of the argument.

(a) First level.   (b) Second level.   (c) Third level.

How far have we got?   Professional game theorists were advised to skip the whole of Volume I. Those who took me at my word will find that the Introduction to Volume II offers a quick survey of the general approach adopted in the first volume, but some guidance on the technical issues covered there, and those still awaiting treatment in Volume II may also be helpful.

After the first chapter, which provides a brief review of the theory defended in the book, most of Volume I is devoted to an attempt to clear the ground of the many misconceptions about game theory that make it so hard to use game theory successfully in moral philosophy. In particular, Chapter 3 patiently reviews numerous fallacies associated with the Prisoners' Dilemma in the hope of persuading the reader that the errors they incorporate are not only wrong but would be impossible to make if proper attention were paid to getting the underlying model right in the first place. Chapter 4 similarly reviews orthodox Bayesian decision theory. This theory is widely misunderstood, even by scholars who routinely make use of it in their work. However, my approach to the interpersonal comparison of utility becomes nonsensical if one insists, for example, that a rational person chooses a over b because he gets more utility from choosing a than b. On the contrary, we assign a larger utility to a than to b because a is chosen when b is available.

Although most of Volume I is expository, it does cover the roots of the theory to be presented in Volume II. These roots are to be found in the social contract theories of Harsanyi [233] and Rawls [417].[1] Both theories employ what Rawls calls the device of the original position to determine what social contract should be deemed to be fair. This device calls for all citizens to imagine themselves bargaining about which social contract to operate behind a veil of ignorance that conceals their current and future roles in society. I call both theories metaphysical, since both appeal to Kant when asked to explain why the terms of the deal reached under such hypothetical circumstances should be thought to impose any obligation on the citizens of a society.

Figure 1.1 of Volume I is intended to clarify the discussion of the metaphysical theories of Harsanyi and Rawls, which commences in Chapter 1 of Volume I and concludes after many interruptions in Chapter 4 of the same volume. A more elaborate version of the figure is reproduced below.

Route 1 represents the defense of egalitarianism offered by the younger[2] Rawls [417], showing how he traces his philosophical ancestry through Locke, Rousseau, and Kant—although it is the influence of the last that is most obviously in evidence.

Somewhat idiosyncratically, I divide Harsanyi's [233] defense of utilitarianism into teleological and nonteleological variants. A teleological theory takes as its foundation stone the existence of an *a priori* Good toward which society ought to strive. The problem for a metaphysician propounding such a theory is then to characterize the nature of the Good. A nonteleological theory takes the *procedure* by means of which a society makes its decisions as fundamental. A common good function that arises in such a theory has no causal role. One simply asserts that to use the procedure under study is to behave *as though* maximizing a particular common good function. In his own metaphysical version of such a nonteleological theory, Rawls describes himself as deducing the Good from the Right—with the device of the original position serving to operationalize Kant's categorical imperative.

Route 2 shows the teleological variant of Harsanyi's theory, which lies squarely in the tradition of Bentham and Mill. Route 3 shows the nonteleological variant, which I have classified as a piece of Kantian constructivism by assigning it the same philosophical ancestry as Rawls' Route 1.

What separates Route 1 and Route 3 is the different manner in which Rawls and Harsanyi model rational choice under uncertainty. Harsanyi [233] is entirely orthodox on this subject whereas Rawls [417] is wildly

---

[1]Although it should be noted that neither cares much for the freedom I allow myself in reinterpreting their ideas; Harsanyi is not even willing to be called a contractarian.

[2]Rawls [417, 419] has passed through a number of metamorphoses over the years. His recent *Political Liberalism* essentially abandons the project of producing a "moral geometry" to which he devoted *A Theory of Justice*. When Rawls is mentioned without further qualification, I always have in mind the views of the younger Rawls.

heterodox in his insistence on using the maximin criterion (Section 4.6 of Volume I). Unless one is to follow Rawls in his iconoclastic rejection of Bayesian decision theory, one must therefore regard Route 3 as a corrected version of Route 1. That is to say, the Kantian constructivism adopted by Rawls should have led him to utilitarianism.

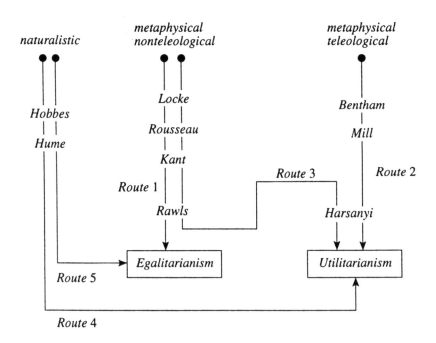

Rival social contract theories.

Routes 1, 2, and 3 are labeled as metaphysical because they assume that citizens are mysteriously bound to the terms of their social contract, without any mechanism being proposed to enforce the postulated commitment. Sometimes attempts are made to justify the necessary commitment assumptions by making appeals to Kant, but it is of the utmost importance to appreciate that one cannot respect Kant's views on the nature of rationality and simultaneously believe that game theory is worthwhile. As far as I can see, Kant's rationality arguments escape derision only because they are expressed in such obscure language that nobody can follow them.

My own approach is not at all metaphysical. I follow David Hume in taking an entirely naturalistic view of morality, according to which a moral code is something that a society grows as it accumulates experience. Examining the consequences of adopting such a Humean stance is the major task of Volume II. Fortunately, throwing over Kant in favor of Hume does not mean that all the lessons learned in studying Routes 1, 2 and 3 need be lost. In fact, Harsanyi's nonteleological Route 3 provides a foundation stone for the two new routes to be studied in Volume II.

Route 4 is studied in Chapter 2 of the current volume. It differs from Route 3 in abandoning any metaphysical pretensions at providing a reason why the terms of a hypothetical bargain reached in the original position should bind the citizens of a society. Instead, a commitment mechanism is postulated in the form of a benign and incorruptable government, whose officers are assumed to lie outside the society they administer. Attention can then center on how social evolution determines the manner in which societies make interpersonal comparisons of utility.

Route 5 represents my own considered view. It is studied in Chapter 4 of the current volume. From Chapter 3 onwards, commitment assumptions are abandoned altogether on the grounds that the terms of a stable social contract must include the officers of society along with its rank and file. But with no outside enforcement agency, the only viable social contracts must be self-policing. The implications of applying this principle systematically are far-reaching. Instead of being led as in Route 4 to a utilitarian conclusion, we are led to an egalitarian outcome of the type that Rawls was originally seeking to defend in his study of Route 1.

Acknowledgment

The author gratefully acknowledges the financial support of the Leverhulme Foundation and the Economic and Social Research Council.

Just
Playing

The scene being set in this most famous of William Blake's drawings is on a somewhat larger scale than the scene setting of the coming introductory chapter.

Introduction:

Setting the Scene

> Early mankind soon reached the grand generalization that everything
> has its price, everything can be paid for. Here we have the oldest and
> most naive canon of justice, of all *fair play*. ... Justice, at this level,
> is good will operating among men of roughly equal power, their readi-
> ness to come to terms with one another, to strike a compromise—or,
> in the case of others, to *force* them to accept such a compromise.
>
> Friedrich Nietzsche [386, p.202]

0.1 Whither Away?

This is the second of two volumes in which I try to put the case for approaching social contract questions from a game-theoretic perspective. The
current introductory chapter sets the scene for new readers.

Readers who are familiar with game theory are advised to skip the material that follows Section 0.2. Those who are less confident of their game-
theoretic expertise will find it useful to devote some time to the overview
of the subject that begins with Section 0.4. Only professional philosophers
will gain much from Section 0.3, where I try to relate my work to more
orthodox approaches.

0.2 The Art of Compromise

Diplomacy is supposedly the art of compromise. If so, then von Clausewitz
[125] teaches us that war is simply an attempt to reach a compromise by
violent means. Far from being shocked by such realpolitik, I do not think
von Clausewitz goes far enough. *All* stable human relationships, without
exception, represent a compromise of sorts—including the *modus vivendi*

1

established between the victor and the vanquished after a war. The balance of power may be very unequal in such a relationship. But even a slave is not totally powerless. Only a stupid master does not eventually learn to be sparing with the whip in seeking to persuade a slave to labor efficiently on his behalf. Those who fail to curb their brutal impulses find themselves sitting on a powder keg.

It may seem odd that a liberal like myself should begin a book on justice and fair play by expressing such Machiavellian sentiments. But those of us who aspire to rethink the thoughts of the founding fathers of the American Republic in the light of the lessons of history cannot afford to be squeamish. Whether we refer to ourselves as classical liberals or modern whigs,[1] our aim is to revive the coolly rational approach to reform that the world owes to the glorious days of the Scottish Enlightenment. We betray this tradition immediately if we allow sentiment to color our perception of the nature of human nature. Everybody wishes that human beings were more noble than evolution has made us, but it is irrational to allow our preferences to influence our beliefs about matters of objective fact.

Like Madison [215, p.268], modern whigs believe that justice is the end

---

[1]The Whigs were originally a British political party that arose in opposition to the Tories of the seventeenth century. The modern Conservative party is a direct descendant of the Tories. The Whigs were eventually outflanked on the left by the modern Labour party and squeezed into insignificance. Their remnants survived as the Liberal party, which now continues as the Liberal Democratic party after a merger with the Social Democrats who splintered from the Labour party at a time when the latter lurched dangerously to the far left. However, in recent years, Labour has perhaps become even more whiggish than the Liberal Democrats.

American history also boasted a Whig party, broadly similar in character to its British counterpart. It was vocal in its opposition to Andrew Jackson's authoritarian innovations in the use of the presidential veto. Before joining the newly emergent Republican party, Abraham Lincoln was a Whig, but modern Republicans seem largely to have forgotten their whiggish roots.

What did the Tories and the Whigs represent? Etymology does not help, since a Tory was originally an Irish bogtrotter and a Whig a Scottish covenanter. Nor does it assist to observe that Edmund Burke was the Whig credited with being the founder of modern conservatism, or that David Hume, whose ideas are the inspiration for my own brand of whiggery, was held to be a Tory by his contemporaries, since he famously confessed himself able to shed a tear for Charles I. It is more informative to observe that the Whigs are traditionally associated with the Glorious Revolution of 1688, in which the Catholic and authoritarian James II was expelled in favor of the Protestant and constitutionally minded William III. However, even here the water is muddy, since the coup could not have been bloodless without the cooperation of most Tories. But politics makes the facts of history less important than the mythology they engender. According to popular wisdom, we whigs are for economic and political freedom, thrift, self-help, and equality of opportunity. Our enemies are either the advocates of big-spending government intent on creating a lickspittle citizenry, or else the corrupt backers of arbitrary government and ancient privilege.

of government and civil society. As he says, justice has ever been and ever will be pursued until it is obtained, or liberty lost in the pursuit. But we will not achieve a just society if we are unrealistic about what people are like. As Madison [215, p.259] puts it: "A nation of philosophers is as little to be expected as the race of philosopher-kings wished for by Plato."

0.2.1 Nonsense upon Stilts

I don't like it any more than anybody else, but it is idle to insist that Nature has made us members of one another. The brotherhood of man is strictly a human invention. Insofar as Nature endowed us with the milk of human kindness, she intended that we suckle only our near and dear. Nor is the right to hold property somehow written in our genes. Nor yet the right to publish our thoughts without fear of reprisal. Nature knows nothing of so-called natural rights to life, liberty and the pursuit of happiness.[2] Jeremy Bentham [53], the founder of my college, put it very bluntly: "Natural Rights is simple nonsense: natural and imprescriptible rights, rhetorical nonsense—nonsense upon stilts."

As the founding fathers did not need to be told, power is held in check in human societies only by some countervailing power. We deceive ourselves if we think that it is our dedication to abstract moral ideals that prevents civilization from lapsing into chaos. Civilization survives because those

---

[2]Which is not to say that we are not talking about something real when, for example, we say that Adam has a right to exclude Eve from his property. However, it is a mistake to take such statements literally—just as it would be a mistake to take literally the proposition that Nature abhors a vacuum. Maynard Smith [347], for example, describes games played by animals over the possession of territory. After a confrontation, it is usually the newcomer who backs down when neither player has any obvious fighting advantage. If the animals were human, we might explain what happens by saying that the first animal on the scene has a right to possess the territory—perhaps through a Lockean mixing of his labor with the land—and that the intruder backs down because he recognizes this right. But the truth is more mundane. Each animal is programmed to maximize its biological fitness. When both optimize simultaneously, the result is necessarily an equilibrium of the game they are playing. But the game we are concerned with here has several equilibria. One equilibrium calls for the newcomer to back down, as we usually observe. However, there is a corresponding equilibrium in which it is the animal who was first on the scene who backs down. Through the workings of evolutionary forces, the animals have become programmed to coordinate on the first equilibrium. However, if they were capable of examining their own behavior, they would be unlikely to hit upon this explanation at their first attempt. They might well come up with an explanation in terms of squatters' rights, just as our ancestors did. If evolution had succeeded in coordinating behavior on the second equilibrium (Burgess [108]), they would, of course, invent some other story—perhaps about the rights of newcomers to displace any squatters they may find on a piece of land. The point is that such explanations may succeed in describing what actually happens with great success, but it does not follow that they come anywhere near describing *why* it happens.

with some power to influence matters are willing to compromise with their competitors. To behave morally is not to transcend the need for compromise. On the contrary, morality evolved in our species because a successful human society needs to *institutionalize* the compromises that allow it to operate.

Neither the left nor the right is willing to bite this particular bullet. Socialists think it moral to insist on reforms that will necessarily fail in the long run because they neglect the realities of power. Conservatives resist reform because they imagine that morality consists in honoring rules that evolved in the past to sustain power structures that are no longer relevant. Both sides are only too willing to discredit their opponents by calling them immoral. But whigs need not be disturbed by such abuse. Neither the left nor the right have much in their history of which they can be proud. Even by their own faulty standards, everything of lasting worth in our societies has been brought about by reformers following the whiggish principles from which Adam Smith [500, 501] drew the inspiration for his two great books, *The Wealth of Nations* and *The Theory of Moral Sentiments*.

0.2.2 Social Contracts

In game-theoretic terms, the mistake made by both those on the left and those on the right is to confuse the morality game played by a particular society with the Game of Life. The Game of Life may change over time, but it is beyond our power to alter its rules at will. Its rules are determined by the laws of physics and biology; by geographical and demographic facts; by technological and physiological constraints; and by whatever else sets unbreakable bounds on our freedom of action.

On the other hand, the rules of a morality game are very definitely human constructs, or else the product of social evolution. We have changed them in the past and we will change them in the future. Moreover, whigs should delight in their malleability. If it were not possible to alter the rules of a morality game, all our whiggish reforming zeal would necessarily come to naught.

But we cannot plan to reform the morality game our society currently operates unless we first understand what function it currently fulfills in keeping our society going. Here, I join Hume [267], and modern scholars like Axelrod [31], Bicchieri [60], Gibbard [205], Hammond [218], Kliemt [307], Schotter [467], Skyrms [499] and Sugden [511], who see Von Neumann and Morgenstern's [538] theory of games as the appropriate vehicle within which to express Hume's ideas.

The folk theorem of repeated game theory tells us that one must expect the Game of Life to have *many* equilibria. Such equilibria arise when each citizen acts in his own enlightened self-interest on the assumption that others will do the same. In this context, each equilibrium represents a compromise of sorts. But to say that self-interest will lead people to a compromise is far from being the end of the story. Some compromises are better for everybody than others.[3] In such circumstances, the compromise that everybody prefers is said to be a Pareto-improvement on its rival. For example, it is an equilibrium in the Driving Game if we all independently toss a coin each morning to decide whether to drive on the left or the right. But it is better for everybody if we all drive on the right. However, to sustain such a Pareto-improving equilibrium, we all need to be party to a suitable *coordinating convention*.

I identify the set of all the commonly understood coordinating conventions operated by a society with its *social contract*. Nobody is bound by the terms of such a social contract. It serves only to coordinate behavior on an equilibrium in the Game of Life. The survival of the social contract does not therefore depend on its being backed up by some external enforcement mechanism. In a well-ordered society, each citizen honors the social contract because it is in his own self-interest to do so, provided that enough of his fellow citizens do the same.[4]

To carry out their part in sustaining a social contract, the citizens of a society need rules to regulate their conduct. Some of these rules are

---

[3]Hobbes [251], for example, argued that the equilibrium represented by an absolutist state is better for everybody than the equilibrium that pertains in the state of nature. Much as I admire Hobbes, I think that he was mistaken, both about the state of nature and about the range of alternative equilibria.

[4]In taking this line, I am agreeing with the answer that Hume [267, p.280] had in mind when he asked: "What theory of morals can ever serve any useful purpose, unless it can show that all the duties it recommends are also the true interest of each individual?" This Humean position is anathema for more traditional moralists. Gauthier [196], for example, begins his book with an explicit rejection of Hume on this subject. Others even go so far as to claim that a rule can only properly be described as moral if it calls for behavior that actually runs counter to one's interests. Such extreme views presumably assume a much narrower conception of the nature of self-interest than I intend. But, even if traditional moralists were to broaden their horizons on this score, I would still be at odds with them. Insofar as their intuitions are derived from observing human behavior, I suspect that they make the mistake of confusing what people say with what they do. The rules that govern our actual moral behavior are much more complex and subtle than our conscious minds readily appreciate. Indeed, I think that traditional folk wisdom on moral issues overlooks much of our moral behavior altogether, because the moral rules in actual use succeed in coordinating our behavior so smoothly that we fail even to notice that there was a risk of a coordination failure at all. Traditionalists tend to notice that a morality game is being played only when its rules are misapplied and used in situations for which it is not adapted (Section 2.3.3 of Volume I). In seeking to extrapolate its rules from such situations, it is then inevitable that they come up with the same type of distortion that would result if one sought to deduce the rules of formal etiquette from watching the Simpson's cartoon series.

enshrined in law. Some are seen as being part of the society's moral code. Others are classified as being matters of fashion or convention. Yet others are so taken for granted that they are not noticed at all. I take the totality of all such rules as defining the morality game for a particular society. It is vital for the purpose of this book to recognize that a society's morality game is a human artifact. As in Chess, we obey its rules, insofar as we do, because we *choose* to obey them. Just as it is actually within our power to move a bishop like a knight when playing Chess, so we can steal, defraud, break promises, tell lies, jump lines, talk too much, or eat peas with our knives when playing a morality game. But rational folk choose not to cheat for much the same reason that they obey traffic signals. Like traffic signals, a morality game is a coordinating device. If we play *as though* bound by its rules, then we are led to coordinate on an equilibrium—not only in our society's morality game, but also in the *Game of Life*.

0.2.3 Reform

By tradition, whigs are practical reformers. We are therefore sometimes confused with those intelligent conservatives who grudgingly accept reform as the price of holding on to power. It is true that we share the conservative distaste for classical utopias like New Atlantis or Plato's Republic. Perhaps the City of the Sun or the Christianopolitan Republic dismissed by Burton [110, p.85] as witty fictions were more jolly, but I doubt it. Utopians say that we deny the practicality of such castles in the air because we don't *want* them to be practical—thereby revealing their characteristic inability to see that it is irrational to allow one's beliefs about what is feasible to be influenced by what one would like to be optimal.

It is easy to laugh off accusations of misanthropy from people whose idea of philanthropy is to invent definitions of collective rationality which supposedly show that individuals do best to cooperate in the one-shot Prisoners' Dilemma (Chapter 3 of Volume I). But the misery created when utopian schemes based on such wishful thinking gain widespread support is not at all funny. There is no point in proposing social contracts that are not equilibria in the Game of Life. It is *inevitable* that any meddling with our current social contract that neglects this truth will only lead back, sooner or later, to some new balance of power—to some new equilibrium in the Game of Life. And history is not at all encouraging about the period of disequilibrium that ensues after a society is jolted out of one equilibrium and has yet to find its way to another.

But it does not follow that we should therefore timidly refrain from all attempts to improve our society. To be sure, caution must be exercised when reforms are contemplated. It is not enough to advocate moving to a

Pareto-improving equilibrium. Reformers must also explain how the reform
is to be implemented. If the change is not carefully planned, the system
may slip out of equilibrium during the transition. One way of avoiding
this difficulty is to restrict attention to new social contracts that can be
achieved by moving slowly but surely through a sequence of intermediate
equilibria. Others may propose more ambitious routes to a better society,
but we whigs need a lot of convincing before agreeing to take risks with our
bourgeois comforts.

Whigs favor reform *by mutual consent* of those with the power to prevent
the implementation of the proposed reform.[5] Such an approach calls for
the art of compromise to be practiced on a grand scale. However, the type
of diplomacy required differs from the explicit bargaining that takes place
when, for example, an author bargains with his publisher about the size of
his advance. Everybody cannot have his say when a whole society is to be
shifted from one equilibrium to another.

Fortunately, stereotyped methods of reaching a compromise without
the need for explicit bargaining, have evolved along with our species. The
rules of our current morality game provide for settling a whole range of
issues efficiently without any need for anyone to say anything. For example,
when a dish in short supply is shared at a polite dinner party, there is
seldom any verbal dispute. If things go well, and they usually do, the
dish gets divided without any discussion or intervention by the host. When
questioned, everybody will agree that each person should take his fair share.

But how do we know what is fair? This is not a simple question. What is
judged to be fair according to our current morality game commonly depends
on a complex combination of contingent circumstances—like who is fat and
who dislikes cheese. Moreover, if we observe what actually happens, rather
than what people say should happen, we will find that it also depends on
how each person at the table fits into the social pecking order. Woe betide
the poor relative sitting at the table on sufferance in the last century who
helped himself to an overgenerous portion of his favorite dish!

---

[5]Do we therefore approve of slavery when this cannot be abolished without the agreement of the slaveholders? Such questions are idle. It does not matter whether one
approves of slavery or not if its abolition really does require the consent of people whose
cooperation cannot be obtained. One might as well ask whether we approve of the in-
verse square law of gravity. To make the question interesting, it needs to be modified
so that abolition can be *forced* on the slaveholders. They will then no longer have the
power to prevent the reform, but they may still have the power to make it costly. The
proposed reform must then be redefined so that the costs of abolition are included in
its specification. Abolishing slavery *efficiently* will then require reaching a compromise
with the slaveholders in which the slaveholders get compensated for losing their slaves.
Notice that one does not need to *like* such a compromise in order to support it. One
needs only to observe that, as with democracy, all the alternatives are worse.

Numerous scholars have tried to make sense of the calculations that people must implicitly perform when they coordinate on an outcome that they afterward describe as fair. It surely can be no accident that the consensus is firmly in favor of some type of do-as-you-would-be-done-by principle.[6] Moralists down the ages have offered numerous arguments that seek to explain why it is morally imperative that each person should follow such a golden rule. But none of these traditional arguments are founded on anything solid. I think we get suckered into taking them seriously because we are too ready to confuse a fairly accurate description of *what* we do in certain situations with an explanation of *why* we do it.

Notice that I take the view that variants of the do-as-you-would-be-done-by principle are *already* firmly entrenched among the instincts and customs that regulate our lives. But I do not accept that our habits and customs survive because we consciously adopt them. On the contrary, I think that most of our habituated behavior is acquired using processes that operate below the level to which our conscious minds have easy access. Like monkeys, we are programmed to imitate the behavior of our more successful neighbors. If those in thrall to a particular habit or custom are perceived as being winners, then their habituated behavior will be copied, without any need for anyone to understand *why* the habituated behavior works well in the currrent social environment. However, such monkey-see-monkey-do explanations are not popular. Just as confidence tricksters flatter naive tourists into buying the Brooklyn Bridge, so we prefer to listen to the tall tales told by those moralists who see *homo sapiens* as a race of potential philosopher-kings seeking to live the classical self-examined life.

0.2.4 The Original Position

One particularly interesting variant of a do-as-you-would-be-done-by principle is described by Rawls [417] in his *Theory of Justice*. Harsanyi [233] and others have independently proposed conceptually similar schemes. Rawls' *original position* is a hypothetical standpoint to be used in making judgments about how a just society should be organized. Each citizen is asked to envisage the social contract to which he would agree *if* his current role in society were concealed from him behind a *veil of ignorance*.

In considering the social contract on which to agree under such hypothetical circumstances, each person will pay close attention to the plight of those who end up at the bottom of the social heap. Devil take the hindmost is not such an attractive principle when you yourself may be at the back of the pack. As Rawls argues, the social contract negotiated in the original

---

[6]And as ye would that men should do to you, do ye also to them likewise—Luke 6:31.

position is therefore likely to generate outcomes that the underprivileged might reasonably regard as fair.

I agree that the device of the original position does generate compromises that would commonly be regarded as fair, but I am not very interested in the Kantian arguments that Harsanyi and Rawls offer when urging its use. The interest of the device for me lies simply in the fact that it represents an attempt to operationalize a stylized do-as-you-would-be-done-by principle that faces up to objections like: don't do unto others as you would have them do unto you—they may have different tastes from yours. It seems to me that evolution has indeed provided us with the capacity to take such complexities into account, and that the device of the original position comes close to describing how we do it.

In approaching the issue from such a perspective, fairness is interpreted entirely in naturalistic terms. The original position is seen as an idealized representative of a class of equilibrium selection criteria washed up on the beach along with the human race by the forces of biological and social evolution. But a whig is not willing to let matters rest where evolution has left them. Like a beachcomber who finds a treasure among the flotsam at the water's edge, he asks, "Why don't we take the device of the original position and apply it, not just to some aspects of our lives, but to the Game of Life taken as a whole?" The defense for such a proposal is entirely pragmatic. Here is a tool supplied by Nature. Let us use it to improve our lives, just as we use whatever tools we find in our toolbox when making repairs around the house.

If reforms are based on the device of the original position, the task of persuading people to move from one equilibrium to a Pareto-improving equilibrium by this familiar method will be much easier than persuading them to move according to principles that are not already rooted in their day-to-day experience of life. But the project will not work at all if we tell ourselves fairy stories about how the Game of Life is currently played.

People pay lip service to the grand principles and utopian aims of traditional moralists, but, as my mother was fond of saying, they know perfectly well that fine words butter no parsnips. This is why I insist on taking such a hard-nosed attitude to the nature of morality games. The plan is to widen the domain in which we make use of the device of the original position to coordinate our behavior. But we shall get nowhere in this enterprise if we refuse to be realistic about how this social tool functions in our current morality game.

The Game of Life. I refuse to wear rose-colored glasses when studying the devices for selecting equilibria that folk wisdom explains as moral imperatives, but my commitment to realism does not extend to modeling the Game of Life in all its vast complexity. On the contrary, I think that one minimizes the risk of allowing sentiment to color one's perception of how morality really works by adopting the simplest possible model of the Game of Life. The model must have some structure, because the game-theoretic explanation of how cooperation is sustained as an equilibrium in the Game of Life depends on fleshing out Hume's insight that the underlying mechanism is *reciprocity* (Section 1.2.3 of Volume I). It is therefore important not to follow the political philosophers who have sought to model the Game of Life as a one-shot Prisoners' Dilemma.[7] It is impossible for reciprocity to emerge in a one-shot game, because Eve cannot threaten to stop scratching Adam's back if he stops scratching hers without *time* entering the picture. Nor is it enough to model the Game of Life as a finitely repeated game. To succeed in supporting cooperation as an equilibrium phenomenon, our model of the Game of Life must be at least as complicated as an indefinitely repeated game (Section 2.2.6 of Volume I).

Not only do I simplify by modeling the Game of Life as an indefinitely repeated game, I also usually keep the number of players down to two. The setting for the discussion is then the Garden of Eden, where Adam and Eve are considering the terms of a fair marriage contract. Sometimes, one can see Adam and Eve as representatives of two homogeneous classes. For example, when questions of sexual inequality are at issue, Adam might be regarded as a representative man and Eve as a representative woman. In a Marxist discussion, Adam and Eve might perhaps personify Labor and Capital. However, I am unable to offer more than pious hopes for the future when the situation is genuinely multilateral. Even when treating Adam and Eve as delegates of mutually exclusive coalitions, it is necessary to be very cautious, because coalitions cannot be expected to have the same single-minded and enduring preferences as an individual player.

The Game of Morals. Recall that the original position is to be modeled as a natural device that evolved to help us coordinate on equilibria in some of the games we play. The blatant injustices that surround us amply testify to the fact that we do not currently employ the device to the Game of Life as a whole. When a game G arises in which we do use the device, players behave *as though* they were constrained, not only by the rules of G, but also by a set of socially determined rules whose addition to the naturally

---

[7]It cannot be rational for Adam to choose the dovelike strategy in the Prisoners' Dilemma because the hawkish strategy gives him a higher payoff whatever strategy Eve may choose (Section 2.2.4 of Volume I). Chapter 3 of Volume I is devoted to identifying the errors in some of the many fallacies that have been invented to justify claims to the contrary.

determined rules of G creates a morality game M.

In my model, the rules of the artificial game M require that Adam and Eve pretend that each round of G is preceded by an episode during which each player can call for the veil of ignorance to be lowered. They then bargain in the original position about which equilibrium is to be operated when the veil is lifted. The rules of M then require that the actions specified by this equilibrium be implemented in future play—unless Adam or Eve demand that the veil of ignorance be lowered again for the agreement to be renegotiated. An equilibrium of the natural game G is said to be *fair* if its play would never give a player reason to appeal to the device of the original position under the rules of the morality game M.[8]

This model is an attempt to provide a stylized description of what actually happens in the world when we settle coordination problems using a fairness norm. Doubts about the validity of the predictions generated by such a descriptive interpretation of the model can be resolved in principle by running laboratory tests. My basic prediction—that Aristotle [18] was right to argue that people evaluate fairness in proportional terms—was confirmed by social psychologists long before I set pen to paper (Section 4.4). I was heartened when I later came across this work, but nobody else is likely to be convinced by empirical work that preceded the writing of the theory. It is therefore important for the theory that sharper tests based on the comparative statics of Section 4.8 be attempted. Even if the theory passes these tests, its predictions will doubtless prove wrong at some level since it is most unlikely that I have succeeded in making the right modeling judgments at every stage. However, perhaps others will think that I have got enough right to make it worthwhile to produce more sophisticated versions of the theory that are consistent with the data.

Such a concern with empirical work is alien to the mainstream of moral philosophy, but we whigs are practical people who think that the building blocks of any proposed reform should be tested to destruction before any attempt is made to implement the reform. Once we properly understand how fairness norms actually work within the domain in which we routinely use them, we can cast off our laboratory coats and go forth into the world to preach the advantages of using fairness as the major organizing principle for a social contract.

---

[8]Chapter 2 considers the utilitarian implications of following Harsanyi [233] or Rawls [417] in assuming that the rules of M can be made as unbreakable as the rules of G. An agreement reached behind the veil of ignorance at the very outset of the game can then be assumed to bind the players at all future times. No provision need then be made for renegotiation. Nor need agreements be confined to equilibria of G. Everything is thereby greatly simplified, but at the expense of introducing the indefensible assumption that the rules of morality game cannot be broken.

If my theory were to pass all the necessary empirical tests unscathed, the attainable utopia to be preached would simply require that the game G be replaced by the *entire* Game of Life. When discussing this whiggish utopia, I cease to call M a morality game and refer to it instead as the *Game of Morals*. However, one can always deflate the grandiose aspirations I express in this prescriptive mode by reinterpreting the Game of Morals as a morality game associated with some small subsidiary game G of the Game of Life. The real-life pumpkin from which I evoke a fairy coach to take Cinderella to the ball will then be revealed.

Sympathy and empathy. Hume [267] emphasized the importance of sympathetic identification for human moral behavior. His pupil, Adam Smith [500], made this insight central to his whole system of moral philosophy. But neither David Hume nor Adam Smith succeeded in refining their concept of sympathy sufficiently for it to become a serious tool of analysis.

Davis [150] is typical of modern social psychologists in substituting the word "empathy" for the complex of ideas that Hume and Smith had in mind when speaking of sympathy. Experimental work by social psychologists[9] suggests that the notion can be separated into many strands, but I follow Wispé [562, 563] in focusing only on two primary strands, reserving the word *sympathy* for one and *empathy* for the other.[10]

Adam sympathizes with Eve when he so identifies with her aims that her welfare appears as an argument in his utility function. Section 2.5.2 argues that we instinctively sympathize with our kin. The extreme example is the love a mother has for her baby.

Adam empathizes with Eve when he puts himself in her position to see things from her point of view. Empathy is not the same as sympathy because Adam can identify with Eve without caring for her at all. For example, a gunfighter may use his empathetic powers to predict an opponent's next move without losing the urge to kill him. To quote Wispé [563]: "The object of sympathy is the other person's well-being. The object of empathy is understanding."

If Eve has *sympathetic preferences* that incorporate a concern for Adam's welfare, she has reason to give him a fig leaf if he is sufficiently more modest than she.[11] On the other hand, Eve's *empathetic preferences* can provide

---

[9]Hoffman [255, 256] is particularly instructive.

[10]Section 3.8.2 argues that the "Adam Smith problem" arises because Smith [500] defined sympathy to be something close to what Wispé calls empathy, but operated in practice with what I call sympathy.

[11]Kant's [298, p.52] reworking of the myth of Adam and Eve makes the fig leaf represent: "the inclination to inspire others to respect ... by concealing all that might arouse low esteem"!

no such motivation. Eve is expressing an empathetic preference if she says that she would find it less distressing to be naked if she were Adam than if she were Ichabod (Section 4.3.1 of Volume I). But such a preference provides her with no reason to give a fig leaf to Adam, because she is never actually going to find herself in either his body or Ichabod's.

Everybody is able to formulate empathetic preferences, but what are they for? Why has evolution gifted us with the apparently useless capacity to understand them?

I suspect that we are able to express empathetic preferences because we need them as inputs when using the device of the original position— and for no other reason. In order to evaluate a social contract behind the veil of ignorance, a person *must* be able to compare being Adam in a particular social contract with being Eve. Since we never need to make such comparisons outside a fairness context, the benefit of being able to use fairness norms in our evolutionary past must have been very high in order to outweigh the cost of carrying the apparatus for expressing empathetic preferences around in our heads.

0.2.5 Bargaining

The device of the original position calls on people to predict what deal would be negotiated if everybody concerned were unaware of their current role in society. To understand how this social tool works, it is necessary to have a view on how people bargain. Chapter 1 argues in favor of using a bargaining model for this purpose that was first analyzed by Ariel Rubinstein [445].

The defense for using Rubinstein's bargaining model, rather than one of the many alternatives that have been proposed, is that Rubinstein's model mimics the cut and thrust of bargaining that we actually observe in real life. That is to say, the protagonists simply trade proposals and counterproposals until an agreement is reached. One can complain that Rubinstein is less than realistic in modeling the bargainers as members of the imaginary species *homo economicus*,[12] but one cannot reasonably deny

---

[12]But not so unrealistic as the authors who advocate the use of *homo ethicus* or *homo behavioralis* or some other invented hominid who can be relied upon to give suckers an even break. It is true that *homo economicus* is not a carbon copy of *homo sapiens*. But the discrepancies quoted by critics usually involve deviations from rationality that cost very little, or else occur only rarely (Section 1.2 of Volume I). The fact that people tip in restaurants that they will never visit again is one such picayune example. However, in a social contract discussion, we care about what people do in *important* situations that occur *frequently*. In such situations, I work on the assumption that *homo economicus* is not a bad approximation to *homo sapiens*. This is not to argue that *homo sapiens* commonly solves problems by thinking them through in the style attributed to *homo economicus*. Insofar as *homo sapiens* succeeds in behaving like *homo economicus*, it is

that Rubinstein's model of the bargaining procedure itself is very realistic. Unlike other social contract theorists who have taken the bargaining issues involved in a social contract discussion seriously, this commitment to realism is very important to me. I claim that we *already* use variants of the device of the original position when coordinating on equilibria in games that are subsidiary to the Game of Life. But how could this be if the type of bargaining hypothetically employed were outside the experience of those who supposedly use it?

As a preliminary to modeling the implicit bargaining that is used in Rawls' original position, I therefore think it necessary follow Rubinstein in studying the simpler case of explicit bargaining in which people negotiate face-to-face—as in an oriental bazaar. The fact that we are able to do a workmanlike job of this in the two-person case is one of the small triumphs of modern game theory. However, the ideas have yet to percolate fully even into labor economics, where mistaken notions continue to be widely used in predicting the outcome of wage negotiations. I hope I will therefore be forgiven for devoting nearly all the next chapter to a slow and careful review of bilateral bargaining theory.

The multiperson case.   No promises have been made about exploring *multilateral* bargaining theory because I do not believe that game theory has advanced sufficiently far in this area for it to be a reliable tool. The simple mathematics I employ would extend directly to the n-person case if it were legitimate to ignore potential coalitions of m people when $1 < m < n$. But I am not ready to follow Harsanyi [233] or Rawls [417] in making such an implicit assumption about the structure of a society. The mythology of the right pictures society as nothing more than a collection of atomized households ($m = 1$), while the mythology of the left sees it as a monolithic structure that acts like a single individual written large ($m = n$). But a whig sees no need to choose between these two misconceptions. Whigs understand that real societies consist of a complex of overlapping subsocieties that interact in subtle ways. Modeling such complex systems adequately is a very difficult problem. I certainly have no idea how to solve it. But, to paraphrase Socrates, the first step on the road to understanding is to recognize what one does not understand.

Some cooperative game theorists are willing to claim much more than I am willing to concede about the nature of coalition formation in a large society of rational optimizers. But, even if I were less skeptical of their claims, the problem of interpersonal comparison of utility in such a society

---

because his culture incorporates lessons that his ancestors painfully learned by trial-and-error experimentation long ago.

would still leave me stumped. I follow Harsanyi [233] in deducing such comparisons from the players' empathetic preferences, which I see as being an artifact of a society's culture. But each of the interlocking subsocieties that together make up a large modern society will necessarily have its own subculture, and hence its own internal pattern of empathetic preferences. Each citizen will therefore necessarily have a whole wardrobe of hats that he will don in different circumstances when making fairness judgments. In consequence, what is regarded as fair in one subsociety need not be regarded as fair in another—and what is fair in society as a whole may depend in a complicated way on how fairness is construed in each different subsociety.

I appreciate that it is disappointing to be offered a theory in which only two-person examples are studied. However, I do not feel very apologetic. It may be that my slow and cautious approach will take us only a little way down the road—but those who seek to run before they can walk seldom get anywhere at all.

0.3 Moral Philosophy

This section tries to find the right philosophical pigeonhole for my theory. Readers who are unfamiliar with the language of metaethical debate are advised to skip forward to Section 0.4 without delay. Philosophers will perhaps be more patient as I rummage among their dirty linen looking for a place where my own piece of laundry won't look too disreputable.

Scientific or hortatory? An immediate classification problem is created by the fact that the model of this book admits both a *descriptive* and a *prescriptive* interpretation. Reichenbach [421] would say that I was contributing to *descriptive ethics* when using the first interpretation, and to *volitional ethics* when using the second. I never wear my descriptive and prescriptive hats simultaneously, but I fear that confusion is inevitable if one fails to notice when I take off one hat and put on the other. Sidgwick [492] would not have approved at all. As he wisely observes:

> A treatment which is a compound between the scientific and the hortatory is apt to miss both the results it would combine; the mixture is bewildering to the brain and not stimulating to the heart.

In my descriptive mode, the game G represents only the small segment of the Game of Life to which we currently apply the Game of Morals. In this setting, I see myself as a scientist doing my best to formulate propositions capable of being refuted in the laboratory.

As explained in Section 2.4, I claim no authority for the reforms I propose when wearing my prescriptive hat. I merely express my own personal preference for the kind of society that would result if we were to apply the Game of Morals to the *entire* Game of Life. Unlike most utopian projects, following such a road to a just society requires nothing outside our capacity to deliver. We know that we don't need to be saints or eggheads to operate the Game of Morals, because we already use it when solving small-scale coordination problems.

0.3.1 Traditional Philosophical Categories

In philosophy, a *consequentialist* is not merely someone who judges acts by their consequences: he maintains the existence of an *a priori* Good toward which mankind should strive. A *deontologist* maintains to the contrary that we have an unconditonal obligation to follow the rules determined by whatever *a priori* notion of the Right he favors.

I am not a consequentialist, nor any other kind of teleologian (Broome [102]). According to Snare [503] and other textbook writers, this should make me a deontologist. But I no more believe in metaphysical notions of the Right than in metaphysical notions of the Good. When wearing my descriptive hat, I am a *naturalist*—someone who tries to be realistic about the ethical systems that actually operate in different societies. More importantly, I am a moral *relativist*, since it seems undeniable that different societies have operated different ethical systems at different times and places.[13] As Epicurus [174] explains:

> Justice is the same for all, a thing found useful by men in their relations with each other; but it does not follow that it is the same for all in each individual place and circumstance.

Finally, I am also a moral *realist* of a kind, since I believe that statements about the morality of a particular society are objectively true or false, and a *psychological egoist,* because the model of man that seems to me to provide the most useful first approximation to the data is *homo economicus.*[14]

---

[13]Relativism is sometimes held to be vulnerable to a piece of sophistry used by Socrates in demolishing Euthyphro. According to relativists, people think an action is right or good because this is generally held to be the case in their society. But how does one tell that a view is generally held? By observing that this is what most people think! But the circularity in this argument is no more fatal than the circularity in the argument used in defending a Nash equilibrium. Why did Adam do that? Because Eve did this. Why did Eve do this? Because Adam did that.

[14]Psychological egoism is supposedly vulnerable to a knock-down argument of the philosopher Butler [111]. In its modern form, it amounts to the claim that one can only ultimately defend psychological egoism by reducing it to a tautology (Snare [503]).

When I replace my descriptive hat by a prescriptive hat, I become a *subjectivist*—someone who simply offers his opinions about the way he would like society to be organized. Tradition holds that a subjective naturalist must translate the sentence "Lying is wrong" into "I disapprove of lying."[15] I suspect that this insistence lies at the root of a confusion between moral relativism and subjectivism that has unnecessarily discredited naturalistic approaches to morality for many years (Section 3.8.4).

If I say that lying is wrong, I am wearing my descriptive hat to make an objective assertion about a particular society. Within the society under discussion, it is then objectively true that nobody ought to lie. These statements are relative, since they only genuinely apply in certain primitive societies. Our own social contract would collapse if we began to pay more than lip-service to the principle that the truth should always be told!

I never say that *I* disapprove of lying, because it seems to me like asking someone to listen to the sound of one hand clapping. Only *societies* can have moral values, because a moral value must be common knowledge among the players of the Game of Life before it can coordinate behavior on the equilibrium that serves as their social contract (Section 3.8.1). Individuals have *preferences* rather than values. With my prescriptive hat on, I am very ready to say that I would like our society to replace its current social contract by one in which politicians caught out in major lies were required to resign their offices in disgrace. This is a subjective view that other people may or may not share, depending on the extent to which their history of experience resembles mine. But such a subjective expression of my likes and dislikes needs to be firmly distinguished from the objectively false assertion that it is wrong for politicians to lie. Our current social contract not only endorses the lies of our great statesmen; it treats oddball politicians who tell truths unhelpful to their cause as unfit for office.

0.3.2 Fin de Siècle

To mark the end of the century, Darwell, Gibbard and Railton [142] offer a classification of currently active lines in moral philosophy. How does my theory fit into this up-to-the-minute survey from Michigan?

Section 2.2.4 of Volume I argues that this claim is false—but if it were true that psychological egoism were a tautology like $2 + 2 = 4$, why would that be an argument against psychological egoism?

[15] According to A. J. Ayer's [35] popular definition, an emotivist is a species of subjective naturalist who claims that to make a moral judgment that lying is wrong is simply to *say* something equivalent to "I disapprove of lying." But why on earth should we want to say that someone who approves of lying, and therefore says that he disapproves of lying, has made a moral judgment that lying is wrong?

Even the prescriptive interpretation of my model would seem to fall within the scope of the theories they classify as scientific, since one can always ask how nature and nurture combined to equip me with the cultural prejudices I reveal when expressing my preferences over different social contracts. However, I am unenthusiastic about being classified either as a scientific *reductionist* or as a scientific *nonreductionist*.[16]

Section 2.7.1 explains why I think the debate between holists and reductionists is unproductive, but this isn't why I decline to be labeled either as a reductionist or a nonreductionist in this context. It is because I want to disassociate myself from the tradition that the role of a naturalist is to provide naturalistic interpretations of the words and phrases that people use when discussing moral questions. What if physicists had similarly seen their role as analyzing the language of folk physics?

Consider the folk principle "Nature abhors a vacuum" (Section 1.2 of Volume I). Aristotle's thoughts on the nature of the void were apparently confirmed when Strato found that a partially evacuated vessel will suck up water through a narrow neck. Only when Torricelli experimented with mercury instead of water was the principle refuted. But how would physics have advanced if Torricelli had devoted his life to studying how some omnipotent being might have developed a distate for vacua? If physicists had treated language as their primary data in this kind of way, we would still be struggling to make sense of whatever misconceptions about atmospheric pressure were built into the folklore of pre-Aristotelian times.

It seems to me that the data for a scientific study of the social contracts of different societies need to be based primarily on what people *do* rather than what they *say*. Modern psychology is rife with examples which demonstrate that our explanations of our own behavior are often nothing more than *post hoc* rationalizations. What could be more likely than that the same holds true of our moral behavior as well?

Fortunately for my claim to be a moral scientist, the Michigan survey recognizes a third scientific category to accommodate the work of Philippa Foot [184].[17] Since any naturalist must acknowledge a major debt to Aristotle, I am pleased to be able to call myself a *neo-Aristotelian*. However, I fall into this category only to the extent that the same is true of David Hume. To learn something worthwhile about my relationship with contemporary moral philosophy, it is necessary to turn to the categories that the Michigan survey deem to be discontinuous with science.

---

[16]The second category is actually postpositivist nonreductionist, but I assume that everybody agrees that philosophical positivism is passé.

[17]Although she has recently repudiated her characterization of morality as a system of hypothetical imperatives.

How is it possible that a scientific approach to moral philosophy might be partly inspired by approaches that deny the relevance of the scientific method? I don't know where metaphysical moral philosophers think their intuitions about what is Good or Right come from, but a naturalist has no difficulty in attributing their intuitions to the same source as his own— namely, our knowledge and experience of the way actual social contracts work. The fact that such intuitions are not reliable is demonstrated by the fact that different people offer mutually inconsistent explanations of the same phenomena. Nevertheless, a good beginning for a naturalist is to follow Strato in examining the substance of the available intuitive explanations to see how well they accord with the data. Approaching the nonscientific categories of the Michigan survey in this spirit, I find that the substantial debts I owe to modern moral philosophy are categorized under the headings of *practical reasoning* and *constructivism*.

Kant [299] distinguishes between pure and practical reasoning. Pure reasoning is used to determine what is true, and practical reasoning to decide what to do. However, like all scientifically trained people, I think that the principles of reasoning are the same wherever they are applied, and that the practical reasoning appropriate for studying problems in moral philosophy is embodied in modern decision theory. Since Kant denies the Humean principle that reason is the slave of the passions on which modern decision theory is based, I therefore part company with the majority of moral philosophers who still carry his standard. But Darwell *et al* [142] do not restrict their practical reasoning category to Kantians, they also recognize a Hobbesian variety of practical reasoning, which they associate with Baier [36] and Gauthier [196]. I believe that the difficulties moral philosophers have perceived in their work can be eliminated by adopting a game-theoretic approach that takes on board the difficulty of making decisions in an environment in which other people are simultaneously making decisions.[18]

I see myself as a constructivist because my principal source of inspiration is the work of Rawls [417] and Harsanyi [233], who both undertake to construct a concept of the good from more primitive concepts. Instead of simply asserting the nature of the good after which we should strive in common, they describe a procedure which we ought to follow in making moral decisions. They then argue that our notions of the good are derived from the fact that people who follow this procedure will behave as *though* maximizing the common good they construct. I go further down this road

---

[18]My differences with Gauthier stem from a fear that his idiosyncratic proposals for solving multilateral decision problems will be thought to represent the orthodox approach of game theory (Sections 1.2.9 and 3.2.2 of Volume I).

by repudiating the Kantian arguments that Harsanyi and Rawls give to explain why people are morally obliged to follow their procedures. Instead of constructing the good from the right, I therefore find it necessary to construct both the good *and* the right from more primitive assumptions. These assumptions are entirely naturalistic in that they depend on the biological and social facts of human history.

The remaining categories in the Michigan survey are *noncognitivism* and *sensibility*. Gibbard [205] emphasizes the importance of appreciating the role of morality in coordinating behavior on an equilibrium in the Game of Life. I also agree very much that the practical reasoning that mediates such coordination is seldom operated by the consciously Machiavellian parts of our brains. It is instead largely controlled by emotional responses that evolution has developed for the purpose (Section 3.5.2). I therefore have something in common with noncognitive and sensibility theorists. But I see no future in pursuing the relevant issues by philosophical means. It seems to me that the time has come to turn our attempts to analyze human emotive responses over to the new subject of evolutionary psychology.

By the standards of contemporary philosophy, it therefore seems that I am a constructive neo-Aristotelian who takes a Hobbesian approach to practical reasoning without being committed to cognitivism. With a foot in so many camps, perhaps I have a chance of seeming to be all things to all people!

0.4 Noncooperative Game Theory

The remaining sections of this intoductory chapter are intended for readers whose knowledge of game theory is derived largely or exclusively from reading the first volume of the book. The most immediate need for such newcomers to game theory is an appreciation of the ideas of cooperative game theory offered in Section 0.5. However, I plan to begin by reviewing noncooperative game theory, since I follow Nash [380, 381, 382] in thinking that the two branches of game theory have to be understood together or not at all. Since the whys and wherefores of noncooperative game theory were discussed at length in the first volume, its reappearance here signals that I am often shamelessly repetitive when there is a risk of being misunderstood.

Books on game theory sometimes give the misleading impression that game theory consists of a single organic whole whose branches differ only in the emphasis that they accord different issues. However, the truth is much less satisfactory. Some branches of game theory are built on very shaky foundations indeed. Others are mutually incompatible because the aims of the theorists who invented them are diametrically opposed. Yet other branches are closely interwined but must be carefully distinguished when exact reasoning is necessary, since they use the same words with different meanings. Eliminating all the possible sources of confusion would be a task worthy of Hercules. However, I hope at least to demonstrate that game theory is a house with many mansions, before I settle on the approach known as the *Nash program*, which will serve as the foundations stone for the rest of the book.

0.4.1 The Ultimatum Game

The theory of games came into being with the publication of Von Neumann and Morgenstern's *The Theory of Games and Economic Behavior* in 1944. Their approach is more than a little schizophrenic, in that the style of analysis in the first part differs markedly from the second. This schizoid tendency is preserved in modern game theory by the distinction drawn between cooperative game theory and noncooperative game theory.

Noncooperative game theory calls for a complete description of the rules of the game to be given so that the *strategies* available to the players can be studied in detail. The aim is then to find a suitable pair of *equilibrium* strategies to label as the solution of the game. All the game-theoretic discussion of Volume I was devoted to noncooperative ideas. We met both the *extensive form* of a game and the *strategic form* of a game. In the former, the rules are laid out in full detail by drawing a game tree. To illustrate these ideas, a simple bargaining game will be considered.

Bargaining between two individuals is worthwhile when an agreement between them can create a surplus that would otherwise be unavailable. Much negotiation in real life is concerned with sharing information with a view to estimating the size of the potential surplus. But I shall neglect all such informational questions in this book. Thus, if you have a fancy house to sell that is worth $2m to you and $3m to me, then I shall assume that it is common knowledge that a surplus of $1m is available for us to split. The archetypal version of this problem is called *dividing the dollar*. The story is that a philanthropist has offered you and me a dollar to divide between us provided that we can agree on a division. If we are unable to agree, he takes his dollar back.

The simplest possible negotiation rules that might be employed in dividing the dollar apply when an object is sold in a Western store. The seller writes a take-it-or-leave-it price on the object to be sold. The buyer then decides whether to pass or buy. A game tree for the bargaining problem faced by such a buyer and seller is shown in Figure 0.1(a). This is the extensive form of the Ultimatum Game.

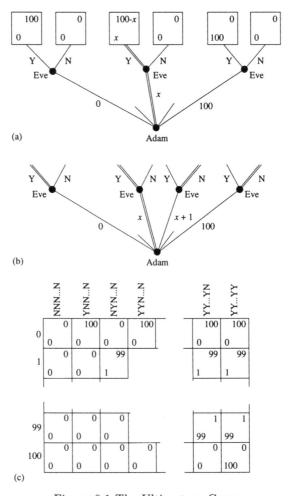

(a)

(b)

(c)

Figure 0.1 The Ultimatum Game.

Adam is the seller. He begins by making a choice at the first move of the game, which is represented in Figure 0.1(a) by the node labeled Adam at the root of the tree. Each line emanating from this node represents an ultimatum that he can put to the buyer. Each ultimatum is labeled with the number of cents that Adam demands as his share. Eve is the buyer. After Adam has moved, she has two possible responses, yes or no. In the diagram, each of Adam's demands is accordingly followed by a node corresponding to the final move by Eve. Two lines labeled Y and N emanate from each such node to represent her two choices at this move. A play of the game consists of Adam's choice of an ultimatum followed by Eve's choice of a response. One such play is shown in Figure 0.1(a) by doubling appropriate lines. In this play, Adam demands x and Eve responds with Y. The doubled path through the tree leads to a box that shows the *payoffs* that each player derives from this play. Adam's payoff of x appears in the southwest of the box. Eve's payoff of $1 - x$ appears in the northeast of the box.

Pure strategies. A *pure strategy* for a player in a game consists of a plan of action which is so complete that it specifies what the player would do under all possible contingencies that might arise. A pure strategy for Adam in the Ultimatum Game is just a number x chosen from the set $\{0, 1, 2, \ldots, 100\}$. A pure strategy for Eve is more complicated. It must specify what her response would be to all Adam's possible demands. One such pure strategy is for her to plan to accept when Adam offers her the whole dollar and to refuse otherwise. This greedy strategy may be denoted by $YN \ldots NN$. The Y in $YN \ldots NN$ indicates that a demand of 0 cents is to be accepted. The first N indicates that a demand of 1 cent is to be refused; and so on. The number of "words" of 101 letters that can be made using only Y or N is 2^{101}. Eve's set of pure strategies is therefore enormously large.

The strategic or normal form of the Ultimatum Game is a table in which the rows correspond to Adam's pure strategies and the columns to Eve's pure strategies, as illustrated in Figure 0.1(c). The entries in a cell of the table are the payoffs that the players will receive if Adam and Eve happen to play the row and column in which the cell lies. As always, Adam's payoff lies in the Southwest of each cell and Eve's payoff lies in the Northeast.

Von Neumann and Morgenstern felt that the normal procedure in analyzing a game should be to codify the rules as a suitable extensive form, and then pass immediately to its strategic form for the purposes of analysis. Modern writers are more cautious. The Ultimatum Game is not exceptional in having a large and unwieldy strategic form that is much harder to analyze than the extensive form from which it is derived. Moreover, as we shall see in discussing the concept of subgame-perfection, vital issues concerning time and information get hidden in the shadows rather than being floodlit as they deserve, when one passes from the extensive to the strategic form.

Payoffs. If each player is risk neutral, the entries in the strategic form of the Ultimatum Game—the players' payoffs—can be identified with how much money each player gets at the end of the game. To say that a player is risk neutral simply means that his Von Neumann and Morgenstern utility for \$x can be taken to be x utils (Section 4.6 of Volume I). If a player were not risk neutral but cared only about money, it would be necessary to take his payoff for \$x to be u(x), where u is his Von Neumann and Morgenstern utility function. If he cared not only about ends but also about the means by which an end is achieved, it would be necessary to write his payoffs in the form u(s,t), where s is Adam's pure strategy and t is Eve's pure strategy.

The point in being careful about such matters is to ensure that no doubts exist about the motivation of the players (Section 2.2.4 of Volume I). If payoffs are always measured in Von Neumann and Morgenstern utils, then we can be sure that a rational player will necessarily act as though seeking to *maximize his expected payoff*.

Von Neumann and Morgenstern utility functions were introduced for the first time in an appendix to *The Theory of Games and Economic Behavior*. The idea has since proved a vital foundation stone for the economic theory of risk. It is equally vital for the subject matter of this book. I hope that readers who are rusty on this subject will review Chapter 4 of Volume I. Von Neumann and Morgenstern's thoughts on this subject are often badly misunderstood—especially when it comes to the vexed question of how Adam's utils are to be compared with Eve's.

Nash equilibrium.   The part of Von Neumann and Morgenstern's book that was devoted to noncooperative game theory dealt only with a special class of games—two-person, zero-sum games. All the outcomes of such a game result in payoffs to Adam and Eve that sum to zero.[19] Von Neumann and Morgenstern argue that the solution to such a game will often require that the players use mixed strategies.

A *mixed strategy* requires a player to randomize over his pure strategies in order to keep the opponent guessing.[20] But with what probabilities should each pure strategy be played? In the noncooperative part of their book, Von Neumann and Morgenstern [538] propose that a player should choose these probabilities in order to guarantee that his expected payoff in the game can never be less than his *security level* whatever the opponent may do. His security level is the largest amount that can be so guaranteed.

---

[19]Sometimes this is thought to involve the implicit assumption each of Adam's utils is worth precisely one of Eve's utils. Readers who do not see why no such interpersonal comparison of utility is implied are strongly recommended to review Section 4.2.2 of Volume I.

[20]If it seems odd that rational play might require randomizing, consider the game of Poker. It is obvious that one must sometimes bluff in Poker. If one only bet high with a big hand, the opponents would soon learn to fold unless holding an even bigger hand. It is equally obvious that one should not bluff in a predictable way. Randomizing the times when one chooses to bluff therefore makes good sense—provided that the frequency with which one bluffs is chosen with great care.

As Section 4.6 of Volume I explains, a player computes his security level in a game by first calculating the minimum expected payoff he could conceivably get from using each of his mixed strategies. His security level is then the maximum of all such minima. Von Neumann and Morgenstern's recommendation for playing two-person, zero-sum games is therefore called the *maximin criterion*.

Although Von Neumann and Morgenstern did not suggest that the maximin criterion has anything to recommend it outside the context of a two-person, zero-sum game, it has nevertheless somehow become a candidate as a criterion for making decisions in general. For example, Rawls [417] assumes that people behind the veil of ignorance will use the maximin criterion, even though the decision problem they face bears no resemblance to a zero-sum game. It is therefore not surprising that he should be led to recommend the use of the maximin criterion when goods are distributed in a just society. Nor is it surprising that those trained in rational decision theory see no reason why his arguments on this score should carry any weight. The use of the maximin criterion is rational in two-person, zero-sum games because Von Neumann's [537] celebrated Minimax Theorem[21] asserts that Adam and Eve's security levels sum to zero in such a game. Neither can therefore get *more* than his security level unless the other gets less. On the other hand, neither player need end up with *less* than his security level, because he is always free to use the maximin criterion and so guarantee his security level or better. Thus rational play in two-person, zero-sum games must result in each player getting exactly his security level.

This conclusion is very far from being true in games that are not zero-sum. Indeed, the generalization of Von Neumann's theorem to games that are not zero-sum is disappointingly weak.

It is easy to prove that the use of the maximin criterion by both players in a two-person, zero-sum game necessarily results in each player choosing a strategy that is an optimal reply to the strategy choice made by his opponent. A pair of strategies with this property is called a *Nash equilibrium*. Nash [381] showed that all games with a finite number of strategies have at least one equilibrium, provided that mixed strategies are allowed.

The idea of a Nash equilibrium is basic to noncooperative game theory. An authoritative game theory book cannot possibly recommend a strategy

---

[21]Why minimax rather than maximin? The theorem is usually stated in the form $\max_p \min_q \Pi(p, q) = \min_q \max_p \Pi(p, q)$, where $\Pi(p, q)$ is the expected payoff to Adam if he uses mixed strategy p and Eve uses mixed strategy q. The left hand side of this equation is Adam's security level. The right hand side is the negative of Eve's security level. (Her payoff in a two-person, zero-sum game is the negative of Adam's payoff. But $\max_p \min_q -\Pi(p, q) = -\min_p \max_q \Pi(p, q)$.) In consequence of Von Neumann's Theorem, people often carelessly refer to the maximin criterion as the "minimax criterion".

pair as the solution to a game unless the pair is a Nash equilibrium. If the book recommends a strategy to me that is not a best reply to the strategy it recommends to my opponent, then I will not follow its recommendation if I believe that my opponent will. The proper way to extend Von Neumann and Morgenstern's theory of two-person, zero-sum games does not therefore lie in a mindless attempt to apply the maximin criterion to all decision problems. The proper extension simply says that, if a noncooperative game has a solution, then it lies among the Nash equilibria of the game.

Why did Von Neumann and Morgenstern not formulate this extension themselves? My guess is that they recognized that it is often not very helpful to say that the solution of a game must be a Nash equilibrium. Interesting games typically have many different Nash equilibria. For example, any split of the dollar is a Nash equilibrium outcome of the Ultimatum Game—including the split in which Adam offers the whole dollar to Eve. Figure 0.1(b) illustrates a pair of strategies in which Adam demands x and Eve accepts $100 - x$ or more but refuses anything less than $100 - x$. Each is optimizing given the choice of the other. For each value of x, including $x = 0$, this pair of pure strategies is therefore a Nash equilibrium.

Which of all such Nash equilibria should be selected as the solution of the game? For a two-person, zero-sum game, the answer to this *equilibrium selection problem* is irrelevant, because all the Nash equilibria in such a game are equally satisfactory. I think that Von Neumann and Morgenstern saw that the same is not true in general and therefore said nothing at all rather than saying something they perceived as being unsatisfactory.[22]

Subgame-perfect equilibrium.   Game theorists have made progress with the equilibrium selection problem in recent years, but much remains controversial. The approach that has become traditional was pioneered by Reinhard Selten [471]. As explained in Section 2.5.2 of Volume I, Selten introduced the notion of a *subgame-perfect equilibrium*. For any subgame of the original game, whether it is reached or not during the play of the game, a pair of subgame-perfect strategies requires that the players plan to use a Nash equilibrium in that subgame.

Since the whole game counts as a subgame of itself, a subgame-perfect equilibrium is a Nash equilibrium of the whole game with additional properties. If we restrict attention to subgame-perfect equilibria, we therefore make a selection from the class of all Nash equilibria. But why should we

---

[22]As they would have put it, the defense given above of the Nash equilibrium concept is entirely negative. It only says that nothing other than a Nash equilibrium can be the solution of a game. But a player needs positive reasons for choosing one strategy rather than another.

be interested in restricting attention to subgame-perfect equilibria? To see the reason, we need to ask why players would stay on the equilibrium path that their original choice of strategy specifies.

A rational player stays on the equilibrium path because of what *would* happen if he *were* to deviate. But how does he know what would happen if he were to deviate? This information is supposedly embodied in each player's original choice of a strategy. (Recall that a pure strategy tells us what the players would do under *any* contingency that might arise in the game, not just those which actually do occur.) But why should rational players stick by their original choice of a strategy after a deviation has led them to a subgame that would not otherwise have been reached? The reason that the players were satisfied with their original choice of strategies was that these make up a Nash equilibrium for the whole game—so that each player's strategy is an optimal reply to the strategy choice of his opponent. If the players are not to abandon their original strategy choices after a deviation has led them to an unanticipated subgame, then Selten [471] argues that the same conditions must hold in the subgame—that is, the original strategy choices must induce a Nash equilibrium in each *subgame*, as well as in the whole game.

As explained in Section 2.5 of Volume I, subgame-perfect equilibria in simple games can be computed by *backward induction*. One begins by looking at the smallest subgames—those which occur at the end of the game. In the Ultimatum Game of Figure 0.2(a), these smallest subgames are the one-player games in which Eve must answer Y or N after the delivery of an ultimatum by Adam. One argues that, if such a subgame were reached, then the players would play it rationally and hence use a Nash equilibrium in the subgame.

The Nash equilibria in the one-player subgames of the Ultimatum Game are easy to characterize. They simply require that Eve choose an optimal action in the subgame. These optimal actions have been indicated by doubling the optimal choices for Eve in Figures 0.2(a) and 0.2(b). In each case when Adam demands less than the whole dollar, she is left with a choice between something and nothing—and so takes the former. When Adam demands the whole dollar, she has two optimal choices,[23] since both an acceptance and a refusal will then leave her with nothing Figure 0.2(a) shows her refusing when offered nothing, and Figure 0.2(b) shows her accepting.

Having dealt with the smallest subgames, the backward induction principle requires us to look at the next largest subgames. In the Ultimatum Game, there is only one such game—the whole game itself. Taking the be-

---

[23]If we allow mixed strategies, then Eve has many more optimal choices, since any mixture of Y and N is also optimal.

havior already specified in the smaller subgames as given, we now look for a Nash equilibrium in what remains. After player II's behavior in the smaller subgames of the Ultimatum Game has been fixed as in Figures 0.2(a) and 0.2(b), we are left with another one-player game to consider, but this time Adam is the protagonist. In Figure 0.2(a), his optimal response to the behavior attributed to Eve is to demand 99 cents. This demand will be accepted, but a demand of 100 cents will be refused. In Figure 0.2(b), his optimal response to the behavior attributed to Eve is to demand 100 cents, since this will be accepted.

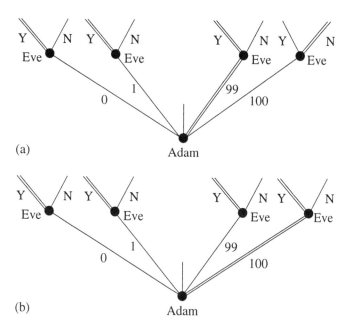

Figure 0.2 More Ultimatum Games.

An analysis of the Ultimatum Game by backward induction therefore reveals two subgame-perfect equilibria in pure strategies. One of these is illustrated in Figure 0.2(a) and the other in Figure 0.2(b). Both yield nearly the same outcome—Adam gets the whole dollar, give or take or a penny. A subgame-perfect equilibrium analysis therefore captures the intuition that a person who is on the receiving end of a take-it-or-leave-it demand has no bargaining power at all.

This conclusion appears in a purer form if we modify the Ultimatum Game so that Adam is allowed to demand any fraction of the dollar, rather than being restricted to demands in whole numbers of cents. In this case, the strategy pair illustrated in Figure 0.2(a) ceases to be subgame-perfect, because Adam could do even better by demanding 99.9 cents, which Eve would accept because even 0.1 of a cent is better than nothing. In fact, when *any* fraction of the dollar is admissible as a demand, the only subgame-perfect equilibrium requires that Adam demand the whole dollar and Eve plan to accept whatever demand is made.[24] This provides the simplest example of a principle that will be useful when we come to consider multistage bargaining games in which no restrictions are placed on how the surplus can be divided among the protagonists: In a subgame-perfect equilibrium, whoever is currently making a proposal always plans to offer the responder an amount that will make the responder indifferent between accepting and refusing. In equilibrium, the responder always plans to accept such an offer or better, and to refuse anything worse.

Since the equilibria of Figures 0.2(a) and 0.2(b) are subgame-perfect for the original version of the Ultimatum Game, they are also necessarily Nash equilibria. But they are by no means the only Nash equilibria of the Ultimatum Game. As observed earlier, any split of the dollar is a Nash equilibrium outcome—including the highly counter-intuitive outcome in which Adam offers the whole dollar to Eve.

The important idea to carry away from this discussion of the Ultimatum Game is the reason why the Nash equilibrium illustrated in Figure 0.1(b) is *not* subgame-perfect.

If Adam were to demand more than x, then Eve's strategy requires that she refuse. But, if she actually were to be faced with a demand of anything less than the whole dollar, such behavior would not optimal. Eve might threaten Adam before the play of the Ultimatum Game by announcing that she was committed to refusing insultingly small offers, but Adam would regard her threat as incredible for a rational player. Once the offer has been made, only an irrational player would insist that nothing is better than a penny.

0.4.2 Anomalies?

One reason for using the Ultimatum Game as an example in introducing the basic ideas of noncooperative game theory is because it has served as a focus in the literature for doubts about the extent to which such ideas have any relevance to human behavior at all. Thaler [519] has been particularly widely quoted. The unrest derives from a much replicated laboratory experiment.

---

[24]Chapter 5 of my *Fun and Games* (Binmore [70]) discusses this point in greater detail.

Selten [471], the inventor of subgame-perfect equilibrium, proposed to Güth , Schmittberger and Schwarze [214] that they test the subgame-perfect prediction for Ultimatum Game in the laboratory. Selten did not expect the prediction to be confirmed, and he was correct. Numerous later experimental studies show that Adam would be unwise to offer Eve less than a third of the money available, since the probability of her refusing then exceeds one half.

This experimental work commonly provokes one of two reactions, both of which I believe to be mistaken. The first reaction comes from old-style economists who insist that *homo economicus* is the right model of man whatever the context. They therefore squeeze all behavior into an optimizing shoe, no matter how tight the fit. Since an Eve who refuses something positive in an Ultimatum Game experiment is clearly not maximizing money, they therefore argue that she must be maximizing something else. Exotic utility functions are therefore proposed that incorporate a spite component, or a taste for fairness, or something else that overwhelms Eve's greed for money.

The second reaction comes from New Age economists who think that strategic considerations are largely irrelevant to human behavior. They prefer *homo sociologicus* to *homo economicus* as their model of man, and hence see human behavior as being governed by social norms. In a bargaining context, they believe that fairness norms are invoked whose strength outweighs any selfish inclinations to which the subjects may be prone.

Although I use *homo economicus* as my model of man throughout this book, I am anxious to distinguish myself from the naive positivists who argue that humans *always* optimize. In my own experimental work, which is largely concerned with testing game-theoretic predictions in bargaining games, I stress that it would be unreasonable to expect the theory to be successful unless:

- The game is easy to understand;
- Adequate incentives are provided;
- Sufficient time is available for trial-and-error learning.

Early experimental work often neglected all three criteria. As a consequence, experimental studies of games like the Prisoners' Dilemma typically found that game theory came nowhere near predicting the subjects' behavior, if their aim was taken to be maximizing money. However, as Ledyard [313] documents, more careful modern work confirms that, when the three criteria are respected, there is no need to attribute exotic motivations to players in order that their behavior be consistent with game-theoretic predictions in games like the Prisoners' Dilemma, which model the private provision of public goods.

Nevertheless, anomalies certainly do exist. Their existence forces us to ask some difficult questions. How simple does a game have to be before is "easy" to understand? Has the game been framed in a "misleading" way? How large do the incentives have to be before they are "adequate"? How much experience is "sufficient" for trial-and-error learning to converge? It is the last question that I think vital for understanding the experimental results on the Ultimatum Game.

Binmore, Gale and Samuelson [76] distinguish between a number of time spans in discussing experimental game theory: the short run, the medium run, the long run, and the ultralong run. These terms are not used here in the sense of Section 1.3.7 of Volume I, and it is important to stress that later chapters of the current volume revert to this earlier usage.

The short run refers to the behavior triggered when the subjects enter the laboratory by the manner in which the experiment is framed. This short-run behavior may well be optimal in the circumstances for which it evolved. But, as in the case of Lorenz's baby jackdaw mentioned in Section 2.3.3 of Volume I, behavior that has evolved for one purpose may be utterly unadapted to the circumstances under which it is elicited in the laboratory. Kahneman and Tversky [286] have demonstrated this fact repeatedly for the short-run behavior of human subjects.

In the medium run, subjects begin to adapt to their novel circumstances. In the long run, such adaptive behavior may lead them to an equilibrium of the game they are playing. Which equilibrium they reach in the long run will generally depend on where they started—that is, on their short-run behavior.

If one hopes to observe a history-independent equilibrium selection in the laboratory, one must turn to the ultralong run, as studied by Young [573], Kandori, Rob and Mailath [292], Binmore *el al* [87] and numerous others. Adaptive behavior will necessarily be noisy. If one waits long enough, the random shocks that perturb the learning process will eventually bounce the system out of the basin of attraction of the equilibrium to which it first finds its way, into the basin of attraction of another equilibrium. If one waits a great deal longer, this mechanism will establish a probability distribution over the set of all equilibria of the game. Often, this distribution puts nearly all its mass on just one of the equilibria, which can then be said to be selected in the ultralong run.

How long is the long run? Theoretical studies of simple stylized adjustment rules show that the answer to this question depends very strongly on the game being played. In simple auction games, convergence tends to be very fast, and so it is not surprising that microeconomic predictions traditionally work out well in such games. However, in the Ultimatum Game, convergence is very slow. Roth and Erev [440] suggest that the observed results in the Ultimatum Game are therefore simply snapshots taken before the learning process has had a chance to do much converging. Such a medium-run explanation of the data has a lot going for it. However, Binmore *et al* [76] argue that, even in the long run, one should not necessarily expect convergence to the subgame-perfect equilibrium in the particular case of the Ultimatum Game. If the adjustment process is modeled using a noisy version of the Darwinian replicator dynamics, they find that the system typically converges to a Nash equilibrium of the game in which Adam offers Eve about one quarter of the available money! The subgame-perfect equilibrium would doubtless be selected in the ultralong run, but the expected waiting time would be so long that the subjects would probably all be drawing their old-age pensions before the experiment was over.

Keeping the logic straight.   Given the preceding discussion of the Ultimatum Game, why do I persist in using the idea of a subgame-perfect equilibrium as an analytical tool in this book? A major motivation lies in the fact that the theory of subgame-perfect equilibria provides a logically coherent framework within which to reason. Whatever else the history of ethical thought may teach, one lesson at least is clear—the temptation to fudge the logic is very strong. Section 2.2.4 of Volume I therefore emphasizes that utilities in this book are to be strictly interpreted according to the tenets of the theory of revealed preference. This is the official orthodoxy of neoclassical economics—although I have to admit that heretics are not in short supply, especially within welfare economics. As Section 2.2.5 of Volume I points out, the backward induction principle behind the notion of subgame-perfect equilibrium then becomes a tautology.[25]

At least two problems arise in taking such a strict revealed preference line. The first problem is terminological. As emphasized repeatedly in Volume I, revealed preference theory reverses the causality implicit in the English expressions that it reinterprets. For example, when we say that Eve cares only about maximizing money in the analysis of the Ultimatum Game given in Section 0.4.1, we are not offering a reason why she will act as indicated in Figure 0.2(a) or 0.2(b). We are simply summarizing the way that she is assumed to act in a convenient shorthand form. If it is also assumed that Adam is also a maximizer of money who knows that Eve will maximize money no matter what ultimatum he chooses, we are then led to restrict our attention to the subgame-perfect equilibria in which Eve is

---

[25] I certainly do not agree with Aumann [26] that common knowledge of rationality implies that the subgame-perfect equilibrium outcome will necessarily be observed in finite games of perfect information when the payoffs are not determined by the theory of revealed preference (Binmore [72, 74]).

offered at most a penny.

What mathematically naive critics find hard to understand in such stories is that nothing substantive is being said at all. One is simply pointing out a logical consequence of the assumptions. The real issue for a critic is whether a theory based on such assumptions can genuinely be applied in the circumstances for which it was constructed. In the case of the Ultimatum Game, experimental work in the laboratory suggests otherwise. One may seek to rescue the theory by modeling the preferences revealed by the players with exotic utility functions of one kind or another, but my own view is that the problems arise at a deeper level. The difficulty is not so much that people necessarily react to simple situations in an intrinsically complicated way, but that their behavior tends to be confused and inconsistent in unfamilar environments until they have learned the ropes. We therefore have reason to doubt the *consistency* assumptions of revealed preference theory in such situations.

Section 1.2.2 of Volume I glibly justifies the consistency requirements of revealed preference theory using an evolutionary argument whose effect is to identify utility with *fitness*. Such airy appeals to evolution paper over many cracks. However, as an Oxford philosopher might say, the evolutionary argument holds together "tolerably well" when used to defend the notion of a *Nash* equilibrium. But the abstract form of the argument comes under strain when applied to *subgame-perfect* equilibria, because this notion depends on postulating that players would behave consistently even under circumstances that never actually occur. The orthodox neoclassical notion of consistency then parts company with an evolutionary story, since irrational plans can only be selected against if they are sometimes brought into play.

Game theorists commonly adopt Selten's [471] approach to this problem by assuming that players always make decisions with a trembling hand, so that all branches of the game tree are reached with some minimal probability, including those that rational players with steady hands would never visit. If the hands making decisions at different nodes all tremble independently, Selten shows that only subgame-perfect equilibria are viable in finite games of perfect information—of which the Ultimatum Game is an example.

Like the poor, trembling-hand errors will doubtless always be with us. However, as Section 3.4.1 of Volume I argues, it is a mistake to regard such a game-theoretic equivalent of a typo as the only kind of error that may perturb a real-life game. We need to face the fact that thinkos are likely to be much more important, even though such a source of noise is much harder to model. One might loosely describe a result of Samuelson [457] by saying that the frequency of typos needs to be enormously greater than the frequency of thinkos before typos can have a significant effect on the evolutionary process. I have already mentioned that perturbed versions of the Darwinian replicator dynamics do *not* necessarily converge to a subgame-perfect equilibrium of the Ultimatum Game (Binmore *at al* [76]). This result survives the introduction of trembling-hand errors. It therefore seems to me that Selten's trembling-hand defense of subgame-perfect equilibrium cannot be sustained in the context of this book.

Fortunately, it is not necessary for our purpose to argue that evolution will ensure that the theory of revealed preference will eventually be applicable to *all* games in the long run. We need only defend subgame-perfect equilibrium outcomes for games that are of interest for a social contract discussion. With the approach to social contract ideas adopted in this book, attention therefore has to be focused on two classes of games. The first class contains the bargaining games of Chapter 1, for which subgame-perfection is important only in the case of the Alternating Offers Game of Section 1.7.1. The second class contains the indefinitely repeated games studied in Chapter 3.

In the case of indefinitely repeated games, the most important result is the folk theorem of Section 3.3. This remains valid whether we allow all Nash equilibria or restrict attention to subgame-perfect equilibria. Subgame-perfect equilibria only become important when we ask why a player should be deterred by the threat of being punished if he deviates from the equilibrium path. For such a threat to be credible, the person acting as policeman must find it optimal to carry out the punishment should his threat be ignored. This may require that the policeman should also be credibly threatened with punishment if he fails to punish the original deviation. The idea of a subgame-perfect equilibrium allows such threat-chains to be extended without limit. One could doubtless get by with a less severe concept that only extended such chains a few stages, but the gain in realism would only be marginal.

The Alternating Offers Game of Section 1.7.1 presents more of a challenge. Section 1.7.2 observes that Binmore *et al* [81] are able to duplicate Rubinstein's [445] uniqueness conclusion in the symmetric case by studying evolutionarily stable strategies capable of being played by finite automata. But the asymmetric case remains problematic. Until more is known, a question mark therefore hangs over my use of Rubinstein's theorem to defend the Nash bargaining solution. However, critics are free to redo the analysis of this book using whatever alternative bargaining solution strikes their fancy.

In summary, it is pointless to attack the *logic* of revealed preference theory, but its consistency assumptions are certainly questionable as a description of human behavior—especially when used to defend the notion of a subgame-perfect equilibrium. Indeed, a study of the Ultimatum Game and its cousins shows that the theory sometimes fails hopelessly to predict what actually happens in laboratories. However, the consensus among economists remains that these situations should be regarded as anomalous. I am not so complacent myself, but nevertheless agree that the orthodox theory is the best available for a wide variety of applications. Above all, the theory has the inestimable virtue of being logically watertight.

Homo sociologicus. This fictional hominid is governed by the social norms that characterize his society. Such norms provide examples of the *memes* proposed by Dawkins [151] as the social equivalent of genes (Section 1.2.2 of Volume I). Proponents of *homo sociologicus* therefore have a ready explanation of the Ultimatum Game data. The story is that the manner in which the game is framed in the laboratory triggers a fairness norm in the heads of the participating subjects, who then simply "play fair". Except insofar as it is relevant to framing questions, the strategic structure of the game is seen as being unimportant in determining the subjects' behavior.

Personally, I have no quarrel with *homo sociologicus* as a model of man. He is admittedly not very useful for theoretical purposes as he stands, since, by equipping him with a suitably exotic set of norms, he can be used to rationalize almost any behavior observed in the laboratory. *Homo sociologicus* is of interest for me because I see him as a stepping stone to *homo economicus*. The argument is that social evolution is likely to operate on the norms that govern *homo sociologicus* until he ends up acting like a maximizer. As observed in Section 1.2.2 of Volume I, this claim becomes almost a tautology if the maximand is taken to be whatever makes the bearer of a norm the locus for its replication to other heads.

Such an attitude to *homo sociologicus* differs sharply from that adopted by economic psychologists like Thaler [519] or Kahneman *et al* [285], who seem to me to confuse the species with *homo behavioralis* of Section 2.4.2 of Volume I. The behavioral patterns of *homo behavioralis* are hardwired into his head and hence are not subject to alteration. This leads his advocates into the error of supposing that short-run human behavior is likely to be a good guide to long-run human behavior.

My own experimental work points in a quite different direction. In a two-stage Ultimatum Game (Binmore *et al* [88]), just two trials turned out to be enough to shift Adam's modal offer very dramatically from fifty : fifty to the subgame-perfect offer.[26] If the explanation for subjects playing fair

---

[26]Thaler [519] dismisses this result by claiming that the experimental instructions told the subjects what to do. If so, why didn't they do what they were told at the first trial? Experiments which offer learning opportunities to the subjects are similarly dismissed

at the first trial is that people simply operate fixed fairness norms, how came they didn't "play fair" at the second trial? Similarly, if the failure of the subjects to use the subgame-perfect equilibrium at the first trial is to be explained by their having exotic preferences, how come they played the subgame-perfect equilibrium at the second trial?

Later experiments with more complicated bargaining games were also strongly supportive of the learning hypothesis (Binmore *et al* [90, 89, 79]). Moreover, when debriefed after the experiment, subjects have a strong tendency to say that what actually happened in their experiment is "fair". Binmore *et al* [91] obtained particularly sharp results in the case of the smoothed Nash Demand Game analyzed in Section 1.6.1.

Different groups of subjects were first conditioned[27] to coordinate their demands at the outcomes specified by symmetric versions of the four different bargaining solutions described in Section 1.3. Thirty trials against randomly chosen human opponents from the same group then followed. By the end of the experiment, each group was sharply concentrated at an exact Nash equilibrium of the game,[28] but different groups converged to different equilibria. However, although different groups found their way to different equilibria, in the computerized debriefing that followed their session in the laboratory, subjects showed a strong tendency to assert that the outcome reached by their own group was the "fair" outcome of the game. In fact, the median of the final demands actually made by a group of subjects turned out to be a remarkably sharp predictor of the median of the demands said to be "fair" by members of that group.

If one thinks of the subjects in one of these experimental groups as the citizens of a minisociety, and the computerized bargaining game they played as an artificial Game of Life, then the convergence of the group on a Nash equilibrium corresponds to the evolution of a social contract. Whatever this social contract turns out to be, subjects are ready to call it "fair". How should we react to this data?

My own view is that the result simply reflects our confusion over the concept of fairness when it is applied outside the domain for which it evolved.

---

on the grounds that subjects can be "taught" to do anything whatever. My advice is to take nothing on trust in these matters. Read the original papers and form your own opinion on whether the subjects are learning for themselves or being taught!

[27] The conditioning was achieved by having the subjects play against robots programmed to converge on the bargaining solution chosen for that group. Only ten trials with robot opponents sufficed to condition the subjects.

[28] As Section 1.6.1 explains, a smoothed Nash Demand Game normally has a unique Nash equilibrium. But computer technology forces the use of a discrete approximation which necessarily has a large number of Nash equilibria. (Exact Nash equilibria are emphasized in the text because everything on the Pareto-frontier in the discrete game is an ϵ-equilibrium with ϵ less than one dime.)

For example, Elster [173] distinguishes twenty different fairness criteria that are supposedly employed under various different circumstances.

Everybody knows that people care about "fairness", and so it figures large in the rhetoric of public debate. But nobody has any clear idea what it means. We chatter all the time about fairness without appreciating at all why the concept has such a powerful hold on our imaginations. Seldom has anyone ever got anything so wrong as Wittgenstein when he said that we have no choice but to remain silent when faced by ideas that we are unable to express.[29] On the contrary, the less we understand something, the more ready we are to talk endlessly about it. Linguistic philosophers therefore set themselves a hopeless task when they try to distill the essence of fairness from a study of what people have to say on the subject. Our culture is so confused about the nature of fairness that we merrily contradict ourselves even in passing from one sentence to the next.

This book tries to sidestep the confusion by offering a stylized explanation of the *origins* of the fairness concept. If we can achieve an understanding of how it works in the situations for which it evolved, I believe that there is some hope that we might begin to be able to use the fairness concept successfully for more than rhetorical purposes on a larger stage. As Heraclitus [149] observes:

> Talk about everything according to its nature, how it comes to be and grows. Men have talked about the world without paying attention to the world or to their own minds, as if they were asleep or absent-minded.

Section 2.3.3 of Volume I explains why I feel similarly about moral questions in general. I believe that the rules that govern our moral behavior are simultaneously more subtle and less grandiose than the folk wisdom built into our language allows. In situations for which the rules are well-adapted, we tend not to notice them at all. Our intuitions about them are therefore primarily founded on observing their use outside the domain for which they evolved, when they stick out like a sore thumb. It is therefore no wonder that what we say about morality is such a bad guide to the way that morality actually works.

---

[29] *Wovon man nicht sprechen kann, darüber muss man schweigen.* This celebrated final sentence from Wittgenstein's [564] *Tractatus Logico-Philosophicus* is a tautology if translated literally. My rendering is more than a little mischievous since Wittgenstein had nothing so mundane in mind. His subtext is that the ultimate questions of philosophy are incapable of verbal expression, but can somehow be appreciated by mystics whose understanding of the *Tractatus* is sufficiently deep that they are able to recognize that Wittgenstein is talking nonsense when he speaks of them!

0.5   Cooperative Game Theory

Cooperative game theory takes a more freewheeling attitude than its stay-at-home sister. It is concerned with those situations in which players can negotiate before the game is played about what to do in the game. It is standard to assume that these negotiations culminate in the signing of a *binding* agreement. Under these conditions, so it is argued, the precise strategies available in the game will not matter very much. What matters most is the *preference* structure of the game, since it is this structure that determines what contracts are feasible.

0.5.1   Games in Coalitional Form.

The literature on cooperative game theory mostly follows the second half of Von Neumann and Morgenstern's [538] book in concentrating on coalition formation in games with many players. To study this issue, Von Neumann and Morgenstern proposed a further simplification of the structure of a game, which has come to be known as the *coalitional form* in recent years. The coalitional form of a game is specified by a *characteristic function V*, which assigns a value $V(S)$ to each possible coalition S of players. Von Neumann and Morgenstern interpreted $V(S)$ as the amount that would be available for the coalition S to share among its members if it should form.

It is not hard to criticize the idea that a coalition must necessarily have a well-defined value. Only in certain games is it true that what a coalition S can get is independent of the behavior of the players excluded from S.[30] However, the games in this class are certainly not without interest since they include most trading games. If the members of a coalition S choose to trade only among themselves, then the surplus they can jointly achieve is independent of any trading that may take place in the rest of the economy.

However, even if we restrict ourselves to such trading games, the notion of the value of a coalition remains troublesome, since it takes for granted

---

[30]Von Neumann and Morgenstern [538] were not satisfied to restrict themselves to this class of games. They began by studying multiplayer, zero-sum games. The formation of a coalition S in such a game will presumably induce the players excluded from S to get together to form the complementary coalition CS. The two rival coalitions can then be treated as two ordinary players in a two-person, zero-sum game. Von Neumann's theorem then tells us that the value $V(S)$ of the coalition S must be its security level. So far so good, but Von Neumann and Morgenstern [538] insisted on interpreting $V(S)$ as a security level even in games that are not zero-sum. Although this tradition is not yet dead, the consensus is now that Von Neumann and Morgenstern went astray on this particular point. (To deal with an n-player game that is not zero-sum, they invented an extra mythical player whose payoffs are carefully chosen so that the $(n+1)$-player game created by his introduction is zero-sum. But why should an analysis of this invented game be relevant to an analysis of the original game?)

that utils can be passed unaltered from one player to another. Von Neumann and Morgenstern introduced this metaphysical notion of "transferable utility" without any serious attempt at justification. The idea is that Adam may need to offer a bribe to Eve in order to persuade her to sign an agreement to cooperate. Such an inducement is called a *side payment*. If they receive payoffs of x_A and x_E as a result of their cooperation in the game and the value of the agreed side payment is t, then the final payoff to Adam will be $x_A - t$ and the final payoff to Eve will be $x_E + t$.

Of course, if the payoffs are in dollars and the players are risk neutral, then there is no difficulty with making such side payments, and the value $V(S)$ of a coalition S can simply be taken to be the sum of the dollars earned by each player in the coalition as a result of their jointly agreed course of action. But when the payoffs cannot be identified with dollars, how meaningful is it to speak of trading utils? It has to be faced, albeit reluctantly, that the answer is almost never. As Chapter 4 of Volume I makes clear, it is hard even to justify *comparing* Adam and Eve's utils. To go one step further and to suggest that utils can be *traded* is to introduce the straw that breaks the camel's back. However, facing up to the fact that transferable utility is an unrealistic assumption has the unpleasant consequence that we can no longer treat $V(S)$ as a number. Instead, it must be treated as a set.

Von Neumann and Morgenstern use the term *imputation* to describe a vector $x = (x_1, x_2, \ldots, x_n)$ of payoffs, one for each player, that corresponds to one of the possible outcomes of a game. In a game without transferable utility, we can define $V(S)$ to be a *set* of imputations by requiring that an imputation x lies in $V(S)$ if and only if the coalition S has available a joint course of action which ensures that each member of the coalition receives a payoff at least as large as that assigned to him by x. As in the case of transferable utility, this definition makes most sense for games in which the opportunities available to a coalition do not depend on the actions of players excluded from the coalition.

The core. Two cooperative solution concepts will be mentioned to provide an indication of how Von Neumann and Morgenstern intended that the coalitional form of a game be used. Recall that cooperative game theory assumes that the players negotiate on how they will play before the game begins. After negotiating an agreement, they are irrevocably bound by that agreement once play commences. A cooperative solution concept tells us something about the agreement that rational players might reach under such circumstances. Cooperative solution concepts sometimes nominate a unique imputation as the one and only agreement possible for rational players, but usually the aim is less ambitious and the concept simply excludes imputations that lie outside a given set of possible agreements.

An imputation y that lies in some $V(S)$ is an *objection* to an imputation x if each player in S prefers y to x. The *core* of a game is the set of all imputations to which no objection can be made. The idea is that nothing outside the core can be viable as the outcome of preplay negotiation by rational players, because the proposal of an imputation x outside the core would result in a coalition S getting together to enforce an objection y.

Like much else in cooperative game theory, this idea seems compelling until one realizes that the behavior it attributes to players during the preplay negotiations is highly myopic. In many games, for example, the core is empty. That is to say, an objection can be found to each and every imputation. If the reasoning that goes with the core were to be applied in such games, the players would never be able to agree on anything! But rational players would not allow themselves to be caught in such an elementary trap. The members of a coalition S would look ahead and see that the mere existence of an objection y to x is not enough to motivate their getting together to veto x. What is the point of vetoing x in favor of y if a new coalition T will then immediately veto y?

The core is much studied in economic textbooks because, under mild conditions, the core allocations in a market game with a large number of players approximate the Walrasian equilibria of the game. Since economics is largely concerned with the study of Walrasian equilibria in perfectly competitive markets, it is not surprising that economists have embraced this result with open arms. It seems to say that they have devoted their lives to studying exactly the outcome on which rational players would agree if they had the opportunity to do so. However, as observed in Section 2.5 of Volume I, just as an argument is not refuted by showing that it leads to an unwelcome conclusion, so it is not confirmed by showing that it leads somewhere attractive. I do not doubt that economists are right to devote much of their time to studying Walrasian equilibria, but I think they deceive themselves when they offer a core analysis of a market game as a reason for doing so. The market-clearing price in large markets is demonstrably *not* achieved as a consequence of rival coalitions registering objections to this or that proposed allocation. Nor would it be in a world inhabited by farsighted superbeings.

Stable sets. In spite of its current popularity, Von Neumann and Morgenstern presumably shared similar doubts about the core, since the cooperative solution concept they proposed does not assume that the players are myopic. Their notion is nowadays known as a Von Neumann and Morgenstern   stable set. They argued that forward-looking players will not cooperate in vetoing an imputation x unless the objection y they have in mind is itself viable as a possible outcome of the preplay negotiations. To be coherent, they suggested that the set Y of such viable imputations must satisfy the following criteria:

- An objection inside Y can be found to everything outside Y.
- No objection inside Y can be found to anything inside Y.

Von Neumann and Morgenstern were aware that a game may have many such stable sets.[31] They regarded different stable sets in the same game as representing different "standards of behavior", but were not very forthcoming on how such a standard of behavior might come to be established. Presumably, a society might operate one standard of behavior rather than another in much the same way that Americans choose to drive on the right while Japanese drive on the left.

As an example, consider the case of a three-player, divide-the-dollar game in which the amount assigned to each player is decided by majority vote. If the players are risk neutral, it makes sense to treat this as a game with transferable utility. The coalitional form of the game is then specified by assigning a number $V(S)$ between 0 and 1 to each possible coalition S. The coalitions containing just one player are $\{1\}$, $\{2\}$, and $\{3\}$. These coalitions can guarantee nothing at all for their members and so $V(1) = V(2) = V(3) = 0$. The remaining nonempty coalitions are $\{1, 2\}$, $\{2, 3\}$, $\{3, 1\}$, together with the *grand coalition* of all three players $N = \{1, 2, 3\}$. These latter coalitions can all guarantee that the whole dollar be divided among their members, and so $V(1, 2) = V(2, 3) = V(3, 1) = V(1, 2, 3) = 1$.

The core of this game is empty.[32] The game has an infinite number of stable sets, of which only $Y = \{(\frac{1}{2}, \frac{1}{2}, 0), (0, \frac{1}{2}, \frac{1}{2}), (\frac{1}{2}, 0, \frac{1}{2})\}$ treats each player symmetrically. No imputation in Y is an objection to any other imputation in Y. Moreover, given any x outside Y, there is an objection inside Y to x. When $x_1 \leq x_2 \leq x_3$, the objection is $y = (\frac{1}{2}, \frac{1}{2}, 0)$.[33]

I have criticized the core, but I do not want to spend time itemizing the criticisms that can be directed at Von Neumann and Morgenstern stable sets[34] or at other traditional cooperative solution concepts based on the coalitional form of a game—like the Shapley value or Aumann and

---

[31]But not that some games have no stable sets at all (Lucas [332]).

[32]Suppose that the coordinates of an imputation x satisfy $x_1 \leq x_2 \leq x_3$. Then $x_1 + x_2 < x_1 + x_2 + x_3 = 1$. An imputation y can then be found such that $x_1 < y_1$ and $x_2 < y_2$. The imputation y is an objection to x for the coalition $\{1, 2\}$.

[33]If it is false that $x_2 < \frac{1}{2}$, then $\frac{1}{2} \leq x_2 \leq x_3$ and so $x = (0, \frac{1}{2}, \frac{1}{2})$.

[34]The reader will already have noticed that, although the players in the stable set story are less myopic than in the core story, they still only look two steps ahead. If they looked further ahead, they might see that it could be in their interests to destabilize an

Maschler's notion of a bargaining set (Chapter 14 of Osborne and Rubinstein [397].) Some of these concepts admittedly capture some of the considerations that matter, and hence are capable of generating intuitively satisfying conclusions when applied to particular games for which other considerations happen to be irrelevant. However, I think it fair to say that the consensus is that none of the traditional cooperative solution concepts come close to capturing *everything* that matters.

It is only with great reluctance that a game theorist gives up the opportunity to prove the general theorems for multiplayer societies that would be possible if an appeal to traditional cooperative game theory were permissible. However, although I find it necessary to make heroic assumptions about the way society works in order to bring game theory to bear, I think it important that the game theory to which appeal is made should itself be free of any heroic tendencies. In the rest of this book, the coalitional form of Von Neumann and Morgenstern will therefore only receive an occasional mention in passing.

0.6 Nash Program

Cooperative game theory presupposes a preplay negotiation period during which the players come to an irrevocably binding agreement on how the game is to be played. Sometimes people speak of a bull session in the bar the night before the game when referring to such a preplay negotiation.

Nash observed that any negotiation is itself a species of game. The moves are the proposals and statements that the players may interchange while bargaining and the outcomes are the possible agreements on which they may finally settle. If all the possible moves that can be made during a preplay negotiation were to be specified formally, the result would be an enlarged game. This enlarged game would then need to be studied *without* presupposing any preplay negotiation—such a preplay negotiation having already been incorporated into its rules. I have referred elsewhere to a game that is to be analyzed without presupposing any preplay interaction at all between the players as a *contest* (Binmore [65]).

Predicting bargaining outcomes. Within the Nash program, the analysis of contests is fundamental. If noncooperative game theory had advanced to a stage which allowed us to write down the "solution" of any contest, then we would also have solved all the problems of cooperative

---

imputation y in Y, because the flurry of objections and counterobjections that follow might eventually end in a new imputation in Y that they prefer to y.

game theory. The appropriate cooperative solution concept for any game would simply be the "solution" of the contest obtained by prefixing the original game with an appropriately formalized negotiation period.

However, to attempt such a frontal attack on the problem would be absurdly ambitious. The set of bargaining ploys to which people have been known to resort when bargaining for real is bewilderingly large. How could one possibly construct an extensive-form game that is sufficiently general to capture each twist and turn that a negotiation might conceivably take?

Nash recognized that a frontal attack on the bargaining problem is not practical. In spite of the beliefs of some authors, the Nash program therefore does *not* boil down to the imperative that only noncooperative bargaining models should be used in predicting bargaining outcomes. Nash [380, 381, 382] did not regard cooperative and noncooperative game theory as rival approaches to the problem of analyzing bargaining problems. On the contrary, he saw them as providing complementary insights, with the strengths of one serving to buttress the weaknesses of the other. In particular, he advocated the use of cooperative solution concepts when seeking to predict a bargaining outcome.

The reason for using a cooperative solution concept for predictive purposes, rather than analyzing a noncooperative model of the actual bargaining procedure, is that the latter will necessarily incorporate all kinds of fine detail about which the modeler is unlikely to be fully informed and which are probably irrelevant to the final outcome. For example, in the grand old game of Cricket, the traditional rules require that the players wear white. But nowadays it is common for one team to wear red and the other green without any suggestion being made that the colors worn by the two teams have an effect on the outcome of the game.

Attention to detail. Appealing to a cooperative solution when seeking to predict a bargaining outcome cuts through all the tiresome fuss about which details matter and which do not. When passing to the coalitional form of a game, all the fine details present in the extensive form are swept away and so cannot affect a prediction derived from a cooperative solution concept.

One sometimes hears a particularly silly criticism of noncooperative bargaining models based on this last point. It is said that predictions derived from cooperative solution concepts are superior to those obtained by analyzing the equilibria of noncooperative bargaining models *because* the former do not depend on fine details of the bargaining procedure. This would only be true if *none* of the fine details of the bargaining procedure were relevant to the final outcome. In Cricket, for example, a layman might

think it unimportant whether the bowler is allowed to roughen the ball a
little on one side from time to time. However, it turns out that that it
makes an enormous difference to the amount of swing that he can put on
the ball. Very slight changes in the rules concerning the degree to which
the bowler is allowed to interfere with the ball by polishing or scratching it
may therefore make an enormous difference to how the game is played. To
insist on predicting the outcome of a game of Cricket without paying any
attention to the details of this particular rule would therefore be absurd.

Similarly, there are certain issues to which bargaining outcomes are
highly sensitive, whose importance is not immediately apparent to the untrained intuition. For example, in the Rubinstein [445] bargaining model,
who gets how much is determined by the relative impatience of the two
bargainers. A cooperative solution concept that takes no account of such a
matter of detail therefore cannot possibly predict the agreement that will
be reached when the Rubinstein procedure is used. The class of bargaining
procedures for which such a cooperative solution concept predicts successfully must therefore exclude the Rubinstein procedure—and doubtless other
procedures as well.

Gedanken experiments. It is a blessing that a cooperative solution
concept should ignore the many details of a bargaining procedure that are
irrelevant to the bargaining outcome—but a curse in the case of those few
details which actually are significant. Of course, if one hits upon a cooperative solution concept that just happens to give the right prediction for the
current configuration of significant details, then the curse loses its sting.
But how is one to know whether one has hit upon the right cooperative
solution concept?

Nash's advice on this subject is not easy to implement, but it constitutes
the hallmark of the Nash program. Instead of taking a proposed cooperative
solution concept on trust, Nash proposed that it be *tested* with the help
of noncooperative bargaining models constructed to capture the essence of
the bargaining procedures whose outcome the cooperative solution concept
supposedly predicts.

Whenever an applied worker proposes the use of a particular cooperative solution concept to predict a bargaining outcome, the Nash program
dictates that we ask the applied worker what he knows about the bargaining institutions in use within his field of application. We then perform a
thought experiment by constructing a bargaining game in extensive form
that satisfies these institutional requirements. Next the equilibria of this
bargaining game are computed. If we are able to solve the equilibrium selection problem, we will then know the agreement on which rational players

would settle if they were bound by the rules of the noncooperative bargaining game studied. This agreement may or may not coincide with the prediction obtained using the cooperative solution concept. If the two do not coincide, then the applied worker's confidence in his cooperative solution concept cannot be sustained. The thought experiment will have refuted his theory.

Windtunnel models. Using untested cooperative solution concepts to predict bargaining outcomes has been likened to strapping wings on your back and jumping off a high building in the hope that you will fly. This is a particularly apt simile for the Nash program, since there is a close analogy between the noncooperative bargaining games used to test cooperative solution concepts and the model airplanes used in windtunnel studies. In constructing a model for a windtunnel experiment, an engineer does not aim to produce an exact replica of its full-scale counterpart. Indeed, he will usually have only a rough-and-ready idea of what the final airplane will look like. However, he will take enormous pains to ensure that the aerodynamic properties of his models be as close as possible to the aerodynamic properties of the full-scale airplane under design. Once this aim has been achieved, the simpler the model can be made in its other characteristics, and the simpler it will be to interpret the windtunnel data.

Just as an aeronautical engineer will only have a limited amount of information about the design of his final product, so an applied worker seeking to predict a bargaining outcome is unlikely to be well informed about the detailed structure of the bargaining procedure in use. Even the bargainers themselves would probably be hard put to enumerate all the twists and turns that their negotiation might take. However, just as an aeronautical engineer needs only a sketchy notion of the seating arrangements in an airliner when constructing a windtunnel model, so we can make do with very fuzzy information on many matters of detail when constructing a noncooperative bargaining game to be used within the Nash program. In fact, the less we know about the bargaining procedure in use, the easier it becomes to refute the claim that a given cooperative solution concept will necessarily predict its outcome.

The aim is always to begin by constructing the *simplest* noncooperative bargaining games consistent with the known facts—and the simpler the game we have to analyze the less onerous the task of analysis becomes. Of course, if we are entirely ignorant even of those details which materially affect the bargaining outcome, then we shall be wasting our time altogether in seeking to employ the Nash program—just as it is a waste of time to run windtunnel experiments unless one already has some clear thoughts on the aerodynamic features of the airplane being designed.

Rational bargainers. The task of constructing windtunnel bargaining games is simplified by the necessity of working with stripped-down models of the players as well as stripped-down models of the bargaining procedure. I do not want to repeat the arguments of Section 1.2.2 of Volume I in favor of using *homo economicus* as our model of man rather than one of the varieties of *homo behavioralis* or *homo ethicus* that are sometime proposed. Economists know perfectly well that the *homo economicus* paradigm captures only some small part of the full richness of the human species. However, this book is not concerned with the full range of human experience. Decisions on social contract issues are not taken with the same careless abandon with which we pay inflated prices for the privilege of overloading our supermarket carts with items we finally consign to the trash can. As the old form of the English marriage service puts it, decisions on social contract questions are to be taken "advisedly and soberly". One must therefore look for the heroes and heroines of this book in the works of Jane Austen rather than those of the Brontë sisters.

Fortunately, *homo economicus* is a simpler beast than *homo sapiens.* When two members of the latter species interact, much of what is going on consists of attempts by each to exploit the folly of the other. But rational players will regard the insults, the flattery, the appeals to fairness and all the other bargaining bombast hallowed by time and custom as so much idle chatter. Rational players pay attention only to things that matter. For example, in neighborhood Poker games, people play in all kinds of crazy ways. When playing in such nickel-and-dime games, the interest lies largely in seeking to fathom and exploit the psychological weaknesses of your opponents. But when the players are known to be rational, their psychological makeup ceases to be an issue. Indeed, when Cutler [140] kibitzed two players using the optimal strategies that he had calculated for two-player, pot-limit Poker, he remarked that it was as entertaining as watching paint dry! Similarly, watching two rational players bargain is unlikely to be a bundle of laughs. However, it is the factors that determine how two such players bargain which correspond to the aerodynamic considerations that are crucial in designing useful windtunnel models.

Significant details. It is not hard to list some of the factors that will be crucial when rational players bargain. The bargaining outcome will depend on what is being bargained about and the preferences the players have over the various agreements they might reach. It will also depend on how the players feel about the consequences of a disagreement. On both counts, a player's attitude to risk-taking or delay will be relevant. It may also be of strategic importance to take note of the manner in which disagreements can occur. For example, a breakdown in the negotiations may be caused by one of the bargainers walking off to take up his best outside option. Or it could be that the negotiations have to be abandoned as a result of the intervention of some external event beyond the players' control. Each possibility will affect the final outcome in different ways.

These and other similar factors will be discussed in Chapter 1. Of the other considerations that may affect the bargaining outcome, only two will be mentioned here. The first concerns what the players do or do not know about the rules of the game and the preferences of their opponents. As documented in Binmore, Osborne and Rubinstein [80], the bargaining outcome is more than sensitive to such informational questions—they are utterly crucial. Moreover, the literature on bargaining with incomplete information is an almighty mess, because the models typically pose the equilibrium selection problem in a particularly intractable form. This book therefore systematically sidesteps informational questions in bargaining by restricting attention to games of complete information. In such games, everything of interest to the players at the beginning of the game is taken to be common knowledge.

The second consideration concerns the commitment possibilities available to the players. For game theorists, a commitment is an *irrevocable* threat or promise—from which it is not possible to retreat if later events lead one to regret having been so rash. As Section 1.2.4 of Volume I explains, it is sometimes claimed that rational individuals somehow have the capacity to commit themselves now to an action at some future time t that they definitely would not choose at time t if they had been left with any freedom to choose. Such a power to make commitments can be a highly effective instrument for getting your way when bargaining. For example, if Eve could advertise a commitment to accept no less than 99 cents before playing the Ultimatum Game, then Adam would be forced to restrict his demand to one penny. However, the work of Schelling [464] has taught game theorists to treat the commitment issue very gingerly. Claims that players are able to make forward commitments in the absence of any enforcement mechanism are treated with great suspicion. When commitment power *is* to be attributed to the players, each commitment opportunity is formally modeled as a move in the game—which is then analyzed without any further commitment assumptions being smuggled into the analysis.

I make a fuss about the commitment question in this chapter for two reasons. The first is that Nash [380, 382] wrote before the modern attitude to commitment assumptions developed. He was therefore very free in building commitment opportunities into the rules of his bargaining games.

By modern standards, the noncooperative bargaining games he studied are therefore of doubtful value for the purposes of the Nash program. A second reason is that the bargaining results are to be applied in a social contract context. I believe that both Harsanyi [233] and Rawls [417] made an error of judgment in assuming that the citizens in the societies they consider are somehow committed to the hypothetical deal on which they would agree if they were to bargain in ignorance of their current and future roles in society.[35] As Hume [267, p.306] put it: "We are not surely bound to keep our word because we have given our word to keep it." Still less are we bound to keep a word that would have been pledged under certain hypothetical circumstances!

Self-policing rules. This discussion of commitment brings us to the final point that needs to be made about constructing windtunnel models for the Nash program. A rational player always chooses whatever available action maximizes his expected utility. He will therefore not honor a promise or a threat made in the past unless it is expedient for him to do so.

The same also goes for the rules of the game he is playing. It would be most unusual to find anyone acting as a referee when two people bargain. The players can therefore cheat on the rules without the risk of being punished. The rules that are proposed for a bargaining game therefore need to be *self-policing*. If the rules are honored, it will be because the players find it expedient to honor them.

A similar point was made in Section 1.2.4 of Volume I about formulating the Game of Morals. What really constrains the players are the rules of the Game of Life. If they honor the rules of the Game of Morals, it must therefore be because they see it as being in their interests to do so. The same applies when formulating a bargaining game. We waste our time if we do not accept that the rules we invent must be consistent with the players' incentives.

However, this is not all bad news for the problem of formulating windtunnel models. One must often face the task of choosing among many alternative models, and it can be exceedingly helpful to have a criterion that allows one to dismiss many of the possibilities.

For example, in the Rubinstein [445] bargaining model, the players take turns in making proposals. How long should a player have to wait after rejecting a proposal before he is allowed to make a counterproposal?

---

[35]However, neither Harsanyi nor Rawls would wish to join that merry band of brothers who seek to use use commitment arguments to show that cooperation is rational in the one-shot Prisoners' Dilemma. The fallacies on this subject listed in Chapter 3 of Volume I provide an awful warning that one cannot afford to relax one's guard for a single instant on the commitment question.

Whatever counterproposal he plans to make, it would be better for him to make it with as little delay as possible. Unless the rules are somehow enforced by an external agency, there is therefore no point in studying a game in which a year must go by between each proposal. Players will simply ignore such a rule. In Chapter 1, we therefore study the limiting case of the Rubinstein model in which the interval between successive proposals is allowed to converge to zero. The advantage a player can get by making his proposal too soon then becomes vanishingly small.

Unfinished business. If the Nash program were fully realized, all cooperative solution concepts would come accompanied with examples of the noncooperative bargaining games whose outcomes they succeed in predicting. An applied worker would then be able to compare what he knows about the bargaining institutions in his field of application with the available examples of noncooperative bargaining games in the hope of finding a match.

Unfortunately, matters have not advanced anywhere near this far. Even in the extensively studied case of the Nash bargaining solution, to which the next chapter is largely devoted, the practice on the ground is disappointing. We know a great deal about when and how to apply this particular cooperative solution concept. We also know how to relate the parameters with which it comes equipped to the bargaining environment in which it is to be applied. However, this knowledge is largely neglected by labor economists who use the Nash bargaining solution in seeking to predict the outcome of wage negotiations. I hope that Chapter 1 will persuade at least some labor economists that it isn't perhaps so hard after all to use the Nash bargaining solution in the manner that the Nash program recommends.

0.7 Implementation

Within the Nash program, noncooperative game theory is fundamental. Cooperative solution concepts are used to predict the outcomes of negotiations between rational players, but the ultimate test of the value of such a cooperative solution concept is whether the predictions it generates coincide with the relevant equilibrium outcomes of appropriate noncooperative bargaining games. Unfortunately, the simplicity of this viewpoint is sometimes obscured by a failure to distinguish between the Nash program and implementation theory.

The Judgment of Solomon. The Judgment of Solomon provides an early example of an implementation problem. When confronted by two women disputing the motherhood of a baby, King Solomon famously proposed that the baby be sliced in two so that each claimant could have half. The false mother agreed to the judgment, but the Bible tells us that the true mother's "bowels yearned upon her son" so that she begged for the baby go to her rival rather than being hacked in two (1 Kings: 3: 26). King Solomon then knew the true mother and awarded her the baby.

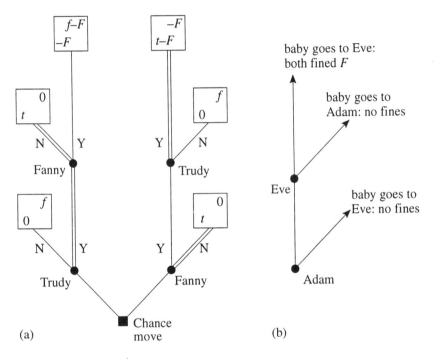

Figure 0.3 The Judgment of Solomon.

Actually, the biblical story does not support Solomon's proverbial claim to wisdom particularly well. His scheme would have failed had the two women been more strategically minded. Solomon would have done better to use the game of Glazer and Ma [207] illustrated in Figure 0.3(a) in seeking to assign the baby to its true mother. In this game, the true mother is Trudy and the false mother is Fanny. The game begins with a chance move that determines which of Trudy and Fanny is first asked whether she is the mother. If the first woman questioned denies that the baby is hers, it is given to her rival. If the first woman claims the baby, then the same question is put to the second woman. If the second woman denies being the mother, the baby is given to the first woman. If both women claim the baby, it is given to the second woman, but Solomon imposes a fine of F on both for trying his patience. It is in the choice of F that the famed wisdom of Solomon is required most. He needs to ensure that $f < F < t$, where f is the payoff that Fanny gets if she is awarded the baby and t is the payoff that Trudy gets if she is awarded the baby.

The doubled lines indicate the result of applying backward induction to the game. The unique subgame-perfect equilibrium that results always awards the baby to Trudy without any fine needing to be paid.

Designer games. The Judgment of Solomon is an example of an implementation problem. A more general formulation requires the terminology introduced in Section 2.3.3 of Volume I. Let $\mathcal{P}$ be the set of all possible preferences that a rational individual might hold over a set S of social states. If the society contains only Adam and Eve, then their attitudes to these social states are built into their preference relations $\preceq_A$ and $\preceq_E$. Each possible set of attitudes can therefore be identified with a preference profile $(\preceq_A, \preceq_E)$ in the set $\mathcal{P} \times \mathcal{P}$. A *social welfare function* $F : \mathcal{P} \times \mathcal{P} \to \mathcal{P}$ maps each such preference profile into a communal preference $\preceq = F(\preceq_A, \preceq_E)$.

A social welfare function may be envisaged as summarizing a particular conception of the "common good". Most attempts to put ethics on a quantitative footing for application to public policy issues, are *teleological* in that they take some such conception of the common good as fundamental (Section 2.3.5 of Volume I). Teleological theories proceed as though we had *a priori* knowledge of the proper aims and purposes of a society, and so devote themselves only to the question of how such aims and purposes can best be achieved.

My own view is that approaches to ethics based on such teleological foundations are unacceptably metaphysical. Nor do I care for political parties with paternalistic platforms. We whigs are deeply suspicious of authoritarianism, even in its most benign forms. However, in the current section, it is necessary to put aside such prejudices in order to adopt the attitude of a social reformer who has decided what is optimal for society and is now considering how best to implement the reforms he believes appropriate.

Section 2.3.4 of Volume I describes Arrow's paradox for social welfare functions. Arrow [19] showed that only social welfare functions that give one of the citizens dictatorial power satisfy his apparently innocent requirements. Section 4.2.4 of Volume I describes one way of evading this paradox. Others involve lessening the "collective rationality" requirements that the use of a social welfare function imposes on a society. In particular, a preference relation in $\mathcal{P}$ is necessarily transitive (Section 2.2.4 of Volume I). But do we always want to insist that a society exhibits the same standards of rationality as an individual? I think the answer must be no, for the reasons given when the notion of a "common good" was discussed in Section 2.3.6 of Volume I.

If this judgment is accepted, we can turn our attention away from social welfare functions to social decision functions $G : \mathcal{P} \times \mathcal{P} \to S$. Notice that $s = G(\preceq_A, \preceq_E)$ is *not* a communal preference relation like $F(\preceq_A, \preceq_E)$. A social decision function G maps a preference profile into a *social state* rather than a preference profile. For each set of attitudes that a society might hold, G specifies which social state should be implemented. A social decision rule is a little more sophisticated. Rather than mapping a preference profile into a single social state, a *social decision rule* $g : \mathcal{P} \times \mathcal{P} \to 2^S$ maps a preference profile into a set $T = g(\preceq_A, \preceq_E)$ of social states.[36]

The implementation problem for a reformer equipped with a social decision rule g is to design rules[37] for a noncooperative game whose solution set coincides with $g(\preceq_A, \preceq_E)$ whenever Adam and Eve have the preferences $\preceq_A$ and $\preceq_E$. Of course, if the reformer knew Adam and Eve's preferences, he could simply impose the social state $g(\preceq_A, \preceq_E)$ without the need to design any game. However, Adam and Eve's preferences are their own private information. The reformer's purpose in designing the game is to compensate for his ignorance by inventing a system of incentives that persuades the players to bring about the reformer's goal, although each is pursuing only his own individual self-interest.[38]

In the case of the Judgment of Solomon, we can use the label Adam for the first of the two women to be questioned and Eve for the second. A social state is a pair $s = (W, v)$, in which W is either A for Adam or E for Eve, depending on who is assigned the baby, and v is the fine to be imposed on both Adam and Eve. Solomon needs to consider two preference profiles. The first of these profiles corresponds to the case when Trudy is Adam and Fanny is Eve. Adam's preferences are then described by a utility function u_A^1 satisfying $u_A^1(A, v) = t - v$ and $u_A^1(E, v) = -v$. Eve's

---

[36]The notation Y^X denotes the set of all functions from X to Y. A standard abuse of notation identifies 2^X with the set of all subsets of X. (In formal set theory, the natural number $n + 1$ is *defined* to be the set $\{0, 1, 2, \ldots n\}$. In particular, $2 = \{0, 1\}$. Thus 2^X is strictly the set of all functions defined on X that take the values 0 or 1. Each such function ϕ can be identified with the subset T of X defined by $T = \{x : \phi(x) = 1\}$.)

[37]In my *Fun and Games* (Binmore [70]), I refer to such a set of rules as the *script* for a game. It is usual to use the more prosaic term, *game-form*. A script or game-form becomes a game when the players' preferences are specified.

[38]Economists refer to this as a principal-agent problem. The reformer is the principal who must act through agents who know things hidden from the principal. He therefore seeks to create an incentive-compatible scheme of inducements that will bend them to his purpose in spite of his ignorance of their private information.

preferences are described by a utility function u_E^1 satisfying $u_E^1(A, v) = -v$ and $u_E^1(E, v) = f - v$. In the second case, the two roles are reversed and so Adam's utility function u_A^2 satisfies $u_A^2(A, v) = f - v$ and $u_A^2(E, v) = -v$. Eve's utility function u_E^2 satisfies $u_E^2(A, v) = -v$ and $u_E^2(E, v) = t - v$.

Solomon can be taken to be a utilitarian for the purposes of this example. In each case i, he wants to maximize $u_A^i(s) + u_E^i(s)$. His social decision function therefore selects social state $(A, 0)$ in case $i = 1$ and social state $(E, 0)$ in case $i = 2$. If we are prepared to identify the solution set of a noncooperative game with its set of subgame-perfect equilibria, then the rules of the game that Solomon can use to implement his social decision function are shown in Figure 0.3(b). (Notice that Figure 0.3(a) just consists of Figure 0.3(b) written twice—once with Trudy as Adam and Fanny as Eve, and once with Fanny as Adam and Trudy as Eve.) Whichever of the two cases applies, a backward induction analysis of the game of Figure 0.3(b) results in no fine being levied and the baby being assigned to Trudy. In the jargon of the trade, the game provides an incentive-compatible mechanism for inducing the players to reveal their types.

In assessing such a result, it is important not to lose track of the assumptions made along the way—especially those assumptions which have been made only implicitly. Note in particular that Solomon must not only be able to *enforce* the rules of his implementation game, he must be able to *commit* himself not to change the rules if things do not go as he anticipates. In equilibrium, either Adam or Eve will say no when asked if the baby is hers. But if both Adam and Eve were to say yes, Solomon would like to scrap the whole scheme and start again. He certainly would not like being forced to assign the baby in ignorance of its mother while simultaneously levying a heavy fine. However, if the women thought it possible that Solomon might change the rules in such an eventuality, then they would not believe they were playing the game he originally proposed and the whole story would collapse.

Implementability criteria. A practical reformer will regard it as a waste of time to propose unworkable reforms. He will therefore confine his attention to *implementable* social decision rules. There is a substantial literature on the characteristics of such rules, the highlight of which is the Gibbard-Satterthwaite theorem.

As Chapter 3 of Volume I testifies, many fallacious arguments against identifying (*hawk, hawk*) as the noncooperative solution of the one-shot Prisoners' Dilemma have been proposed. However, game theorists regard the principle that a player will not use a strongly dominated strategy in a noncooperative game as the weakest of all conceivable rationality criteria.

Gibbard [203] and Satterthwaite [460] therefore studied implementation in strongly undominated strategies. Can rules for a noncooperative game be found so that the outcomes remaining after strongly dominated strategies have been deleted always coincide with the outcome set selected by a given social decision rule? To this question, the Gibbard-Satterthwaite theorem provides an answer reminiscent of the Arrow paradox. Implementation in these terms is possible if and only if the social decision rule is dictatorial.[39]

What of implementation by Nash equilibria? Maskin [345] has shown that any Nash-implementable social decision rule must be monotonic.[40] Since the social decision rule attributed to Solomon in the Bible is not monotonic, it follows that no rules for a game that he might invent will always result in his wishes being carried out if the women claiming the baby are rational.[41]

The idea of a monotonic social decision rule will be illustrated by an example. Suppose that Adam's preferences over the set $S = \{a, b, c, d\}$ are given by $a \sim_1 b \prec_1 c \prec_1 d$. If his preferences change to $a \prec_2 c \prec_2 b \sim_2 d$, then b is ranked at least as high in his second set of preferences as in the first. A *monotonic* social decision rule has the property that a state s in its outcome set remains there when the preference profile changes so that everybody ranks s at least as high as before. Thus, if a monotonic social decision rule includes b in its outcome set when the preference profile is $(\preceq_1, \preceq_1)$, then b must also be in the outcome set for the preference profile $(\preceq_2, \preceq_1)$.

Implementation and the Nash program. Implementability criteria seem relevant to the Nash program. A social decision rule is not the same

---

[39] As in Arrow's paradox, the proof requires that $\mathcal{P}$ contain *all* preference profiles and that S have at least three elements. In addition, for each social state s in S, the outcome set of the social decision rule must coincide with s for some profile of preferences.

[40] Suppose that g is an implementable social decision rule for two players. For any s in $g(\preceq_A, \preceq_E)$, we can then find a pair of strategies (α, β) which is a Nash equilibrium that generates the outcome $s = \pi(\alpha, \beta)$. To say that (α, β) is a Nash equilibrium means that $\pi(\alpha, \beta) \succeq_A \pi(a, \beta)$ and $\pi(\alpha, \beta) \succeq_E \pi(\alpha, b)$ for all strategies a and b. These inequalities remain valid if Adam and Eve's preferences change in such a way that both players rank s at least as high as before. Thus s remains a Nash equilibrium in the new situation and so must be in the outcome set of g for the new preference profile. Thus g is monotonic.

[41] In the biblical case, the available social states are A, in which Adam gets the baby, E, in which Eve gets the baby, and H, in which each woman gets half the baby. When Trudy is Adam and Fanny is Eve, their preferences are $H \prec_A E \prec_A A$ and $A \prec_E H \prec_E E$. In this case, Solomon wants to implement A. When Fanny is Adam and Trudy is Eve, their preferences are $E \prec_A H \prec_A A$ and $H \prec_E A \prec_E E$. In this case, Solomon wants to implement E. But A ranks at least as high in everybody's estimation in the second case as in the first. Thus A would be in the outcome set of Solomon's decision rule in the second case if his rule were monotonic.

thing as a cooperative solution concept, but the two ideas are closely related. Obviously, a cooperative solution concept is of no interest at all for the purposes of the Nash program if it is not implementable. One might therefore hope that implementation theory would provide some criteria to restrict the class of cooperative solution concepts that need be considered. However, with regard to social decision functions for three or more players, it turns out that anything whatever can be *approximately* implemented by subgame-perfect equilibria (Abreu and Matsushima [2]).[42]

It follows that implementation theory is of doubtful relevance to the Nash program and vice versa. Nevertheless, the two are often inexplicably confused. Part of the reason is historical. In the fifties and sixties, the noncooperative half of Von Neumann and Morgenstern's book received little attention from economists, since zero-sum games have few economic applications. Attention was focused instead on cooperative game theory. Following Nash's [380] axiomatic characterization of the Nash bargaining solution (Section 1.4.2), it became fashionable to invent ever more exotic systems of axioms to characterize a longer and longer list of cooperative solution concepts. However, Nash's idea that noncooperative bargaining models should be used to test the domain of applicability of a cooperative solution concept was so far ahead of its time that it was almost entirely overlooked.

In the seventies and eighties, noncooperative game theory staged a comeback. Economists came to appreciate that Nash's other major contribution to game theory—the idea of a Nash equilibrium—could be used much more effectively in economic applications than had previously been realized. With this revival of noncooperative game theory came a better understanding of the Nash program. As a consequence, the flow of new cooperative solution concepts was stemmed for a time. But the appearance of implementation theory has put the industry back into business. However, nowadays it is not thought not enough to promulgate some axioms and then to describe the cooperative solution concept that they characterize—one must also invent the rules of a game that implements the new cooperative solution concept. This would be a harmless species of intellectual endeavor if some of its practitioners were not under the impression that they are furthering the aims of the Nash program. Let me therefore emphasize that the Nash program and implementation theory are very different animals.

I noted earlier that implementation theory forms part of a *teleological*

---

[42]In the sense that, given any $\epsilon > 0$, it is possible to formulate the rules of a game which implements the social decision function with probability at least $1 - \epsilon$ for all preference profiles. Abreu and Sen [6] call this *virtual* implementation. Abreu and Matsushima [2] actually show that all social decision functions for three or more players can be virtually implemented via the iterated deletion of strongly dominated strategies.

approach to social welfare. In such an approach, the axioms that characterize a social decision rule or a cooperative solution concept are fundamental. They represent *a priori* goals that a reformer takes as given. The rules of the noncooperative game that he invents to implement these goals are secondary. By contrast, the Nash program has no teleological pretensions at all. Its purpose is to assist in predicting actual bargaining outcomes. These are determined by the bargaining game actually in use. It is this complicated and ill-understood noncooperative game that is therefore fundamental. A cooperative solution concept is an auxiliary device that is intended to assist in predicting what will happen when it is played.

It is true that both the Nash program and implementation theory require the analyst to invent rules for noncooperative games. However, the Nash program imposes severe constraints on an analyst's powers of invention. In order for his analysis to provide a useful thought experiment within the Nash program, the rules of the game that he invents must not only be self-policing, they must also capture the strategic essence of the game that is actually used in practice. No such constraints bind an analyst working in implementation theory. His rules do not need to be self-policing. It is assumed that the reformer is able to enforce any rules whatever. Nor do the rules need to follow any preestablished pattern. They can be as bizarre as the imagination of their inventor allows.

Far from being candidates as windtunnel models for use in the Nash program, the sets of rules that the conventions of the implementation literature admit sometimes define mathematical objects whose admissibility as games is at best doubtful. Consider for example the two-player "game" in which a player loses unless he shouts louder than his opponent. This game has no Nash equilibrium. If my shout is a best reply to your shout, then I am shouting louder than you. But then your shout cannot be a best reply to mine, and you would do better to shout louder.

Such existence problems are standard in decision problems with an infinite action space that is not compact. The crudest case occurs when I am asked how many dollars I am to be given. Whatever I say, it cannot be optimal because I could always ask for one dollar more. However, such difficulties are artificial. They arise only in problems for which an inappropriate mathematical formalism has been adopted.[43]

---

[43]The safest procedure is to stick with finite games. But sometimes the mathematics required to solve a finite game are much harder or less transparent than the mathematics required to solve a limiting version of the game. For example, the infinite version of the Ultimatum Game in which Adam can choose any demand from the compact interval [0, 1] has only one subgame-perfect equilibrium and hence is simpler than the realistic version in which Adam can only choose whole numbers of cents. Nevertheless, the infinite version is an acceptable surrogate for its finite cousin because the subgame-perfect equilibria of

In practice, there is a limit to how loud someone can shout or how many dollars could be available for distribution. Nevertheless, rules with a shouting loudest component are a staple of the implementation literature (Maskin [345], Moore [365]). Such unrealistic rules have no place within the Nash program. The same goes for most "bargaining games" invented for implementation purposes. It seems to me an abuse of language to call something a bargaining game unless it is an idealization of a game that actual people might conceivably use while bargaining. In practical terms, this means as a minimum that we need to restrict our attention to infinite games that approximate a relevant finite game in a sensible topology.

---

the latter are all approximated by the unique subgame-perfect equilibrium of the former. But an infinite version of the Ultimatum Game in which Adam chooses his demands from the open interval $(0, 1)$ has no subgame-perfect equilibria at all, and hence is useless as a surrogate for the finite case. Such infinite games seem of little interest to me because they cannot be approximated by games that real people could actually play.

William Blake's drawings for Wodehouse's *Aunts Aren't
Gentlemen* deserve wider recognition. The negotiation
illustrated here didn't go at all well for the hero. As he
puts it, "When I suggested twenty-five, a nicer-looking
number than his thirty, he shook his grey head regretfully,
and so we went on haggling, and he haggled better than
me, so that eventually we settled on thirty-five."

Chapter 1

Nuances of Negotiation

Yet will I distress Ariel, and there shall be heaviness and sorrow.

Isaiah 29:2

1.1 Realistic Bargaining Models

The pragmatic reason for advocating the use of the original position in reforming society is that we already tacitly use it in settling picayune fairness questions. It therefore has a strong intuitive appeal when proposed for use on a grand scale. However, if we fail to use it for large-scale purposes in the same way that it is currently used for small-scale purposes, its appeal will evaporate. It is particularly important to be realistic about the bargaining hypothesized as taking place behind the veil of ignorance. If the hypothesized bargaining were outside the experience of the players using the device of the original position, how could they be using it at present—albeit only for such petty matters as deciding who should wash tonight's dishes?

Progress therefore depends on having access to a sound economic theory of bargaining. Not only must such a theory be realistic in the rules it postulates to govern how the bargainers negotiate, it must also put aside the temptation to attribute motives to a player beyond the urge to advance his own self-interest. Fairness criteria therefore have no place in the bargaining theory to be adopted. Aside from other considerations, there would be no point in proposing the device of the original position as a final court of appeal on fairness questions if it could not itself be used without making fairness judgments. This is not to say that theories of fair bargaining are meaningless or devoid of interest. On the contrary, one of the main objectives of this book is to defend a particular view of what constitutes a fair

59

bargain. It is simply to insist that such a view should emerge as one of the outputs of the theory rather than being subsumed among its inputs.

Volume I is full of references to Nash equilibria of noncooperative games. In this chapter, Nash's name will appear on nearly every page. We shall hear of Nash bargaining problems, the Nash bargaining solution and the Nash Demand Game. Above all, the underlying ethos will be that of the Nash program. I have to admit to being more of a grasshopper than an ant when it comes to the reading of introductions. But fellow grasshoppers who skipped the account of the Nash program in the Introduction to this volume may wish to look again at Section 0.6 if the methodology of this chapter proves puzzling.

It is also necessary to warn noneconomists that the chapter contains a substantial amount of technical material. This is usually not at all deep, but some of it needs to be assimilated if the following chapters are to be properly understood. One possible strategy would be to skim the chapter in the first instance, taking note only of the properties of the Nash bargaining solution, but it will then be necessary to return to the chapter repeatedly when seeking to follow the later arguments.

1.2   Bargaining Problems

Attention will be restricted to two-player bargaining problems with *complete information*. This means that the parameters of the bargaining problem are common knowledge among the players. In real-life bargaining, we are seldom so lucky. For example, when a union bargains with a firm about the wage rate, it may not know the firm's revenue with any precision. Part of its problem will then be to infer what it can about the amount of surplus available for division from the manner in which the firm conducts the negotiation. Nash [380] suggested that "haggling" be used as a substitute for "bargaining" in such cases of incomplete information. This suggestion has not been taken up in the literature, but all references to bargaining in this chapter will follow Nash in taking for granted that only the complete information case is intended.

1.2.1   Payoff Regions

Two people bargain until they reach an agreement or the negotiations break down. An agreement consists of an undertaking by each party to carry out some action. In the general case, the set of agreements available to the bargainers can therefore be identified with the set of strategy pairs in a certain game G. It follows that we can view any negotiation as a discussion about how some game G should be played.

Noncooperative payoff regions. Figure 1.1(a) shows a simple candidate for G in strategic form. Adam's two pure strategies are the rows, *top* and *bottom*. Eve's two pure strategies are the columns, *left* and *right*. If the players stick to pure strategies, there are therefore four ways the game can be played: $(top, left)$, $(top, right)$, $(bottom, left)$ and $(bottom, right)$. The payoff pairs $P = (5, 1)$, $Q = (1, 5)$, $R = (3, 4)$ and $S = (2, 3)$ that correspond to these four pure strategy pairs are shown in Figure 1.1(b). However, the players are not restricted to using pure strategies. They can also use mixed strategies.

If the game were being analyzed as a contest—which means without any preplay communication at all—the randomizing activities of the two players would necessarily be independent. The shaded region V of Figure 1.1(b) shows the set of payoff pairs that can be achieved if Adam uses one mixed strategy and Eve *independently* uses another. One might call V the noncooperative payoff region for the game G.

The line segment labeled ℓ shows the set of payoff pairs that are possible once Adam has settled on the mixed strategy in which he plays *bottom* with probability ℓ, but Eve has yet to decide what to do. The line segment labeled m shows the set of payoff pairs that are possible once Eve has fixed on the mixed strategy in which she plays *right* with probability m, but Adam remains undecided. If Adam chooses ℓ and Eve chooses m, the resulting outcome is therefore the point of intersection of the line segments labeled ℓ and m in Figure 1.1(b). The set V is the collection of all such points as ℓ and m vary over all possible mixed strategies for Adam and Eve respectively.

In Figure 1.1(b), it turns out that if Adam chooses $\ell = \frac{4}{5}$, then the corresponding line segment is horizontal. This guarantees that Eve will get the same payoff in reply to ℓ whatever she does. All her replies to ℓ are therefore best replies. Similarly, if Eve chooses $m = \frac{2}{3}$, then the corresponding line segment is vertical. This guarantees that Adam will get the same payoff whatever he does. All his replies to m are therefore best replies. It is therefore a Nash equilibrium for Adam to use the mixed strategy $\ell = \frac{4}{5}$ and for Eve to use the mixed strategy $m = \frac{2}{3}$.

The game G has no other Nash equilibria. Its unique Nash equilibrium outcome therefore consists of the payoff pair $N = (3\frac{2}{5}, 2\frac{1}{3})$, which lies at the point of intersection of the line segments ℓ and m drawn in Figure 1.1(b). If we were analyzing G as a contest, N would therefore be our only candidate for a solution outcome.

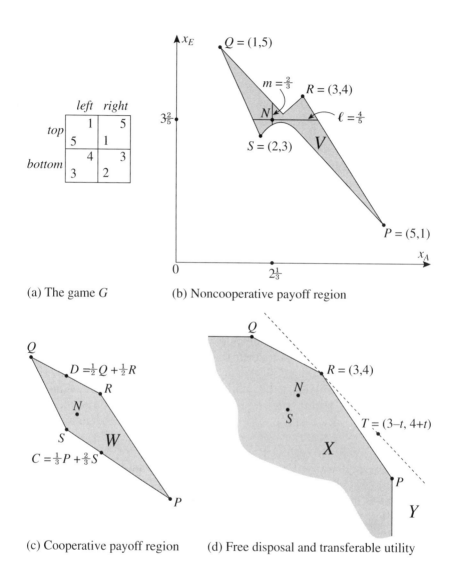

(a) The game G (b) Noncooperative payoff region

(c) Cooperative payoff region   (d) Free disposal and transferable utility

Figure 1.1: Payoff regions for the game G.

Cooperative payoff regions. However, we are definitely not interested in analyzing G as a contest. Far from wishing to forbid any preplay communication, we shall now adopt the conventions of cooperative game theory and assume that the players are allowed to negotiate *irrevocably binding* contracts about how G is to be played. When such preplay interaction is permitted, the players are no longer forced to randomize independently. If they so choose, they can use the *same* randomizing device when mixing among their pure strategies. In soccer, for example, the two team captains decide who will kick off by agreeing to abide by the toss of a single coin.

In the case of the game G, Adam and Eve might perhaps agree to implement the outcome P with probability $\frac{1}{3}$ and the outcome S with probability $\frac{2}{3}$. Adam's expected utility for this agreement is $\frac{1}{3}P_A + \frac{2}{3}S_A = \frac{1}{3}{\times}5 + \frac{2}{3}{\times}2 = 3$. Eve's expected utility is $\frac{1}{3}P_E + \frac{2}{3}S_E = \frac{1}{3}{\times}1 + \frac{2}{3}{\times}3 = 2\frac{1}{3}$. We can therefore identify their agreement with the convex combination

$$C = \tfrac{1}{3}P + \tfrac{2}{3}S = (\tfrac{1}{3}P_A + \tfrac{2}{3}S_A, \tfrac{1}{3}P_E + \tfrac{2}{3}S_E) = (3, 2\tfrac{1}{3})$$

of P and S shown in Figure 1.1(c). Notice that C is located two thirds of the way along the line segment joining P and S. Similarly, the point $D = \frac{1}{2}Q + \frac{1}{2}R$ is halfway along the line segment joining Q and R.

When all the agreements corresponding to such convex combinations[1] have been included, the result is the *convex set* W shaded in Figure 1.1(c).[2] The set W is called the *convex hull* of the points P, Q, R and S because it is the smallest convex set that contains them.

For the same reason that V is referred to as the noncooperative payoff region for the game G, W is called its *cooperative payoff region.*

Free disposal. However, W is not necessarily the end of the story. Adam and Eve can often expand the set of feasible agreements from W to a larger set X.

Consider any payoff pair w in the set W and suppose that $x \leq w$.[3] Then Adam and Eve can achieve x by agreeing that w will first be implemented, and then each player will burn just enough money to reduce the utility that he gets from w to the utility he gets from x. Economists speak of *free disposal* when admitting the possibility that goods may be costlessly

---

[1] A convex combination of two vectors x and y is an expression of the form $z = \alpha x + \beta y$, in which $\alpha \geq 0$, $\beta \geq 0$ and $\alpha + \beta = 1$. As α increases from 0 to 1, z sweeps out the straight line segment joining x and y.

[2] Recall from Section 1.2.5 of Volume I that a convex set W has the property that, whenever two points belong to W, so does the line segment joining them.

[3] Which means that $x_A \leq w_A$ and $x_E \leq w_E$.

thrown away in such a manner. Rational players will obviously not behave in such a crazy way, but they could if they were so inclined. We should therefore expand the set W to the set X shaded in Figure 1.1(d). The set X is *comprehensive*, which means that, whenever it contains x, it also contains all points y satisfying $y \leq x$.[4] Since X is the smallest comprehensive set containing W, one might call X the comprehensive hull of W.

Transferable utility. When working with "transferable utility", one can go on to expand the set X of feasible agreements to a still larger set Y. Adam and Eve could agree that the point $R = (3, 4)$ will first be implemented, and then that Adam will pay Eve a side payment of t. The result of this agreement will be the payoff pair $T = (3 - t, 4 + t)$. By varying the amount of the side payment t, Adam and Eve could thereby achieve any payoff pair on the line of slope -1 that passes through R. With transferable utility, the set Y of possible agreements therefore consists of all payoff pairs on or below the broken line marked in Figure 1.1(d).

However, let me repeat the refrain of Section 4.1 of Volume I, that the metaphysical notion of transferable utility has no place in an analysis of the kind being attempted in this book. We shall therefore be content to identify the set of possible preplay agreements for the game G with the set X of Figure 1.1(d). Such an understanding does not preclude the possibility that Adam and Eve might make side-payments to each other. However, we shall then insist that the game G be expanded so that the making of side-payments is included in its strategy set.

1.2.2 Nash Bargaining Problems

For the purposes of cooperative game theory, the set of possible preplay agreements for the game G is the set X of Figure 1.1(d). In formulating an abstract representation of a bargaining problem, Nash [380] short-circuited the manner in which such an agreement set X was constructed from an underlying game G in Section 1.2.1. Instead, he simply admitted any set X of the type illustrated in Figure 1.2(a) as a possible *agreement set*. The formal requirements to be imposed are:

- The set X is closed and bounded above.
- The set X is convex.
- The set X is comprehensive.[5]

---

[4]A comprehensive set is *strictly* comprehensive if its boundary contains no horizontal or vertical line segments.

[5]Nash [380] himself did not assume that X is comprehensive.

Nash recognized that the agreement on which rational players settle will necessarily depend on what would happen if the negotiations were to fail. In addition to an agreement set X, he therefore also required that a *status quo* point ξ inside X be given. The *status quo* ξ is understood to be the pair of payoffs that the players will receive if they are unable to reach an agreement. Formally, a *Nash bargaining problem* is therefore a pair (X, ξ).

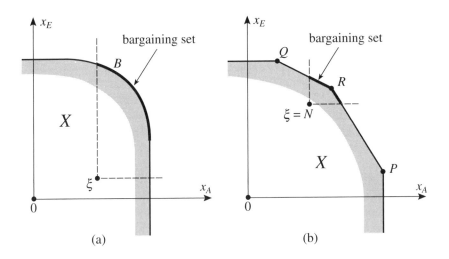

Figure 1.2: Nash bargaining problems.

Suppose, for example, that the negotiations between Adam and Eve on how to play the game G of Figure 1.1(a) were to fail. While bargaining, both Adam and Eve will doubtless have accompanied their proposals with threats or warnings about the consequences of a refusal. Such threats may sometimes have the status of commitments, as in Section 1.5.2. However, it will usually be more realistic to assume that a preplay negotiation which fails to end in a binding agreement will leave the players free to choose whatever strategy for G they like. That is to say, after the negotiations have failed, Adam and Eve will play G as a *noncooperative* game. Since the point N of Figure 1.1(b) is the unique Nash equilibrium outcome of G, it is therefore the natural candidate as the *status quo* for the problem.

Figure 1.2(b) shows the set X of Figure 1.1(d) together with the *status quo* point N. The pair (X, N) is then an abstract representation of the bargaining problem faced by Adam and Eve when they negotiate over how the game G should be played.

Complex disagreements. Nash's *status quo* is sometimes called the disagreement point or the threat point for the bargaining problem. I prefer Nash's original terminology because it is more neutral about how ξ is to be interpreted. In general, there may be many ways in which a negotiation can fail to end in an agreement, each of which results in the players receiving different payoffs. Only in special cases will it be true that we can follow Nash in restricting attention to just one of these.

At the very least, we should distinguish between the deadlock point and the breakdown point. The *deadlock point* is the payoff pair d that would result if the players were to sit at the negotiation table indefinitely without ever reaching an agreement. Of course, rational players will not get trapped in such an impasse, but it does not follow that we can neglect what would happen if they did. As so often in game theory, the wise course of action is determined by what *would* happen if someone *were* to act foolishly.

The *breakdown point* is the payoff pair b that would result if the negotiations were somehow broken off at their outset. For example, it might be that Adam simply walks off to take up his best outside option b_A, leaving Eve to make do with her best outside option b_E. Or it may be that the negotiations have to be discontinued as a result of the intervention of some external factor. For example, on learning that a firm and its workers are in dispute over the wage rate, the firm's customers may choose to take their business permanently elsewhere, leaving no surplus to be divided.

Other more complex scenarios easily come to mind. However, my guess is that most wage-bargaining problems can be fitted into a scheme with just two disagreement points. Even with just b and d, we have something more complicated than a Nash bargaining problem. Appendix C calls the triple (X, b, d) a *generalized* Nash bargaining problem. However, for the moment, attention will be confined to the simple case considered by Nash. One may think of this as the case in which the players' outside options are sufficiently bad that sitting at the negotiation table for ever is no worse than opting out.

1.2.3   The Bargaining Set

As mentioned in Section 1.2.8 of Volume I, Von Neumann and Morgenstern [538] did not neglect the two-player, Nash bargaining problem. However, they added little to what was then the traditional view among economists that the bargaining problem is indeterminate. They decided that the bargaining outcome reached by rational players must necessarily lie in what they called the bargaining set, but thought it impossible to narrow down the field of possible agreements any further without information about the "bargaining skills" of the players. The world had to wait until Nash [380, 382] came along for a more sophisticated approach to the issue.

Unless ξ is Pareto-efficient,[6] the *bargaining set*[7] B for a Nash bargaining problem (X, ξ) is the set of all Pareto-efficient points of X that are Pareto-improvements on the *status quo* ξ (Section 1.2.8 of Volume I).

The set of Pareto-efficient points of the agreement set X is called its Pareto-frontier. If X is strictly comprehensive, its Pareto-frontier F coincides with its boundary. When X is not strictly comprehensive, as in Figures 1.2(a) and 1.2(b), the Pareto-frontier does not coincide with the boundary of X. Given two points lying in the same horizontal or vertical line segment on the boundary of X, one must be a Pareto-improvement on the other.

Sometimes, as in Section 1.3.3, it is necessary to replace B by a weak version of the bargaining set, in which Pareto-efficiency is replaced by weak Pareto-efficiency. Recall that y is a Pareto-improvement on x if a movement from x to y makes someone strictly better off without making anyone else worse off. For y to be a *weak* Pareto-improvement on x, a movement from x to y must make everybody strictly better off. Just as a Pareto-efficient point of X has no feasible Pareto-improvements, a *weakly* Pareto-efficient point of X has no feasible weak Pareto-improvements. The weak Pareto-frontier of a comprehensive set X always coincides with its boundary. Given two points lying in the same horizontal or vertical line segment on the boundary of the set X illustrated in Figure 1.2(b), neither is a weak Pareto-improvement on the other.

If ξ is not Pareto-efficient, the *weak* bargaining set B_0 for a Nash bargaining problem (X, ξ) is defined as the set of all *weakly* Pareto-efficient points that are Pareto-improvements on the *status quo* ξ.

Although it dates back at least as far as Edgeworth [169], the idea that rational bargainers negotiating costlessly will necessarily achieve at least a weakly Pareto-efficient outcome has become associated with the name of Ronald Coase [127]. However, the Coase theorem does little more than to reiterate Edgeworth's observation that rational agents unhampered by any transaction costs would not be satisfied with an agreement x as long as a feasible, weak Pareto-improvement y remains available. The only agreements on which rational players might settle must therefore be weakly Pareto-efficient, because these are the only points which have no feasible, weak Pareto-improvements. Most economic textbooks are less cautious and replace the conclusion that rational agreements will necessarily be *weakly*

---

[6] In which case the bargaining set is defined to be the single point ξ.

[7] Von Neumann and Morgenstern's notion of a bargaining set differs from that of Aumann and Maschler [27] mentioned in Section 0.5.1.

Pareto-efficient with the simple requirement that they be Pareto-efficient. Since nothing of any importance hangs on the issue, I follow the latter convention wherever possible.

Some authors regard the idea that rational players will necessarily bargain to a Pareto-efficient outcome as a "collective rationality" principle.[8] However, when such authors speak of "individual rationality", they are seldom contrasting the rationality of a particular individual with the rationality of the collective to which he belongs. Instead, the term *individual rationality* is used to indicate that no player will agree to settle for less than his security level.

In a Nash bargaining problem, a player's security level is taken to be his *status quo* payoff, because a player can guarantee at least this much simply by never agreeing to anything at all. Thus an agreement x is individually rational for Adam if $x_A \geq \xi_A$. It is individually rational for Eve if $x_E \geq \xi_E$. It is individually rational for both Adam and Eve if $x \geq \xi$. The individually rational points of X therefore coincide with the points of X that are Pareto-improvements on ξ, together with ξ itself. The bargaining set B for the problem (X, ξ) is therefore the set of all individually rational, Pareto-efficient points of X.

The core. For a point x of X to lie in the bargaining set B for a Nash bargaining problem (X, ξ), it must be both collectively and individually rational. It must be collectively rational to the extent that it is Pareto-efficient, and individually rational to the extent that it is a Pareto-improvement on ξ. These considerations link B to the core, which was defined in Section 0.5.1.

In a two-player problem, the grand coalition is just the set $N = \{1, 2\}$. The set of agreements that can be guaranteed by N is $V(N) = X$. The other nonempty coalitions are $\{1\}$ and $\{2\}$. The set $V(1)$ consists of all points on or left of the line $x_1 = \xi_1$. The set $V(2)$ consists of all points on or below the line $x_2 = \xi_2$. The grand coalition has an objection to anything in $X = V(N)$ that is not Pareto-efficient. The coalition $\{1\}$ has an objection to any x in $V(1)$ satisfying $x_1 < \xi_1$. The coalition $\{2\}$ has an objection to any x in $V(1)$ satisfying $x_2 < \xi_2$. After all imputations x to which some coalition has an objection have been removed, we are left with the Von Neumann and Morgenstern bargaining set B for the problem (X, ξ).

---

[8]Aside from this generalized use, the term *collective rationality* is also used to indicate that a collective preference relation is to be assumed transitive.

1.2.4 Dividing the Dollar

Section 0.4.1 introduced the problem of dividing the dollar. Recall that a philanthropist donates a dollar to Adam and Eve, provided that they can agree on how to divide it. If they cannot agree, he takes the money back. Many bargaining problems have this simple structure. For example, wage negotiations often reduce to a dispute over how the surplus created by the joint efforts of a firm and its workers should be divided. If no agreement is reached, then there will be no surplus.

Contracts typically specify the action that each party agrees to implement. One can therefore always see a contract as an agreement to play a game in a certain way. In this section, the underlying game is the *Nash Demand Game*. Unlike the Ultimatum Game of Section 0.5.1, the rules of the Nash Demand Game specify that each player must *simultaneously* make a monetary demand to the philanthropist, who then pays each player his demand if the two demands add to a dollar or less. Otherwise, both players get nothing.

Section 1.6.1 discusses the Nash Demand Game in more detail. It is needed for the moment only so that an agreement can be identified with a pair (m_A, m_E) of monetary demands for which $m_A + m_E \leq 1$. Any other pair of demands counts as a disagreement, which each player regards as being equivalent to ending up with nothing.

In some cases, it may be that Adam cares not only about how much money he gets, but also about how much Eve receives. However, here we simplify by assuming that each player is concerned only with his own payment.[9] We can then describe Adam's preferences with a Von Neumann and Morgenstern utility function $u_A : \mathbb{R} \to \mathbb{R}$ that assigns $u_A(m_A)$ utils to the agreement (m_A, m_E). Similarly, Eve's preferences can be described with a Von Neumann and Morgenstern utility $u_A : \mathbb{R} \to \mathbb{R}$ that assigns $u_E(m_E)$ utils to the agreement (m_A, m_E).

The discussion of dividing the dollar in Section 0.4.1 assumed that each player is risk neutral. The current section relaxes this assumption, requiring only that each player be risk averse. As explained in Section 4.6 of Volume I, Adam is *risk averse* if his Von Neumann and Morgenstern utility function u_A is concave.[10] Figure 1.3(a) illustrates the problem of dividing the dollar

---

[9] Recall that game theorists do not refer to m_A and m_E as *payoffs*. Payoffs are always measured in utils. The payoff pair $x = (x_A, x_E)$ that corresponds to the pair $m = (m_A, m_E)$ of monetary payments is defined by $x_A = u_A(m_A)$ and $x_E = u_E(m_E)$.

[10] Chords drawn to the graph of a concave function always lie on or below the graph. Risk-loving players have convex utility functions. Chords drawn to the graph of a convex function always lie on or above the graph. Risk-neutral players have affine utility functions. Affine functions have straight-line graphs (Section 4.6 of Volume I).

in the case when Adam is strictly risk averse and Eve is risk neutral. The figure has four separate parts. The southwest part of the diagram shows the problem in monetary terms. The southeast part shows the concave graph of the equation $x_A = u_A(1 - m_E)$ that expresses Adam's payoff in utils in terms of *Eve's* payment in money. The northwest part shows the straight-line graph of the corresponding equation $x_E = u_E(1 - m_A)$ for Eve. As explained in Section 4.2.2 of Volume I, the zero and the unit on a Von Neumann and Morgenstern utility scale can be freely chosen to be whatever is convenient. In this case, it seems easiest to require that $u_A(0) = u_E(0) = 0$ and $u_A(1) = u_E(1) = 1$.

The northeast part of Figure 1.3(a) shows the problem of the dividing the dollar in utility terms. Notice how the pair m of monetary demands in the southwest part of the diagram corresponds to the payoff pair x in the northeast part of the diagram. Similarly, the set M of monetary agreements corresponds to the set X of agreements in utility terms. Since disagreement corresponds to the payoff pair $\xi = (0, 0)$, the northeast part of the diagram represents the problem of dividing the dollar for risk-averse players as a Nash bargaining problem (X, ξ).

 Negotiating over payoff flows. A divide-the-dollar story treats Adam and Eve's relationship as transient. They meet to decide the division of a single dollar and never interact again. But people often bargain over the division of income *flows*. For example, while a firm is operating normally, its revenue can be measured as a flow of dollars per week. After nonlabor costs have been subtracted, the firm and the union have the remaining surplus left to divide. In bargaining over how to split this surplus, they will then be negotiating over the payoff flow that each is to receive.

A payoff flow for Adam is a function $f : T \to \mathbb{R}$, where T is the set of relevant times, and $f(t)$ is the payoff that Adam receives at time t. Attention is restricted to the case when T is a set of discrete times t_0, t_1, t_2, ... satisfying $0 = t_0 < t_1 < t_2 < \cdots$. If time is measured in weeks and $t_k = k$, then t_k will occur k weeks from now. In the case of the constant payoff flow in which $f(t)$ is always 2, Adam receives two utils now and at weekly intervals thereafter.

A payoff flow is distantly related to the idea of a lottery, with the payoff $f(t)$ available at time t in the former serving as an analogue of a prize in the latter. Section 4.2.2 of Volume I describes the axioms used by Von Neumann and Morgenstern to characterize a rational player's preferences over lotteries. In the same spirit, axioms have been proposed that characterize a player's preferences over payoff streams (Fishburn and Rubinstein [182]). A relatively mild set of conditions suffices to ensure that Adam's

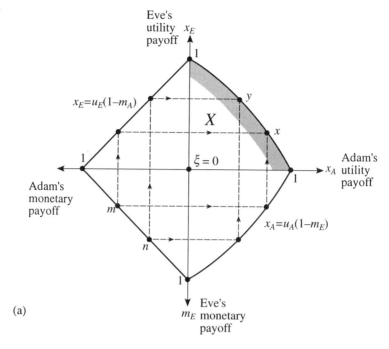

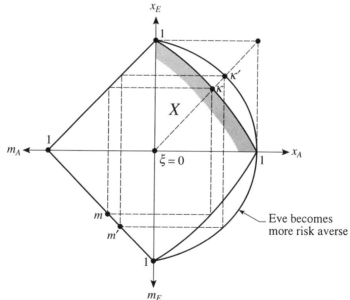

Figure 1.3: Dividing the dollar.

Von Neumann and Morgenstern utility function u_A over payoff flows takes the form

$$u_A(f) = f(t_0)\delta^{t_0} + f(t_1)\delta^{t_1} + f(t_2)\delta^{t_2} + \cdots, \qquad (1.1)$$

where the constant δ satisfies $0 < \delta < 1$. We call δ Adam's *discount factor*. It measures his *impatience*. The smaller δ gets, the more impatient Adam becomes.

The utility $u_A(f)$ is Adam's *present discounted value* for the payoff flow f. It is easily computed when the weekly payoff flow is constant. If $f(t) = x$ for all t, then

$$u_A(f) = x\{1 + \delta + \delta^2 + \delta^3 + \cdots\} = x/(1 - \delta). \qquad (1.2)$$

If Adam and Eve reach an agreement in which Adam is to receive a constant flow of x utils per week, he will therefore evaluate the deal as being worth $x/(1 - \delta)$ utils in total.

Converting a constant flow payoff of x into the corresponding payoff $x/(1 - \delta)$ for the total flow therefore simply involves multiplying by the constant $1/(1 - \delta)$. In consequence, bargaining problems over the division of a flow can often be expressed directly in terms of the flow payoffs without any explicit mention of the present discounted value of the whole flow. For example, the problem of dividing the dollar can be reinterpreted as a problem about how to divide an income flow of one dollar per week.

When the payoff pairs in the agreement set are interpreted as pairs of payoff flows, one must be careful to evaluate disagreements according to the same criterion. For example, in a wage negotiation over the weekly wage rate, one will typically wish to identify the deadlock point with the pair of flow payoffs that the firm and the union receive *during a strike*. The breakdown point is the pair of payoff flows that each would receive if both were to abandon the negotiations in favor of their best outside options. For a worker, his best outside option may be to go on the dole. His payoff flow after the negotiations have been abandoned will therefore be determined by the size of his social benefit check. If the law allows, a firm may have the opportunity to replace its workforce when the negotiations break down. If so, its breakdown payoff flow is determined by its anticipated weekly profit after negotiating a new deal with the new workforce.

Interest rates. Discount factors arise in a more concrete way when evaluating flows of dollar payments. One must then pay close attention to the current *rate of interest*. If the weekly interest rate is r, then a payment of $\$x$ that I receive now and put in the bank will have increased to $\$(1 + r)x$ after a week has passed. It follows that a cast-iron promise to pay me $\$y$ next week is worth only $\$y/(1 + r)$ to me now. Similarly, a promise to pay

$y two weeks hence is worth only $y/(1 + r)^2$ now. In general, the present value of a promise to pay $y after a delay of t weeks is $y/(1 + r)^t$.

The current worth in dollars of a flow of weekly dollar payments is therefore found by computing its present discounted value using the discount factor $\delta = 1/(1 + r)$. This is therefore the right discount factor to attribute to a risk-neutral player with access to a rock-solid banking system which borrows and lends at a fixed interest rate r. In applications, δ is therefore often set equal to $1/(1 + r)$ without further ado. However, the discount factors that appear in this book are generally meant to be interpreted less narrowly.

We have been measuring time in weeks, but what if payments are made at monthly, yearly or daily intervals? Let τ be the period between successive payments. If time is measured in weeks and payments are made daily, then $\tau = \frac{1}{7}$. To find the present discounted value of a constant payoff flow of x paid at intervals of length τ, write $t_k = \tau k$ in (1.1). We then obtain (1.2), but with δ replaced by δ^τ.

The interest rate $r(\tau)$ corresponding to the discount factor δ^τ is given by the equation $\delta^\tau = 1/(1+r(\tau))$. Thus $r(\tau) = -(1-\delta^{-\tau})$. The *instantaneous* rate of interest ρ is the rate of interest charged per unit time as τ becomes vanishingly small. When τ is very small, the rate of interest $r(\tau)$ is therefore approximately $\rho\tau$, where

$$\rho = \lim_{\tau \to 0} \frac{r(\tau)}{\tau} = -\lim_{\tau \to 0} \frac{1 - \delta^{-\tau}}{\tau} = -\log\delta,$$

by L'Hôpital's rule.[11] The discount factor δ and the instantaneous rate ρ of interest are therefore connected by the formula

$$\delta = e^{-\rho}. \tag{1.3}$$

The parameter ρ defined by (1.3) is often useful even when it does not admit an interpretation as the instantaneous rate of interest. It is then referred to as the *discount rate* corresponding to δ. It is easy to confuse discount rates and factors, but it is important remember that impatient economic agents have low discount *factors* but high discount *rates*.

1.2.5 Edgeworth Box

Edgeworth's [169] quaintly titled *Mathematical Psychics* of 1891 anticipated a number of the themes that have preoccupied modern economic theorists.

---

[11]L'Hôpital's rule says that $f(x)/g(x) \to f'(\xi)/g'(\xi)$ as $x \to \xi$, provided that $g'(\xi) \neq 0$. I follow most mathematicians in using $\log x$ to denote the *natural* logarithm of x. Thus $y = \log x$ if and only if $x = e^y$.

Bowley's [97] adaptation of his representation of the two-player bartering problem in terms of the Edgeworth box has been particularly fruitful.

Adam has brought W_A bushels of wheat and F_A pounds of fish to market. His initial commodity bundle is therefore (W_A, F_A). Eve has brought W_E bushels of wheat and F_E pounds of fish to market. Her initial commodity bundle is therefore (W_E, F_E). The total amounts of wheat and fish available for redistribution are therefore $W = W_A + W_E$ and $F = F_A + F_E$. The bundles (W_A, F_A) and (W_E, F_E) are said to be Adam and Eve's initial *endowments*. Assuming that no other trade opportunities exist, the problem is to predict the commodity bundles with which Adam and Eve will leave the market.

Figure 1.4(a) shows a box $\mathcal{E}$ of length W and height F. A point (w, f) in this *Edgeworth box* represents the possible trade in which Adam gets the bundle (w, f) and Eve gets the bundle $(W - w, F - f)$. The pair $e = (W_A, F_A)$ is called the *endowment* or *no-trade* point. If Adam and Eve fail to reach agreement, then Adam will be left with the bundle (W_A, F_A). Eve will be left with the bundle $(W_E, F_E) = (W - W_A, F - F_A)$.

We also need information about Adam and Eve's preferences. These are described by Von Neumann and Morgenstern utility functions u_A and u_E. An *indifference curve* for Adam is the set of points (w, f) that satisfy the equation $u_A(w, f) = c$, for some constant c. Since each bundle on an indifference curve is assigned the same utility, Adam is indifferent between any pair of bundles on the curve.[12]

It will be assumed that u_A is continuous and increasing.[13] It will also be assumed that Adam is risk averse so that u_A is concave. Figure 1.4(b) illustrates some indifference curves for Adam when u_A has the properties specified. The arrows indicate the direction of Adam's preferences.

Eve's utility function u_E will be assumed to have the same properties as Adam's. However, her indifference curves in the Edgeworth box of Figure 1.4(b) have a different shape. When Adam gets the bundle (w, f), Eve gets the bundle $(W - w, F - f)$. An indifference curve for Eve is therefore the set of all points (w, f) that satisfy the equation $u_E(W - w, F - f) = d$ for some constant d. Note also that the arrows indicating Eve's direction of preference do not point in a northeasterly direction like Adam's. Other things being equal, she prefers trades in which Adam gets less wheat and less fish. Her arrows therefore point in a southwesterly direction.

---

[12]Mathematicians say that an indifference curve is a *contour* or a *level curve*. The graph of a function $f : \mathbf{R}^2 \to \mathbf{R}$ is a three-dimensional surface. A contour $f(x) = c$ consists of all points $x = (x_1, x_2)$ at which the height of this surface above the (x_1, x_2) plane is precisely c. As in geography, a contour map is often the best way to represent a three-dimensional surface using a two-dimensional picture.

[13]To say that u_A is increasing means that $x \leq y$ implies $u_A(x) \leq u_A(y)$.

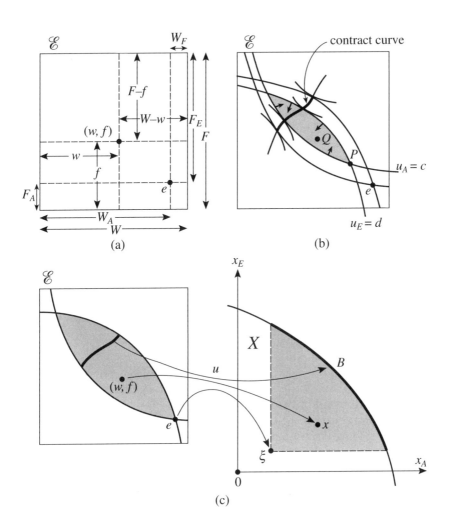

Figure 1.4: The Edgeworth box.

Aside from Adam and Eve's indifference curves, Figure 1.4(b) also shows what Edgeworth called the *contract curve*. This is the set of Pareto-efficient, individually rational trades in the Edgeworth box.

In locating the set of Pareto-efficient points in the Edgeworth box, it is helpful to notice that an interior point of the box where two indifference curves cross cannot be Pareto-efficient. For example, all of the points like Q in the canoe-shaped region enclosed by the indifference curves through the point P of Figure 1.4(b) are Pareto-improvements on P. The only interior points of the Edgeworth box that represent Pareto-efficient trades are therefore those where the indifference curves are tangent.

As explained in Section 1.2.3, individual rationality is the requirement that no player gets less than his security level. In the two-player bartering problem, each player is free to refuse to trade at all and hence can guarantee coming away with at least the endowment with which he arrived. Individual rationality therefore requires that any trade that takes place be on or above Adam's indifference curve through the endowment point e. Similarly, any such trade must be on or below Eve's indifference curve through e.

Figure 1.4(c) shows how the Edgeworth box transforms into a Nash bargaining problem (X, ξ) under the action of the function $u : \mathcal{E} \to \mathbb{R}^2$ defined by

$$u(w, f) = (x_A, x_E)\,,$$

where $x_A = u_A(w, f)$ and $x_E = u_E(W - w, F - f)$. The function u therefore maps the trade (w, f) into the pair (x_A, x_E) of utilities that Adam and Eve will receive if the trade (w, f) is implemented.

Notice how the contract curve maps into a Von Neumann and Morgenstern bargaining set B. It is sometimes helpful to think of the function u as first folding the Edgeworth box along the contract curve and then stretching and compressing the result until it coincides with the set X.

The preceding discussion takes for granted that the *status quo* point ξ should be identified with the payoff pair corresponding to the endowment point e. However, it is of some importance to note that the endowment point actually corresponds to the breakdown point introduced in Section 1.2.2, and nothing in the story told so far offers any indication where the deadlock point might be. This difficulty in identifying a *status quo* point disappears if the commodities are reinterpreted as *flows* rather than stocks of goods. To say that Adam is endowed with the bundle $(1, 2)$ then means that he receives one bushel of wheat and two pounds of fish *per week*. This will remain true even if Adam and Eve are deadlocked and so sit at the negotiation table forever without reaching agreement. The deadlock point and the breakdown point will then coincide.

1.3 Bargaining Solutions

A confusingly large number of cooperative solution concepts have been invented for use with multiplayer games in coalitional form. The situation is equally desperate in the case of cooperative solution concepts invented specifically for use with Nash bargaining problems. Such cooperative solution concepts are called bargaining solutions.

Let $\mathcal{B}$ be the set of all possible Nash bargaining problems (X, ξ). A *bargaining solution* is then a function $f : \mathcal{B} \to \mathbb{R}^2$ with the property that $\sigma = f(X, \xi)$ lies in the set X. The payoff pair $\sigma = (\sigma_A, \sigma_E)$ is to be interpreted as the solution outcome for the Nash bargaining problem (X, ξ).

Section 1.2.3 explains why one would anticipate that rational players will settle on an agreement that is simultaneously weakly Pareto-efficient and individually rational. Only bargaining solutions with this property will be considered. The assumption that $\sigma = f(X, \xi)$ lies in the agreement set X will therefore always be strengthened to the assumption that σ lies in the weak bargaining set B_0 of the problem (X, ξ). Wherever possible, the weak bargaining set B_0 is replaced by the regular bargaining set B. In fact, only in the case of the proportional bargaining solution of Section 1.3.3 is it necessary to work with B_0 rather than B.

Since pretty much everything else in this chapter is a Nash this-or-that, it will come as no surprise that the first and most famous of the bargaining solutions is the Nash bargaining solution. The Nash bargaining solution will be our fundamental tool when the time comes to analyze what goes on when rational players negotiate a social contract in the original position behind the veil of ignorance.

The chief rival to the Nash bargaining solution is the bargaining solution of Kalai and Smorodinsky. For two-player bargaining problems, this coincides with the bargaining solution invented by Gauthier [196] for his social contract theory. It will therefore be worthwhile to discuss at some length why the Kalai-Smorodinsky solution will not suffice for our purpose.

1.3.1 Nash Bargaining Solution

Recall that Von Neumann and Morgenstern [538] argued that nothing could be said about the deal that rational players would reach when faced with the bargaining problem (X, ξ), beyond the fact that it must lie in the bargaining set. Nash [380] was the first to challenge this counsel of despair by proposing the Nash bargaining solution as a candidate for a *unique* rational solution to the bargaining problem.

Given the social welfare function

$$W_N(x) = (x_A - \xi_A)(x_E - \xi_E),\tag{1.4}$$

the *Nash bargaining solution* for the bargaining problem (X, ξ) is the point ν in X at which $W_N(x)$ is maximized, subject to the requirement that $x \geq \xi$.

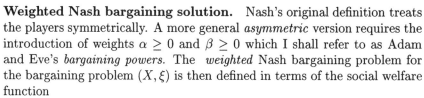 **Weighted Nash bargaining solution.** Nash's original definition treats the players symmetrically. A more general *asymmetric* version requires the introduction of weights $\alpha \geq 0$ and $\beta \geq 0$ which I shall refer to as Adam and Eve's *bargaining powers*. The *weighted* Nash bargaining problem for the bargaining problem (X, ξ) is then defined in terms of the social welfare function

$$W_{wN}(x) = (x_A - \xi_A)^\alpha (x_E - \xi_E)^\beta.\tag{1.5}$$

It is the point n in X at which $W_{wN}(x)$ is maximized subject to the requirement that $x \geq \xi$. Figure 1.5(a) illustrates the definition.

The location of n depends only on the *ratio* of α and β. By changing this ratio, the weighted Nash bargaining solution can be made to coincide with any point in the bargaining set B for the bargaining problem (X, ξ).[14] When $\alpha = 0$, Adam gets only his *status quo* payoff of ξ_A. One might then reasonably say that all the bargaining power lies with Eve. Equally, when $\beta = 0$, Eve gets only her *status quo* payoff of ξ_E and all the bargaining power lies with Adam.

In Nash's [380] original definition of the Nash bargaining solution, Adam and Eve are assumed to have equal bargaining powers, so that $\alpha = \beta$. When necessary to distinguish it from the weighted versions, the original solution will therefore be referred to as the *symmetric* Nash bargaining solution.

Nash [380] first defended his symmetry assumption in terms reminiscent of Von Neumann and Morgenstern [538] by arguing that the players should be taken to have equal "bargaining skills". However, he explicitly corrected himself in a later paper (Nash [382]), in which he points out that there can be no question of one rational player being more skilled at bargaining than another. Just as a perfectly rational chimpanzee will play chess at least as

---

[14]A silly criticism of the weighted Nash bargaining solution deduces that it therefore has no predictive power. The same criticism can be directed at Newton's law of gravity on the grounds that, by varying the gravitational constant, it too can be made to predict that the force acting at any point is anything whatever! However, in applying Newton's law of gravity, one uses known data to estimate what the gravitational constant would have to be if the law were correct. This estimate is then used in predicting unknown data. Similarly, in using the weighted Nash bargaining solution for predictive purposes, one first employs known data to estimate the ratio of the bargaining powers.

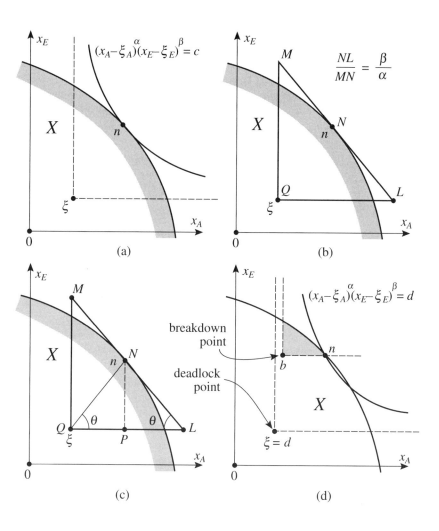

Figure 1.5: The Nash bargaining solution.

well as Alekhine,[15] so Adam will necessarily bargain at least as skillfully as Eve, and vice versa. The bargaining powers in a weighted Nash bargaining solution should therefore not be interpreted in terms of bargaining skills. Bargaining powers are determined by the strategic advantages conferred on players by the circumstances under which they bargain. The leading example is provided by the Rubinstein bargaining model to be studied in Section 1.7.1.

Other characterizations.   It is not always easy to use the Nash products of (1.4) and (1.5) to find the location of a Nash bargaining solution. Other characterizations can then be valuable. Two such characterizations are shown in Figures 1.5(b) and 1.5(c).

In Figure 1.5(b), LM is a tangent[16] to the frontier of X, which has been chosen so that its point of contact with X satisfies $LN/MN = \beta/\alpha$. The weighted Nash bargaining solution for the bargaining problem (X, ξ) then lies at N.[17]

Some elementary Euclidean geometry[18] suffices to deduce the criterion illustrated in Figure 1.5(c) for locating the *symmetric* Nash bargaining solution of a Nash bargaining problem (X, ξ). The symmetric Nash bargaining solution N is found by drawing the tangent LN to X at N that makes the same angle θ to the vertical as the ray QN from the *status quo* point Q to the point N.

Nash solution with outside options.   Section 1.2.2 observes that the manner in which disagreement arises in real bargaining problems can sometimes be too complex to be summarized using a single *status quo* point ξ. The simplest generalization involves two disagreement points: a deadlock point d and a breakdown point b. A bargaining problem then becomes a triple (X, b, d).

Adam's breakdown payoff b_A is what he would get if the negotiations were to break down leaving him with no alternative but to take up his best outside option. Applied workers typically look no further than the pair b of outside option payoffs when seeking to identify a *status quo* to use with the Nash bargaining solution, but identifying ξ with b is likely

---

[15]Alekhine is reported to have said, "Position, what does position matter! It is my will that counts." However, in playing a perfectly rational chimpanzee, nothing but the current position would matter, even to the genius who invented Alekhine's Defense.

[16]If the frontier is not smooth, LM becomes a supporting line to the convex set X.

[17]See Section 5.5.6 of my *Fun and Games* for the mathematical details (Binmore [70]).

[18]The tangents of the angles QNP and LNP in Figure 1.5(b) are QP/NP and LP/NP respectively. These are equal in the symmetric case, because the proportional division theorem then tells us that $1 = LN/MN = LP/QP$.

to be a mistake unless there is some reason to suppose that that Nash's [382] variable-threat theory of Section 1.5.2 is relevant to the problem. However, the commitment power that this theory attributes to the players will usually be absent. This leaves Rubinstein's Alternating Offers Game as the leading candidate for a model of the bargaining procedure. As we shall see in Section 1.7.1, when the Alternating Offers Game implements the Nash bargaining solution, it places the *status quo* at the deadlock point d. This represents the payoffs that the players would receive if they bargained forever without reaching an agreement.

What role is then left for the breakdown point b? Simple accounts of the Alternating Offers Game assume that the players have no outside options at all. However, when outside options are included in the model,[19] they do not figure in the appropriate Nash product at all. Instead, they appear as *constraints* on the set over which the Nash product is maximized.

To be more precise, the definition of the weighted Nash bargaining solution for the bargaining problem (X, ξ) has to be modified in the case of the bargaining problem (X, b, d) so that the ξ in the definition of the social welfare function W_{wN} is replaced by d and the ξ that appears in the constraint $x \geq \xi$ is replaced by both b and d. Thus the weighted Nash bargaining solution for (X, b, d) is the point n in X at which

$$W_{wN}(x) = (x_A - d_A)^\alpha (x_E - d_E)^\beta$$

is maximized subject to the constraints $x \geq b$ and $x \geq d$. Figure 1.5(d) illustrates the definition in the case when the constraint $x \geq b$ is *active*.[20] When the constraint $x \geq b$ is inactive, the bargaining outcome is identical with the weighted Nash bargaining solution for the problem (X, d).

A formal defense of this definition requires appealing to the model of Section 1.7.1 as in Appendix C. However, it is not hard to see that, when the constraint $x \geq b$ is inactive, so that $n_A > b_A$ and $n_E > b_E$, it cannot possibly make any difference if Adam's outside option is increased to c_A provided that $b_A < c_A < n_A$. In seeking to exploit what seems to be an advantageous change in his bargaining position, Adam might threaten Eve that he will opt out unless she concedes more than n_A. But Eve can safely ignore the threat and bargain precisely as before. At each occasion when Adam previously had the opportunity, Adam chose not to opt out because $n_A > b_A$. In the new situation, opting out is less unattractive than it was before, but he will nevertheless always continue negotiating when the

---

[19]See Binmore, Shaked and Sutton [90], Binmore, Rubinstein and Wolinsky [83], or Binmore, Osborne and Rubinstein [80]. The first of these papers includes an experimental comparison of b and d as candidates for the *status quo* in an alternating offers scenario. The latter overwhelmingly outperforms the former.

[20]Which means that $n_A = b_A$ or $n_E = b_E$.

opportunity to opt out arises, because it is still true that $n_A > c_A$. Only if $c_A \geq n_A$ does the situation change. Eve must then concede at least c_A to Adam to make it worth his while to do a deal at all.

Although this is a simple argument that lay folk readily understand, it seems to be largely ignored by labor economists. In modeling situations where bargaining theorists would use the *symmetric* Nash bargaining solution for the bargaining problem (X, b, d), labor economists frequently use a *weighted* Nash bargaining solution for the problem (X, b) instead. The value of the ratio α / β is then estimated from the available data.

Since a weighted Nash bargaining solution has a free parameter, the result sometimes fits the facts fairly well. Using the correct model may even result in a worse fit—just as a Newtonian, two-body model may fail to predict the movement of Mars as accurately as a Ptolemaic model with enough epicycles. However, this is not a good reason for using the wrong theory. It is an argument for moving to a version of the correct theory that abstracts less away. For example, in the Martian case, one might move to a three-body Newtonian model that takes account of the whereabouts of Jupiter.

1.3.2   Kalai-Smorodinsky Solution

The *Kalai-Smorodinsky bargaining solution* for the Nash bargaining problem (X, ξ) can be defined in terms of the social welfare function

$$W_{KS}(x) = \min \left\{ \frac{x_A - \xi_A}{U_A - \xi_A}, \frac{x_E - \xi_E}{U_E - \xi_E} \right\}, \qquad (1.6)$$

where Adam's utopian payoff U_A is the most he can get given that Eve is to receive at least her *status quo* payoff of ξ_E, and U_E is the most that Eve can get given that Adam is to receive at least ξ_A. The Kalai-Smorodinsky solution selects the point K at which $W_{KS}(x)$ is maximized subject to the requirement that $x \geq \xi$. Figure 1.5(b) of Volume I illustrates the definition.

The geometric characterization of Figure 1.6(a) is easier to appreciate. It will not usually be possible for both Adam and Eve to receive their utopian payoffs simultaneously—a fact reflected in Figure 1.6(a) by placing the utopian point U outside the agreement set X. Having located the utopian point U, join this to the *status quo* point Q with a straight line. The Kalai-Smorodinsky bargaining solution K is located where this straight line crosses the Pareto-frontier of X.

Section 1.2.9 of Volume I explains why Gauthier [196] thinks the Kalai-Smorodinsky solution is relevant to rational bargaining. Kalai and Smorodinsky [289] were much more tentative in the claims they made for their solution. Their aim was simply to point out that axiom systems other than

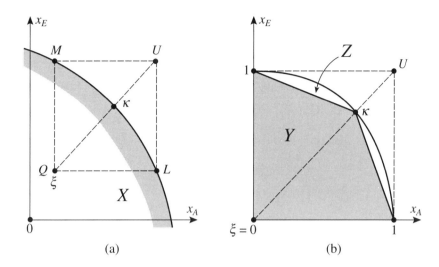

Figure 1.6: The Kalai-Smorodinsky solution.

Nash's [380] can lead to bargaining solutions that are no less elegant from the mathematical point of view than the Nash bargaining solution. Once this has been pointed out, even mathematicians have no choice but to ask interpretive questions about the *meaning* of the rival axiom systems—an issue that is taken up in Section 1.4.1.

1.3.3 Bargaining with Interpersonal Comparison

To decide whether a bargaining outcome is fair in some sense or another, one needs to be able to compare the utils that Adam and Eve get from the deal. But neither the Nash bargaining solution nor the Kalai-Smorodinsky solution depends on Adam's utils being comparable with Eve's. In contrast to the proportional bargaining solution or the utilitarian solution to be introduced shortly, neither the Nash bargaining solution nor the Kalai-Smorodinsky solution therefore makes any sense as an *ethical* concept.

This is a point on which it is necessary to insist, since Raiffa [413] muddied the waters at an early stage by considering the merits of a number of bargaining solutions as fair arbitration schemes. As a result, the idea that the Nash bargaining solution should be seen only as a candidate for a fair arbitration scheme has taken on a life of its own. Even now, it is not unusual to come across authors who think it appropriate to reject the Nash bargaining solution because it has no merit as an ethical concept.

However, nobody who understands the Nash bargaining solution would wish to claim that it has any merit whatsoever as an ethical concept. If it did, its use behind the veil of ignorance would preempt the role of the original position in *determining* what is to be regarded as fair. The Nash bargaining solution is intended to predict the outcome of a negotiation between rational players with different personal preferences. Kantians hold that strategic issues will then be irrelevant because true rationality requires setting aside one's personal preferences in order to do what is "right". But such romanticism has no place in the theory being developed here. Rationality does not lie in playing fair—it lies in seeking the best available compromise, given the current balance of power. But nothing says that the best available compromise need be fair.

Although the Nash bargaining solution and the Kalai-Smorodinsky solution operate without any comparison of Adam and Eve's utils being made, there are other bargaining solutions that do presuppose that utils can be compared. Those which matter for the purposes of this book are the weighted utilitarian bargaining solution and the proportional bargaining solution. Little is said about these ideas in the current chapter, but this is not because they are of no interest. On the contrary, a cooperative game theorist might well characterize this whole book as being nothing more than an attempt to extend the Nash program to such concepts.

Weighted utilitarian solution.   Section 2.2.4 of Volume I defines a utilitarian to be consequentialist with an additively separable common-good function. The *utilitarian bargaining solution* for the Nash bargaining problem (X, ξ) is accordingly defined in terms of the social welfare function

$$W_H(x) = x_A + x_E \, .$$

It is the point H in X at which at which the social welfare function is maximized, subject to the requirement that $x \geq \xi$. Figure 1.5(b) of Volume I illustrates the definition.

The *weighted utilitarian bargaining solution* allows a social planner to choose his own rate of comparison between Adam and Eve's utils. If he regards V of Adam's utils as being equivalent to U of Eve's, then he replaces the utilitarian social welfare function by a weighted version

$$W_h(x) = U x_A + V x_E \, ,$$

in which the weights U and V are predetermined nonnegative constants. As always with such weights, only the ratio U/V is significant. In Figure 1.7(a), the weighted utilitarian solution is located at the point h.

Proportional bargaining solution. This is a notion that has been extensively studied by Raiffa [413], Isbell [281] Kalai [288], Myerson [373], Roth [438], Peters [403], and others. The *proportional bargaining solution* with weights $U \geq 0$ and $V \geq 0$ for the Nash bargaining problem (X, ξ) is located at the point ρ in Figure 1.7(a). It is found by drawing the ray with slope U/V through the *status quo* point ξ. The point ρ lies at the point of intersection of this ray with the frontier of X. Roth [438] emphasizes that the proportional bargaining solution may fail to be Pareto-efficient when X is not strictly comprehensive. It is therefore the only bargaining solution of those to be considered that may fail to lie in the bargaining set B. However, since it is always *weakly* Pareto-efficient when X is comprehensive, it always lies in the weak bargaining set B_0.

When X is strictly comprehensive, the proportional bargaining solution for the Nash bargaining problem (X, ξ) can also be defined in terms of the social welfare function

$$W_\rho(x) = \min \{U(x_A - \xi_A), V(x_E - \xi_E)\} . \tag{1.7}$$

It is the point ρ in X at which at which the social welfare function is maximized, subject to the requirement that $x \geq \xi$.

Characterizing the proportional bargaining solution in terms of the social welfare function W_ρ makes it easy to see its relation to the Rawlsian maximin criterion. If we write $y_A = U(x_A - \xi_A)$ and $y_E = V(x_E - \xi_E)$, then a maximizer of the social welfare function W_ρ will not be satisfied until he has attained a welfare level of

$$\max_y \min\{y_A, y_E\} .$$

The proportional bargaining solution therefore leads to exactly the same outcome as the Rawlsian maximin criterion, provided that Adam and Eve's utility scales are suitably recalibrated. Rawls is uninterested in rescaling the players' utils because he believes that one can evade problems of interpersonal comparison using his concept of a primary good. Section 1.2.5 of Volume I explains why I think that such an evasion of this crucial issue is not viable. Much of this book can be read as an attempt to plug this gap in Rawls' reasoning by justifying why attention should be concentrated on one particular recalibration of Adam and Eve's utility scales.

When worlds collide. Figure 1.7(b) reproduces Figure 1.6(a) of Volume I in showing three different bargaining solutions: the *symmetric* Nash bargaining solution ν, the *weighted utilitarian solution* h, and the proportional

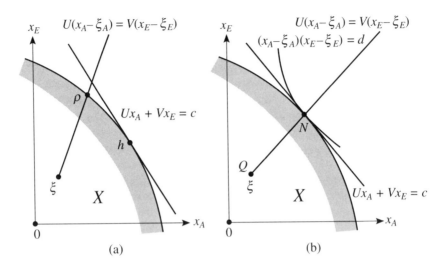

Figure 1.7: Proportional and utilitarian solutions.

bargaining solution ρ. As Yaari [570] has shown, if the weights U and V are chosen to make two of these solutions coincide, the third solution is then located at the same point as the other two.

This result is easy to check using the characterization of the Nash bargaining solution given in Figure 1.5(c). The case of most interest to us in later chapters will be that in which the two bargaining solutions known to coincide are the symmetric Nash bargaining solution and the proportional bargaining solution. The common location of ν and ρ in Figure 1.7(b) is the point N. Since $N = \rho$, the slope of the line QN is U/V. Since $N = \nu$, the slope of the line QN is the negative of the slope of the line PN (because QN makes the same angle to the vertical as PN). It follows that the slope of PN is $-U/V$. The weighted utilitarian solution h must therefore also lie at the common location of ν and ρ.

Section 4.6.7 explains why I think that the link between these three different bargaining solutions is of significance for social contract issues. In particular, I suspect that it may go some way toward explaining why thinkers like Harsanyi and Rawls, whose ethical positions are apparently so far apart, nevertheless end up advocating similar social reforms.

Fairness and justice. Is utilitarianism fair? Does justice require the use of the proportional bargaining solution? I care a great deal about such questions. But it would be premature to ask them at this stage because I think that fairness and justice are concepts without an *a priori* meaning. Like Mackie [338], I believe that such ethical concepts are human artifacts rather than Platonic forms whose true definition can be found by adopting the posture of Rodin's thinker and waiting for inspiration. Nor is a full and comprehensive definition to be found by analyzing the way that words like "fair" and "just" are used in the English language. Indeed, the idea that our ancestors hid a coherent theory of morality in the structure of our language seems to me quite bizarre. The words available to us when we try to talk about moral questions originally evolved for use in small communities faced by problems quite different from those afflicting modern societies. It is therefore not surprising that contemporary debates about justice and fairness are confused and inconclusive. In spite of all the rhetoric about fairness that accompanies wage negotiations, there is no consensus whatever about how a fair wage should be calculated. Each side typically proposes whatever criterion seems expedient at the moment and claims that it would be unfair for them to be forced to accept less.

Moral philosophers *invent* their definitions of fairness and justice, and I think it important not to fudge this point. One is not free, of course, to ignore the dictionary definitions of such words altogether. The definitions need to be consistent with the usage of the words when applied to the small-scale phenomena for which they evolved and where their meaning is relatively uncontroversial. I believe that Rawls' [417] device of the original position is successful in providing such a definition for the case of fairness. I am less enthusiastic about his identification of fairness and justice, since it seems to me that "justice" carries the further connotation that due process is to be observed. However, since I have little or nothing to say about legal matters, it seems harmless for the purposes of this book to treat the words "fair" and "just" as synonyms.

Where I differ from Rawls is in my belief that we need to look at the forces of social evolution if we are to understand why the device of the original position accords with our intuitions in those small-scale situations where a broad consensus exists about the nature of fairness. Adopting such an approach makes it impossible to separate a study of the character of fairness judgments from the question of how interpersonal comparisons of utility are made. Indeed, I believe that the two issues are so inextricably entangled that one *cannot* sensibly discuss the weighted utilitarian solution or the proportional bargaining solution without simultaneously discussing the interpersonal comparison question. Any analysis of these bargaining solutions will therefore have to await progress on the problem of what it means to say that Adam and Eve's utils are comparable.

1.4 Characterizing Bargaining Solutions

The tradition in cooperative game theory is that solution concepts should come equipped with a set of characterizing axioms. It is widely believed that the virtues and vices of a cooperative solution concept can be evaluated simply by comparing such axioms with our gut feelings about what ought to be true. The Greeks felt the same about physics, and hence invented the Aristotelian principles which remained unchallenged for a millenium. Indeed, surveys demonstrate that most people still hold to the Aristotelian doctrine that heavy objects fall faster than their lighter brethren!

If our untutored physical intuition is bad, how much worse are our gut feelings about the way a society of rational individuals would organize itself? Those with doubts on this score are invited to contemplate the traditional preeminence of Kant on rational ethics (Section 2.4 of Volume I). If Kant's writings on this subject are what seem reasonable to the untutored intuition, then our intuitions are badly in need of tutoring! As when learning to ski or to helm a dinghy, we need to recognize that our gut feelings need to be tested and discarded when found wanting.

Our minds are often amazingly adept at inventing excuses why some particularly cherished notion need not be abandoned, but the Nash program provides a discipline whose strictures are hard to evade. Section 0.6 describes the Nash program at some length. It is designed to assist in bringing down to earth the castles we so readily build in the air when formulating systems of axioms.

When a solution concept from cooperative game theory is proposed for use in predicting the deal that will be reached after a negotiation between rational players, the Nash program calls for testing its effectiveness in this role by constructing a detailed model of the bargaining procedure supposedly in use. Such a model will constitute an extensive-form game that is open to analysis using the methods of noncooperative game theory. If the equilibrium selection problem can be solved for this game, we will then know what the outcome of a rational negotiation would be if the players were to use the modeled bargaining procedure. If the cooperative solution concept fails to predict this outcome, then it will have proven useless in its proposed role—provided that the bargaining procedure used in the test genuinely captures the strategic essentials of the problem faced by the bargainers.

I shall be insisting on this final proviso a great deal. After all, who cares if a cooperative solution concept fails to predict the results of using a bargaining procedure that nobody would ever think of using in real life?

1.4.1   The Kalai-Smorodinsky Axioms

The dangers of relying on untrained intuition in evaluating axioms that characterize a cooperative solution concept will be exemplified using the Kalai-Smorodinsky bargaining solution. The same example will also serve to illustrate how the Nash program is sometimes misapplied.

Wherever possible, it is assumed that the solution outcome $f(X, \xi)$ obtained by applying a bargaining solution $f : \mathcal{B} \to \mathbb{R}^2$ lies in the bargaining set $\mathcal{B}$ for the problem (X, ξ). In addition, Kalai and Smorodinsky [289] proposed the following axioms:

- Independence of Utility Calibration
- Symmetry
- Individual Monotonicity[21]

These axioms will be considered one by one below.

Utility calibration axiom. This axiom is based on the idea that the contract the bargainers sign will remain the same in real terms even though Adam or Eve's utility scale may be recalibrated by altering the location of its zero point or changing the units in which utils are measured. Such a shift from one set of utility scales to another should therefore have no impact on the bargaining solution beyond requiring that it be rewritten in terms of the new utils along with everything else.

As explained in Section 4.2.2 of Volume I, measuring utility in utils is like measuring temperature in degrees. The water remains the same temperature whether the thermometer is graduated in degrees centigrade or degrees fahrenheit. Similarly, Adam and Eve's preferences remain the same whether we use the Von Neumann and Morgenstern utility functions u_A and u_E or the equivalent Von Neumann and Morgenstern utility functions $v_A = au_A + c$ and $v_E = bu_E + d$, in which $a > 0$, $b > 0$, c and d are constants. The utility function v_A describes the same preferences as u_A, but relocates the zero and the unit of the utility scale being used to measure the preferences.

For example, in the problem of dividing the dollar, a contract will specify that Adam is to get m_A cents and Eve is to get m_E cents. The pair $m = (m_A, m_E)$ of monetary payments corresponds to a payoff pair $x = (x_A, x_E)$

---

[21]Monotonicity means different things to different people. To mathematicians, a monotonic function is either increasing or decreasing, but an economist usually means that the function is increasing. Sometimes he means that it is *strictly* increasing. Monotonic social decision rules appeared in Section 0.7. A monotonic bargaining solution is not the same thing at all. It is usually understood to be a solution that assigns more to both players when the agreement set expands. (The Nash bargaining solution is not monotonic in this sense.)

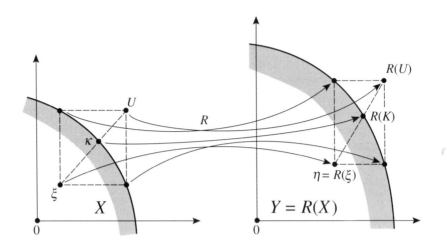

Figure 1.8: Recalibrating the utility scales.

with $x_A = u_A(m_A)$ and $x_E = u_E(m_E)$. The original problem is thereby transformed into a Nash bargaining problem (X, ξ), as illustrated in Figure 1.3(a). If we recalibrate the utility scales by replacing u_A by $v_A = au_A + c$ and u_E by $v_E = bu_E + d$, then the original problem will be transformed instead into a different Nash bargaining problem (Y, η). As Figure 1.8 illustrates, the second bargaining problem is the image of the first under the action of the function $R : \mathbb{R}^2 \to \mathbb{R}^2$ defined by

$$R(x_A, x_E) = (ax_A + c, bx_E + d).$$

In brief, $(Y, \eta) = (R(X), R(\xi))$.

If the calibration of utility scales is irrelevant to the bargaining outcome, then the solution $f(X, \xi)$ of the first problem must correspond to the same monetary payments to Adam and Eve as the solution $f(Y, \eta)$ of the second problem. It follows that $f(Y, \eta) = R(f(X, \xi))$. This leads to a formal statement of the requirement that a bargaining solution be independent of how the utility scales are calibrated. For any recalibration function R we require that

$$f(R(X), R(\xi)) = R(f(X, \xi)). \tag{1.8}$$

Figure 1.8 shows the Kalai-Smorodinsky solutions for both (X, ξ) and (Y, η). Notice that the Kalai-Smorodinsky solution for the second problem does indeed coincide with $R(\kappa)$, where κ is the Kalai-Smorodinsky solution for the first problem.[22] Thus the Kalai-Smorodinsky solution is indeed independent of how the utility scales are calibrated. The same can easily be demonstrated for the Nash bargaining solution using the geometric characterization of Figure 1.5(b).

As a consequence of these ideas being misunderstood by authors who have only a limited grasp of Von Neumann and Morgenstern's theory of utility, the issue of utility calibration is perhaps the prime source of confusion in cooperative bargaining theory. Some game theorists, notably Osborne and Rubinstein [397, p.299], therefore discuss the Kalai-Smorodinsky and the Nash bargaining solutions directly in terms of Adam and Eve's preference relations $\preceq_A$ and $\preceq_E$, thereby avoiding any mention of utility at all. Unfortunately, this option is not open to us since later chapters discuss the utilitarian and proportional bargaining solutions, which *do* depend on how the players' utility scales are calibrated. If there are readers who remain confused, I can only recommend a rereading of Section 4.2.2 of Volume I before going any further.

It remains to observe that, since the Kalai-Smorodinsky bargaining solution is independent of how utility scales are calibrated, it cannot depend on comparisons of the relative value of Adam and Eve's utils. The same goes for the Nash bargaining solution, which also satisfies Independence of Utility Calibration.

Symmetry axiom. A Nash bargaining problem (X, ξ) is symmetric when it looks exactly the same to each player. A symmetric problem (X, ξ) is therefore characterized by two properties. The first is the requirement that $\xi_A = \xi_E$. The second is the requirement that (y, x) must lie in X whenever the same is true of (x, y). If the bargaining procedure in use also treats Adam and Eve symmetrically, then it is natural to postulate that the bargaining solution will also be symmetric. The symmetry axiom formalizes this postulate by requiring that $f(X, \xi)$ assign Adam and Eve the same payoff whenever (X, ξ) is a symmetric bargaining problem.

Individual monotonicity axiom. Suppose that (Y, ξ) and (Z, ξ) are two different bargaining problems with the same *status quo* ξ and the same utopian point U. Suppose also that Y is a subset of Z but $f(Y, \xi)$ is a Pareto-efficient point of Z, as illustrated in Figure 1.6(b). Then Individual Monotonicity says that $f(Z, \xi) = f(Y, \xi)$.

Once Individual Monotonicity has been added to the other two axioms, it is easy to demonstrate that the three axioms together characterize the

---

[22]Since R is affine, it maps straight lines to straight lines. In particular, it maps the straight line through ξ and U to the straight line through $R(\xi)$ and $R(U)$.

Kalai-Smorodinsky solution. First recalibrate the utility scales for the problem (X, ξ) so that the problem $(Z, 0)$ which results has its *status quo* at $(0, 0)$ and its utopian point at $(1, 1)$. Take Y to be the kite-shaped region illustrated in Figure 1.6(b). Symmetry tells us that the solution of the problem $(Y, 0)$ must be K. Individual Monotonicity then says that K must also be the solution of the bargaining problem $(Z, 0)$. Thus $f(Z, 0)$ is the Kalai-Smorodinsky solution for the problem $(Z, 0)$. Now apply Independence of Utility Calibration. This tells us that the bargaining solution of (X, ξ) can be obtained from K by reverting to the original utility scales. Since the Kalai-Smorodinsky solution is invariant under such recalibrations of the utility scales, it follows that $f(X, \xi)$ must be the Kalai-Smorodinsky solution of the bargaining problem (X, ξ).

The defense of Individual Monotonicity given by Kalai and Smorodinsky [289] notes that, for each possible outcome y in Y, there is an outcome z in Z that assigns Eve more than y_E without assigning Adam less than y_A. They argue that the solution payoff for Eve should therefore be larger in the problem (Z, ξ) than in the problem (Y, ξ). Since the same goes for Adam, it follows that $f(Z, \xi) \geq f(Y, \xi)$. But when $f(Y, \xi)$ is a Pareto-efficient point of Z, this implies that $f(Z, \xi) = f(Y, \xi)$.

How good is this defense? Why should we suppose that Eve's bargaining position must necessarily be at least as strong after some change has expanded the agreement set so that, for each payoff Adam might get, there is more left for her? Even if we accept this principle, why should it be applied only when the two situations being compared have the same utopian point?

In the literature on cooperative games, such questions are seldom asked. Axioms are stated in formal terms with only a sentence or two of motivation, since the serious business is thought to be the proving of theorems that follows. The attention of trained mathematicians is then diverted to the details of the proofs, while lay folk read only popularized versions of the work because they are intimidated by the algebra. But perhaps the current discussion of the Kalai-Smorodinsky bargaining solution will suffice to show that one can sometimes get by with almost no algebra at all, but that it never makes sense to leave the axioms unquestioned.

Bargaining and risk-aversion. The problem of dividing the dollar will be used as an example in exploring the plausibility of the individual monotonicity axiom.

Figure 1.3(b) shows how the agreement set Y can be expanded to an agreement set Z by making Eve more risk averse. I hope that the diagram makes it evident why an increase in a player's risk-aversion never makes him worse off when the Kalai-Smorodinsky solution is used (Khilstrom *et al* [304]). But is this a sensible property for a bargaining solution? Why should it be a strategic asset to a player that he has become less willing to take risks? An increase in his risk-aversion will presumably mean that some of the strategic ploys that seemed attractive before will cease to be tempting because of the high level of risk they involve. Handicapping Eve in this way ought to be to Adam's advantage rather than Eve's. That is to say, an increase in Eve's risk-aversion should benefit *Adam* rather than Eve—as indeed it does in the case of the Nash bargaining solution.

The story which accompanies Individual Monotonicity therefore has only a surface plausibility. In particular, it fails to hold water in the case of dividing the dollar. It remains conceivable that increasing a player's risk-aversion may be advantageous in other bargaining problems, but it is enough to produce one counterexample to demolish a general claim.

Implementing the Kalai-Smorodinsky solution. The Nash program requires that a cooperative solution concept be tested by comparing its predictions with the equilibrium outcomes of noncooperative bargaining models. How does the Kalai-Smorodinsky bargaining solution fare when tested in this way?

Moulin [366] has invented the rules for a noncooperative game that implements the Kalai-Smorodinsky bargaining solution. In this game, Adam and Eve begin by simultaneously naming a probability. The player who wins this auction by bidding the highest probability p then gets to propose a feasible agreement s. If his opponent accepts this proposal, a referee then organizes a lottery that yields s with probability p and the *status quo* ξ with probability $1 - p$. If the proposal is refused, the opponent makes a counterproposal t. If the first player accepts this proposal, the referee then organizes a lottery as before, using the *same* probability p. If the counterproposal is refused, the *status quo* ξ is implemented for certain.

This result is sometimes quoted to show that the Kalai-Smorodinsky bargaining solution meets the requirements of the Nash program no less well than the Nash bargaining solution. However, before accepting such an interpretation of the result, one should ask a number of questions. Who organized Moulin's game and why? Where did the referee come from? What constrains the players to obey the rules? Most importantly, can Moulin's game really be said to model a bargaining process?

I hope it is obvious that Moulin's game was never meant to survive such a line of questioning. It is not a bargaining model, and hence has no relevance to the relative merits of the Nash and Kalai-Smorodinsky bargaining solutions. It is a contribution to implementation theory.

1.4.2   The Nash Axioms

By discussing the axioms that characterize the Kalai-Smorodinsky bargaining solution, Section 1.4.1 sought to encourage a skeptical attitude to relying on the axiomatic approach alone. The next set of axioms to be considered characterizes the symmetric Nash bargaining solution. I do not plan to prove this fact, since I shall sketch a proof of what I believe to be a superior characterization in Section 1.4.3.[23] However, the historical development of the Nash bargaining solution makes it impossible to proceed without some mention of Nash's Independence of Irrelevant Alternatives axiom.

Although we shall find that Nash's [380] axioms stand up much better to criticism than those of Kalai and Smorodinsky [289], the time to relax our guard will only arrive in Section 1.7.1 after studying some of the noncooperative bargaining models that implement it.

The first of Nash's [380] axioms is the requirement that the bargaining solution be Pareto-efficient. Actually, when the other axioms are in place, one can get by with the assumption that the bargaining solution is individually rational (Roth [438], Binmore [65]). However, such niceties become irrelevant in the face of our blanket assumption that, wherever possible, bargaining solutions are assumed to lie in the bargaining set B for a problem (X, ξ). The remainder of Nash's axioms are:

- Independence of Utility Calibration

- Symmetry

- Independence of Irrelevant Alternatives

Independence of Utility Calibration and Symmetry were discussed in Section 1.4.1. Symmetry is only required in characterizing the *symmetric* Nash bargaining solution. The other axioms characterize the weighted Nash bargaining solution (Kalai [287], Roth [438], Binmore [65]). The vital axiom for Nash's formulation is the Independence of Irrelevant Alternatives.

Independence of Irrelevant Alternatives.   This axiom is usually defended as a "collective rationality" principle. As such, it is best expressed as the requirement that, if the players sometimes agree on s when t is feasible, then they never agree on t when s is feasible. For example, suppose that Adam and Eve are choosing a dish to share in a Chinese restaurant where the menu lists chow mein, chop suey and egg foo yung. After much

---

[23]After reading the proof that Kalai and Smorodinsky's [289] axioms characterize their solution, most readers will find it easy to reconstruct a version of Nash's proof. If not, my *Fun and Games* contains an accessible proof (Binmore [70, p.186]).

discussion, they choose chop suey. The waiter then appears with the information that chow mein is actually unavailable. If this leads Adam and Eve to change their decision from chop suey to egg foo yung, then they will have violated the Independence of Irrelevant Alternatives, with chop suey in the role of s and egg foo yung in the role of t. Chow mein plays the role of an irrelevant alternative. Since it is not going to be chosen anyway, whether it is available or not shouldn't matter.

The same principle can be expressed in a form that makes it into a kind of opposite to the individual monotonicity axiom required in characterizing the Kalai-Smorodinsky bargaining solution.[24] Suppose that (Y, ξ) and (Z, ξ) are two different bargaining problems with the same *status quo* ξ. Suppose that Z is a subset of Y but $f(Y, \xi)$ lies in Z. Then the Independence of Irrelevant Alternatives says that $f(Z, \xi) = f(Y, \xi)$.

The Nash solution and risk-aversion. After observing that the Irrelevance of Independent Alternatives reverses the logic of Individual Monotonicity, it should come as no surprise that the Nash bargaining solution, whether symmetric or asymmetric, responds in the opposite way to the Kalai-Smorodinsky solution when one of the players becomes more risk averse (Crawford and Varian [137], Roth [439]).

The payments that each player receives when either bargaining solution is applied to the problem of dividing the dollar depend entirely on how risk averse each player is. This is what determines the shapes of graphs of the utility functions in Figure 1.3(a). These in turn determine the shape of the agreement set X.[25] However, when Eve becomes more risk averse, the Nash bargaining solution awards her less money than she got before, whereas the Kalai-Smorodinsky solution awards her more money than before. Section 1.4.1 comes down firmly in favor of the former being the more plausible of the two possibilities.

Implementation. Although the Independence of Irrelevant Alternatives has a defense as a "collective rationality" principle, it is not a defense

---

[24]It is not entirely an opposite for two reasons. The first is that the Independence of Irrelevant Alternatives says nothing about the location of the utopian point. However, as observed in Section 1.4.1, there seems no particular reason why the utopian point should be assigned any special significance. The second reason is that nothing is now said about $f(Y, \xi)$ being a Pareto-efficient point of Z. However, since $f(Y, \xi)$ is Pareto-efficient in Y, it must also be Pareto-efficient in any subset Z to which it belongs.

[25]What if the players are not risk averse, but risk-loving? The set X then fails to be convex and so must be replaced by its convex hull. Any symmetric bargaining solution then requires that Adam and Eve toss a fair coin to decide who gets the whole dollar. Each then takes a risk of ending up with nothing, but risk-loving folk like taking risks.

that accords comfortably with the use of the Nash bargaining solution in
the Nash program. Players of a noncooperative bargaining game have no
interest in principles of collective rationality. Each is interested only in
seeking to maximize his own utility function.

Personally, I think that speaking of collective rationality at all creates
more confusion than it is worth. It tempts one to treat a group of people like a single individual written large, with coherent aims and purposes
of its own. Sometimes, groups do indeed manage to act in this way for
certain purposes. But Section 2.3.6 of Volume I argues that one cannot
always expect this to be the case. However, the Independence of Irrelevant
Alternatives has another defense as an *implementation criterion*.

Maskin's [345] monotonicity criterion was discussed in Section 0.7. It
provides a necessary condition for a social decision rule to be implementable.
For a given *status quo* ξ, the criterion can be applied to bargaining problems
by regarding the set of all payoff pairs $x \geq \xi$ as a set S of social states. With
each agreement set Y, one can then associate a pair of preference relations
$\preceq_A$ and $\preceq_E$. Adam's preference relation is defined so that he is indifferent
between all infeasible payoff pairs x outside the set Y and the *status quo*
ξ. Given two feasible payoff pairs inside Y, he prefers whichever gives him
the higher payoff. Eve's preference relation is defined similarly. Given a
bargaining solution f, one can then define a social decision function g by
writing

$$g(\preceq_A, \preceq_E) = f(Y, \xi) \,.$$

Such a social decision function differs from a social decision rule only in
always assigning a *unique* social state to each preference profile.

Now suppose that Y is replaced by a subset Z which still contains
$\sigma = f(Y, \xi)$. Then Adam and Eve's preference relations change in such a
way that σ is ranked just as high as before. Maskin's monotonicity condition
then says that σ must therefore remain the social state selected by g. Thus
$\sigma = f(Z, \xi)$, as required by the Independence of Irrelevant Alternatives.

1.4.3   Renegotiation Axioms

Section 1.5 begins to examine how well the Nash bargaining solution predicts the bargaining outcome when the players have unlimited commitment
power. We shall need to appeal to this work when studying Harsanyi's defense of utilitarianism in Chapter 2. However, my own theory goes to the
opposite extreme and denies any but the most transient powers of commitment to the players.

As we shall see in Section 1.6, a player may then demand a renegotiation
of the agreement halfway through its implementation. Rational players will

predict that such demands for a renegotiation of the terms of the deal will be made, and so will restrict their attention to agreements which have the property that any call for a renegotiation will simply renew the original agreement. This important idea can be used as the basis for a more satisfactory axiomatic characterization of the Nash bargaining solution. Among other advantages, it makes it clear why the Nash product of Figure 1.5(a) enters the picture.

The underlying structure envisaged by the new axioms splits the negotiation into two stages. First a partial agreement is signed, followed by a full agreement. The signing of the partial agreement is meant to ease the later negotiation of a full agreement. However, the implementation of the partial agreement is contingent on full agreement being reached eventually. If a full agreement is not reached, all bets are off and the players are left with their *status quo* payoffs. If a full agreement is reached, the provisions of the full agreement must be consistent with those of the partial agreement. After full agreement has been reached, the bargaining can then be implemented in two stages without fear of a destabilizing call for renegotiation after the first stage.

The Independence of Irrelevant Alternatives already captures something of this idea. The effect of signing a partial agreement is to reduce the set of feasible final agreements from the original set X to a subset Y. The requirement that $f(X, \xi) = f(Y, \xi)$ can therefore be seen as a condition requiring that the full agreement be consistent with the partial agreement.

The weakness of this new defense of the Independence of Irrelevant Alternatives is its assumption that the bargaining game that comes into existence after the signing of the partial agreement can be solved independently of the terms of the partial agreement. Why should the terms reached at the first stage of the bargaining process not alter the bargaining solution to be used at the second stage?

As an example of a case when the second-stage bargaining problem can sensibly be solved independently of the first-stage problem, suppose that Adam and Eve have been left a thousand French francs and a hundred German marks without any instructions on how to divide their inheritance. They may perhaps begin with a partial agreement on who gets how many francs. A full agreement then awaits a joint decision on the division of the marks. To see that the latter decision can be separated from the problem of dividing the francs, observe that Adam and Eve can now delegate the negotiations over the division of the marks to their lawyers without telling them who got how many francs. The lawyers would then have all the information they need to negotiate sensibly on behalf of their principals. However, this would certainly not be the case if Adam and Eve had been left a house and a car, and the preliminary partial agreement were concerned with the disposal of half the house and half the car. The lawyers at the second stage would need to know who was awarded what at the first stage in order to know how Adam and Eve value the remaining half-shares in the house and the car.

Such reasoning suggests introducing a weakened form of the Independence of Irrelevant Alternatives that restricts its application to situations in which the second-stage bargaining problem is genuinely independent of the first stage bargaining problem. For the purposes of this book,[26] the replacement axiom for the Independence of Irrelevant Alternatives in the list of Section 1.4.2 that characterizes the Nash bargaining solution is called:

- Renegotiation Consistency

One can think of it as an axiom that operationalizes the word "irrelevant" in the Independence of Irrelevant Alternatives.

Separable bargaining problems. The axiom of Renegotiation Consistency begins by restricting its area of application to separable bargaining problems. To be separable, a bargaining problem must satisfy three requirements. The first asks that the players' preferences allow the original bargaining problem to be split into two subsidiary bargaining problems that can meaningfully be negotiated separately. The second asks that any agreement available in the original problem be equivalent to a pair of agreements in the pair of subsidiary bargaining problems. The third insists that a partial agreement obtained in either of the subsidiary bargaining problems should be worthless unless a full agreement is eventually reached.

Renegotiation Consistency then requires that the same result be obtained by applying the bargaining solution $f : \mathcal{B} \to \mathbb{R}^2$ to a separable bargaining problem as is obtained by applying it separately to the two subsidiary bargaining problems into which the original problem separates.

The first and second separability requirements are satisfied in the example of bargaining over francs and marks, but the third is much more stringent. It purpose is to ensure that the completion of the negotiations at the first stage gives neither player any extra bargaining power that he can bring to bear at the second stage. Concern over the strength of this assumption is misplaced, since the stronger the requirements placed on the definition of a separable bargaining problem, the weaker the Renegotiation Consistency axiom becomes.

Formalizing Renegotiation Consistency. With the notation of Section 4.2.2 of Volume I, let **L** and **M** be two lotteries. Create a new lottery

---

[26]In Binmore [65, 62], the new axiom was called "convention consistency".

$(\mathbf{L}, \mathbf{M})$ from this pair, in which the pair of prizes $(\mathcal{D}, \mathcal{E})$ is available with probability pq if and only if the prize $\mathcal{D}$ is available in $\mathbf{L}$ with probability p and the prize $\mathcal{E}$ is available in $\mathbf{M}$ with probability q. The lottery $(\mathbf{L}, \mathbf{M})$ is meant to capture the result of simultaneously running the two lotteries $\mathbf{L}$ and $\mathbf{M}$ independently. A preference relation $\preceq$ over such joint lotteries is *separable* if and only if

$$(\mathbf{L}, \mathbf{M}) \prec (\mathbf{L}, \mathbf{M}') \quad \Rightarrow \quad (\mathbf{L}', \mathbf{M}) \preceq (\mathbf{L}', \mathbf{M}'),$$
$$(\mathbf{L}, \mathbf{M}) \prec (\mathbf{L}', \mathbf{M}) \quad \Rightarrow \quad (\mathbf{L}, \mathbf{M}') \preceq (\mathbf{L}', \mathbf{M}'),$$

for all relevant $\mathbf{L}$, $\mathbf{L}'$, $\mathbf{M}$, and $\mathbf{M}'$.

If player i has separable preferences and his Von Neumann and Morgenstern utilities for the lotteries $\mathbf{L}$ and $\mathbf{M}$ are x_i and y_i, then his Von Neumann and Morgenstern utility for the lottery $(\mathbf{L}, \mathbf{M})$ is

$$a x_i y_i + b x_i + c y_i + d,$$

where a, b, c, and d are nonnegative constants.[27] All of these constants can be eliminated by choosing the utility scales conveniently. Normalize all bargaining problems so that the *status quo* lies at $(0, 0)$ and the utopian point lies at $(1, 1)$. An appeal to the third requirement in the definition of a separable bargaining problem then yields that $b = c = d = 0$. The only nonzero constant is $a = 1$.

This result allows Renegotiation Consistency to be expressed in terms the idea of a direct product. The direct product of two payoff payoff vectors x and y is defined by $x \otimes y = (x_A y_A, x_E y_E)$. The set $X \otimes Y$ is the set of all $x \otimes y$ with x in X and y in Y. If $(X, 0)$ and $(Y, 0)$ are normalized bargaining problems, then Renegotiation Consistency reduces to the requirement that

$$f(X \otimes Y, 0) = f(X, 0) \otimes f(Y, 0), \tag{1.9}$$

whenever $(X \otimes Y, 0)$ is a normalized bargaining problem.[28] It is trivial to use the characterization illustrated in Figure 1.5(a) to check that any weighted Nash bargaining solution satisfies (1.9). That is to say, if λ, μ and ν are respectively the weighted Nash bargaining solutions with bargaining powers α and β for the bargaining problems $(X, 0)$, $(Y, 0)$, and $(X \otimes Y)$, then $\nu = \lambda \otimes \mu$.

To show that (1.9) is sufficient to replace the Independence of Irrelevant Alternatives in the characterization of a weighted Nash bargaining solution, the first step is to observe that, if x is in X and y is in Y, then $x \otimes y = \nu$

---

[27]Use the fact that Von Neumann and Morgenstern utility functions are unique up to a strictly increasing, affine transformation (Binmore [65, 62]).

[28]$X \otimes Y$ may not be convex in the general case.

implies that $x = \lambda$ and $y = \mu$.[29] The second step requires considering the normalized bargaining problem $(\Delta, 0)$, in which Δ is the unit simplex.[30] Since $\nu = f(\Delta, 0)$ lies in the bargaining set of the problem $(\Delta, 0)$, we can find bargaining powers α and β that make ν a weighted Nash bargaining solution for $(\Delta, 0)$. The third step requires picking any normalized bargaining problem $(X, 0)$ and observing that we can then find another normalized bargaining problem $(Y, 0)$ such that $X \otimes Y = \Delta$ (Binmore [65, 62]). But Renegotiation Consistency then implies that $f(X, 0) \otimes f(Y, 0) = f(\Delta, 0) = \nu$. From the first step of the argument, it then follows that $f(X, 0) = \lambda$. That is to say, the solution to the bargaining problem $(X, 0)$ is the weighted Nash bargaining solution with bargaining powers α and β.

It remains to extend the result to the case of unnormalized bargaining problems by appealing to the Independence of Utility Calibration. To obtain the regular Nash bargaining solution, the Symmetry axiom must be invoked as well to ensure that $\alpha = \beta$.

1.5 Bargaining with Commitment

Harsanyi [233] and Rawls [417] assume that players are committed to the deal hypothetically reached in the original position. In studying their approaches to the social contract, consistency therefore requires that it be assumed that the players are also able to make commitments when they bargain behind the veil of ignorance. Fortunately, Nash [380] took the same freewheeling attitude to commitment as Harsanyi and Rawls when he formulated the first noncooperative bargaining game ever studied. In presenting my reconstruction of the ideas of Harsanyi and Rawls in the next chapter, it will therefore be possible to follow Nash's approach without any need for modification.

1.5.1 Nash Demand Game

The noncooperative bargaining game studied by Nash [380] is known as the Nash Demand Game.[31] We met this game briefly in Section 1.2.4 while

---

[29]For any x in X and y in Y, $x_A^\alpha x_E^\beta \leq \lambda_A^\alpha \lambda_E^\beta$ and $y_A^\alpha y_E^\beta \leq \mu_A^\alpha \mu_E^\beta$. If $x \neq \lambda$ or $y \neq \mu$, then at least one of these inequalities is strict. But then $(x_A y_A)^\alpha (x_E y_E)^\beta < (\lambda_A \mu_A)^\alpha (\lambda_E \mu_E)^\beta$, and so $x \otimes y \neq \nu$.

[30]The convex hull of the points $(0, 0)$, $(0, 1)$, and $(1, 0)$.

[31]Section 2.3 of Volume I considers some of the simple games that have been proposed as microcosms of society in social contact discussions. The one-shot Prisoners' Dilemma is the most popular choice by far, but Chicken and the Stag Hunt Game also have their followers. My own view is that society is much better modeled as a *repeated* game, but if the need to keep things simple requires that we restrict our attention to one-shot games,

studying the problem of dividing the dollar. However, in this section, we shall follow Nash in assuming that the players make demands directly in utils rather than cents.

The Nash Demand Game associated with the Nash bargaining problem (X, ξ), requires that each player simultaneously make a demand. If Adam demands a payoff of x_A and Eve demands a payoff of x_E, then these demands are compatible if the pair $x = (x_A, x_E)$ lies in the set X. In this case, each player receives his demand. If the demands are incompatible, each player receives his *status quo* payoff.

The Ultimatum Game captures the essence of bargaining with commitment when one of the players is gifted with the opportunity to speak first. Suppose, as in Section 0.4.1, that the first player is Adam. His optimal strategy is to accompany his opening proposal with a commitment to treat any response except an immediate acceptance as ending the negotiation. That is to say, it is advantageous for Adam at his first move to limit his future freedom of maneuver to the maximal extent. The resulting situation is modeled by the Ultimatum Game, in which Adam ends up with everything.

In contrast, Nash [380] focused on the case in which the strategic opportunities available to the two players are identical. If commitment is allowed, both players will then race to be the first to register a take-it-or-leave-it opening demand. The Nash Demand Game captures the essence of procedures in which such a race is preordained to be a draw, so that it is common knowledge that both players will submit take-it-or-leave-it demands *simultaneously*. Indeed, given the opportunity, a player does best to commit himself right now to take-it-or-leave-it proposals for each bargaining problem with which he might conceivably be faced in the future, thereby ensuring that he is at no risk of losing any commitment races. However, in doing so, he will recognize that the same ploy will be employed by any rational opponent. If it is common knowledge that such advance commitments will be made for all bargaining problems, then the effect will

---

then there is a good case for using the Nash Demand Game rather than other one-shot games that have been proposed. The Nash Demand Game has a continuum of Nash equilibria, just as the folk theorem assures us is the case for repeated games. As in real life, the Nash Demand Game therefore offers us many choices for a viable social contract. I do not pursue this point in the text for fear of diluting the emphasis I give later to the importance of modeling society as a repeated game. Nor have I taken this point very far in writing on the Nash Demand Game elsewhere (Binmore [66, 64, 68]). However, other workers may perhaps find some of these writings of interest. In particular, the connection between an incomplete information version of the Nash Demand Game and Harsanyi and Selten's [236] generalization of the Nash bargaining solution to the case of incomplete information is studied in Binmore [64]. Evolutionary considerations are explored in Binmore [68] and Skyrms [499]. Young [572] has particularly interesting things to say on this latter topic.

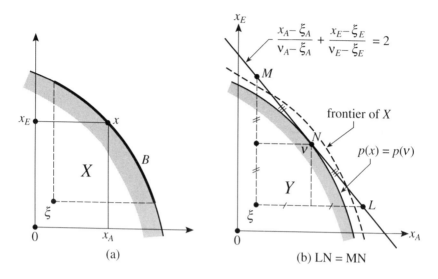

Figure 1.9: Nash Demand Game.

be the same as if the bargainers chose their strategies simultaneously at the beginning of each bargaining session.

Every x in the bargaining set B of the bargaining problem (X, ξ) is a Nash equilibrium for the Nash Demand Game. As illustrated in Figure 1.9(a), if Adam demands x_A, Eve's best response is to soak up the available surplus by demanding whatever payoff x_E makes x Pareto-efficient—unless it happens that $x_E < \xi_E$. If so, her best reply is to make a demand which is incompatible with Adam's so that she gets her *status quo* payoff ξ_E. Aside from the Nash equilibria in the bargaining set, we therefore have to take note of a large set of uninteresting Nash equilibria x characterized by $x \geq U$, where U is the utopian point for the problem (X, ξ).

Nash [380] dealt with the equilibrium selection problem for the Nash Demand Game using a method that has been recently generalized by Carlsson and Van Damme [114] to a wider class of games. Nash modified his demand game by introducing some uncertainty about whether a pair of demands x will prove to be compatible or not. The new game is equipped with a function $p : \mathbb{R}^2 \to [0, 1]$. The value of $p(x)$ is interpreted as the probability that x_A and x_E will turn out to be compatible. If $p(x) = 1$ for x in X and $p(x) = 0$ otherwise, then the new game coincides with the original Nash Demand Game. However, Nash assumed p to be differentiable, so that

$p(x)$ changes smoothly as x crosses the frontier of X. For this reason, the modified game is sometimes called the smoothed Nash Demand Game.

Under what circumstances does a pair ν of demands constitute a Nash equilibrium for the smoothed game? If Eve demands ν_E and Adam demands x_A, he gets an expected payoff of

$$x_A\, p(x_A, \nu_E) + \xi_A \left\{ 1 - p(x_A, \nu_E) \right\} . \tag{1.10}$$

Adam's best reply to Eve's demand is found by maximizing this quantity.[32] Similarly, Eve's best reply to Adam's demand of ν_A is found by maximizing

$$x_E\, p(\nu_A, x_E) + \xi_E \left\{ 1 - p(\nu_A, x_E) \right\} .$$

If ν is a Nash equilibrium, then ν_A and ν_E must be the respective solutions to the two maximization problems. Since p can be differentiated, necessary conditions for ν to be a Nash equilibrium are therefore

$$(\nu_A - \xi_A) p_{x_A}(\nu) + p(\nu) = 0 ,$$
$$(\nu_E - \xi_E) p_{x_E}(\nu) + p(\nu) = 0 ,$$

where p_{x_A} is the partial derivative of p with respect to x_A, and p_{x_E} is the partial derivative of p with respect to x_E.[33]

The set of all pairs x that satisfy $p(x) = p(\nu)$ defines a contour of the function p. The tangent line to this contour at the point ν is given by

$$p_{x_A}(\nu)(x_A - \nu_A) + p_{x_E}(\nu)(x_E - \nu_E) = 0 .$$

On substituting for $p_{x_A}(\nu)$ and $p_{x_E}(\nu)$ from the necessary conditions for a Nash equilibrium, the tangent equation reduces to the pleasingly simple form

$$\frac{x_A - \xi_A}{\nu_A - \xi_A} + \frac{x_E - \xi_E}{\nu_E - \xi_E} = 2 ,$$

as illustrated in Figure 1.9(b). On comparing this diagram with Figure 1.5(b), we find that a Nash equilibrium ν for the smoothed Nash demand game must be the symmetric Nash equilibrium for the Nash bargaining problem (Y, ξ), in which Y is the shaded set in Figure 1.9(b) consisting of all x that lie on or beneath the contour $p(x) = p(\nu)$.

The Nash Demand Game was smoothed to shed light on the equilibrium selection problem posed by the original game. However, motivating the smoothed game in this fashion puts the cart before the horse. In real life,

---

[32]Without loss of generality, one can simplify by taking $\xi = (0, 0)$.

[33]Thus $p_{x_A}(x_A, \nu_E) = \partial p(x_A, \nu_E)/\partial x_A$. We then find $p_{x_A}(\nu)$ by evaluating the partial derivative $p_{x_A}(x_A, \nu_E)$ at the point $x_A = \nu_A$.

there is always some uncertainty about everything—even death and taxes. The smoothed game is therefore truer to life than the unsmoothed version.[34] A justification for abstracting away the true-to-life details of the smoothed version in favor of the unsmoothed version would be that the details which are lost along the way are unimportant. But the argument shows that the suppressed details do actually matter. Abstracting them away creates a spurious equilibrium selection problem. However, we can have the best of both worlds by treating the unsmoothed Nash Demand Game as the limit of the smoothed version when the uncertainty level becomes vanishingly small. In particular, we can use the limiting value of the Nash equilibrium[35] of the smoothed game as our selection from the class of all Nash equilibria for the unsmoothed game.

In order that the unsmoothed Nash Demand Game appear as the limit of the smoothed version, we shall insist that each contour $p(x) = c$ with $0 < c < 1$ approximates the Pareto-frontier of X as indicated in Figure 1.9(b). We can then study the limit as the band of such contours shrinks down to the frontier. The set Y of Figure 1.9(b) then converges on X. Because the Nash equilibrium n of the smoothed game is the symmetric Nash bargaining solution for (Y, ξ), its limiting value is therefore the symmetric Nash bargaining solution for (X, ξ).[36]

Since our equilibrium selection criterion selects the symmetric Nash bargaining solution from the set of all Nash equilibria for the original Nash Demand Game, the symmetric Nash bargaining solution has survived the kind of test required by the Nash program—albeit only for the case when the players are free to make whatever commitments take their fancy.

1.5.2 Fixed and Variable Threats

Nash's [380] formulation of a bargaining problem as a pair (X, ξ) is sometimes called his *fixed-threat* theory to distinguish it from the *variable-threat* theory which will be discussed next. I dislike this terminology because it takes for granted that the only bargaining procedures for which it makes sense to apply the Nash bargaining solution are those which ascribe unlimited commitment power to the players. However, Schelling [464] has convincingly argued that this will normally be a highly unrealistic assumption.

A technical advantage of working with the case of unlimited commit-

---

[34]This is an overly glib assertion. In modeling only one source of uncertainty, the implicit assumption is that other sources are negligible in comparison.

[35]I plan to ignore the exceptional cases in which there may be more than one Nash equilibrium for the smoothed game (Binmore [66]).

[36]The Nash bargaining solution is continuous in the Hausdorff metric (Roth [438]).

ment is that it resolves the problem of where to put the *status quo*. In some cases, locating a *status quo* creates no difficulty. For example, if Adam and Eve are bargaining over how to play the one-shot Prisoners' Dilemma, it clearly belongs at the payoff pair that results if both players choose their hawkish strategy, whether or not we assume that the players can make forward commitments. However, in a game like G of Figure 1.1(a) it matters a great deal whether forward commitments are permitted.

Section 1.2.1 notes that G has only one Nash equilibrium outcome, the point N in Figure 1.1(b). If the players bargain on how to play G without being able to make commitments, it would therefore be natural to place the *status quo* at N. However, if they can make forward commitments while bargaining, Nash [382] thought it equally appropriate to assume that the players can also make commitments about what action they would take if the negotiations were to break down.

The idea of a subgame-perfect equilibrium rests on the assumption that players will ignore threats unless they are credible (Section 0.4.1). Adam might threaten Eve that he will burn down the family home unless she agrees to let him watch his favorite television program, but she will be unimpressed because she knows his threat is empty. If he doesn't get his way over the television program, why should he then damage himself further by destroying his home? However, if he could credibly commit himself to carrying out the threat, then things would be very different. After he had made the threat, Eve would then be faced with a choice between missing her preferred television program or seeing her house burned down. Assuming that she prefers the former, Adam would therefore win the argument over which television program should be watched. Perhaps we should therefore be grateful that commitments are actually so hard to make in real life, since it is hard to see a marriage surviving long if all disputes were settled in this way!

Of course, when Adam threatens to turn to arson if he fails to get his way, he does not expect to carry out the threat. He knows that Eve is rational and that he would only have to implement his threat if she were to behave irrationally. Gauthier [196] therefore goes astray in observing that:

> Maximally effective threat strategies ... play a purely hypothetical role ... since Adam and Eve do not actually choose them ... But if Adam and Eve would not choose these strategies, then they cannot credibly threaten with them. Maximally effective threat strategies prove to be idle.[37]

---

[37]The original refers to Ann rather than Eve, and arises in a discussion of an approach to bargaining due to Zeuthen [576] that was revived by Harsanyi [233], but then rendered obsolete by the Rubinstein [445] model of Section 1.7.1. I am not clear why Gauthier

It is certainly true that neither Adam nor Eve will actually carry out their maximally effective threats. But this is not because the strategies that call for such threats to be made are ineffective. On the contrary, the threats are not actually carried out because the strategies are so effective that the opponent always buckles under. Similarly, one seldom sees people run over by cars. But it does not follow that the threat of being run over by a car can be ignored. Insofar as people avoid accidents, it is precisely because they keep this threat very much in mind.

What are maximally effective threat strategies? As usually told,[38] the story that goes with Nash's variable-threat theory supposes that the players simultaneously open the negotiations by making a commitment to the action they will take in the game G if they are unable to reach an agreement on how it should be played. These opening commitments then determine a *status quo* for an application of the symmetric Nash bargaining solution.[39] The problem for the players is to choose these opening commitments in a maximally effective manner.

In the game G of Figure 1.1(a), each player has an infinite number of mixed strategies available as possible threats. Recall from Section 1.2.1 that the label ℓ is used for the mixed strategy in which Adam uses *bottom* with

---

thinks that Nash's threat theory is relevant in such a context. Nor do I feel that his views on such matters are clarified in a recent article (Gauthier [198]), in which he distinguishes between "assurances" and promises or threats. He thinks it rational, for example, to drink the toxin in Kavka's [302] "toxin puzzle" after offering an assurance that one intends to do so. However, it seems to me that Gauthier is simply dressing up his version of the Transparent Disposition Fallacy of Section 3.2 of Volume I in new clothes. We are told that it can be rational to choose not to deliberate rationally. But this is just another way of saying that one can commit oneself to being disposed to behave in a certain type of way. Similarly, his notion of an assurance seems no more than a revival of his idea that rational folk can be supposed to have transparent skulls.

[38] Owen [398] contains a very clear account. My *Fun and Games* (Binmore [70, p.261]) discusses only a simplified version of the theory.

[39] Although this story simplifies the computation of the bargaining outcome in the case of variable threats, it is not consistent with the attitude to commitment adopted earlier. When the bargaining opens, Adam and Eve will not only make commitments about what they will do should the ensuing negotiations fail, they will also seize the opportunity to make commitments that limit their future freedom of maneuver during the bargaining itself. They will therefore not only choose mixed strategies ℓ and m for G as described in the text, thereby determining a *status quo* $\xi(\ell, m)$; they will simultaneously make take-it-or-leave-it demands x_A and x_E as in the fixed-threat case. If Eve commits herself to $m = \mu$ and $x_E = \nu_E$, then Adam needs to choose ℓ and x_A to maximize (1.10) with ξ_A replaced by $\xi_A(\ell, \mu)$. Adam can proceed by first finding the optimal value of x_A for each fixed value of ℓ. Assuming that Eve has already optimized, we know from our study of the fixed-threat case that, in the limit, as the level of uncertainty becomes vanishingly small, the optimal value of x_A lies at the symmetric Nash bargaining solution for the problem with *status quo* $\xi(\ell, \mu)$. It now remains for Adam to calculate the optimal value of ℓ. This he does precisely as described in the text.

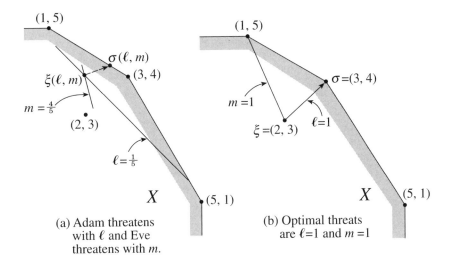

(a) Adam threatens
with ℓ and Eve
threatens with m.

(b) Optimal threats
are $\ell=1$ and $m=1$

Figure 1.10: A Nash threat game.

probability ℓ. Similarly, the label m denotes the mixed strategy in which
Eve uses *right* with probability m. After Adam has committed himself to ℓ
and Eve has committed herself to m, let the corresponding outcome of the
game G be $\xi(\ell, m)$.

When the symmetric Nash bargaining solution is used with $\xi(\ell, m)$ as
status quo, the result is a payoff pair $\nu(\ell, m)$ as indicated in Figure 1.10(a).
We are therefore faced with a game in which Adam chooses ℓ, Eve chooses
m, and the resulting outcome is $\nu(\ell, m)$. To find each player's maximally
effective threat strategy, we need to examine the Nash equilibria of this
game of threats.

Fortunately, this problem is less hard than it seems at first sight, because
all the outcomes lie on the Pareto-frontier of the set X of Figure 1.10(a).
The threat game is therefore strictly competitive, because Adam and Eve
have strictly opposed preferences over the outcomes. Although the threat
game is not zero-sum, the principles that Von Neumann and Morgenstern
used to analyze zero-sum games nevertheless apply. In particular, the Nash
equilibria are found by applying the maximin criterion. Figure 1.10(b)
shows the maximin choices $\ell = \lambda = 1$ and $m = \mu = 1$.

When Adam and Eve bargain over how to play the game G of Figure
1.1(a), Adam will therefore threaten to play *bottom* if the negotiations
break down. Eve will threaten to play *right*. When the symmetric Nash
bargaining solution is calculated using the resulting *status quo* $\xi(\lambda, \mu) =$

$(2,3)$, the final outcome will be an agreement on $\nu(\lambda,\mu) = (3,4)$.

This is such an elegant piece of work that it is an enormous pity to have to say that it is has little or no relevance to practical questions. To echo Schelling [464] yet again, it is simply not realistic to attribute unlimited commitment power to the players.

1.6   Trustless Transactions

So far, this chapter has followed the standard convention of cooperative game theory by assuming that the players are irrevocably committed to any agreement that they reach during a preplay negotiation. The preceding section goes further and assumes that the players can also make any commitments that take their fancy during the negotiation itself. The first of these assumptions is easier to defend that the second. Bilateral agreements are supposedly enforceable under the law of contract, whereas the legal position on the circumstances under which unilateral commitments are unconditionally binding is much less certain. A player may perhaps post a bond with a third party to be forfeited if he backs down from the commitment he is attempting to register, but such attempts to reduce a unilateral commitment to a bilateral contract are rare, presumably because the courts cannot always be relied upon to enforce them.

Even the existence of a standard bilateral contract is no guarantee that the agreement it embodies will actually be carried out. If Adam fails to honor his contractual obligations, it falls to Eve to decide whether to sue. But she will go to law only if the probability of a favorable ruling is high enough to make it worth her while to incur the huge costs that the legal profession levies on those unfortunate enough to require its services.

A legally binding contract therefore does not bind in the strict sense. If Adam or Eve disregard the terms of a contract they have signed, they must then play a legal game—which has many chance moves and is played for stakes that are sometimes much higher than those of the original contract. Rational players will therefore not treat contractual obligations as sacrosanct. When Adam and Eve append their signatures to a contract, they know perfectly well that they are only exchanging one game for another.

We therefore do not escape the problems of noncooperative game theory altogether by adopting the methodology of cooperative game theory. We simply exchange the problem of computing the equilibria of the original game for that of computing the equilibria of various legal games. However, sometimes the latter problem can be trivialized. For example, when the legal precedents are favorable, one may be able to write such large penalties for default into the contracts that compliance becomes a strongly dominant strategy. However, no such simplifying expedient is possible when we are discussing *social* contracts.

Social contracts are not contracts in the legal sense, since there is nobody outside the Game of Life to whom one can appeal for redress if someone cheats. Everybody—up to and including the Lord High Executioner—is a player in the Game of Life. However, the fact that social contracts are not legally binding does not imply that the methods of cooperative game theory must be abandoned altogether. Indeed, applications are simplified in one respect because the rules of the game that has to be played after someone cheats are always the same, whereas the rules of the legal game that is played after a legal contract is broken naturally depend on the terms and conditions written into the contract. However, in most respects matters become more difficult when there are no policemen to whom one can appeal, because only *self-policing* agreements are then worth the paper on which they are written. As Section 1.2.5 of Volume I explains, such a self-policing agreement reduces to an understanding between the players about the equilibrium in the Game of Life on which they plan to coordinate.

1.6.1 Repeated Games

Removing the assumption that preplay agreements are binding makes an enormous difference to what should be taken as the agreement set X for the game G of Figure 1.1(a). The only self-policing preplay agreement for G is the game's unique Nash equilibrium, marked N in Figure 1.1(d).[40] However, the game G is even less appropriate as a model of the Game of Life than the Prisoners' Dilemma or Chicken, whose suitability for this purpose came under question in Section 2.3 of Volume I. In this book, the Game of Life is modeled as a *repeated* game.

Chapter 3 discusses the folk theorem of repeated game theory. This theorem demonstrates that there is a close connection between the set of equilibrium outcomes of an infinitely repeated game and the cooperative payoff region of the game that is repeated.

To understand the connection, it is necessary to begin by explaining how

---

[40]If G had several equilibria, the players might agree to toss a coin to decide on which equilibrium to coordinate. The game theory literature usually observes that neither player then has an incentive to cheat on the agreement unless he believes his opponent will cheat. However, this is only strictly true if the players can commit themselves not to interact again after the fall of the coin. After all, if no opprobrium will result, a soccer captain disadvantaged by the fall of a coin tossed to decide who kicks off could repudiate the agreement and insist that the coin be tossed again. Whether this is a good idea will depend on what would then happen next. If this seems a trivial issue, perhaps I may recommend a rereading of the discussion of the Game of Morals in Section 1.2.4 of Volume I.

payoffs are measured in repeated games. As in Section 1.2.4, we assume that Adam evaluates a payoff flow $x_0, x_1, x_2, \ldots$ by calculating its present discounted value. If his discount factor[41] is δ, the total payoff he derives from the payoff flow is $x_0 + x_1\delta + x_2\delta^2 + \cdots$. A constant payoff flow $x, x, x, \ldots$ correspondingly results in a total payoff of $x + x\delta + x\delta^2 + \cdots = x/(1 - \delta)$. For a flow of $x_0, x_1, x_2, \ldots$ to be equivalent to a constant per-game payoff of x it is therefore necessary that

$$x = (1 - \delta)(x_0 + x_1\delta + x_2\delta^2 + \cdots).$$

In repeated game theory, the payoff to a player is always given on such a per-game basis.

The reason for refusing to use one-shot games to model the Game of Life is that they allow no opportunity for long-term relationships to develop. If one believes that *reciprocity* is the mechanism that keeps society's wheels turning, one therefore has to model the Game of Life with something at least as complicated as a repeated game. Even in a repeated game, reciprocal arrangements will not be feasible if the players' discount factors are too small. The gains available to Adam from betraying Eve today will then outweigh the advantages to be gained from maintaining their relationship. We therefore study the limiting case in which the players' discount factors are allowed to approach unity. The payoff to a player then becomes

$$x = \lim_{\delta \to 1}(1 - \delta)(x_0 + x_1\delta + x_2\delta^2 + \cdots).$$

It turns out that it must then also be true that

$$x = \lim_{N \to \infty} \frac{x_0 + x_1 + \cdots + x_N}{N},$$

when this limit of the means exists.[42] One can therefore think of the per-game equivalent of a payoff flow in a repeated game as being its long-run average.

---

[41] Section 2.2.6 of Volume I discusses indefinitely repeated games rather than infinitely repeated games. In the former, play does not continue forever. Instead it ends as soon as a weighted coin tossed after each repetition shows heads. Suppose that the probability of heads on any toss is p, where $0 < p < 1$. A pair of strategies that would result in Adam receiving a payoff flow of $x_0, x_1, x_2, \ldots$ if the coin always fell tails, then has an expected utility of $x_0 + x_1\delta + x_2\delta^2 + \cdots$, where $\delta = 1 - p$. Thus $1 - p$ serves as a discount factor when evaluating a payoff flow in an indefinitely repeated game. For many purposes, it therefore does not matter whether one is dealing with an infinitely repeated game or an indefinitely repeated game. Indeed, the reason that nature has programmed us to discount the future at all is presumably because the real Game of Life is full of chance moves that may prevent our being unable to cash in on future opportunities.

[42] Mathematicians will recognize this result as saying that Abel summability implies Césaro summability. It seems a long time ago that I wrote a thesis on such arcane matters!

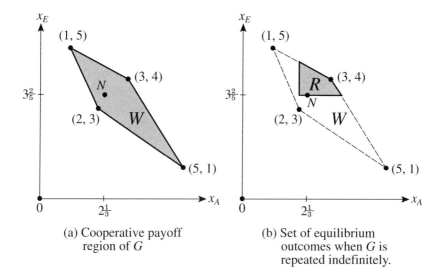

(a) Cooperative payoff region of G

(b) Set of equilibrium outcomes when G is repeated indefinitely.

Figure 1.11: Equilibrium outcomes in a repeated game.

The outcome of an infinitely repeated game between Adam and Eve can therefore be thought of as a pair of long-run average payoffs. Figure 2.4 of Volume I shows the set R of all Nash equilibrium outcomes for the infinitely repeated versions of the Prisoners' Dilemma, Chicken, and the Battle of the Sexes. Figure 1.11 compares the set R of Nash equilibrium outcomes of the repeated version of the game G of Figure 1.1(a) with its cooperative payoff region. Since we shall be working with a *status quo* point ξ inside R, the differences between the two sets are irrelevant.[43]

The folk theorem of repeated game theory therefore has the very striking conclusion that all contractual arrangements on how to play the one-shot game G that Adam and Eve might make are also realizable when G is to be played infinitely often—but without any need that agreements in the latter case be enforceable by an outside agency. In the infinitely repeated case, the agreements to coordinate on an equilibrium are self-enforcing.

Even though my own social contract theory never assumes that the agreements Adam and Eve make are binding, the folk theorem will nevertheless allow us to borrow much of the apparatus of cooperative game theory and use this unchanged when studying preplay bargaining over in-

---

[43] As explained in Section 1.2.1 of Volume I, my theory places the state of nature for a social contract discussion at the social contract currently being operated by a society. Since I assume that social contracts can only survive if they correspond to equilibria in the Game of Life, the state of nature must therefore lie in R.

finitely repeated games. Indeed, I shall always assume that the set of equilibrium outcomes of an infinitely repeated game satisfies all the properties attributed to an agreement set X in Section 1.2.2. The set X is therefore be assumed to be closed, bounded above, convex, and comprehensive.

It is particularly important that the folk theorem guarantees that the set of equilibrium outcomes in a repeated game is necessarily *convex*. Even when contracts can be written, the claim that players can always agree to toss a coin to decide a difficult issue is often open to debate. What union boss would be able to explain to his members that they ended up with a low settlement in a wage negotiation because he lost a coin toss! However, in the context of an infinitely repeated game, the convexity of the set of equilibrium outcomes is achieved without any need to randomize. If two teams play soccer every Saturday, they do not need to toss a coin each time to decide who kicks off—they can take turns.

The set of equilibrium outcomes is assumed comprehensive only to keep things simple. If anything important turned out to hinge on this assumption, it would be necessary to abandon it and to start all over again.

1.6.2   Transitional Arrangements

Suppose that Adam and Eve are currently operating an agreement that results in a payoff flow of ξ_A to Adam and ξ_E to Eve. However, they agree to shift from ξ to a Pareto-improvement x. If ξ and x are both equilibrium outcomes of whatever game is being played, then there will be no enforcement problems while they are operating under the old arrangement. Nor will enforcement problems arise once the new arrangement is in place. But what of the transitional period while they are shifting from one equilibrium to another?

Sometimes the transition can be managed overnight, as when Sweden switched from driving on the left to driving on the right. After the fall of the Soviet Empire, some East European countries attempted a similarly quick shift from a centrally planned, command economy to a Western-style market economy, with varying degrees of success. Such attempts to move fast are sometimes criticized as rash and brutal, especially when things go wrong. However, critics seldom recognize that moving fast can minimize the risk of a regime being destabilized altogether while the system is in an out-of-equilibrium transitional phase.

The alternative is to attempt to move slowly and gradually from the original state to the planned Pareto-improvement through a long sequence of neighboring equilibria. This is the route to reform that one would expect to see employed by a rational society, since it removes the possibility of control being lost while the reforms are being implemented. Gorbachev's attempt to reform the Soviet Union along these lines is admittedly not a good advertisement for such a cautious approach. On the other hand, China seems to be managing a similar step-by-step shift without yet sliding into chaos, although I imagine that few Western economists would regard the route to reform being followed as ideal, since it involves so much suffering by so many people. In a rational society, no uncompensated suffering should be necessary at all. One of the tasks awaiting us is to confirm this intuition by showing that Adam and Eve have a safe and painless route from the state of nature, where their society begins, to the Pareto-efficient social contract that is their final destination.

The Centipede Game and continuous exchange. At this point, I want to study a much simpler transition problem. Two criminals agree to exchange a quantity of heroin for a sum of money. Adam is to end up with Eve's heroin and Eve with Adam's money. But how is this transition to be engineered, if each is free to walk away at any time carrying off whatever is currently in his or her possession? In real life, matters are complicated by the threat of physical violence, but we shall assume that Adam nor Eve cannot be forced to act against their will. Nor are they fearful of losing their reputations or their lives should either go back on their word.

Obviously, there is no point in Adam handing over the agreed price and waiting for the goods. Somehow the criminals have to arrange a flow between them, so that the money and the drug change hands *gradually*.[44] The problem of modeling trustless cooperation in terms of movement along a continuous path will be handled by assuming that both Adam and Eve are to be trusted in the case of small enough transactions, and then going to the limit.

Suppose that Adam has $100, each dollar of which is worth only one cent to him if not spent on heroin. Each grain of heroin that Adam buys for one dollar at the agreed rate of exchange is worth only one cent to Eve if not sold to Adam. Figure 1.12 shows a game with alternating moves that represents a procedure which the players might use to facilitate the exchange. At each decision node, the player whose turn it is to move can choose "across" or "down". To choose "across" is to make a gift to the other player which is worth one cent to the donor, but one dollar to the

---

[44]Biology offers the more exotic example of sex among the hermaphroditic sea bass (Dugnatov [163, p.45]). Eggs are expensive to produce, but sperm is cheap. When two sea bass mate, each therefore takes turns in laying *small* batches of eggs for the other to fertilize. A sea bass that laid all its eggs at once would be outperformed by an exclusively male mutant that fertilized the eggs and then swam off without making an equivalent investment in the future of their joint children.

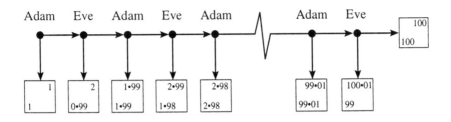

Figure 1.12: The Centipede Game.

recipient.[45] To choose "down" is to cheat on the arrangement by exiting with what one currently has. Cognoscenti will recognize that this game of exchange has the same structure as the much studied Centipede Game[46] (Rosenthal [437], Binmore [63], McKelvey and Palfrey [349]).

The Centipede Game has only one subgame-perfect equilibrium. This requires that both players *always* cheat. No trade would then take place. To see this, consider what is optimal in the subgame that arises if the rightmost node is reached. The seller must then choose between 100.01 and 100, and so cheats by choosing the former. In the subgame that arises if the penultimate node is reached, the buyer will predict this behavior by the seller. He must therefore choose between 99.01 and 99, and so cheats by choosing the former. Since the same backward induction argument works at every decision node, the result of a subgame-perfect analysis is that both players plan always to cheat.

Does this mean that exchange is not possible without trust? To draw such a conclusion would be to put more weight on the mathematical model proposed above than it can bear. The real world is imperfect in many different ways. The Centipede Game takes account of the imperfection that money is not infinitely divisible. But real people are even more imperfect than real money. In particular, they are not infinitely discriminating. What is one cent more or less to anybody? Following Radner [412], one may seek

---

[45]The models of trustless transactions studied in this section assume that the players do not discount time. Since each model has a finite time-horizon, one would obtain the same conclusions if discount factors were introduced and then allowed to approach unity after the analysis.

[46]So called because it has a hundred legs.

to capture such an imperfection by looking at ϵ-*equilibria*, in which players are satisfied to be within ϵ of the optimal payoff. Provided that the relevant trading unit (one cent in Figure 1.12) is chosen smaller than ϵ, it will then be an ϵ-equilibrium to honor the deal.

One can now idealize by considering versions of the Centipede Game with smaller and smaller trading units. This will allow us to study ϵ-equilibria with smaller and smaller values of ϵ. When the limit as $\epsilon \to 0$ is taken, a continuous model of exchange results. If the smallest unit of currency is made substantially smaller than ϵ in the Ultimatum Game of Section 0.4.1, then numerous ϵ-equilibria will be found, since Adam's ultimatum must take account of the fact that Eve may not accept a proposal which assigns her ϵ or less. However, in the limit as $\epsilon \to 0$, all these equilibria converge on the unique subgame-perfect equilibrium of the continuous version of the Ultimatum Game.[47] But it is not true that all ϵ-equilibria of the Centipede Game coalesce as $\epsilon \to 0$. In particular, honoring the deal is a limit of ϵ-equilibria. Cheating is also the limit of ϵ-equilibria, but it is not a limiting equilibrium on which the agents would wish to coordinate.

Modeling in continuous time. Sometimes philosophical difficulties are raised about the approach to continuous exchange surveyed above. The fact that similar difficulties are raised about the Rubinstein bargaining model of Section 1.7.1 is the chief reason for being so pedantic about the details of the argument.

It is sometimes argued that one need not bother with the fussy details of taking limits because it is more sound to "model directly in continuous time" by writing down whatever differential equation is relevant. An analysis then reduces to finding an appropriate solution to the equation. Time *is* continuous, so the story goes, and it is therefore a mistake to treat the case when time moves in discrete jumps as fundamental.

It is not easy to respond in a manner that such critics can understand. In particular, drawing attention to the way physicists do things cuts no ice with critics who have little or no scientific training. Still less does it help to suggest that the metaphysical nature of time is open to debate. Usually it is thought that a joke is being made. However, I have sometimes found it effective to point out that a derivative is *defined* as a limit. In writing down a differential equation, one is therefore making a statement about the manner in which certain limits are related. If one wants to get these relationships right, one therefore has little choice but to look at the limiting procedures that led to them.

This is the crucial issue. For example, in the pursuit-evasion games

---

[47]In the continuous version of the game, Adam can demand any real number in $[0,1]$.

studied by the methods of differential game theory, it is often assumed
that the hunter and his prey move at constant speed. But it makes no
sense to assume that they can choose the direction in which they move
at each separate instant of time. Indeed, it makes no sense to think in
terms of an extensive-form game in continuous time at all.[48] One cannot
ask how a player will alter his behavior between one instant and the next,
because there is no such thing as a next instant when time is treated as a
continuous variable. One therefore chooses a discrete sequence of points in
time at which decisions are envisaged as being made and then computes the
limiting values of the quantities involved as the distance between successive
points in this grid is allowed to become vanishingly small. However, in
pursuit-evasion games, these limits will not exist unless constraints are
imposed on how much a player can alter his direction between one of his
decision nodes and the next. After limits have been taken, such constraints
will translate into assumptions about the radius of a player's minimum
turning circle. But to provide a proper defense for the appearance of such
a quantity in the final differential equation, one cannot avoid following the
limiting process through from beginning to end.

The same holds for the continuous version of the Centipede Game. Just
as the players of a pursuit-evasion game are assumed to be imperfectly agile,
so Adam and Eve are assumed to be limited in their power to discriminate
between small sums of money. This assumption is first imposed in a discrete
model and maintained when the limit is taken. The existence of a powerful
theory of differential equations does not allow one to dispense with the need
to fuss about the relative magnitudes of small imperfections that the use of
such a methodology entails. On the contrary, one has no choice but to pay
close attention to such matters if one is to have any hope of ending up with
the right differential equation. If, as always in this book, the analytical
problems that remain after completing the necessary fuss about the details
of the model are too easy to require any appeal to the theory of differential
equations, so much the better. When "modeling directly in continuous
time" seems to lead to different conclusions, always ask why the differential
equation studied is thought to be relevant. Usually, no coherent answer
will be forthcoming.

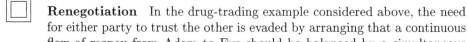

Renegotiation   In the drug-trading example considered above, the need
for either party to trust the other is evaded by arranging that a continuous
flow of money from Adam to Eve should be balanced by a simultaneous
flow of heroin from Eve to Adam. When the Centipede Game is used to

---

[48]Shubik's [491] drawing of a game played in continuous time belongs with Escher's
pictures of impossible landscapes!

model the rules of exchange, neither party has any incentive to cheat at any time, because all that will be achieved is to terminate the transaction before it is complete, to the disadvantage of both parties.

The Centipede Game is a particularly simple example for at least two reasons. In the first place, the rules of the game do not allow the transaction to be resumed if the exchange is interrupted. In the second place, the terms of trade are fixed in advance: either grains of heroin are traded for dollars at a rate of one for one, or no trade takes place at all. In a more realistic model, it would be necessary to make allowance for the fact that a player might call a halt to the exchange with a view to renegotiating the terms of trade. After such a renegotiation, the transaction could then possibly continue at a different rate of exchange.

When bargaining without trust, Adam and Eve therefore have to remember that the implementation of the agreement they reach at time zero may be interrupted at any time if either party thinks he can thereby gain an advantage. The deal will then be renegotiated and the exchange continued on different terms. The player who loses out will naturally feel aggrieved— just like Eve in the Ultimatum Game when she is offered a derisory share of the dollar. However, just as it is rational for Eve to swallow her pride and accept what she is offered on the grounds that even a cent is better than nothing, so the aggrieved party after a renegotiation will see that nothing is to be gained by refusing to accept that the balance of power has shifted to his disadvantage.

The possibility of renegotiation means that Adam and Eve will not bother to consider agreements that they know are not viable, because someone will renege on the deal before it has been fully implemented. It is therefore necessary to restrict attention to agreements that are proof against renegotiation,[49] because any renegotiation will simply reproduce the arrangements specified originally. The case of the drug-trading example is particularly simple.

The set X of Figure 1.13(a) shows the set of all payoff pairs that the drug dealers can jointly achieve. The deadlock point d is the payoff pair that will result if the traders are unable to agree, and so hang around indefinitely on some street corner dickering, at constant risk of attracting the attention of the ever-vigilant forces of law and order. As in Section 1.3.1, it is necessary to distinguish between deadlock and breakdown. If the negotiations break down immediately and Adam and Eve then both

---

[49]Renegotiation-proofness is a notoriously intricate and controversial topic in game theory. However, Section 3.7.3 explains why I think it appropriate to ask for social contracts to be proof against renegotiation only on the equilibrium path. Such a relaxation of the orthodox requirements trivializes the issues that make renegotiation-proofness a hot potato in the literature.

depart to take up their best outside options, the result will be the payoff
pair $b(0) > d$. The players are therefore initially faced with the bargaining
problem $(X, b(0), d)$ of Section 1.3.1. However, since they are bargaining
without trust, it is not enough to propose some payoff pair n as the solution
to their bargaining problem. It is necessary to find a continuous trading
path through X that describes the rates at which they are to trade heroin
for money at each point of time until the transaction is complete. The
point on this curve reached at time t is labeled $b(t)$, since the result of the
negotiation breaking down at this time is that each player will depart with
the amount he has accumulated so far during the exchange. The transaction
concludes when the curve reaches the Pareto-boundary of X. Attempts to
negotiate further will inevitably lead to an immediate breakdown, because
neither can improve his position without making the other worse off. If it
takes time T to complete the transaction, the final deal will therefore be
represented by $b(T)$.

In Section 1.3.1, we saw that taking account of the bargainers' outside
options forces us to impose constraints on the range over which a proposed bargaining solution is valid. Renegotiation considerations have the
potential to impose further constraints. For example, if it is proposed that
$b(T)$ should be the symmetric Nash bargaining solution for the problem
$(X, b(0), d)$, then the proposal needs to be supplemented by specifying a
continuous path through X from $b(0)$ to $b(T)$ along which $b(T)$ is also
the symmetric Nash bargaining solution to each problem $(X, b(t), d)$ with
$0 \leq t \leq T$. If such a path can be found, neither player will have an incentive at any time during the exchange to demand a renegotiation, because
the result of such a renegotiation would simply be to reproduce the original
agreement. As Figure 1.13(a) shows, such paths are easily found[50] and
hence the lack of trust between Adam and Eve proves to be no handicap
since they are able to reach the same deal they would have reached if they
were bosom buddies.

In the drug-trading example just discussed, the deadlock point was kept
constant during the transaction so as to keep things simple. Figure 1.13(b)
illustrates an example in which the roles of the deadlock point and the
breakdown point are reversed. Imagine a game that is repeated every day
between the workers and the management of a firm. As explained in Section
1.2.4, the payoffs should therefore be interpreted as flows (rather than stocks
as in the heroin-trading example). Each point in X corresponds to the pair
of payoff flows that results from the choice of one of the many equilibria of
the repeated game. Currently the firm is operating the Pareto-inefficient
equilibrium corresponding to $d(0)$. Meanwhile management and workers

---

[50]It is only necessary that $b_A(t)$ and $b_E(t)$ be continuous and increasing on $[0, T]$.

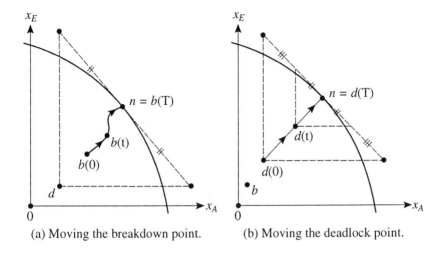

(a) Moving the breakdown point.   (b) Moving the deadlock point.

Figure 1.13: Bargaining without trust.

bargain about how to share the surplus that would result if agreement could be reached on a move to a more efficient equilibrium. Thus $d(0)$ is the deadlock point. An inferior breakdown point is located at $b < d(0)$. The breakdown point represents the payoffs to the workers and the management if the workers were to find jobs elsewhere and the management were to hire the best replacements it could locate.

Is it possible for the firm to get from $d(0)$ to the symmetric Nash bargaining solution n of the problem $(X, b, d(0))$ without trust between the two sides? Such a move will involve the gradual surrender of entrenched privileges and practices by both management and workers. Once surrended, these will not necessarily be recoverable in their old form. If the move to n is broken off before it is completed as a consequence of one side or the other insisting on a renegotiation, the firm will lapse into a *new* equilibrium. The point $d(t)$ in Figure 1.13(b) corresponds to such an equilibrium. If there is a call for a renegotiation at time t, then $d(t)$ will serve as a *new* deadlock point while the breakdown point b remains unaltered.

This example therefore differs from the drug-trading example in that the exchange is seen as a moving the deadlock point continuously from $d(0)$ to $d(T)$ instead of moving the breakdown point continuously from $b(0)$ to $b(T)$. As a consequence, it is not so easy to satisfy the renegotiation-proofness requirement that the solution n to the problem $(X, b, d(0))$ also be the solution to each problem $(X, b, d(t))$ with $0 \le t \le T$. In fact, as the characterization of the symmetric Nash bargaining solution illustrated in Figure 1.5(c) shows, the only possible path from $d(0)$ to $n = d(T)$ that will suffice is the *straight line* of Figure 1.13(b). However, since such a straight line does exist, it turns out that trust between the management and its workforce is not necessary for the firm to get to an efficient outcome.

It is this last story which is most relevant to the social contract issues discussed in Chapter 4. However, there is a more general lesson to be learned from such stories, namely that it is pointless to use the Nash bargaining solution, or any other bargaining solution, without a very clear idea of the circumstances in which it is to be applied. In particular, it would definitely be wrong in the management-worker story that goes with Figure 1.13(b) to place the *status quo* at b—the analogue of a Hobbesian state of nature—unless unreasonable commitment assumptions are to be made. This is a point which has been insisted upon by Buchanan [103, 104].

1.7   Bargaining without Commitment

Are rational players able to make unilateral commitments that constrain their future freedom of action? As Section 1.2.4 of Volume I explains, game theorists insist that any commitment opportunities available to the players be built into the rules of the game under study, so that the analysis of the game can proceed without any risk of falling prey to the fallacies catalogued in Section 3.2 of Volume I. The extent to which commitment opportunities should be incorporated into a game's rules is a matter for the judgment of the modeler. My own thoughts on this subject derive from Schelling [464]. Corruption and decay await any large social or political institution that relies on people, particularly its own officers, committing themselves to the goals of the organization at the expense of their own individual self-interest. Realism therefore demands that attempts to model the Game of Life involve building as little commitment into its rules as possible. This is not to say that no commitment opportunities at all should be offered to the players. I am not so disillusioned with my fellow human beings as to think they can never be trusted to keep promises about trifling issues in the near future. Games with commitment opportunities built into their rules must therefore be studied, but I am only comfortable with conclusions derived from such models if they survive when the level of commitment is allowed to become vanishingly small—as in Section 1.6.2.

For the bargaining issues studied in this chapter, commitment problems arise in two places. Are players committed to agreements reached during the bargaining when the time comes for them to be implemented? Are players committed to carry out any threats or promises they may make while bargaining that would constrain their future bargaining behavior? It is standard in cooperative game theory to answer *yes* to the first question. Section 1.2 reviews some of the technical apparatus that has been developed to deal with the implications of such an affirmative answer. But for the social contract issues of this book, the right answer to the first question is *no*. Fortunately, Section 1.6 shows that much of the standard technical apparatus can still be employed if suitably interpreted.

Section 1.5 reviews the implications of answering *yes* to the second question. In spite of the elegance of the theory that results, my own view is that it has little relevance to how people bargain in the real world. When bargaining over something important to both sides, few people are so naive as to take an opponent at his word when he claims to be making a final take-it-or-leave-it offer. The current section therefore studies a model that answers *no* to the second question. It turns out that we are again lucky that answering *no* does not force us to abandon the technical apparatus developed when answering *yes*. In Section 1.5, bargaining with commitment was modeled using the Nash Demand Game. The current section uses the very different Alternating Offers Game of Stahl [507] and Rubinstein [445] to model bargaining without commitment. However, in both cases, we are led to a version of the Nash bargaining solution. It is therefore still true that the more risk averse a player, the smaller the share of the surplus he will receive. But the case of bargaining without commitment is necessarily more complicated: time must be taken into account along with risk, since a patient player can put pressure on an impatient player by employing delaying tactics.

Fortunately, it turns out that this complication can be handled by moving from the symmetric Nash bargaining solution to a weighted version in which the players' bargaining powers reflect how patient they are. Impatient players get smaller bargaining powers and hence are assigned a smaller share of the surplus. Successful bargainers will therefore be both patient and ready to take calculated risks.[51] Our task in this section is to quantify the impact that their attitudes to risk and the unproductive passage of time will have on the share of the surplus they receive.

---

[51] Recall that only bargaining under complete information is to be studied. Matters are much more complicated when the players have different information about each other and the amount of the available surplus. Players may then find it worthwhile to use delaying tactics to signal how tough they are (Binmore, Osborne and Rubinstein [80]).

1.7.1   The Alternating Offers Game

In the noncooperative bargaining models that we have studied hitherto (Sections 0.4 and 1.5), the assumption that players could commit themselves to take-it-or-leave-it offers forced all the action into a single burst of activity at time 0. Without commitment, such take-it-or-leave-it offers cease to be viable. A player can always refuse an offer and still have the opportunity to respond with a counteroffer. Time may therefore pass before agreement is reached. We therefore have no alternative but to take account of the players' attitudes to the unproductive passage of time. As Cross [139] remarks, "If it did not matter *when* the players agreed, it would not matter *whether* they agreed at all."[52]

It will be assumed that a deal s reached at time t results in a payoff of $\delta_A^t u_A(s)$ to Adam and a payoff of $\delta_E^t u_E(s)$ to Eve. Discount factors like δ_A and δ_E were discussed in Section 1.2.4 along with the corresponding discount rates $\rho_A = -\log \delta_A$ and $\rho_E = -\log \delta_E$. It will be assumed that both discount factors lie between 0 and 1 so that the discount rates are both positive.

Recall from Section 1.2.2 that X denotes the set of all feasible payoff pairs in a bargaining problem. Since both Adam and Eve dislike delay, this agreement set is given by

$$X = \{(u_A(s), u_E(s)) : s \text{ is a feasible deal}\}.$$

If time t has passed without an agreement, the set of feasible deals shrinks to the set

$$X_t = \{(x_A \delta_A^t, x_E \delta_E^t) : x \in X\}.$$

To keep things simple, assume that the agreement set X satisfies the requirements of Section 1.2.2. Thus X is assumed to be closed, bounded above, convex and comprehensive. To ensure that X is convex, it is usually necessary to require that the players be risk-averse.

---

[52]Mathematicians may dispute this observation on the grounds that the sequence of sequences which begins $1, 0, 0, 0, \ldots, 0, 1, 0, 0, \ldots, 0, 0, 1, 0, \ldots$ does not converge to the zero sequence $0, 0, 0, 0, \ldots$ in such topologies as ℓ^∞ or ℓ^2. However, the topology one adopts in a model should be determined by practical considerations. The practical consideration here is that someone who delays gratification for sufficiently long risks losing the opportunity for gratification altogether as a consequence of the intervention of some unforeseen circumstance. People therefore have a very good reason for discounting the future. Working with a topology that neglects this fact of life, as is common in the literature on intergenerational equity, seems pointless to me. The claim that $\delta = 1$ is the only fair intergenerational discount factor is sometimes defended by saying that the model is concerned only with *pure* time preferences, considerations of risk being abstracted away altogether. However, such models would seem to have no practical relevance—unless one believes that the human race is certain to survive in its present form forever.

A finite-horizon version of the bargaining model to be studied appears first in the work of Stahl [507]. However, we shall study an infinite-horizon version formulated independently by Rubinstein [445].[53] As its name suggests, the players in this Alternating Offers Game alternate in making offers until an offer is accepted, with Adam making the first offer at time $t = 0$. After an offer has been refused, a time interval of $\tau > 0$ elapses before the next offer is made. In principle, a deadlock could be reached in which all offers are always refused. We assume that such a deadlock results in each player receiving a zero payoff.[54] In the notation of Section 1.2.2, the bargaining problem with which we are concerned is therefore $(X, 0)$.

As in the Ultimatum Game of Section 0.4.1, *any* outcome of the Alternating Offers Game can be supported as a Nash equilibrium. We therefore again turn our attention to subgame-perfect equilibria. Ariel Rubinstein's [445] celebrated result says that:

> The Alternating Offers Game has a *unique* subgame-perfect equilibrium and hence determines a unique solution of the bargaining problem $(X, 0)$.

Proving a special case.   As often happens, the chief difficulty in getting to Rubinstein's conclusion lies in recognizing that the result might be true. Once the mechanism that makes the theorem work is recognized, it then becomes possible to construct quite simple proofs, as in Binmore [67] or Shaked and Sutton [488]. For the special case considered below, the latter paper provides a particularly elegant proof. Appendix C analyzes a very general version of the model.

The first two stages of the extensive form of the Alternating Offers Game are shown in Figure 1.14(a) for the special case of risk-neutral players faced with the problem of dividing the dollar.

The thickened lines in Figure 1.14(a) indicate some strategies for the two players. Adam's strategy requires that he begin by demanding a fraction x of the dollar. This is equivalent to his offering the remainder $1 - x$ of the dollar to Eve. The strategy for Eve shown in Figure 1.14(a) requires her to refuse this offer and to make a counterdemand of y at time $t = \tau$. Figure 1.14(a) shows Adam accepting the $1 - y$ that is then left for him. Because

---

[53]It is sometimes thought adequate to compute the subgame-perfect equilibrium of a finite-horizon model by backward induction, and then to study its properties as the time-horizon is allowed to recede to infinity. In the case of the Alternating Offers Game, this will lead to Rubinstein's result. However, in other situations, the infinite-horizon model will have other subgame-perfect equilibria which have an equal or better claim to be selected than that obtained by considering the limit of the finite-horizon game.

[54]Because $u(s)\delta^t \to 0$ as $t \to \infty$.

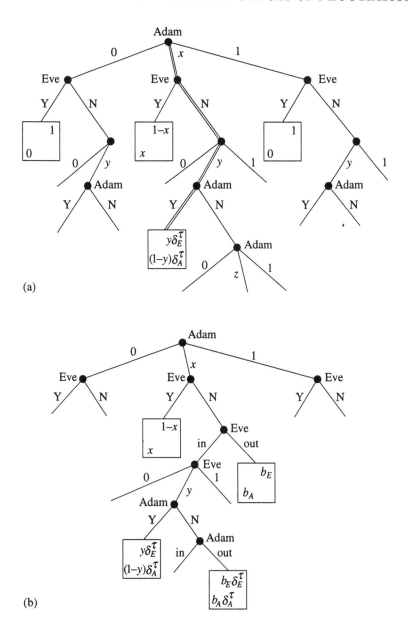

Figure 1.14: Extensive forms for the Alternating Offers Game.

of the delay of one time period in reaching agreement, Adam then receives a payoff of $(1 - y)\delta_A^\tau$ and Eve receives a payoff of $y\delta_E^\tau$.

Our problem is that the game may have many subgame-perfect equilibria. Let m_A and M_A be respectively the minimum and maximum[55] payoffs that Adam can receive at a subgame-perfect equilibrium of the Alternating Offers Game G. Let m_E and M_E be the corresponding quantities for the companion game H in which it is Eve who makes the first move. If Adam's opening demand in G is refused, a subgame of G must then be played which is identical to H except that it begins at time $t = \tau$. The subgame-perfect equilibrium payoffs of this subgame of G are therefore found by discounting those of H by one period. That is, the maximum and minimum subgame-perfect equilibrium payoffs in the subgame that follow her refusing Adam's opening offer are $M_E\delta_E^\tau$ and $m_E\delta_E^\tau$.

The next step in the proof uses this fact to obtain the inequalities:

$$1 - M_E\delta_E^\tau \leq m_A \leq M_A \leq 1 - m_E\delta_E^\tau. \tag{1.11}$$

If Adam's opening demand x in G satisfies $x < 1 - M_E\delta_E^\tau$, then Eve will get $1 - x > M_E\delta_E^\tau$ from accepting and at most $M_E\delta_E^\tau$ from refusing. In equilibrium, she must therefore accept. A contradiction follows if we assume that a subgame-perfect equilibrium exists in which Adam gets $y < 1 - M_E\delta_E^\tau$. To be sure of getting more, he has only to demand x, where $y < x < 1 - M_E\delta_E^\tau$. This proves that $1 - M_E\delta_E^\tau \leq m_A$ in (1.11). The proof that $M_A \leq 1 - m_E\delta_E^\tau$ is easier. A subgame-perfect equilibrium in which Adam gets $x > 1 - m_E\delta_E^\tau$ cannot exist because Eve would then get at most $1 - x < m_E\delta_E^\tau$. But she can guarantee getting at least $m_E\delta_E^\tau$ simply by refusing Adam's opening demand. Finally, $m_A \leq M_A$ because the minimum must be smaller than the maximum.

The same argument is now applied to the game H. Since H is the same as G except that the roles of Adam and Eve are reversed, we are led to the inequalities of (1.11) with the letters A and E exchanged:

$$1 - M_A\delta_A^\tau \leq m_E \leq M_E \leq 1 - m_A\delta_A^\tau. \tag{1.12}$$

We can now use (1.12) to eliminate m_E and M_E from (1.11). A little algebra then generates the inequalities:[56]

$$\frac{1 - \delta_E^\tau}{1 - \delta_A^\tau\delta_E^\tau} \leq m_A \leq M_A \leq \frac{1 - \delta_E^\tau}{1 - \delta_A^\tau\delta_E^\tau}. \tag{1.13}$$

---

[55] Strictly speaking, *infimum* and *supremum* should be written for *minimum* and *maximum*.

[56] First introduce the inequality $M_E \leq 1 - m_A\delta_A^\tau$ from (1.12) into the inequality $1 - M_E\delta_E^\tau \leq m_A$ from (1.11) to obtain $1 - (1 - m_A\delta_A^\tau)\delta_E^\tau \leq m_A$. Rearranging the terms of this inequality leads to the first half of (1.13). The second half is obtained similarly.

It follows that $m_A = M_A$. The use of any subgame-perfect equilibrium in the Alternating Offers Game G therefore results in Adam receiving a payoff of precisely $\xi = (1 - \delta_E^\tau)/(1 - \delta_A^\tau \delta_E^\tau)$.

What does Eve get? If Eve were to refuse Adam's opening demand, she would get her subgame-perfect payoff $(1 - \delta_A^\tau)/(1 - \delta_E^\tau \delta_A^\tau)$ in H, but discounted by one period. However, $\delta_E^\tau(1 - \delta_A^\tau)/(1 - \delta_E^\tau \delta_A^\tau) = 1 - \xi$. The only way that Adam can get ξ is therefore for him to make an opening demand of precisely this amount and for Eve to accept.[57] As in the case of the Ultimatum Game studied in Section 0.4, Eve will then be indifferent between accepting and rejecting, but equilibrium play requires that she accept. Eve therefore gets $1 - \xi$. That is to say, the unique subgame-perfect outcome is Pareto-efficient. The dollar is split immediately, with ξ going to Adam and $1 - \xi$ to Eve.

Rubinstein and Nash. Rubinstein's [445] model seems to capture the cut and thrust of bargaining in the real world. The two sides do indeed alternate in making offers, whether they are bargaining over a loaf of bread in an oriental bazaar or a skyscraper in Manhattan. However, there are difficulties that need to be addressed. How come Adam gets to make the first offer and hence enjoys a first-mover advantage? Why do the players politely wait their turn? Given that it is Eve's turn, why does she wait a period of length τ before responding? Having refused an offer from Adam, it would be optimal for her to make a counteroffer immediately.

These difficulties become less pressing when we consider the limiting version of the model obtained by allowing τ to become vanishingly small. If Adam and Eve are bargaining over a loaf of bread in a bazaar, then neither will regard the loss of some fraction of a second in getting to a deal as significant. In bargaining over a skyscraper in Manhattan, the same will be true of a delay of a day or two. Some commitment to the bargaining conventions of the model must still be postulated, but the level of this commitment becomes progressively smaller as the time period τ is allowed to recede to zero.

The case of dividing the dollar with risk-neutral players was considered when proving Rubinstein's theorem. When $\tau \to 0$, Adam's equilibrium payoff in this special case becomes[58]

$$\lim_{\tau \to 0} \frac{1 - \delta_E^\tau}{1 - \delta_A^\tau \delta_E^\tau} = \frac{\rho_E}{\rho_A + \rho_E}$$

---

[57]If Adam demands anything else and Eve accepts, then Adam will not get ξ. If Adam demands ξ and Eve refuses, then she will eventually get $1 - \xi$. Adam must therefore end up with less than ξ, because their payoffs can sum to 1 only if they agree immediately.

[58]Use L'Hôpital's rule, remembering that $\delta = e^{-\rho}$.

In the limit, Eve's equilibrium payoff is therefore $\rho_A/(\rho_A + \rho_E)$. It follows that they split the dollar in the ratio $\rho_E : \rho_A$, which is the same as splitting it in the ratio $1/\rho_A : 1/\rho_E$.

This result generalizes immediately to the case of the bargaining problem $(X, 0)$. The limiting equilibrium outcome turns out to be the weighted Nash bargaining solution for the problem $(X, 0)$ in which the bargaining powers are $\alpha = 1/\rho_A$ and $\beta = 1/\rho_E$ (Binmore [66]). Moreover, to prove this result, we do not need to begin by recalculating the equilibrium payoffs for this more general case, provided we are willing to take on trust that there is indeed a unique subgame-perfect equilibrium in which the proposer always plans to offer an amount the responder will accept even though indifferent between accepting and refusing.

For some particular $\tau > 0$, let $P = (a, b)$ be the unique subgame-perfect equilibrium outcome in the general Alternating Offers Game G in which Adam is the first mover. Our problem is to relate the payoff pair (a, b) to a suitably weighted Nash bargaining solution of the bargaining problem $(X, 0)$. In this task, it will be helpful as before to consider a companion game H, which is identical to the Alternating Offers Game except that the roles of Adam and Eve are exchanged. Eve therefore enjoys the first mover advantage in the game H, whose unique subgame-perfect equilibrium outcome will be denoted by $Q = (c, d)$.

If Adam's opening offer in G were refused, then a subgame of G would be entered in which Eve is the first mover. This subgame is identical with H except that it begins at time $t = \tau$. Its unique subgame-perfect equilibrium outcome is therefore $(c\delta_A^\tau, d\delta_E^\tau)$. Thus Eve knows that refusing Adam's opening offer will result in her receiving a payoff of $d\delta_E^\tau$ if both players use their subgame-perfect strategies in this subgame. In equilibrium, she must therefore accept any proposal from Adam at time $t = 0$ that would result in a payoff pair (x, y) in X with $y > d\delta_E^\tau$. Similarly, in equilibrium, she must refuse any proposal with $y < d\delta_E^\tau$. Since subgame-perfect play also requires Adam to optimize, his equilibrium proposal must therefore have $y = d\delta_E^\tau$. Since Eve's subgame-equilibrium payoff in the game G is b, it follows that G must begin with this offer being accepted by Eve. Thus $b = d\delta_E^\tau$. Applying the same reasoning to the companion game H, we also find that $c = a\delta_A^\tau$.

Figure 1.15(a) illustrates the two identities

$$b = d\delta_E^\tau \quad \text{and} \quad c = a\delta_A^\tau \tag{1.14}$$

that characterize the unique subgame-perfect equilibrium outcome of the Alternating Offers Game G. Notice that (a, b) and (c, d) must be Pareto-efficient points of the agreement set X, because it is always optimal for

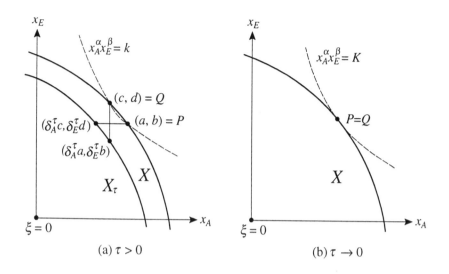

Figure 1.15: Characterizing the outcome in the limiting case.

the proposer to maximize his own payoff after conceding the minimum acceptable amount to his opponent.

Equations (1.14) become a little more friendly after some algebra. Raise both sides of the first equation of (1.14) to the power $\alpha = 1/\rho_E$. Next raise both sides of the second equation to the power $\beta = 1/\rho_A$. Then multiply the two equations that result. We are led to the equation

$$a^\alpha b^\beta = c^\alpha d^\beta ,$$

from which it follows that the points $P = (a, b)$ and $Q = (c, d)$ both lie on the same contour $x_A^\alpha x_E^\beta = k$, as illustrated in Figure 1.15(a).

Figure 1.15(b) shows the limiting case as $\tau \to 0$. The points P and Q converge to a common limit at the point ν where a contour $x_A^\alpha x_E^\beta = k$ touches the Pareto-frontier of the agreement set X. Comparing this diagram with Figure 1.5(a), we see that ν is the weighted Nash bargaining solution for the bargaining problem $(X, 0)$ with bargaining powers $\alpha = 1/\rho_A$ and $\beta = 1/\rho_E$.

Nash program. Rubinstein's result provides a striking example of the use of the Nash program discussed in Section 0.6. The Alternating Offers Game is a realistic noncooperative bargaining model in which the play of the unique subgame-perfect equilibrium results in each bargainer receiving the payoff he is assigned by a weighted version of the Nash bargaining solution.

Not only does the model vindicate the use of the Nash bargaining solution in predicting the deal that rational bargainers will reach when information is complete, it also provides valuable clues on *how* it should be used. In particular, it offers an interpretation of the bargaining powers α and β. As noted in Section 1.7.1, it makes little sense to think of these as reflecting the relative bargaining skills of the players, since ideally rational players will both bargain optimally. In Rubinstein's model, $\alpha = 1/\rho_A$ and $\beta = 1/\rho_E$ measure how patient the players are. An impatient player will have a low discount factor δ and hence a high discount rate ρ. His bargaining power will therefore be low.

The Alternating Offers Game is also instructive on the question of how to interpret the *status quo* ξ when applying the Nash bargaining solution to the bargaining problem (X, ξ). As emphasized in Section 1.2.2, ξ should be placed at the *deadlock point* d. This is the payoff pair that will result if the bargaining continues forever without agreement being reached. Alternatively, in the spirit of Cross's aphorism on the timing of agreements, it is the agreement that makes the players indifferent about when it is implemented. In the simple version of the Alternating Offers Game studied above, $d = (0,0)$. When bargaining about flows, d will be the pair of payoff flows that each player receives while the negotiation is in progress.

Section 1.2.2 also emphasizes the role of the breakdown point b. The payoffs b_A and b_E represent the best available outside opportunities to Adam and Eve if the negotiations are broken off at time $t = 0$. In the Alternating Offers Game studied above, the possibility that the negotiations might be terminated by a player walking off to take up an outside option was not considered. However, it is easy to modify the Alternating Offers Game as in Figure 1.14(b) by allowing a player to decide whether to opt out immediately after refusing an offer. If a player opts out at time t, then the resulting payoff pair is $(b_A \delta_A^t, b_E \delta_E^t)$. If he opts in, then a period of length τ elapses before he makes the next offer.

A formal analysis of a more general model than the Alternating Offers Game with outside options appears in Appendix C. However, the essential ground has already been covered in Section 1.2.2, since the argument given there converts quite easily into a formal proof in the context of the Alternating Offers Game. Matters are particularly simple in the limiting case when $\tau \to 0$. If b lies outside the agreement set X, the players have nothing to bargain about. Each will simply take up their outside options immediately. If $b < \nu$ and τ is sufficiently small, nobody will ever take up an outside option. The bargaining outcome will therefore be precisely the same as if no outside options were available at all. Otherwise the outside options matter. If $b_A \geq \nu_A$, then Adam gets precisely his outside option b_A and Eve gets the rest of the available surplus. If $b_E \geq \nu_E$, then Eve gets precisely her outside option and Adam gets the rest.

In summary, the Alternating Offers Game with outside options has a unique subgame-perfect equilibrium outcome that converges to a weighted Nash bargaining solution for the bargaining problem (X, b, d). As in the case without outside options, the relevant bargaining powers are $\alpha = 1/\rho_A$ and $\beta = 1/\rho_E$.

Let me insist once more that the theory described here does not support the conventional wisdom of the labor economics literature—which applies some version of the Nash bargaining solution to the bargaining problem (X, b) rather than (X, b, d). It therefore neglects the deadlock point altogether and mistakenly locates the *status quo* for the Nash bargaining solution at the breakdown point. It would certainly make sense to model the bargaining problem as (X, b) in the case when $b < d$ if the players were able to commit themselves to take-it-or-leave-it threats, as described in Section 1.5. Each bargainer's optimal threat would then be to take up his outside option. However, let me repeat yet again that threats need to be credible in order to be effective.

1.7.2   How Realistic is Rubinstein's Model?

To appreciate the stir that the publication of Rubinstein's [445] theorem generated, it is necessary to recall that economists up to and including Von Neumann and Morgenstern [538] traditionally held that the bargaining problem was indeterminate. But Rubinstein offered a model of bargaining between rational agents under complete information in which the outcome is entirely determined by the players' attitudes to time and risk.

Other bargaining models available at the time, notably the Nash Demand Game of Section 1.5, also lead to a uniquely determined outcome, but Rubinstein's Alternating Offers Game seems a very much more realistic representation of how real people bargain. But is this impression justified? There is certainly no shortage of critics who express doubts on this score. Often it is possible to respond with a knock down rebuttal, but I have chosen to consider some more subtle criticisms that hinge on fine questions of modeling judgment.

Reaction times. The first criticism is directed less at the model itself than at my interpretation of the players' bargaining powers. After noting that each player will want to wait as short a time as possible before responding to an offer, the criticism assumes that the time intervals between successive rounds in the Alternating Offers Game must therefore be determined by the players' psychological reaction times. Since reaction times are measured in fractions of a second, it would then make even better sense than in the standard interpretation to approximate the model by studying the limiting case as $\tau \to 0$. However, an added problem arises because Eve might be five times as quick as Adam. If both Adam and Eve are equally patient, Eve would then be assigned a bargaining power five times as large as Adam's. But who thinks that being quick on the draw is an important consideration in determining who gets how much in a bargaining encounter?

The first thing to say in response to this line of criticism is that the Alternating Offers Game needs to be modified if we are to fuss about reaction times. Rubinstein's formulation allows no time to elapse at all between an offer being made and its being accepted or refused. If we introduce a delay equal to the responder's reaction time at this point, Eve's advantage disappears, because what matters in determining the bargaining powers is how long each player has to wait after refusing an offer before he gets a response to his counteroffer. This period is the sum of the time it takes the proposer to formulate his proposal and the time it takes the responder to decided whether or not to accept. If Adam's reaction time to all events is τ_A and Eve's is τ_E, then the interval between successive opportunities to accept an offer will therefore be $\tau_A + \tau_E$ for both players. Being quick therefore confers no bargaining advantage.

However, the proper response to the criticism is to deny the relevance of reaction times and the like. It is no accident that the delays observed in real-life bargaining depend on how much is at stake. When bargaining over a loaf of bread in a bazaar, proposals are separated by seconds or less. In real estate deals, delays are measured in hours or days. In the case of international treaties, months can drag by without any action. The delays are therefore endogenously determined. The more important the outcome, the more time we feel that we need to make sure that we are doing the right thing. We also have other things to do with our time than bargain. Eating and sleeping are quite high on my agenda. It is also important to bear in mind that real people seldom bargain with complete information. Almost always we have doubts about how patient or risk averse our opponent may be. In such situations, delays can be used as a strategic signal intended to persuade the opponent that one is bargaining from strength.

Of course, it would be ideal to expand the model so as to include within it all the factors we think may be relevant in determining the bargaining timetable, rather than following Rubinstein in treating the timetable as given. Some first steps along this road have been taken by Reny and Perry [423] but I shall rest content with having drawn attention to the naïveté of the reaction-time interpretation of the delays in Rubinstein's model.

Money comes in whole numbers of cents. The second criticism appears in Myerson's [374, p.404] excellent textbook *Game Theory: Analysis of Conflict* and in papers of van Damme *et al* [529] and Muthoo [369].

The Ultimatum Game was studied in Section 0.4. The version in which the available dollar can be divided arbitrarily finely turned out to have only one subgame-perfect equilibrium, but when the dollar can only be split into whole numbers of cents, there are two subgame-perfect equilibria in pure strategies. A similar result holds for the Alternating Offers Game. Indeed, when the the dollar can only be divided into whole numbers of cents, any division of the dollar is a subgame-perfect equilibrium outcome in the limiting case as $\tau \to 0$. Moreover, there are other subgame-perfect equilibria in which agreement does not occur immediately.[59]

Myerson [374, p.407] deduces that Rubinstein's theorem is therefore no more than a mathematical curiosity. This criticism is easy to shrug off when applying the Nash bargaining solution to social contract problems, since there is then always a continuum of possible agreements. However, this would be a dishonest evasion. As Rubinstein [448] observes in another connection, what matters to people about to play a game is not the brute facts of their predicament, but their *perception* of the facts. The question is therefore not whether the agreement set is actually a continuuum, but whether people *model* it as a continuum when making decisions.

But first it is necessary to deal with a red herring concerning the units in which time and money are measured. Myerson's argument for the existence of multiple equilibria in the Alternating Offers Game requires only that the units used to measure time be sufficiently small compared with those used to measure money—and is it not true that physicists measure time down to the nanosecond while accountants never go below the cent? However, what physicists and accountants do in laboratories or countinghouses has as much do with the case as the flowers that grow in the spring. In everyday situations, real people care as little about one second more or less as they do about winning or losing a penny. Indeed, seconds and cents are presumably the smallest units in which we commonly measure time and money precisely because units of roughly this size are barely perceptible to us. Units an order of magnitude smaller would quickly fall into disuse. Units an order of magnitude larger would equally quickly be broken down into smaller

---

[59]To see why, begin with the case when the only possible agreements are that Adam gets the whole dollar or Eve gets the whole dollar. After proving that there is a subgame-perfect equilibrium in which Adam begins by offering the whole dollar to Eve, the rest will be easy.

fractions by some expedient or another.[60] For example, when bargaining over an indivisible object in circumstances which make it impossible to compensate the loser with money or the like, we may consider deciding ownership by tossing a suitably weighted coin or using some scheme to rotate possession of the object.

Once it is appreciated that the smallest units we commonly use to measure time and money are determined endogenously by the circumstances of our daily lives, it becomes possible to contemplate the possibility that we also usually think of both as continuous variables for most purposes. It is true that Adam may be *many* hours late or that Eve may have lost *many* dollars through careless bidding at the Bridge table. But this is because counting in units is necessarily a discrete operation. However, English grammar forces us to say that Adam's train was *much* delayed and that Eve would be *much* richer if she bid more wisely. Time and money are therefore both treated as continuous variables by our language. The obvious explanation is that this is how we model them when organizing our everyday lives, and it therefore seems entirely reasonable that Rubinstein should also use the same model.

For those who prefer a formal argument, Binmore *et al* [81] offer a model in which evolution is assumed to prune unnecessarily complex strategies in the Alternating Offers Game. Surviving strategies therefore take account of the fine details of the game only to the extent that it pays to do so. In the limiting case as $\tau \to 0$, the modified evolutionarily stable strategies studied get as close to the Rubinstein prediction as the discrete nature of the surplus permits.

Reneging on agreements. The rules of the Alternating Offers Game assume that bargaining will cease as soon as an offer is accepted. But suppose that the players are allowed to renege on agreements after they have been made. Muthoo [368, 371] shows that the players will then sometimes find it advantageous to exploit such an opportunity for strategic reasons. In formal terms, he shows that multiple subgame-perfect equilibria can exist when the Alternating Offers Game is modified to permit the repudiation of agreements.

However, the new subgame-perfect equilibria disappear if the players are assumed to be committed to deals for a period of length $T > 0$, and the interval that separates successive proposals is allowed to become vanishingly

---

[60]Italy provides a good example for both propositions. The lira is too small a unit and so prices are quoted in hundreds of lire. In the other direction, when Italian small change became rare some years ago, it was common to be offered pieces of candy to make up the difference.

small. The uniqueness result therefore survives if the limits are taken in the order $\tau \to 0$ and then $T \to 0$.

The implication is that Rubinstein's result depends on the existence of a small level of mutual trust between the players that can become vanishingly small as we approach the case of continuous time. Since Section 1.6.2 already found it necessary to postulate such vestigial levels of trust when modeling the implementation of an agreement, imposing a minimal trust assumption when justifying the Alternating Offers Game creates no extra difficulty for the purposes of this book.

 Endogenizing the deadlock payoffs. Fernandez and Glazer [181] study a version of the Alternating Offers Game between a firm and a union in which the decision to strike is endogenized. After each refusal, the union decides whether to call a strike for the period of length τ that intervenes before the next offer is made. If there is no strike, the plant keeps operating under the old contract and so both sides receive a positive income. If there is a strike, both sides earn nothing. This model has multiple subgame-perfect equilibria, some of which involve substantive delays in reaching agreement.

But here is an alternative model. The union begins by announcing its strike decision. Its decision remains in force for two periods while first the firm and then the union make their offers. If both offers are refused, the union gets to revise its strike decision and the cycle repeats. This model has a unique subgame-perfect equilibrium outcome. In the limiting case as $\tau \to 0$, the union always strikes if it is relatively worse off than the firm under the old contract. It continues working under the old contract if it is relatively better off. In brief, the union sets the deadlock point at the point most favorable to itself.

Which of these models makes more sense? The question boils down to asking how fast strikes can be switched on and off. I think the second model makes more sense because it assumes that offers can be made faster than strikes can be called. It would make even more sense if the strike decision committed the union for a hundred periods rather than two. If it committed the union forever, we would be back to the Rubinstein model except for the union's opening decision about whether to strike.

 Opting out by telephone. As a second example of a challenge to the robustness of Rubinstein's model, consider the following variant of the Alternating Offers Game with outside options analyzed by Shaked [487]. The standard model of Figure 1.14(b) has a unique subgame-perfect equilibrium. However, Shaked shows that multiple equilibria emerge if the model is changed so that a player's opportunities to opt out come immediately after one of his own offers has been rejected—rather than after he has rejected an offer from his opponent.

Which of these two models makes more sense? The question reduces to evaluating how hard it is for a player to commit himself to taking no further part in the negotiation. In the face-to-face bargaining that takes place in a bazaar, a prospective buyer may convincingly break off the negotiation by walking off to do business with some other seller—but he will not be able to do so without having to listen to one final offer from the stallkeeper as he is walking away. Shaked therefore suggests that his model is more appropriate for bargaining over the telephone, since a player can then terminate the negotiation just by hanging up. But can he? My experience when seeking to break off contact with people who want me to do something important for them is that they keep calling me back no matter how firmly I insist that I prefer to take my outside option. One might perhaps consider cutting off the telephone and refusing to see visitors. But the light plane I recall circling over Ann Arbor pulling a banner saying, "Marry me, Maisie" would seem to suggest that the difficulties involved in avoiding having to listen to unwelcome proposals are well nigh insuperable.

Why subgame-perfection? The final criticism was reviewed at some length while discussing the Ultimatum Game in Section 0.4. Why should we suppose that the players will use a subgame-perfect equilibrium in the Alternating Offers Game?

As Section 0.4.2 explains, I agree with those critics who think that the general arguments given for restricting attention to subgame-perfect equilibria are less than overpowering (in cases when the payoffs have not been chosen to make the validity of the backward induction algorithm tautological). However, an alternative defense of the Rubinstein prediction is available in the symmetric case that operates in the original position. Section 3.3.8 briefly discusses the idea of an evolutionarily stable strategy that has been modified to favor simple strategies that do the same job as more complex rivals. Binmore *et al* [81] apply the idea to the Alternating Offers Game, and are led thereby to the same determinate outcome of the bargaining problem as Rubinstein [445] in the limiting case when $\tau \to 0$. However, the asymmetric case remains problematic.

1.8 Other Approaches to Bargaining

The approach to the bargaining problem defended in this chapter is only one of many that have been proposed over the years. The alternatives are often very seductive in that they can be grasped with very little effort but seemingly have a universal application. The temptation, for example, to seize upon a cooperative solution concept and apply it uncritically to all problems is very strong indeed. Even the Nash bargaining solution will lead us astray if used thoughtlessly—as when labor economists apply it to the bargaining problem (X, b) instead of (X, b, d) (Section 1.3.1). To emphasize the importance of keeping one's feet on the strait and narrow path, stony though it be, this final section briefly reviews two more attempts to short-circuit some of the difficulties of bargaining theory.

1.8.1   The Coase Theorem

The burgeoning subject of law and economics is largely the brainchild of Ronald Coase [126, 127]. Among his many contributions to this and other subjects (Medema [351]), has been to establish that a clear delineation of property rights is vital to the efficient working of a decentralized social system. Coase's work in this field makes constant use of the the claim, familiar from Section 1.2.3, that rational bargainers will necessarily agree on a Pareto-efficient deal if the associated transaction costs are negligible.

Stigler therefore began to refer to this tool of analysis as the Coase theorem.[61] However, as observed in Section 1.2.8 of Volume I, Coase was by no means the first on the scene with this proposition. Its history stretches back through Von Neumann and Morgenstern [538] at least as far as Edgeworth [169]. Nor can it be properly said to be a theorem, since the argument that Coase offered in its defense does not have the status of a mathematical proof. Stigler therefore did Coase no favor in referring to a proposition that is neither Coase's nor a theorem as the Coase theorem.

One can, of course, make the Coase theorem into a genuine theorem by defining a positive transaction cost to be anything that prevents rational bargainers reaching a Pareto-efficient deal. It is frequently asserted, for example, that any asymmetry of information between the bargainers represents a species of transaction cost. Wordplay of this type may sound well, but it is easy enough to expose its hollowness by looking at the following example of Rubinstein [445] (Note 3 of Appendix C.5).

In this example, two perfectly informed bargainers alternate in making offers over the division of a dollar. The players do not discount time as in the standard version of Rubinstein's bargaining theory. Instead, each loses c each time a proposal is refused. The proof used in analyzing the Alternating Offers Game with discounting can be adapted to deal with this

---

[61]Various consequences of the Coase theorem as stated in the text are sometimes quoted in its stead. A totally unrelated proposition concerning the monopoly pricing of durable goods is also sometimes referred to as the Coase theorem.

case also, but it is no longer true that there is a *unique* subgame-perfect equilibrium. The game has many subgame-perfect equilibria, some of which involve delays in reaching an agreement. Equilibria involving such delays cannot be Pareto-efficient because both players would do better to agree immediately to the deal they will eventually reach if they continue to use their equilibrium strategies. In itself, this result creates no problem for the Coase theorem, because one can argue that transaction costs are positive as long as $c > 0$. However, we can study the case of zero transaction costs by considering the limit as $c \to 0$.

It is somewhat of a surprise to discover that the Pareto-inefficient equilibria do not disappear when the limit $c \to 0$ is taken. Far from being Pareto-efficient, one of the limiting equilibrium outcomes calls for the dollar to be split evenly between the players after a delay that costs each player fifty cents. Both therefore end up with nothing! This is a pathological conclusion in that altering almost anything in the model in the direction of more realism restores the uniqueness result of Rubinstein's standard theory. But it shows that the Coase theorem is strictly *false* as a theorem—unless one is prepared to overextend the definition of a transaction cost to the point of absurdity.

Personally, I think that Coasians would do well to accept that all bargaining results—including those on which I rely in this book—are sound only when certain preconditions are met. Insofar as practical applications are concerned, the claim that bargaining outcomes will be approximately Pareto-efficient when transaction costs are small becomes particularly suspect when one of the bargainers has a large outside option.

As we saw in Section 1.7.1, applying Rubinstein's bargaining theory to the problem (X, b, d) does indeed predict that the bargaining outcome will be Pareto-efficient. However, although Pareto-efficient, this deal awards the player with the large outside option no more than he could get by opting out. That is to say, his equilibrium strategy tells him to opt in, but he is actually indifferent between receiving his equilibrium payoff and opting out. To predict that a Pareto-efficient outcome will be observed—even when experienced and well-informed bargainers negotiate over substantial sums of money—is therefore to place too much reliance on a theoretical result that sits on a knife-edge. In fact, laboratory experiments show that the prediction of the standard Rubinstein theory does well in predicting how the surplus will be split in those cases when a deal *is* actually reached, but that the bargainers frequently fail to reach an agreement at all if one of the players has a large enough outside option (Binmore *at al* [90, 79, 82]).

Social optimality? Why is so much importance attached to the Coase theorem? Students are taught that it demonstrates that the state should not intervene in disputes when transaction costs are negligible, since the parties in dispute will come to a socially optimal arrangement if allowed to negotiate privately without constraint. I don't know to what extent students notice that acquiescing in such a claim commits them to the view that "might makes right." After all, as noted in Section 0.2, even a slave and his master have matters over which they can usefully negotiate. But do we want to say that any deal they reach in the absence of transaction costs will be socially optimal?

It seems to me facile to equate social optimality with Pareto-efficiency. Nor are there proper grounds for equating the social welfare of consumers with consumer surplus.[62] To do so is to assume, among other things, that an extra dollar is worth the same to a rich man as it is to a poor man. Still less should we take for granted that total welfare is equal to consumer surplus plus producer surplus plus an externality term. It is true that defining welfare in this way allows us to bring to bear the familiar techniques used in analyzing perfectly competitive markets. But except in isolated patches, the world at large is so far from approaching the ideal of perfect competition that it seems to me a travesty to follow the Coasian tradition of explaining everything in terms of deviations from perfect competition caused by nonnegligible transaction costs.

However, in spite of these and other reservations, it needs to be recognized that my own position has much in common with that of Coase and his followers. I don't agree that the underlying power structure is irrelevant provided that the outcomes are Pareto-efficient. But I do agree that what cannot be altered must be endured. The rational approach to reform is therefore not to rail hopelessly against the rules of what I call the Game of Life, but to follow Coase in looking for equilibria in this game that are Pareto-improvements on the equilibrium that a society currently operates. I again agree that there are no absolute principles that tell us why we should prefer one Pareto-improvement to another. But I do not follow the Coasians in deducing that distributional issues are therefore irrelevant. Different people prefer different Pareto-improvements, and some way of resolving such conflicts must be found. My pragmatic proposal is to harness the fairness criteria that are actually built into our current social contract for this purpose. I do not claim that these criteria are socially optimal; merely that their presence provides us with a useful social tool we would be foolish not to employ.

---

[62]Even for quasilinear utility functions, as explained in Section 4.3 of Volume I.

1.8.2   Gauthier on Bargaining over a Social Contract

An examination of one final source of confusion will serve as a reminder that all this chapter's huffing and puffing about bargaining is merely an attempt at clearing the ground in preparation for an attack on social contract problems.

Rawls [417] proposed that Adam and Eve could settle their differences fairly by employing the device of the original position. That is, each should agree to the deal that would be reached if they were to bargain in ignorance of their current roles in society. Gauthier [196] dispenses with such a Rawlsian veil of ignorance and proposes instead that Adam and Eve reach a compromise directly using the Kalai-Smorodinsky bargaining solution. Section 1.2.9 of Volume I discusses the shortcomings of this approach in general terms. The most telling criticism is that Gauthier's use of the Kalai-Smorodinsky solution telescopes all the ethical content of his procedure into the choice of the *status quo* point ξ. Gauthier argues that ξ should be determined according to Lockean principles, but I see no reason why Gauthier's bourgeois conception of the state of nature should be preferred to such rival proposals as Hobbes' [251] "war of all against all" or Buchanan's [103, 104, 105] "natural equilibrium". Indeed, it is precisely because such questions are controversial that we need the kind of "moral geometry" that Rawls [417, p.121] sought to provide in his *Theory of Justice*. However, the current section puts aside these grand questions to examine the argument that Gauthier offers in defense of the Kalai-Smorodinsky solution as the inevitable outcome of rational negotiation between perfectly informed agents.

It should first be observed that the Nash-Harsanyi-Zeuthen theory which Gauthier [196, p.133] regards as the benchmark against which alternative bargaining theories like his own need to be measured has received no mention in this chapter because the advent of game theory has rendered it obsolete. At about the same time that Von Neumann was first putting together his ideas on zero-sum games, Zeuthen [576] was working on a book in which he presented ideas on bargaining that were way ahead of their time. Indeed, as Harsanyi [233] later showed, the use of Zeuthen's principles leads inexorably to the Nash bargaining solution.

Zeuthen's behavioral assumptions are a tribute to the soundness of his intuition,[63] but are unacceptably *ad hoc* by modern standards. Nowadays, we want to know *why* we should believe that rational players would honor Zeuthen's principles when bargaining. The same goes for the rival behavioral assumptions proposed by Gauthier [196]. Why should we follow

---

[63]Rubinstein, Srafa and Thomson [449].

Gauthier in seeing the essence of rational bargaining as a two-stage process in which each player makes a claim at the first stage and then a concession at the second? Even within this framework, why should we suppose that rational players will follow Gauthier's precepts? In particular, why should we suppose that the vector of initial claims will be the utopian point U of Figure 1.6(a). Without answers to such questions, Gauthier's proposed resolution of the problem of rational contracting lacks any firm foundation.

To explore Gauthier's approach to the bargaining problem, consider the divide-the-dollar problem with two risk-neutral protagonists. There is no particular reason why the players should follow Gauthier's protocol by simultaneously opening the proceedings with claims a and b that neither expects the other to find acceptable. However, we can embroider Gauthier's story by following Muthoo [370, 372] in introducing a cost of backing down from an initial claim. Muthoo assumes that Adam loses $\gamma(a-x)$ if he finally agrees to accept $x < a$ and Eve loses $\gamma(b-y)$ if she finally agrees to accept $y < b$.

A second gap in Gauthier's story concerns the manner in which the players compute the concessions that follow their initial claims (Section 1.2.9 of Volume I). This gap is closed here by assuming that the players follow Rubinstein's procedure after announcing their claims. If their discount rates are equal, their concessions can then be computed using the symmetric Nash bargaining solution.

Without the initial stage at which players make claims from which it is costly to back down, the players would be faced with the Nash bargaining problem $(X, 0)$ of Figure 1.16(a). The symmetric Nash bargaining solution for this problem is $(\frac{1}{2}, \frac{1}{2})$. However, after the players have made their claims a and b, the effective payoff region ceases to be X. For example, if the players finally agree to split the dollar so that Adam gets $x_A < a$ and Eve gets $x_E < b$, then each will then have to pay the cost of backing down from their initial claim. The utility pair that results from such an agreement is $(y_A, y_E) = (x_A - \gamma(a - x_A), x_E - \gamma(b - x_E))$. After the initial claims have been made, the original bargaining problem $(X, 0)$ must therefore be replaced by the bargaining problem $(Y, 0)$ illustrated in Figure 1.16(b) for the case when $a + b > 1$. Since

$$y_A + y_E = (1 + \gamma)(x_A + x_E) - \gamma(a + b) = 1 - \gamma(1 - a - b) = w,$$

the symmetric Nash bargaining solution for $(Y, 0)$ is $(\frac{1}{2}w, \frac{1}{2}w)$, provided that $a \geq \frac{1}{2}w$ and $b \geq \frac{1}{2}w$. Adam's final payoff is therefore a *decreasing* function of a when both a and b are sufficiently large. In this framework, Adam will therefore certainly not want to make a as large as he can in the manner proposed by Gauthier [196]. On the contrary, for each sufficiently

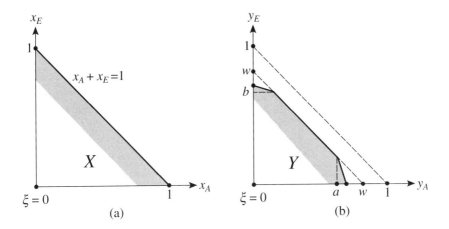

Figure 1.16: Bargaining with revocable commitments.

large b, he will want to decrease a at least until $a = \frac{1}{2}w$. Similarly, Eve will want to decrease b at least until $b = \frac{1}{2}w$.

If $a = b = \frac{1}{2}w$, then $a = b = \frac{1}{2}$.[64] This pair of claims is an equilibrium because the best thing one can do when the opponent claims $\frac{1}{2}$ is to claim $\frac{1}{2}$ for oneself. We have already seen that neither player can gain by claiming more than $\frac{1}{2}$ when the opponent claims $\frac{1}{2}$. If it is not an equilibrium for both players to claim $\frac{1}{2}$, one of the players must therefore be able to get more by lowering his claim. However, the result of one of the players lowering his claim is to create a new bargaining problem $(Z, 0)$ as illustrated in Figure 1.16(b). The Nash bargaining solution for this new bargaining problem remains $(\frac{1}{2}, \frac{1}{2})$ and so a player who lowers his claim from $\frac{1}{2}$ will gain nothing at all.

To summarize, Gauthier [196] argues that players using his bargaining protocol in the problem of dividing the dollar will each begin by claiming the whole dollar. However, according to Muthoo [370], each player will simply claim what he expects to end up with, namely fifty cents.[65] Moreover, this conclusion remains valid no matter how small the cost γ of backing down from a claim is taken to be.

---

[64] Write $a = b = \frac{1}{2}w$ in $w = 1 - \gamma(1 - a - b)$.

[65] It is not hard to show that other equilibria do not exist (Muthoo [370]).

Gauthier [199, p.179] sees no reason to be dismayed by Muthoo's reasoning since it leads to exactly the same final outcome as his own for symmetric bargaining problems. However, if it were possible to confine our attention to symmetric payoff regions in social contract problems, why would we need to appeal to bargaining theory at all? Indeed, what problem would we then be seeking to solve? As Section 1.2.5 of Volume I explains, if the payoff region were symmetric, then the "inequalities of birth and natural endowment" for which Rawls [417, p.100] believes a social contract should provide redress would be absent altogether. But in asymmetric bargaining problems, Muthoo's reasoning does not lead to the same final outcome as Gauthier's. Muthoo is still led to a fifty-fifty split of the dollar even if everything in the set X to the right of the line $x_A = \frac{3}{4}$ is discarded, but this is certainly not true for the Kalai-Smorodinsky solution favored by Gauthier.

Among his many other achievements, William Blake invented the comicbook. Here we see the new medium being used to illustrate the evolutionary history of mankind.

Chapter 2

Evolution in Eden

> He who would understand *baboon* would do more towards
> metaphysics than John Locke.
>
> Charles Darwin

2.1 The Good, the Right, and the Seemly

Traditional moral theories can be loosely categorized into theories of the
Good, the Right, and the Seemly. Those who lean to the left emphasize
the Good. Philosophers call them consequentialists. They maintain the
existence of an *a priori* common good whose advancement takes priority
over our own selfish concerns. Utilitarians and welfarists are the most vocal
members of this lobby.

Conservatives prefer theories of the Right. Philosophers call such theories deontological and see them as the natural opposition to the consequentialist theories of the left. Deontologists maintain the existence of natural
rights, which it is our duty to respect regardless of the consequences. The
right to private property is particularly cherished.[1]

Whigs like myself prefer theories of the Seemly, in which moral rules or
aspirations bind only by habit or custom. Philosophers are not too sure

---

[1]I am not confusing doing something because it is Right with having a right to do
something. In Section 3.2.2, I assume that we have a right to take any action that we
have no duty to shun. Our rights are therefore determined by our duties. But to follow
the Right is simply to do one's Duty.

where we fit into the scheme of things and sometimes forget to include us at all. When we are mentioned, it is typically to make fun of the naturalistic theories of ethics that we espouse.

I have to confess that the works of the great thinkers of the past often defy easy categorization within this scheme. For example, Bentham sometimes says that something is right when he means that it advances the Good. Rawls [417] describes his *Theory of Justice* as an attempt to deduce the Good from the Right. Aristotle's defense of Seemliness is difficult to reconcile with his teleological beliefs. Hume, the great champion of the Seemly, is commonly held to have been a utilitarian of sorts. John Stuart Mill argued that even Kant, the great champion of the Right, sometimes found it necessary to have recourse to utilitarian ideas.[2] Kant might respond by pointing out that Mill himself, the not-so-great champion of the Good, notably fails to reconcile his own views on the liberty of the individual with his basic utilitarian position.

The unprincipled nature of modern political rhetoric throws more dust in the air. Even Marxists now express their calls for reform in terms of newly minted "rights" that Man has supposedly been denied since the dawn of time. At the same time, conservatives bewail the loss of "community spirit", which they somehow contrive to believe can exist independently of the social institutions that enable it to flourish. However, in spite of all such difficulties of classification, I plan to use the scheme of the Good, the Right, and the Seemly as an organizing principle for the next three chapters. The Good will be treated in the current chapter and the Right in the next. But what I have to say will be by way of counterpoint to these themes as they are traditionally developed.

Diogenes [149] remarked that he had seen Plato's cups and table, but had yet to see his cupness or tableness. I feel much the same about the Platonism of traditional theories of the Good and the Right. Bentham [53] described the idea that the Right is more than a human invention as "nonsense upon stilts", and I am ready to apply the same epithet to similar ideas about the Good. But such a denial of the existence of the Good and the Right as moral absolutes does not imply that we are talking nonsense when we distinguish the good from the bad and the right from the wrong in our daily lives. Cupness and tableness are no less human inventions than the Good and the Right, but cups and tables nevertheless exist.

I believe, for example, that the forces of biological and social evolution

---

[2]There is no shortage of ammunition for Mill in Kant's work. For example, Kant [293] tells us that, "We find in ourselves ... a moral teleology ... to do with the reference of our own causality to purposes, and even to a final purpose that we must aim at in the world." Nor do modern scholars necessarily find the accusation bizarre (Hare [230]).

have indeed equipped us with the capacity to make the interpersonal comparisons of well-being that a theory of the Good requires. But I am not prepared to treat this conclusion as axiomatic. In this chapter I ask how and why such a phenomenon might have evolved. Similarly, I believe that the successful running of a society requires that we treat certain privileges as though they were inalienable and certain duties as though they were obligatory. However, as the next chapter explains, I do not believe that the mechanisms that lead us to behave in this way are any different in principle from those which hold baboon societies together. But nobody suggests that baboons behave as they do because they have been gifted with some inborn notion of the Right.

Until recently, theories of the Seemly have been held in very low repute. Evolutionary ethics is still widely regarded as a siren against whose blandishments only the naive or stupid fail to stop their ears with wax. Aside from intellectual considerations, theories of the Seemly do not have the same social cachet as the grandiose theories of the Good and the Right. But perhaps the fact that people like me are moved to write books like this is a sign that the time for Seemliness has come. However, in this book, its hour upon the stage will have to wait until the final chapter, where I try to draw together all the threads that have been spun along the way, in the hope of showing that they can be woven into a logically coherent whole.

2.2 Utilitarianism

The principal concern of this chapter is to say something about evolution and interpersonal comparisons of utility. It is convenient to make the necessary points while mounting the best case for utilitarianism that I think can be made. I believe that this defense has a fatal flaw that disqualifies utilitarianism as a contender in the social contract stakes. However, this is not to say that utilitarianism is without interest. On the contrary, I think that utilitarian principles make a great deal of sense when applied in the circumstances usually faced by a welfare economist, whose remit is not to reform society as a whole but to act as the instrument of a benign but paternalistic government.

My approach is based on Rawls' [417] attempt to use the device of the original position to put flesh on the bones of Kant's categorical imperative. As Section 2.3.7 of Volume I explains, Rawls' analysis of the hypothetical deal that would be reached behind the veil of ignorance leads him to propose two principles of justice. After the rights and liberties of the first principle have been guaranteed, the second principle calls for his difference principle to be used in settling questions of socio-economic inequality. I refer to

Rawls' difference principle as the maximin criterion.

Section 4.6 of Volume I explains at length why I think that Rawls goes astray in arguing that the maximin principle represents the appropriate criterion for individual decision making by rational players in the original position. Although I am a vocal critic of the widespread abuse of Bayesian decision theory, the argument for its use seems to me quite overwhelming in the circumstances of the original position. Indeed, if the orthodox theory is wrong under such circumstances, it must surely always be wrong. It is therefore necessary to classify Rawls as an iconoclast in his views on decision theory. Once this is appreciated, it is natural to ask where Rawls would have been led if he had not been moved to treat Bayesian decision theory as an icon to be consigned to the flames.

Rawls [417] traces his philosophical ancestry through Locke, Rousseau, and Kant, as indicated by Route 1 through the organizational diagram supplied in the Reading Guide. Harsanyi's [233] Route 3 through the same diagram is assigned the same philosophical ancestry because, working quite independently of Rawls, he provided the answer to the question posed at the end of the preceding paragraph before it had even been asked. Harsanyi [233] shows that the use of Bayesian decision theory in the original position leads, not to some form of egalitarianism, but to a version of utilitarianism. If Rawls is correct in seeing the device of the original position as a means of operationalizing the categorical imperative, it then follows that Kantian ethics need to be rewritten to vindicate Mill's suggestion that Kant was something of a utilitarian!

Since I have already recognized Kant as the great champion of the Right, it may seem perverse to argue that his mode of reasoning should have led him to champion the Good. However, I am not alone in my perversity. Recall that Rawls is an avowed Kantian who aims at deducing the Good from the Right—albeit not a utilitarian good. Harsanyi is an even closer ally. By the end of this section, I hope that even dyed-in-the-wool utilitarians will be ready to admit that perhaps they have more in common with Kantians than is commonly recognized. Insofar as matters seem otherwise, I think it is largely because Kant's rhetoric on personal autonomy leads people to suppose that implementing his ideas would lead to a truly free society.[3] Sometimes Kant is even said to be the founder of ratio-

---

[3]Everybody agrees that freedom is a good thing. It is therefore necessary that proposed reforms be shown to increase freedom. In consequence, the history of social debate has left us with a ragbag of conflicting definitions of freedom. Aside from the positive and negative freedoms that he is famous for distinguishing, Berlin [56] counts more than two hundred different senses in which the word "freedom" has been understood. For example, Hayek [240] points out that political freedom is not the same as civil freedom—an Athenian citizen was free to vote but could be banished for no other reason than be-

nal liberalism.[4] However, as Isaiah Berlin [56] explains, this impression is achieved by redefining as slavery what John Doe would regard as unbridled freedom, and then proposing a form of slavery as the essence of true freedom. Once the nature of this conjuring trick has been appreciated, it becomes possible to recognize that the system of ethics proposed by Kant only makes any real sense when the maxims that a rational person would will to be universal laws are enforced by a suitably constituted authority.[5] It is then not such a long step from the resulting paternalistic creed to the paternalism espoused by utilitarian welfare economists.

Although I think that there is a strong case for the seemingly wild claim that Kant should have been a utilitarian, it does not have much relevance to social contract questions, because I believe that both Rawls and Harsanyi go astray in adopting a Kantian foundation for their theories. Most importantly, I see no reason at all to follow Rawls and Harsanyi in adopting the Kantian view that Adam and Eve are somehow committed to the hypothetical deal that they would have reached behind the veil of ignorance. However, my skepticism about the viability of such commitment assumptions will be held in abeyance until the final chapter, where Rawls' faith in egalitarianism will be justified—albeit using a line of reasoning that he would not wish to endorse (Route 5 in the diagram of the Reading Guide). In the main body of the current chapter, I continue to maintain the commitment assumptions of Rawls and Harsanyi without their Kantian foundations by postulating a benign and incorruptible external agency that exists to enforce the terms of a society's social contract (Route 4 in the diagram of the Reading Guide).

---

ing unpopular with a majority of the other citizens. Nor is free will the same as inner freedom or self-mastery. Freedoms from and freedoms to are really concerned with the distribution of goods and power. Indeed, Hayek [240] argues that once freedom has been identified with empowerment, the original sense of the word is in danger of being lost altogether. In the case of Kant, the danger is that individual self-mastery becomes confused with freedom from coercion.

[4] The layman might well suppose this to mean that Kant was the first to put forward rational arguments in support of liberalism, rather than that Kant was the first so-called Rationalist philosopher to emphasize the value of personal autonomy. When taken literally, the claim is painful for modern whigs, who see David Hume as the embodiment of rational liberalism. Kant was moved to write at all only because he thought that Hume was wrong about pretty much everything. Whatever Kant may or may not be, he therefore cannot possibly be regarded as either rational or liberal in any sense acceptable to a whig. In particular, Kant's claim that *rationality* demands that the categorical imperative be obeyed seems to us to be conjured from nothing.

[5] Kant [295, p.vii] was deeply influenced by Rousseau, whom he described as the Isaac Newton of the moral world. I think it no accident that the attempt by the utopians who hijacked the French Revolution with the intention of putting Rousseau's ideas into practice lapsed so quickly into authoritarianism. It needed the Whig, Edmund Burke [101], to explain why their bold experiment was doomed to fail.

2.2.1   Summum Bonum

Francis Hutcheson [277] argued that we should seek the greatest happiness for the greatest number (Scott [470]). In an unguarded moment, Bentham[6] repeated this formula, and hence provided the enemies of utilitarianism with a stick with which its advocates have been beaten ever since. However, unambiguous definitions are easily supplied in the case of a society of fixed size.[7] Section 2.2.4 of Volume I defines a utilitarian to be a consequentialist who has an additively separable common good function. He therefore adds together his estimate of the well-being of each citizen to obtain his measure of the welfare of society as a whole.

Unfortunately, many authors lump every species of consequentialist together, and indiscriminately refer to them all as utilitarians. The confusion is compounded by other authors who believe that the central issue is whether the behavior of *individuals* is compatible with a utility-maximizing model. They therefore assign the label utilitarian to anyone who works with such a model, without seemingly appreciating that a society of utility-maximizing individuals need not be a utility-maximizing society.

This philosophical muddle about what utilitarianism should be taken to be distracts attention from the need to make some careful distinctions, even within the narrow conception of utilitarianism that I adopt in this book. So does the small but persistent controversy among economists over the validity of Harsanyi's [233] utilitarian claims, which I address briefly in Appendix B. The distinction that I borrowed from Rawls [417] between teleological and nonteleological ethical theories can also create confusion if not properly understood (Section 4.2.4 of Volume I and Appendix B). It is especially important to recognize that the defense of utilitarianism to be offered in this chapter has a *nonteleological* character. But before moving to this defense, it is necessary to say something about the more usual teleological approaches that are emphatically not to be defended.

2.2.2   Ipsedixists

Sen [477] denies that Harsanyi [233] is properly to be counted as a utilitarian at all, because his theories are very firmly based on what people want rather than on what they would want if they knew what was good for them. But how do we know what people ought to want? The literature is full of

---

[6]I am grateful to my colleague Fred Rosen for pointing out that Bentham probably derived the formula neither from Hutcheson nor from Leibnitz, but from Beccaria's *Crimes and Punishments* (Shackelton [486]).

[7]As Section 2.5.2 explains, disputes arise about whether total utility or average utility is the appropriate criterion when the population size is variable.

interminable lists of criteria that must somehow be distilled into the one and only true *summum bonum*, but these lists reflect only too obviously the cultural and class prejudices of their compilers.[8] History provides many examples. The Spanish Inquisition tortured and burnt heretics "for their own good". Until comparatively recently, helpless invalids were bled "for their own good". Victorian do-gooders strapped children into specially built anti-masturbation harnesses at night "for their own good".

Will modern do-gooders be judged any less harshly by future generations? My guess is that a particularly bad press awaits those who believe that life is always an unmitigated good, and hence must be inflicted even on those whose suffering is so great that they beg for an easy death. Those who sabotage birth control initiatives for similar reasons have even less prospect of being remembered kindly.[9] In brief, modern do-gooders may be less barbaric then their predecessors, but the source of their inspiration is no more reliable. Mill [358] doubtless overstates the case when he says that all errors that an individual is likely to commit "are far outweighed by the evil of allowing others to constrain him to what they deem his good." However, we whigs see no prospect of stability in a society where such coercion is endemic.

None of this is meant to imply that feeding the hungry, housing the homeless and other similar charitable activities are unworthy of our support. Usually, the hungry want to be fed and the homeless want to be housed. Still less am I suggesting that it is immoral for parents to ignore the whims and fancies of their children in favor of advancing their long-run interests. When the children are older, they will perhaps even be grateful not to have been overindulged in their infancy. I am merely protesting at the arrogance

---

[8] Section 4.6 of Volume I speculates that many scholars find the maximin criterion intuitively attractive because they come from a social class that frowns on risk-taking. But why should this bourgeois prejudice be forced on aristocrats or proles?

[9] One repeatedly hears that the doom-laden predictions from Malthus's [340] *Essay on Population* can be ignored because he recanted in the second edition. It is true that he became less pessimistic about the possibility that a society might eventually learn to keep its population and resources in balance, but his basic contention that a population allowed to grow unchecked must inevitably outstrip the resources necessary to sustain it remains precisely as before. If the birth rate exceeds the death rate, then a population will grow exponentially fast. Since resources can grow only as the cube of time, the population will therefore eventually outstrip its resources—unless the birth or death rate change. We are told that the resource constraint can be evaded by utilizing new technologies and operating more efficiently. To see the error, one need only consider standing room. Even if it were possible to pack the total of human biomass into an ever-expanding sphere in space, the theory of relativity tells us that its radius could only expand at the same rate as time. Its volume, and therefore the maximum population size consistent with the availibility of this particular resource, can therefore grow only as the cube of time. However, the four horsemen will visit us long before the standing room constraint begins to bite.

of those who feel the urge to force their personal conception of the Good on fellow citizens who are perfectly capable of deciding such matters for themselves. They refer to themselves as objectivists, but I prefer to follow Bentham in calling them ipsedixists—those who offer their own prejudices in the guise of moral imperatives for others to follow.

In the idealized Garden of Eden that I use to illustrate the ideas presented in this book, the personal preferences held by Adam and Eve are given only in an abstract form, and so the temptation to argue that disputes between them should be settled by finding in favor of whoever has personal preferences that most accord with one's own prejudices does not easily arise. But even in the imperfect world in which we actually live, history has little positive to say about paternalists who ignore the wishes of those within their power. The more fervent and self-confident do-gooders are, the less likely it is that posterity will endorse their standards of righteousness.

2.2.3   Ideal Observers

A do-gooder of the type we have been discussing sees no reason why he should not seek to impose on others whatever conception of the Good is built into his own personal preferences. For example, Rousseau [444, p.15] insists that the "general will" is not to be confused with the "will of all". As Rousseau [442, pp.255–260] explains, the former is known only to those of "sublime virtue" and so its objectives must be achieved by "bringing all the particular wills into conformity with it."

However, utilitarians belonging to the more mainstream line take for granted that the standard for a common good should be *impersonal*. Instead of offering some dressed-up version of their own prejudices as the standard for a society to follow, they therefore propose that the preferences of some invented ideal observer should determine the goals toward which all members of a society should strive in common.

The notion of an ideal observer was discussed in Section 1.2.7 of Volume I. If an ideal observer is entirely free of personal prejudices, then his preferences must somehow be obtained by aggregating the preferences of all the citizens in the society being studied. He then serves as an embodiment of Rousseau's "will of all". Utilitarians argue that the utility of the ideal observer for a given state should be obtained by simply adding the utilities of each citizen in the state.

Such a view leaves three questions open:

- What constitutes utility?
- Why should individual utilities be added?
- Why should I maximize the sum of utilities rather than my own?

Early utilitarians like Bentham and John Stuart Mill had little to offer in answer to any of these questions. With characteristic frankness, Bentham [52, p.615] says, "That which is used to prove everything, cannot itself be proved." As for the additivity of happiness, this is quaintly described as a "fictitious postulatum".[10]

Mill [360] sometimes endorses this position, but also offers a halfhearted attempt at providing a proof of utilitarianism. The proof consists of a chapter devoted to the claim that what people desire is happiness. Having established this proposition to his own satisfaction, he then rests on his laurels—apparently not feeling the need to tackle the second or third question.[11] No further light is shed by such later utilitarians as Sidgwick [492], from whom we learn that the good is an "unanalyzable notion".[12] Modern moral philosophers writing in the utilitarian tradition seem largely to have lost interest in foundational questions altogether.

Fortunately, Harsanyi [233] is an exception to this rule, along with such followers as Broome [102] and Hammond [219, 221].[13] Recall that Harsanyi wears two hats when writing on utilitarian subjects. In his teleological hat, he offers some foundations for the ideal observer approach to utilitarianism.

Harsanyi answers the first of the three questions posed above by interpreting utility in the sense of Von Neumann and Morgenstern. The players' utilities then simply serve to describe their preferences.[14] The second question is answered by appealing to the intuitions we supposedly have about the nature of the Good, which is conceived of as some pre-existing Platonic form. Like Euclid, with his idealized points and lines, Harsanyi encapsulates his intuitions on this subject in a set of axioms. The axioms include the requirement that the ideal observer, whose individual good is to be identified with the common good, should be as rational as each individual citizen. If his expected utility for any lottery depends only on the expected utilities assigned to the lottery by the citizens in the society he serves, then

---

[10]This addibility of happiness, however when considered rigorously it may appear fictitious, is a postulatum without the allowance of which all political reasonings are at a stand—Bentham [52, p.616].

[11]All we are offered on the second question is the observation: "Each person's happiness is a good to that person, and the general happiness is therefore a good to the aggregate of all persons."

[12]With his usual sharp intelligence, Edgeworth [169] offers an insurance argument in defense of utilitarianism that is sometimes inappropriately bracketed with the writings of Sidgwick and the like. However, I see his argument as a first attempt to give expression to the *nonteleological* approach to which this chapter is devoted.

[13]Hare [228, 229] is another exception following a parallel track.

[14]From the perspective of the theory of revealed preference (Section 2.2.4 of Volume I), it is a tautology that rational players act to maximize utility. Mill's attempt to prove that people desire happiness then becomes redundant if we are willing to identify happiness with utility and to assume that people are rational.

Section 4.2.4 of Volume I shows that the Von Neumann and Morgenstern rationality axioms imply that his utility function can be expressed as a weighted sum of the utilities of all the citizens in society (Appendix B.2).

However, Harsanyi has nothing substantive to say in reply to the third of the three questions. The existence of a common good whose advancement somehow takes priority over our own individual concerns is simply taken for granted. Once we have deduced the properties of this common good from his axioms, Harsanyi sees his task as being over. Nor, as Section 2.3.5 of Volume I observes, do other teleological champions of the Good have anything on offer that might conceivably plug this large gap in their armor. They join the champions of the Right in simply asserting that we have a moral duty to carry out whatever actions are moral according to whatever theory is being defended. Even prophets of the Seemly are not beyond reproach in this regard. However, in the following anticontractarian extract, Edmund Burke [109] is at least honest enough to admit that the source of moral obligation is a total mystery:

> Dark and inscrutable are the ways in which we come into the world. The instincts which give rise to this mysterious process of nature are not of our making. But out of physical causes, unknown to us, perhaps unknowable, arise moral duties, which, as we are able perfectly to comprehend, we are bound indispensibly to perform. Parents may not be consenting to their moral relation: but, consenting or not, they are bound to a long train of burdensome duties towards those with whom they have never made a convention of any sort. ... So it is with men in the community:[15] whether individuals like it or not, they lie under moral obligation to obey the laws and sustain the state.

I think it a major error for utilitarians to fudge the issue of why citizens should pursue the aims of some ideal observer rather than furthering their own individual interests. The question that needs to be decided is whether utilitarianism is a moral system to be employed by *individuals* in regulating their interactions with others, or whether it is a set of tenets to be followed by a *government* that has the power to enforce its decrees. It is understandable that utilitarians are reluctant to argue that their doctrine should be forced down the throats of people who find it hard to swallow. They prefer to imagine a world in which their thoughts are embraced with open arms by

---

[15]The analogy is inappropriate. Nature has built the need to *sympathize* with our infant children into our genes. No social contract is therefore necessary to regulate the relationship between a mother and her baby. But we can only sympathize with those who do not aid in the transmission of our genes to a limited degree. With strangers, we have to rely on our *empathetic* powers when seeking to coordinate our behavior.

all the citizens of a utilitarian state. As with Marxists, there is sometimes talk of the state withering away when the word has finally reached every heart. On the other hand, most utilitarians see the practical necessity of compulsion. As Mill [358] puts it: "For such actions as are prejudicial to the interests of others, the individual is accountable, and may be subjected to social or legal punishment."

My own view is that utilitarians would be wise to settle for the public policy option, which Hardin [224] refers to as *institutional utilitarianism*, and Goodin [209] as *public utilitarianism*. If those in power are inclined to personify the role of the government of which they form a part, then they may be open to the suggestion that its actions should be rational in the same sense that an individual is rational. In a society with egalitarian traditions, Bentham's [53] "everyone to count for one, nobody to count for more than one" will also be attractive. If the powerful are persuaded by such propaganda, then Harsanyi's [233] teleological argument will then require that the government act as though it had the preferences of a utilitarian ideal observer.

Whigs like myself will remain unpersuaded, but not because we are repelled by the idea that citizens need to be coerced into compliance. As long as someone is guarding the guardians, we see coercion as a practical necessity in a large modern state. Who would pay his taxes on time and in full unless compelled to do so? Even Hayek [240, p.153] creates no difficulties on this score. Provided laws are framed impersonally, he is willing to say, "When we obey the laws ... we are not subject to another man's will and are therefore free."[16] There is therefore no reason why enforcement should be a painful issue for teleological utilitarians—provided they are willing to grasp the nettle firmly by facing up to the intrinsic paternalism of their doctrine.

In summary, Harsanyi's [233] teleological defense of utilitarianism applies best to the problem that welfare economics sets out to solve. In its idealized form, a benign but paternalistic government asks what behavior it should enforce on the citizens subject to its authority. Harsanyi's

---

[16]To evaluate Hayek's definition, suppose that we were all constrained by the same strict regimen, specifying a precise ritual that governs how each single instant of the day is to be employed. Since we would all be in the same boat, such a system might be said to be fair, but how could we sensibly be said to be free? Even if the ritual were one we would choose for ourselves if we had the opportunity, the fact that the opportunity is denied means that we are enslaved. Contented slaves we might be, but slaves nevertheless. I think that Hayek [240, p.16] is guilty of the same abuse of language of which he so eloquently accuses others. In the ordinary sense of the word, we have to give up some freedoms when we associate together in a society. Societies therefore cannot be sorted into those which are free and those which are not. At best, one might be able to say that one society offers its citizens more freedom than another.

answer makes sense when the government regards itself as an individual written large who is immune to personal prejudice. However, to understand Harsanyi's nonteleological defense of utilitarianism, it is necessary to adopt a different perspective—the perspective of the philosopher-king.

2.2.4 Philosopher-King

So far, two forms of *teleological* utilitarianism have been considered. In both, an agency outside the system enforces a conception of the Good on those inside the system. The second differs from the first in insisting that the Good be determined in an impersonal manner. I use the word *impersonal* rather than *impartial* to preserve a distinction made by Rawls [417, p.187], which will be important in the discussion of the nonteleological form of utilitarianism that is next on the agenda. Recall that a nonteleological theory treats the *procedure* through which a society makes its decisions as fundamental.[17]

A truly impartial philosopher-king would certainly not want to impose his own *personal* conception of the Good on his subjects—but nor would he be willing to enforce some *impersonal* conception of the Good without a mandate from those he rules. He may perhaps agree with Harsanyi that priority should be given to the requirement that his actions on behalf of the state should be consistent with each other, but why should his personal view prevail? His subjects may well prefer to endorse some irrationality in his decision making should this be necessary to ensure that no citizen is ever treated unfairly.

So what is the role of an impartial philosopher-king? Let us suppose him to be all-powerful but entirely benign. Let us also assume that he has no largesse of his own to distribute, all the productive potential of the state being vested in the hands of his subjects. His role then becomes entirely organizational. First he receives a mandate from the people to pursue certain ends, and then he insists that each person take whatever action is necessary to achieve these ends. In real life, people are only too ready to vote for an end but against the means for attaining it. However, in a rational society, people will accept that working together toward an ambitious goal may require some surrender of their personal freedom. Without a philosopher-king to police their efforts at self-discipline, the citizens would

---

[17]A nonteleological ethical theory can therefore only assert that a society using a certain procedure is operating *as though* it is seeking to maximize a particular common good function. Just as it is a fallacy in modern utility theory to say that a decision maker with utility function u chooses a rather than b because $u(a) > u(b)$, so it is a fallacy in a nonteleological ethical theory to say that a society with common good function G chooses a rather than b because $G(a) > G(b)$.

have no choice but to rely on their own feeble powers of commitment to prevent any free riding. The ends they could jointly achieve would then be severely restricted. But with a philosopher-king to enforce the laws that they make for themselves, the citizens of a society will have a much larger feasible set open for them to exploit.

Political legitimacy. Such an account of the function of a philosopher-king bears a close resemblance to the theory used by modern political parties with liberal pretensions to justify the actions they take when in power.

Edmund Burke, the Whig credited with being the founder of modern conservatism, did not see things this way. In a famous speech to the voters of Bristol, he expounded his version of the *ipsedixist* doctrine that someone voted into power is justified in pursuing his own personal view of what is best for his constituents even if it conflicts with theirs. The Marxist doctrine of the dictatorship of the proletariat is an extreme *ideal observer* theory. The dictator is the ideal observer and the requirement of impersonality ensures that any aggregation of preferences will make him look like a proletarian in societies in which the proletariat is in a large majority. In modern times, neither of these extremes is at all popular. It is perhaps not even controversial to suggest that all political commentators are social democrats now, paying lip service to a theory of power which is essentially that of the *philosopher-king*.

I therefore think it of some importance that this chapter shows that applying Harsanyi's nonteleological ideas to the theory of the philosopher-king leads no less inexorably to utilitarianism than the application of his teleological ideas to the ideal observer theory. Barry [48, p.334] and Broome [102, p.56] are among the many who see no reason why someone unconvinced by Harsanyi's [233] teleological theory of utilitarianism should change his mind when shown his nonteleological theory. However, the admittedly naive classification of political theories offered above suggests an answer that is distinct from the foundational reasons given in Section 2.3.6 of Volume I. In brief, someone who does not lean sufficiently to the left to find an ideal observer theory attractive may still be enough of a social democrat to be convinced by a philosopher-king story.

Kant a utilitarian? A modern Western democracy would want to distinguish itself from other states that claim to be democratic by drawing attention to the freedoms enjoyed by its citizens. But what does it mean to say that a state is free? Nobody is free to withhold his taxes from the state, even under the most liberal of regimes. Nor is it a freedom that anyone would wish to see extended to everyone. This echo of Kant's categori-

cal imperative is no more an accident than the fact that the theory of the
philosopher-king ties in so closely with social democratic theories of political
legitimacy. Nor is it an accident that the role assigned to a philosopher-
king in enforcing the rules that his subjects make for themselves should echo
Rousseau [443, p.19] on the subject of "true freedom". The philosopher-
king serves as the desperately needed master who, in Kant's [294, p.122]
words, "can break man's will and compel him to obey a general will under
which every man could be free."

I think that such abuse of the word *free* in the tradition of Locke,[18]
Rousseau, and Kant serves only to confuse the issues that are supposedly
being clarified.[19] In the absence of some version of a philosopher-king,
this confusion serves as a smoke screen concealing the fact that nobody
in this tradition has any idea why the moral rules they expound should
bind a rational individual.[20] However, in the presence of a philosopher-
king whose subjects have only a limited power to make commitments, this
difficulty disappears and the perverse Kantian understanding of the notion
of freedom admits an interpretation that makes some kind of sense.

A limitation on our power to make commitments can be said to restrict
our "freedom" by imposing constraints on the actions it is feasible for us
to choose.[21] An impartial and benign philosopher-king with unlimited co-
ercive capacity can free us of these constraints by forcing us to honor the
commitments we would have made under our own steam if we had the
power. By enforcing the rules that we make for ourselves, he therefore
offers us a "freedom" that was not available to us before.

But how do we go about making the rules for our philosopher-king to
enforce? For many people, this question has an easy answer—the rules

---

[18]Though this be a State of Liberty, yet it is not a State of Licence—Locke [330].

[19]Earlier I spoke of the philosopher-king restricting the freedom of his subjects by
enforcing laws that decrease their set of available choices. A Kantian says that he is
thereby *increasing* their freedom. More succinctly, Kantian freedom consists in *not*
riding free.

[20]Kant [297] seeks to have it both ways by telling us that obedience to the categorical
imperative is a necessity for a rational individual—but that nothing necessitates his
obedience. Kant [297, p.131] tells us explicitly that the resolution of this apparent
contradiction is intrinsically incomprehensible. The best we can do is to comprehend
its incomprehensibility! I do not know where this leaves Kant's [297] attempt to deduce
the categorical imperative from the crude teleology of the *Groundwork* (Section 2.4.2
of Volume I). The same goes for Kant's [299] later claim that the incentive for obeying
the moral law is provided by a special kind of respect for moral law which is somehow
necessarily induced in rational beings by their knowledge of it!

[21]This is an abuse of the word "free" because it is only by poetic license that anybody
would assert that his freedom is constrained because he cannot perform feats that are
humanly impossible. As Berlin aptly [56] observes, it would be eccentric to say that he
is enslaved or coerced because he cannot understand the darker pages of Hegel.

should be those endorsed by majority vote. But this is a naive response. In the first place, modern democracies have all kinds of checks and balances built into their structure to prevent the kind of folly that led Athenians to vote for an expedition against Syracuse while their city was under siege by the Spartans in the Peloponnesian War. Similar checks and balances serve to protect minorities against exploitation or persecution by the majority. To quote James Madison [215, p.268]:

> In a society under the forms of which the stronger faction can readily unite and oppress the weaker, anarchy can surely be said to reign as in a state of nature, where the weaker individual is not secured against the violence of the stronger.

However, the main argument against treating majority rule as sacred is that part of the enterprise in which we are engaged consists of exploring the extent to which various forms of collective decision making can be justified on more fundamental grounds. It would therefore defeat our purpose to assign a special role to some particular political mechanism. In brief, it ought to be a theorem that some issues should be decided by majority vote rather than an axiom.[22]

Kant would argue that, in a society of truly rational individuals, everybody would agree that the rules to be enforced by the philosopher-king should be the maxims that satisfy the categorical imperative. However, even the most dedicated of Kantians would not claim that the categorical imperative is a precision tool. Rawls [417, p.256] therefore offered the device of the original position as "a procedural interpretation of Kant's conception of autonomy and the categorical imperative". It is on the basis of this claim that I argue that the nonteleological defense of utilitarianism offered in this chapter demonstrates that Kant should have been a utilitarian. Those who reject Rawls' claims to Kantian orthodoxy will have to be satisfied in having it demonstrated that Rawls himself should have been a utilitarian.

2.2.5 The Social Contract Approach

Having upset Kantians by suggesting that the logic of their position should lead them to utilitarian conclusions, it is necessary to remind the reader that I do not myself endorse utilitarianism as a system of public or private morality. I follow neither the Good nor the Right, but the Seemly. In particular, I do not much care whether the original position is or is not

---

[22]However, it will be a theorem proved by someone analyzing much less abstract models of society than I consider in this book.

a legitimate device for implementing Kant's categorical imperative. For me, the original position is merely a stylized representation of a fairness norm that evolved along with the human race for the purpose of resolving small-scale coordination problems. Whigs advocate using the device of the original position on a large scale to generate Pareto-improvements in our current social contract for purely pragmatic reasons.

Not only do whigs see no point in the kind of metaphysical musings with which Kant defends his categorical imperative, we also see no useful role for the idea of an incorruptible philosopher-king when considering constitutional reform. On the contrary, we think it fatal to adopt a viewpoint that abstracts the problem of corruption away. One must always ask: who guards the guardians? The whiggish answer is that we must all guard each other, so that society *as a whole* is in equilibrium. When considering the constitution for a whole society, one cannot afford to treat the officers of the state as convenient social instruments lying outside society. Their roles need to be embedded in a system of incentives that leads them to carry out their planned part in maintaining a social equilibrium without falling prey to the temptations that traditionally bedevil those in power.

Although whigs believe that the idol of utilitarianism has feet of clay, there are several reasons why it remains useful for this chapter to explain why taking the current consensus on political legitimacy seriously leads inexorably to utilitarianism. The first reason is to discredit the consensus, and so open the way to a reconsideration of the social contract ideas that inspired the founders of the American Republic.

The second reason is to provide a preview of how classical social contract ideas need to be updated by considering their implications in the presence of a philosopher-king. Introducing the fiction that the citizens of a society have such a useful instrument in their social toolbox simplifies matters considerably, because we can then neglect the problem of how social contracts are enforced. The problem of what to do when others don't play by the rules is evaded by imagining ourselves back in nursery school with an all-powerful but benign teacher ready to intervene when necessary. Later chapters face up to the need for viable social contracts to be self-enforcing, but maintaining the charade that enforcement is not a problem for a social contract theory allows us to juggle with several fewer balls in this chapter while discussing the evolutionary origins of our notions of fairness.

The third reason is more positive. Both utilitarians on the left and libertarians on the right make the mistake of supposing that their systems are viable as social contracts for a society *as a whole*. But to deny these utopian claims is not to argue that their doctrines have no useful subsidiary role. Like all economists, I am sold on the market mechanism as an efficient and informationally efficient means of allocating resources in many circumstances. But libertarians are wrong to think that the market can operate in a vacuum. Without an appropriate social contract in the background, we wouldn't even understand the concept of private property. Libertarians are equally wrong to think that people don't care about fairness. When it becomes necessary to ask who should take what life-threatening risk, people care very deeply indeed that justice be done. It is then that utilitarianism comes into its own.

For example, the time will soon come when the welfare states of Western Europe will no longer be able to pretend that health care is not rationed. What rationing scheme will a government then implement? The story told in this chapter is ideally suited to answering such subconstitutional questions. A wise government will take on the role of a philosopher-king seeking a mandate from the people obtained under fair circumstances. Economists offer a variety of social welfare functions that might be maximized in such circumstances, but this chapter offers a rationale for selecting the weighted utilitarian social welfare function of Section 1.3.3.

This attitude to utilitarianism is shared by many welfare economists, but my own approach differs from orthodox treatments based on *a priori* notions of the Good. As Section 2.3.6 of Volume I emphasizes, whigs climb Jacob's ladder from the bottom up. That is to say, they reduce the Good to the Seemly by *constructing* the common good to be maximized from more primitive notions. The advantage of this approach is that the weights in the utilitarian social welfare function are assigned in accordance with the way that real people on the ground actually make fairness judgments, rather than being chosen in some supposedly objective manner. The disadvantage for ipsedixists is that they are then left with no opportunity to substitute their own prejudices for those of the people whose health care is being rationed.

2.2.6 Rule-Utilitarianism or Act-Utilitarianism?

Sen and Williams [484, p.2] find it curious that utilitarians are often ambiguous about whether their theories relate to personal morality or public policy, but I don't see any mystery. If utilitarians didn't fudge this issue, they would have to confront the failure of their doctrine to provide an answer to the enforcement question. However, in blurring the boundary between personal morality and public policy, they dig various pits for themselves. One of these is the controversy between act-utilitarians and rule-utilitarians. The former argue that each individual act should maximize the common good. The latter argue that utilitarian principles should be applied to the rules to which we appeal when making decisions.

Harsanyi [233, p.62] is an emphatic rule-utilitarian.[23]   In seeking to discredit act-utilitarianism, he gives an example that is typical of its kind.[24] Adam, who is poor, has borrowed a sum of money from Eve, who is rich. When the due date for repayment arrives, should Adam pay the money back?   An act-utilitarian says no, because the money is worth more to Adam than to Eve.[25] Thus the sum of utilities is maximized by Adam's retaining the money. A rule-utilitarian says yes, because nobody would ever lend money if he couldn't rely on the money being repaid. Honoring the rule of keeping promises therefore makes the overall sum of utilities in society higher than it would otherwise be.

Evaluating this difference of opinion is complicated by the fact that we are not told what standpoint to adopt. Is an appeal being made to our intuitions about how morality works in the world-as-it-is? If so, then the rule-utilitarian would seem to score, because those who break their promises in such circumstances are universally despised. However, as Section 2.2.5 of Volume I argues, people do not keep their promises in the real world for the Kantian reasons taken for granted by the rule-utilitarian argument. They do not ask what would happen if "everybody behaved like that". They keep their promises, insofar as they do, because it is customary to keep promises—and the custom survives because those who break their promises are punished by not being trusted in the future. More generally, as Hume [270] remarks, "All our obligations to do good to society seem to imply something reciprocal."

Is an appeal being made to utilitarianism as an ideal theory of individual morality? If so, then we may ask whether Adam and Eve are both utilitarians. If they are, then we need not be distressed at Adam's promise to repay the debt being broken, because Eve will be only too pleased to

---

[23]Among other things, Harsanyi [235, p.4] correctly argues that rule-utilitarianism cannot achieve a lower level of social utility than act-utilitarianism because one of the rules available is to behave like an act-utilitarian. However, if everybody in a society operating under perfect information were an act-utilitarian taking the same far sighted view of the nature of social utility, the maximum level of social utility would obviously be achieved if all citizens always jointly took whatever step was necessary for this purpose whenever a decision needed to be made. In such an ideal society, there would therefore be no difference between an act-utilitarian and a rule-utilitarian.

[24]I plan to ignore arguments against act-utilitarianism based on the idea that Adam should be treated as a different person at different instants of time, as in Sidgwick's [492] attempt to draw a parallel between the treatment of different people at the same time and the same person at different times (Section 2.5 of Volume I).

[25]As in the traditional story told at the beginning of Section 4.2 of Volume I, the implicit assumption is that utility is a concave function of wealth, so that an extra dollar to Adam is worth more than an extra dollar to Eve. As Bentham [51] puts it: "The quantity of happiness produced by a particle of wealth will be less and less at every particle."

release Adam from his obligation. Indeed, she will shower further wealth upon him until each has the same marginal utility for one dollar more or less. But what if Eve is not a utilitarian? Here we hit the rock that sinks utilitarianism and most other proposals that have been made as a theory of personal morality. What is to be done about those who remain unpersuaded by utilitarian arguments?

The result of applying Kantian reasoning to this problem seems plain enough—although Kantians will not like the answer very much. Adopting the viewpoint of a depersonalized ideal observer, Adam wishes that maximizing the sum of utilities were a universal law. He should therefore act as though it were and break his promise.[26] If the outcome offends our conventional notions of morality, so much the worse for conventional morality. Act-utilitarians are not interested in producing arguments that justify how we currently behave, they are interested in telling us how we ought to behave. As with Kant, their integrity in this respect seems to me entirely admirable.

Such a Kantian rejoinder to the Kantian argument of a rule-utilitarian would seem to hoist the latter by his own petard—provided that we treat utilitarianism as a theory of personal morality. However, if like me, you reject Kantian ideas on morality, you will probably also reject utilitarianism as a viable theory of personal morality. This leaves us with the option of analyzing Adam's dilemma from a public policy perspective. Should a government, whether mandated by an imaginary ideal observer or taking its mandate from from the people like a philosopher-king, enforce contracts? In the case of debts between the kind of rich and poor people to be found in the real world, the answer is so obviously yes that I shall take the reasons for granted.

In summary, if utilitarianism is a theory of personal morality, then a utilitarian can certainly make a good case for act-utilitarianism. However, act-utilitarianism is not a comfortable creed for a bourgeois since it requires him to give away his possessions to the poor. Bourgeois utilitarians who wish to rest easy in their beds are therefore drawn to rule-utilitarianism. However, they then find themselves in the awkward position of having to pretend that a theory designed for the conduct of public policy makes sense as a theory of personal morality. For a really good night's slumber, they would do better to abandon utilitarianism altogether in favor of the species

---

[26]I have not forgotten that keeping promises was one of Kant's own examples of a maxim that Adam would will to be a universal law. But the making of promises would be meaningless in a world where it was a universal law that everyone should act to maximize a single common good. The population would behave like a single individual with a single aim, and there would be no more need of promises between Adam and Eve than between my arms and legs.

of bourgeois liberalism espoused in this book!

2.2.7   The Big Picture

This section is an aside in which I attempt a primitive classification of
political theories with a view to explaining where my own whiggish views
and aspirations stand in relation to utilitarianism. Section 4.10.2 reverts
briefly to this topic after my theory has been presented in detail.

Classifying political attitudes.   It seems to me that one needs at least
two dimensions to come anywhere near capturing the richness of current
political attitudes. Figure 4.16(a) uses two axes to separate the plane into
four regions that I could untendentiously have labeled unplanned centralization, unplanned decentralization, planned decentralization, and planned
centralization. But the language of economics is so dismally dull that I
have translated these terms into neofeudalism, libertarianism, whiggery,
and utilitarianism. A journalist would go further down this road and interpret a utilitarian as a bleeding-heart, big-spending liberal, and so on, but I
prefer to keep my prejudices under slightly firmer control.

In terms of the traditional left-right political spectrum, utilitarianism
sits out on the socialist left and libertarianism sits out on the capitalist
right. The same dichotomy appears in moral philosophy as a split between
the consequentialist followers of the Good and the deontological followers
of the Right. However, I think that these oppositions are orthogonal to
the issues that really matter. The unworkable utopias of both utilitarians
and libertarians have no more relevance to genuine human concerns than
the metaphysical differences on the properties of Absolute Morality that
divide the Good from the Right. Just as we have to assimilate the issues
that trouble consequentialists and deontologists into a single theory of the
Seemly before we can say anything consistent with what evolution has made
of human nature, so we have to separate the feasible from the infeasible
in the aspirations of utilitarians and libertarians before abandoning the
possibility that they may have some common ground on which to stand.

The orthogonal opposition that I think should supersede the sterile and
outdated dispute between left and right contrasts societies in which fairness
is used to coordinate collective decisions with societies that delegate such
decisions to individuals or elites. I use the term *neofeudal* to describe
the latter kind of social contract. All large organized states of historical
times have been neofeudal in character—including those that think of their
prime characteristic as being capitalist or socialist. As is obvious from the
totalitarian regimes operating before the Second World War in Germany
and Japan, and the species of social consensus that has operated so far in

both countries since the war, capitalism does not need libertarian political institutions to flourish. Equally, as is shown by the experience of Britain after the Second World War, a country can ruin its economy by turning to socialism without any need to abandon freedom and democracy along the way. One therefore goes astray in seeking to draw conclusions about the relative merits of the political aspects of the social contracts operating on the two sides of the Iron Curtain in the cold war from the fact that market economies outperform command economies. Insofar as the choice is between the political regimes of living memory, the issue for political philosophers is not how we organize our economies, but whether our children will find more fulfillment in a society in which the same oligarchs rule all the time, or in a society that rotates its oligarchs using a method traditionally regarded as fair.

I guess that nobody likely to be reading this book would prefer the former of these two options. But whigs want to go further and root out entrenched authority altogether, whether its representatives are rotated or not. We differ from the stylized conservatives, socialists, and social democrats of Section 2.2.4 in being unwilling to trust those in power farther than they can be thrown. We see corruption as perhaps the major problem facing the modern state.

Everyone is willing to condemn the straightforward bribery and nepotism which afflict all societies, but whigs are more worried about the corruption of our institutions that arises when their officers cease to operate the institutions to further the purpose for which they were created, but instead imperceptibly and unknowingly come to use the power of their office to advance their own personal goals. To what extent are democratic elections fair now that rich men have learned how to get their puppets elected by employing spin doctors to reduce political debate to an exchange of meaningless advertising slogans? How is justice to be obtained in a law case when the other side has all the money? What poor man seeking his legal rights now expects due process to be respected by the various jacks-in-office who gnaw at the heart of our public institutions? Our institutions were mostly founded with the most benevolent of intentions, but good intentions are not enough to prevent time unraveling the firmest weave if loose ends are left at which it can tug. As the Monty Python catchphrase has it, "Nobody expects the Spanish Inquisition." But this is what corruption eventually made of the institution set up to spread the gospel of Jesus Christ.

Utilitarians and libertarians are as dissatisfied as whigs with our current neofeudal institutions, but respond by making the same mistake as Santayana's *Lucifer*, who rebelled against the *feasibility* constraint in God's decision to create the best of all possible worlds. But there is no point in designing ideal social systems whose workability depends on first changing human nature. Human nature is as it is, and no amount of wishing that it were different will make it so. We therefore have to resign ourselves to living in a second-best society because first-best societies are not stable. Utopians who seek to establish first-best societies are actually condemning us to live in whatever hell evolution eventually makes of their unstable utopia.

The founding fathers of the American Republic understood this point perfectly well when they built a system of checks and balances into the American Constitution in an attempt to confine neofeudalism to the Old World. But what remains of their construction is now hopelessly unfitted to meet the new forms of neofeudalism that have emerged in modern times. My own country does not even have a written constitution or a bill of rights to obstruct the triumphal advance of neofeudalism.

For whigs, the urgent problem for political philosophy is therefore to rethink the thoughts of the classical liberals who wrote the American Constitution as they would think them if they were alive today. We need to find an escape route from the new forms of neofeudalism that have evolved since their time to a new kind of whiggery that places fairness at the heart of a social contract designed to resist corruption.

In summary, we need to cease thinking outdated thoughts about where we would like to locate society on a left-right spectrum. Choosing between utilitarianism and libertarianism makes as much sense as debating whether griffins make better pets than unicorns. We need to start thinking instead about how to move in the orthogonal direction that leads from neofeudalism to whiggery.

Egalitarianism.   How will this book help in promoting a move away from neofeudalism toward whiggery? Not very much, I fear. The best I can hope for is to show how modern advances in game theory can be used to clear away some of the intellectual obstacles that tradition has strewn in our path. Fairness seems to me the issue that most needs to be freed from the confusion that currently clouds rational debate. What are the origins of our intuitions on this subject? How can we predict what people will deem to be fair? Isn't the whiggish concept of a fair society just one more unworkable utopian dream?

To concentrate on fairness is to focus on the *planned* forms of social organization in my crude classification of political attitudes. Most of this book is therefore devoted to utilitarianism and the form of egalitarianism that I call whiggery. One might summarize its conclusions by saying that the book reinterprets Harsanyi's [233] game-theoretic defense of utilitarianism in naturalistic terms, but finds that its logic then leads to an egalitarian conclusion of the type proposed in Rawls' [417] *Theory of Justice.*

However, my feelings about the egalitarian movement that Rawls pioneered are very ambiguous. Although I think that egalitarian intuitions about the way real people actually use fairness norms in their daily lives are sound, I see no future in the way the modern egalitarian movement attempts to incorporate these intuitions into a formal system. Their retreat from utility theory seems particularly retrograde, since they thereby simply reintroduce into moral philosophy all the puzzles and paradoxes that utility theory was invented to eliminate. Readers who know nothing of the current debate between egalitarians and utilitarians may find it ironic that I constantly complain about the errors of modern egalitarians while simultaneously providing egalitarianism with the kind of firm intellectual foundation that only utilitarianism has hitherto enjoyed. However, the real irony is that modern egalitarians have talked themselves into a corner from which they are less able to see the force of the arguments I offer in favor of egalitarianism than our utilitarian opponents.

Modeling unplanned societies? Egalitarianism and utilitarianism are *planned* forms of social organization. One might ask whether game theory also has something to contribute to the study of the *unplanned* forms of social organization in my classification of political attitudes.

The answer is that we have a long way to go before we can say anything at all definitive about the social contract that cultural evolution is likely to select when the choice is left blowing in the wind. Even when a society's Game of Life is as simple as the indefinitely repeated Prisoners' Dilemma, Section 3.3.8 shows that matters are far more problematic than the political science literature imagines. However, although formal studies that demonstrate what time would do to a libertarian utopia would be instructive, there seems little point in theorizing very much about neofeudalism, because we have actual specimens of neofeudal social contracts all around us, awaiting empirical study by political scientists and sociologists who are willing to ask the right kind of questions.

2.3 Fictitious Postulatum?

The previous section distinguished three varieties of institutional utilitarianism: that of the ipsedixist, the ideal observer, and the philosopher-king.

The section also raised three questions that utilitarians need to answer: What constitutes utility? Why should individual utilities be added? Why should I maximize the sum of utilities rather than my own?

When passing from the ideal observer theory to that of the philosopher-king, attention centered on the third of these three questions. It was unnecessary to renew consideration of the first question, because utility is to be interpreted in the sense of Von Neumann and Morgenstern in both types of theory. However, it may have passed unnoticed—as such matters frequently do in this literature—that the answer given to the second question in an ideal observer theory will not suffice for a philosopher-king theory. As shown in Section 4.2.4 of Volume I, the Von Neumann and Morgenstern utility function representing an ideal observer's preferences can be shown to consist of a weighted sum of the utilities of the individual citizens. But a philosopher-king's impartiality disbars him from using his own personal or empathetic preferences as a standard for evaluation. The standard he employs when comparing utils across individuals must derive from the people he rules.

2.3.1   Interpersonal Comparison of Utility

Establishing a standard for making interpersonal comparisons of utility is widely regarded as impossible or hopelessly intractable. Hammond [220] quotes various distinguished economists to this effect, and he could even have quoted Bentham [52]:

> 'Tis in vain to talk of adding quantities which after the addition will continue to be as distinct as they were before, one man's happiness will never be another man's happiness: a gain to one man is no gain to another: you might as well pretend to add 20 apples to 20 pears.

At the time when logical positivism was fashionable, the sound and fury raised against theories of interpersonal comparison reached an almost hysterical pitch. Lionel Robbins [429] was particularly vocal, denouncing interpersonal comparison, among other things, as unscientific on the grounds that statements comparing Adam and Eve's utilities have no empirical content.[27] Echoes of the debate still haunt the economics profession today. Various expedients continue to be invented whose purpose is somehow to allow interpersonal judgments to be made without utils actually being compared (Section 4.3 of Volume I). Meanwhile, a parallel movement within philosophy works hard at making a technical subject out of the notion that ethical values are supposedly incommensurable by their very nature.

---

[27]Harsanyi [233] responds with the example of a man who is unable to go to the opera and has to choose which of his friends would benefit most from a gift of his ticket. I prefer to observe that people reveal their standards of interpersonal comparison whenever they settle a dispute by appealing to a fairness norm.

Personally, I do not see that the causal view of utility taken for granted by Victorians deserves the contempt that continues to be directed at it. Rats apparently forgo all other delights when offered the opportunity to press a lever that stimulates a pleasure center in their brains. Are we so sure that humans would behave any differently? Nor do I see why Victorian ideas should be attributed to those of us who use modern utility theory as the basis for talking about interpersonal comparison.[28] Nor yet do I see how the claim that it is conceptually *impossible* to compare utils across individuals can be taken seriously. It is true that some ways of thinking about utility preclude the possibility that interpersonal comparisons can be made. But there are many possible ways of thinking about utility—as I hope the following brief review of some of the traditional approaches to the question will demonstrate.

Counting perception thresholds. I believe that it was Edgeworth [169] who first proposed observing how far a parameter controlling the environment of a subject needs to be changed before the subject perceives that a change has taken place. If the subject expresses a preference for low parameter values over high parameter values, the number of perceptual jumps he experiences as the parameter moves from one end of its range to the other can then be used as a measure of the intensity of his preference between the two extremes. The psychologist Luce [333] has been a modern exponent of this idea. Rubinstein [447] has also explored its implications.

If workable, such a procedure would provide an objective method of comparing utils across individuals independently of the social mores of their society of origin. But would such a comparison be meaningful? Even if it were, would it be possible to persuade people to regard it as a relevant input when making fairness judgments?

Like many men, I am not only nearsighted, I am also mildly color-blind. At the Poker table, I have to be quite careful when both blue and green chips are in use. Does it therefore follow that I get less pleasure from the use of my eyesight than someone with perfect vision? My hearing is even less reliable than my eyesight. Should those with perfect pitch therefore be assumed to take a keener pleasure in music? I have only the haziest idea of how much I am worth, while others keep accounts that are accurate down to the penny. Is this relevant to how much tax we each should pay?

At the risk of sounding like G. E. Moore (Section 2.3.5 of Volume I), I have to say that attempts of this kind to provide a proper foundation

---

[28]Philosophers sometimes say that utility is interpreted as "preference-satisfaction" in modern theories to distinguish the approach from the hedonic ideas of our Bentham and Mill.

for a Benthamite view of utility seem doomed to founder when confronted
with such questions. As soon as one proposes some absolute definition
of happiness in terms of genuine observables, those who will lose out if
this definition is adopted will say, "But is that the correct definition of
happiness?"[29]

Zero-one rule. The same goes for another proposal that gets a better
press. Hausman [238] even argues that, if there is a "correct" way to compare bounded cardinal utilities, then the zero-one rule is it.

The zero-one rule applies when it is uncontroversial that a person's
individual preferences are to be measured with a cardinal utility function,
usually a Von Neumann and Morgenstern utility function. If the worst
thing available for Adam is $\mathcal{L}$ and the best is $\mathcal{W}$, then the zero-one rule
calls for his utility scale to be recalibrated so that his new Von Neumann
and Morgenstern utility function v satisfies $v(\mathcal{L}) = 0$ and $v(\mathcal{W}) = 1$. The
utility functions obtained after such a recalibration can then be compared
without difficulty.

In Section 4.3.1 of Volume I, Adam and Eve's personal utility functions
are recalibrated in just this way, with $\mathcal{L}$ taken to be hell and $\mathcal{W}$ to be heaven.
However, no attempt is made to use this recalibration as a basis for making
interpersonal comparisons. Eve may well be a jaded sophisticate who sees
$\mathcal{W}$ as only marginally less dull than $\mathcal{L}$, whereas Adam is a bright-eyed
youth for whom the difference may seem unimaginably great. Hausman
[238] quotes Griffin [213], Hammond [220], Rawls [417] and Sen [475], who
all say much the same. In brief, the objections to the zero-one rule are the
same as those which apply to the method of counting perceptual jumps. It
certainly provides a way of comparing utils across individuals, but who is
to say that the comparisons generated are relevant to anything of interest?

Bargaining with transaction costs. My guess is that much of the
reason that the theory of interpersonal comparison has a bad name, is that
it is taken for granted that such a theory must necessarily have the character
of the two approaches considered so far. Let me therefore give an example
of an entirely different kind.

Section 1.7.1 studies the deal that will be reached by rational agents who
bargain according to Rubinstein's [445] Alternating Offers Game. The most
interesting case occurs when the interval τ that elapses between successive

---

[29]However, unlike Moore [364] on the subject of the Good, I do not deduce that
happiness is some transcendental notion we intuitively perceive but which lies beyond
our powers of analysis. I deduce that the concept of happiness as an explanatory entity
is inadequate to describe the phenomena it was invented to rationalize.

proposals is allowed to become very small. The deal is then approximated by an asymmetric Nash bargaining solution with bargaining powers $1/\rho_A$ and $1/\rho_E$, where ρ_A and ρ_E are Adam and Eve's discount rates. As always, the location of the *status quo* for the bargaining problem matters a great deal. In the Alternating Offers Game, the *status quo* must be placed at the deadlock point d, which is the pair of payoffs the players would receive if they were to bargain for ever without reaching an agreement.

In the standard version of the Alternating Offers Game, nothing happens while the players are deadlocked, and so $d = (0, 0)$. But we can elaborate the model so that Adam must pay c_A and Eve must pay c_E for each unit of time that elapses without an agreement. Adam's deadlock payoff is then

$$-c_A\tau\{1 + \delta_A^\tau + \delta_A^{2\tau} + \cdots\} = -c_A\tau/(1 - \delta_A^\tau),$$

which approaches $-c_A/\rho_A$ as $\tau \to 0$. Similarly, Eve's deadlock payoff is $-c_E/\rho_E$. Rubinstein's theory then leads to the conclusion that the deal reached by Adam and Eve when τ is negligible is the Nash bargaining solution with the same bargaining powers as before but with the *status quo* shifted to $d = (-c_A/\rho_A, -c_E/\rho_E)$.

The next step is to consider the case when the players become very patient indeed, so that $\rho_A \to 0$ and $\rho_A \to 0$. However, we still need to keep track of the players' *relative* impatience levels, since this is what matters in determining who gets what. Let us therefore assume that the discount rates converge to zero in such a way that $\rho_A/\rho_E = r_E/r_A$, where r_A and r_E are fixed positive constants. If $r_E > r_A$, Adam will then be more impatient than Eve.

What happens to the Nash bargaining solution in this limiting case? It is an easy task to show that it converges to the weighted utilitarian solution with weights $U = r_A/c_A$ and $V = r_E/c_E$. In brief, very patient players bargaining according to the Rubinstein theory, but who have to pay at a fixed rate for the time that elapses before an agreement is reached, will act *as though* they were utilitarians using the social welfare function W_h defined by

$$W_h(x) = Ux_A + Vx_E.$$

They therefore behave *as though* V of Adam's utils are worth U of Eve's.

Notice that the standard for making interpersonal comparisons of utility established in this way has no *causal* role at all. The agreement on which Adam and Eve settle is not chosen because they are utilitarians who think that one of Adam's utils is worth U/V of Eve's. Nor is there any reason to suppose that they would act as though trading utils at this rate when interacting under other circumstances. A Victorian might therefore feel defrauded when offered a description of their bargaining behavior couched in terms of interpersonal comparisons of utility. However, in this example, the Victorian standpoint has been abandoned in favor of a revealed preference perspective. Revealed preference theory lays no claim to explaining *why* Adam and Eve do what they do. It is content to find a simple description of *what* they do.

Voting. The next example offers a similar lesson, but in a starker context. Adam, Eve and Ichabod are a family who have the opportunity to invite either Hume or Kant to dinner. Eve is unremittingly serious and so votes for Kant rather than the frivolous Hume. Adam and Ichabod have no interest in philosophy, but have heard that Hume understands jokes. They therefore vote for him on the grounds that he is likely to be somewhat less of a bore. As a result, Hume is elected by two votes to one.

It is important to recognize that an interpersonal comparison of sorts is being made whenever voting is used to decide between two candidates in this way. Eve's family made their decision as though they were utilitarians using the zero-one law to make interpersonal comparisons. If p is used to denote the lottery in which Hume is invited with probability p and Kant with probability $1 - p$, then the family chose the value of p *as though* maximizing the utilitarian common good function

$$W(p) = v_A(p) + v_E(p) + v_I(p),$$

where the Von Neumann and Morgenstern utility functions have been normalized so that $v_A(0) = v_E(1) = v_I(0) = 0$ and $v_A(1) = v_E(0) = v_I(1) = 1$.[30] I refer to the use of this welfare function as zero-one utilitarianism.

Following May [346] and Sen [475], it is even possible to give axioms supporting such a utilitarian social decision function. We can represent the preference profile of Eve's family as HKH. Let $f(HKH)$ denote the guest then invited. Pareto-efficiency guarantees that $f(HHH) = H$ and $f(KKK) = K$. For the purposes of contradiction, suppose that $f(HKH) = K$. Then a monotonicity condition like that of Section 0.7 ensures that $f(KKH) = K$. If the labeling of the candidates is irrelevant (neutrality), we can then deduce that $f(HHK) = H$. If the labeling of family members is also irrelevant (anonymity), we are then led to the contradiction that $f(HKH) = H$. It follows that $f(HKH) = H$.[31] The general form of the argument shows that a candidate who commands majority support will always be invited.

---

[30] Then $v_A(p) + v_E(p) + v_I(p) = 1 + p$, which is maximized when $p = 1$.
[31] The logic is "(not P implies P) implies P".

Section 2.3.4 of Volume I mentions Condorcet's Paradox, which illustrates the difficulties that can arise when pairwise majority voting is used to settle disagreements involving more than two issues. After observing that majority voting over two issues is equivalent to zero-one utilitarianism, one might argue that zero-one utilitarianism is therefore a natural extension of majority voting to multi-issue problems. Dhillon and Mertens [160] strengthen the case by providing a weakened version of Arrow's axioms that characterize zero-one utilitarianism uniquely.[32]

As in the previous example, a frequent reaction to introducing interpersonal comparison into a discussion of majority rule is that the resulting story is fraudulent. Voters seldom think of themselves as utilititarians when casting their ballot. There is therefore no sense in which their joint decision is *caused* by their making interpersonal comparisons according to the zero-one rule. But revealed preference theory makes it a fallacy to assert that the utilities assigned to choices necessarily have a causal relationship to the choices actually made. Instead, utilities are used simply to describe the choice behavior of a rational decision maker—just as we can describe the institution of majority voting over two issues in terms of zero-one utilitarianism.

It is admittedly true that making interpersonal comparisons in such a way is highly *context-dependent*. One must look to some other approach if one hopes to obtain measures of interpersonal well-being that are universally applicable. Zero-one utilitarianism is particularly badly tailored for applications to fairness questions. After all, it is hardly fair that Eve's deep longing to meet Kant should be weighed in the same balance as the shallow reasons Adam and Ichabod have for preferring Hume. As we all know only too well, voting procedures often ride roughshod over such fairness considerations.

Nor is this any less true after axioms have been given to characterize zero-one utilitarianism. As in bargaining theory, axiom systems always seem to make good sense when contemplated in the abstract. To uncover their hidden secrets, one has to worry them like a fretful terrier. The secret hidden in the axiom system of Dhillon and Mertens [160] is that the implied criterion for comparing utils is essentially the same as that embodied in majority voting.

Cost-benefit analysis. The preceding argument shows that the one man, one vote principle implicitly defines a standard for making interpersonal comparisons of utility. Economists who use cost-benefit analysis are more explicit in the criterion they employ. Instead of employing the Ben-

---

[32]They refer to zero-one utilitarianism as relative utilitarianism.

thamite principle that each person is to count the same, they employ the principle that each dollar is to count the same—whether it is the billionth dollar of a rich man in his castle or the only dollar of the poor man at his gate.

Varian [534, p.169] expands on the use of this principle in the case when the ith consumer has a quasilinear utility function $u_i(x, y) = U_i(x) + y$, where x perhaps represents a quantity of apples and y represents dollars. Each consumer will then always be willing to exchange one util's worth of apples (measured according to his utility function U_i) for a dollar. If all dollars are to be counted the same, then the utils with which consumers evaluate apples must therefore also be counted the same. In this special case, one can therefore justify the standard procedures of cost-benefit analysis on utilitarian grounds. In more general cases, one can then choose to follow Varian [534, p.167] in defending cost-benefit analysis as a rough-and-ready procedure that is only fully justifiable in the ideal case when consumers have quasilinear preferences.

I do not want to deny the usefulness of making cost-benefit analyses[33] However, it is pointless to deny that they implicitly incorporate a standard for making interpersonal comparisons of utility that makes it socially neutral to take a dollar out of the pocket of a poor man and put it into the pocket of a rich man. But such a standard is patently unfair. As Paul Samuelson famously explains, the reason that a rich man takes a taxicab when it rains while a poor man gets wet is that the poor man values the fare more than the rich man.

The Shapley value. Yaari [570] has offered an ethical evaluation of Shapley's [489] proposal for comparing utils across individuals and so something has to be said on this subject.

The Shapley value is based on the coalitional form of a game with "transferable utility" (Section 0.5.1). In the case of the problem of dividing the dollar, it calls for the dollar to be split equally between two risk-neutral protagonists.[34] But suppose we are faced with a game in which utility is

---

[33]I think the major role for social choice theory will eventually turn out to be the provision of critical evaluations of public policy based on the descriptive devices that the theory provides. For example, those who wish to argue that women are disadvantaged by a particular measure may be able to argue that the government is acting as though it were a zero-one utilitarian with Adam weighted much more heavily than Eve. A cost-benefit analysis is similarly a useful tool for those who wish to argue that a government is not getting value for its money.

[34]With n players, first list each of the $n!$ routes by means of which the grand coalition of all players can be built up by adding one player at a time. For each such route, compute the amount that a particular player adds to the value of the coalition as he joins it. The Shapley value assigns him the average of all such contributions. In the

not transferable? Shapley [489] showed that the players' utility scales can then be recalibrated so as to create an imputation that would have been the Shapley value of the game if utility were transferable. This recalibration then serves as a basis for making interpersonal comparisons of utility.

The situation is easiest in the case of the two-player problem of dividing the dollar. Shapley's proposal then leads to the selection of the symmetric Nash bargaining solution, provided that the players' *status quo* payoffs are identified with their security levels. The necessary recalibration of utility scales requires that the points M and L of Figure 1.5(b) be adjusted until the line LM has slope -1. If the payoffs the players receive at the Nash bargaining solution N after this symmetrization were transferable, then N would be the Shapley value of the new configuration.

The bottom line is that Shapley's proposal for comparing utils in the two-player case reduces to recalibrating Adam and Eve's utility scales so that the payoff increments the players receive in passing from the *status quo* to the symmetric Nash bargaining solution are equal when measured in new utils.

I have never been able to understand why the Shapley value has received so much attention in the transferable utility case. I therefore see no virtue at all in making it the basis for a theory of interpersonal comparison in the multiplayer case when utility is not transferable. However, matters are more interesting in the two-player case, since the Shapley value then reduces to the symmetric Nash bargaining solution if the *status quo* is suitably placed. In fact, I shall be making a proposal similar to Shapley's for those cases in which the forces of social evolution have led Adam and Eve to regard the Nash bargaining solution as fair.

Such a proposal seems paradoxical unless one accepts that interpersonal comparisons need have no causal role and may differ radically between contexts. As emphasized in Section 1.3.3, no interpersonal comparisons are required to locate the Nash bargaining solution. It therefore has no merit as an ethical concept. If Adam and Eve come to regard it as fair, all ethical content will therefore have been eroded from their dealings with each other. But this prevents neither them nor us from continuing to use the language of fairness. If a standard of utility comparison has evolved that results in Adam and Eve being perceived as receiving the same number of utils when they pass from the *status quo* to the symmetric Nash bargaining solution,

---

two-player problem of dividing the dollar, there are two routes to the grand coalition, namely 12 and 21. Along the first route, the first coalition to form is {1}. Player I contributes nothing when he joins this coalition because a player is worth nothing acting alone. The next coalition to form is {1, 2}, which is worth 1. Thus player II contributes 1 along the route 12. The players' contributions along the other route 21 are reversed, and so each ends up with a Shapley value of one half.

then matters will be *as though* the symmetric proportional bargaining solution had been used. In such circumstances, Adam and Eve would see their transaction as fair—just as the subjects in the bargaining experiments reported in Section 0.4.2 found no difficulty in regarding whatever behavior happened to evolve as fair.

An onlooking analyst has two ways of describing the situation. He can say that Adam and Eve are using the proportional bargaining solution and thereby revealing a particular standard for comparing utils. Such a story will have descriptive value whether Adam and Eve explain their own behavior in such terms or not. If they don't, then the story will have no causal role at all. If they do, then one can point to the fairness norms in their heads as the immediate cause of their behavior. However, as Section 0.4.2 emphasizes, such an explanation of their behavior is superficial. A fuller explanation will have something to say about the reasons that one fairness norm evolved rather than another.

Empathetic preferences. Section 1.2.6 of Volume I traces my own approach to interpersonal comparison back to Hume [267] and Adam Smith [500]. Their ideas can be formalized using the modern theory of extended sympathy preferences, as developed by Suppes [513], Arrow [20], Sen [475] and Harsanyi [233]. The risk of confusing an extended sympathy preference with the more straightforward idea of a sympathetic preference is very real. I therefore always speak of an empathetic preference rather than an extended sympathy preference. This usage reflects the distinction made by Wispé [562], de Waal [539] and others between the concepts of sympathy and empathy. As explained in Section 0.2.4, for Adam to sympathize with Eve, he must care directly about her well-being, but he can empathize with her while remaining indifferent to her plight.

Section 4.3.1 of Volume I discusses the algebra of empathetic preferences at length. Much more will be said about the concept later in the current chapter. Those who plan to study this coming material in detail should therefore skip forward to Section 2.4.

If I say that I would rather be Adam wearing a fig leaf than Eve eating an apple, then I am expressing an empathetic preference. Under suitable assumptions, a utility function representing my *empathetic* preferences can be expressed as a weighted sum of the Von Neumann and Morgenstern utility functions that represent Adam and Eve's *personal* preferences. The weights in this sum serve as a standard for comparing Adam and Eve's personal utils.

Such a standard for making utility comparisons is *intra*personal rather than *inter*personal. It is my own idiosyncratic standard and need not be shared by anyone else. But it does not help very much that I have my own private intrapersonal standards for comparing utils if I am unable to agree with others on a common interpersonal standard to be applied when joint decisions are made. Nor is social choice theory very encouraging about the possibility of aggregating our intrapersonal standards. Hylland [279] has shown that a version of Arrow's Paradox applies, and so the only aggregation methods that meet the usual social choice criteria are essentially dictatorial. That is, one must pick some citizen and impose his standards on society. Such a procedure fits comfortably into an ipsedixist philosophy. The ideal observer approach can also be accomodated if the ideal observer is treated as an imaginary citizen. But this chapter is devoted to exploring the notion of a philosopher-king who must derive *all* his standards from the people.

In Section 2.3.4 of Volume I, it was observed that the assumptions required for Arrow's Paradox are overly strong. In this particular case, it is Arrow's condition of Unrestricted Domain that I think most needs to be challenged. This requires that the aggregation method work for all possible profiles of empathetic preferences. However, I agree with Harsanyi [233, p.60] that persons from the same sort of cultural background do in fact show a broad degree of consensus on how utils should be compared across individuals. If so, then Arrow's Paradox is not a problem for us. But, if some degree of consensus on interpersonal comparison exists, how did this come about?

Harsanyi [233] offers a Kantian-like answer. Behind his very thick veil of ignorance, people forget the empathetic preferences they have in real life and so find it necessary to adopt new empathetic preferences. He then appeals to a principle that has become known as the Harsanyi doctrine. The doctrine asserts that rational individuals placed in identical situations will necessarily respond identically.[35] The empathetic preferences that Adam and Eve adopt in the original position will therefore be the same and the problem of interpersonal comparison is solved. Since Adam and Eve will be led to the same standard for comparing utils, no difficulties can arise in taking this as their common standard.

Section 3.4.1 of Volume I comments at length on the Paradox of the Twins in an attempt to highlight the dangers of such reasoning. However, even if the argument could be adequately defended, how would poor mortals like ourselves be able to predict the empathetic preferences that rational superbeings would adopt behind the thick veil of ignorance envisaged by Harsanyi? Even if we could, why should we substitute these empathetic preferences for those we already have?

---

[35]Kolm [309] proposes a similar principle.

My own approach is quite different from Harsanyi's. I think that we have empathetic preferences at all only because we need them as inputs when using rough-and-ready versions of the device of the original position to make fairness judgments in real life. Insofar as people from similar cultural backgrounds have similar empathetic preferences, it is because the use of the original position in this way creates evolutionary pressures that tend to favor some empathetic preferences at the expense of others. In the medium run, the result is that everybody in a society tends to have the same set of empathetic preferences. In fact, although it is ostensibly devoted to utilitarianism, my chief purpose in writing this chapter is to offer a highly stylized account of the underlying mechanisms that I see as operating to this effect.

2.4   Evolutionary Ethics

This section begins a long aside from the reworking of Harsanyi's nonteleological defense of utilitarianism to which this chapter is largely devoted. The aside is necessary to introduce the naturalistic ideas with which I replace the Kantian foundations on which both Harsanyi and Rawls base their theories.

Until we return to the topic of nonteleological utilitarianism in Section 2.6, the notion of a philosopher-king can therefore be forgotten. He is, after all, only a fiction with whom I shall dispense altogether when presenting my own theory in Chapter 4. But the evolutionary ideas that follow would nevertheless remain valid if Nature had actually been kind enough to supply us with a philosopher-king to enforce our agreements. They will therefore be freely used when he reappears in Section 2.6.

Gott mit uns?   It will be evident from Section 2.3 that the interpersonal comparison controversy is an intellectual minefield. Almost anything one might say on the subject is liable to trigger some misconception or other. But matters get worse when evolutionary considerations are introduced. The mere juxtaposition of the words "evolution" and "ethics" is enough to make many philosophers foam at the mouth. Those of us who follow Darwin [143] in regarding the development of ethics as part of the natural history of mankind are either denounced as sinister eugenicists or else mocked as woolly minded relativists.

G. E. Moore [364] is still quoted as an authority for the claim that evolutionary ethics necessarily maintains that "we ought to move in the direction of evolution simply because it is the direction of evolution." It is then a short step to misrepresenting evolutionists as asserting that it is

Right that the weak should go to the wall because only the fit should survive. Some Social Darwinists of Victorian times held such views[36] (Section 2.2.1 of Volume I), but modern social Darwinists[37] like myself do not even believe that the teleological questions to which Moore thinks ethics should provide answers make any sense.

Evolutionary ethics denies that human societies exist to fulfill some great purpose. Evolutionists simply seek to understand why some types of human organization survive better than others. We may use this information in arguing that some reforms have a better chance of bringing about a lasting change in a society than others, but evolutionary ethics offers no authority whatsoever to those who wish to claim that some moral systems are somehow intrinsically superior to others. Evolutionary ethics is a theory of the Seemly—and what is seemly in one society may very well be unthinkable in another. Herodotus's [245, p.229] story of the Greeks and Indians brought together at the court of Darius to compare their funeral customs says everything that needs to be said on this subject. The Greeks were horrified to learn that the Indians ate their dead fathers. The Indians were no less horrified to learn that the Greeks burned theirs!

Such frank moral relativism leads some critics to complain that evolutionary ethics is entirely empty. As de Laguna [152] puts it, "It has been demonstrated again and again that the Darwinian theory will lie down with almost any variety of ethical faith."[38] Of course it will! How could an evolutionary explanation of the origin of ethical systems fail to be consistent with any ethical system that has actually evolved? Like everybody else, social Darwinists often do not *like* the mores of societies with aspirations very different from their own, but evolutionary ethics teaches us not to label a society as Wrong or Bad according to some supposedly absolute standard.

The basic misunderstanding is that traditionalists think that to refuse to label a Society as Wrong or Bad is to say that all societies are equally Right or Good. But a relativist finds no more meaning in the claim that

---

[36]Although I think that Dennett [158, p.393] is too responsive to modern critics of sociobiology when he describes the views of Herbert Spencer as "odious" and "heinous". Spencer was certainly not free of the prejudices of his time, and his scientific ideas are often both quaint and muddled, but the ideals revealed in his later writings seem much like those of other Victorian liberals. It is therefore a pity that he should be denied his proper place as the father of evolutionary ethics on the grounds that he was some kind of Nazi. Even his Lamarckism is not so crazy when viewed from the perspective of social evolution. To argue that memes are replicated from head to head by imitation is to take for granted that acquired characteristics can be transmitted.

[37]I know that calling myself a social Darwinist lays me open to accusations of being a follower of Adolf Hitler. Let me therefore observe that I take a modern social Darwinist simply to be someone who believes in applying Darwinian ideas to social questions.

[38]See Farber's [178] *Temptations of Evolutionary Ethics* for a litany of similar howlers.

two societies are equally Good than he does in the claim that one is Better than the other. The incomprehension deepens when traditionalists hear relativists like myself expressing our aspirations for the kind of society in which we would like our children to live. Where do we get our opinions from? What is the source of our authority? Why do we bother at all if "nothing matters"?

The answers are simple. Like everyone else, social Darwinists are just mouthpieces for the memes that have successfully replicated themselves into our heads. These memes seek to replicate themselves into other heads, because they wouldn't have survived if they didn't have this property. Being relativists, social Darwinists know perfectly well that other memes would be using us as instruments for their replication if we lived in different times or places. But why should our willingness to recognize such an obvious fact be thought to disqualify us from advocating reform? The opinions of those who claim the moral high ground are no less artifacts of their culture than ours. Nor are the underlying reasons that they seek to convert others any different.

When advocating reforms, traditionalists differ from social Darwinists only in claiming a spurious authority for their views. In seeking to persuade others, social Darwinists are constrained by the neo-darwinian meme to making pragmatic appeals to the self-interest[39] of our fellow citizens. But traditionalists tell themselves and others much more exciting stories in which our mundane arguments are trumped by appeals to Moral Intuition or Practical Reason or some other modern successor to Mumbo-Jumbo and the other gods who served as sources of authority for our ancestors. Sometimes, as when traditionalists invoke the Sanctity of Life as a reason for inflicting suffering on babies who have yet to be conceived, it is tempting to fight fire with fire by inventing new sources of authority, like Humanism or Gaia or Science—but the meme that is pushing this pen wants so much to be replicated that it won't even let me appeal to Intellectual Honesty in arguing against the creation of such new graven images.

Sociobiology and reductionism. Critics of social Darwinism who find the notion of a nonteleological ethics inconceivable, are only half the story. One must also beware of the enemies of reductionism, who wilfully misrepresent evolutionary theories in order to denounce them as simplistic. Such critics argue that evolutionary ethics is an attempt to reduce human morality to the crudest of biological imperatives. So successful has this campaign been that one must now refer to much of what used to be called sociobiology as evolutionary psychology or behavioral ecology if one hopes

---

[39]Broadly interpreted, as in Section 1.2.2 of Volume I.

to evade knee-jerk abuse.[40] But not even the most reductive sociobiologist ever held that evolution has written the rules of correct moral conduct into our genes. Sociobiologists believe in *co*-evolution—according to which our biologically transmitted capacities and propensities evolved *in tandem* with the socially transmitted customs and conventions that govern primitive societies (Wilson [559, 560], Lumsden and Wilson [336]). In brief, the genes and memes relevant to prehistoric moral behavior evolved *together*, and so cannot sensibly be studied in isolation.

The evolution of language seems a particularly compelling example of this phenomenon. Everybody knows of Chomsky's [120, 121, 122] claim that all human languages share a common deep structure. Pinker's [405] magnificent *Language Instinct* puts the case for the stronger claim that an innate capacity for language is actually carried in our genes. This is not to argue that Frenchmen carry genes for speaking French or that an American baby adopted by Japanese parents would find it any harder to learn Japanese than the natural children of his adoptive parents. Just as our immune system is not just a stockpile of specific antibodies, but a piece of biological hardware that enables our bodies to create antibodies as and when required, so the language instinct is a hardwired learning device that makes it relatively easy for toddlers to learn languages structured according to certain innate principles. But this mental hardware is very far from tying down all the details of a language. On the contrary, French and Japanese are such different languages because our genes leave a great deal to be determined by *social* evolution.

Having made a powerful case for the existence of a human language instinct, Pinker [405, p.420] joins Cosmides and Tooby [136] and numerous others in speculating about the extent to which the *tabula rasa* theory of human nature is mistaken on other counts. His claim that the human sexual drive is innate is presumably uncontroversial. He is also doubtless right in asserting that we are genetically programmed to shun contaminated foods, or to flee from danger in certain situations, or to turn and fight in others. The robotic character of our behavior when in the grip of such strong emotions as love, disgust, fear, or anger seems too transparent for serious doubt to arise on such questions.[41]

---

[40]See Tooby [522] or Barkow *et al* [43] for an account of the genuinely distinct features of evolutionary psychology. For behavioral ecology, see Barash [42], Hughes [263] or Krebs and Davies [310].

[41]Although a recognition of this fact of life is often combined with a belief that our capacity to feel emotion somehow makes us more than machines. But who has not observed the behavior of a love-sick teenager without being reminded of the mechanical behavior of a duckling relentlessly following whatever moving object first met its gaze on being hatched? Nor does age seem to disengage the autopilot. In short, the fact that we have an emotional life is entirely consistent with our being essentially machines.

It is more controversial to suggest that at least some of our capacity to empathize with our fellow men is instinctive. Plato defined man as a broad-nailed, featherless biped[42] with a gift for politics—and modern studies of chimpanzee or baboon societies would seem to confirm that Plato was right to bracket our status as a political animal with the genetically determined characteristics of our anatomy. Critics of sociobiology may mock such claims by treating them as equivalent to the political theory of Gilbert and Sullivan's *Iolanthe*, according to which ...

> Every boy and every gal
> That's born into this world alive,
> Is either a little Liberal,
> Or else a little Conservative.

But nobody at all thinks that we are genetically programmed to operate any *specific* political system—any more than we are programmed to call an apple a *pomme* or stack all the verbs at the end of a sentence. Like the language instinct, the political instinct operates at a deep structural level. As Hume [267] explains, it is natural that our societies should be governed by "natural laws", but the specific "natural laws" that operate in a particular society are not determined by Nature. What moral philosophers call "natural laws" are either human inventions or else the product of *social* evolution.

The moral sense. I have been stressing the importance of the concept of coevolution to sociobiology, in order to prepare the way for a discussion of the most controversial of the items on Pinker's [405, p.420] list of instinctive human traits. I agree with Pinker [405], James Wilson [561] and numerous others that we probably come into the world equipped not only with an intuitive grasp of the essentials of human psychology, but also with a hardwired sense of justice. This is not to say that our genes determine what will be regarded as fair in any particular society, only that they determine or constrain the algorithm that a society uses in deciding what is fair. But such an algorithm cannot operate without some input to chew on. I suspect that the necessary inputs are almost entirely socially determined. Since different societies have different social histories, it follows that they

---

Indeed, it is precisely when we are hopelessly in love or beside ourselves with anger that our behavior is most robotic. To experience such emotions is not to transcend the mechanical. On the contrary, to be hopelessly in love is to know how it feels to be a certain kind of stimulus-response machine!

[42]Broad nails were apparently added to the definition after Diogenes appeared with a plucked chicken.

will make different fairness judgments.[43] Even if Pinker is correct about the existence of a justice instinct, it therefore does not follow that ethics can be reduced to biology. On the contrary, with some important exceptions, I believe that almost everything in dispute among moral philosophers is determined by *social* evolution.[44]

Section 2.5 seeks to clarify where the boundaries lie between biological evolution and social evolution in the model I use to study social justice. I know perfectly well that the considerations that lead to the model are highly speculative and that the model itself is much too simple to come near capturing the richness of the human predicament. But, to repeat my standard excuse, one has to begin somewhere.

2.5   Evolution and Justice

The early part of the road to an understanding of the evolutionary origins of human morality is by now so well trodden that there seems little point in my attempting to do more than indicate some of the signposts along the way. Darwin [143] and Huxley [278] are more than a little lame on the evolution of ethics, but modern pathfinders like Alexander [11], Boyd and Richerson [98], de Waal [539], Hamilton [216, 217], Maynard Smith [347], Nitecki and Nitecki [387], Power [409], Trivers [524, 525], Williams [555], Wilson [559, 560], and Zahn-Waxler *et al* [574]. have stories to tell that are theoretically coherent and strongly supported by the evidence from the field. However, those new to the subject might do better to begin with the more popular books of Dawkins [151], Dennett [158], Cronin [138], Ridley [425], or Wright [567].

Paradoxically, it is only when such pilgrims on the evolutionary road get near their ultimate destination of modern man that their path begins to degenerate into a weed-strewn track of uncertain direction. It seems that we are able to observe other species and primitive human societies with a dispassionate objectivity that eludes our grasp when we turn our attention

---

[43]Elster [173] and Young [573] see different fairness algorithms using similar social inputs being used in different contexts. My guess is rather that similar fairness algorithms are being used with different social inputs.

[44]The exceptions mostly concern sex and the family. One ought also to include relationships within small close-knit groups that seemingly trigger the same biological mechanisms that evolution has provided for use within the family. Without noticing that humans all have pretty much the same genes, political philosophers sometimes debate such matters as whether we share a common human nature, but the biological underpinnings of our capacity to relate to the rest of the human race are more usually taken entirely for granted. Philosophically minded amoebas in slime mold colonies doubtless find it equally hard to envisage social organizations based on different biological realities.

upon ourselves. However, just as ontogeny is said to recapitulate phylogeny, so I think that there is hope of sorting out our confused intuitions about human morality by tracing their origins back to their humble beginnings. Rather than making attempts at squaring the circle of human cooperation as in Chapter 3 of Volume I, we need instead to follow the title of Singer's [495] pioneering work on evolutionary ethics by *Expanding the Circle* as illustrated in Figure 2.1.

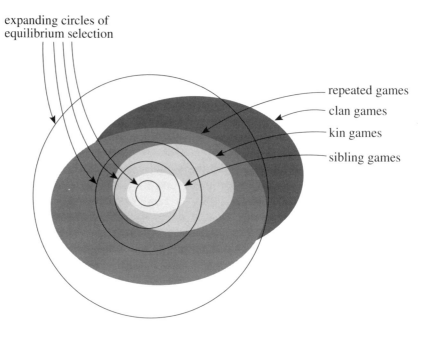

expanding circles of
equilibrium selection

repeated games

clan games

kin games

sibling games

Figure 2.1: Expanding the circle.

Singer traces the expansion of human moral horizons through the circles of kin selection, reciprocal altruism and group selection. Although I adopt his expressive metaphor, I differ from him in seeing the advance of human moral horizons in terms of the evolution of progressively more elaborate *equilibrium selection devices*. In spite of their importance, neither reciprocal altruism nor kin selection is concerned with how equilibria are *selected* in the Game of Life. Reciprocal altruism is about how equilibria in the

Game of Life are *sustained*.[45] Kin selection is about the manner in which *payoffs* should properly be calculated in family games.[46] As Figure 2.1 illustrates, reciprocity and kinship therefore form part of the ground on on which the expanding circles of an evolving morality need to be drawn. My circles therefore all lie in the domain of *group selection*—a subject on which Singer [495] is more than a little tentative because of a major controversy in biology over the use of this term.

Because of this controversy, it is necessary for me to stipulate that I share the orthodox view of Williams [555] that the definition of group selection proposed by Wynne-Edwards is based on a major misunderstanding about the way evolution operates. Wynne-Edwards [569] writes, "In the case of social group-characters, what is passed from parent to offspring is the mechanism, in each individual, to respond correctly in the interests of the community—not in their own individual interests—in every one of a wide range of social situations." In this definition, Wynne-Edwards overlooks the fact that groups whose members do not put their personal welfare before that of the group are vulnerable to invasion by individuals who do. However, one only has to read the title of Dawkins' [151] *The Selfish Gene* to realize that it is obvious that evolution must either operate at the level of the gene or not at all. The reason is that genes are what actually get replicated. As long as a mutant gene replicates faster than a normal gene, it will therefore take over whether or not the species as a whole benefits. In game-theoretic terms, Wynne-Edwards asks us to believe that Nature is one of the would-be circle-squarers castigated in Chapter 3 of Volume I for inventing fallacious arguments that supposedly show that it is rational to cooperate in the Prisoners' Dilemma.

But although Wynne-Edwards [569] was mistaken in the explanation he offered of the group phenomena his book describes, it does not follow that such phenomena do not evolve as a result of selection operating among groups.[47] Indeed, such mechanisms are routinely studied in the behavioral ecology literature.[48] But the mention of group selection nevertheless

---

[45]Since the altruism involved is then illusory, I prefer simply to speak of *reciprocity*. Insofar as true altruism is meaningful in a biological context, it is discussed under the heading of kin selection.

[46]The term kin *selection* is therefore misleading, and I prefer simply to speak of kinship. When players are related, the Game of Life they play makes their equilibrium selection problem easier to solve, but the mere fact of kinship doesn't solve the problem in itself. Hamilton [217] prefers to speak of inclusive fitness rather than kin selection.

[47]Nor, as Pollock [406] documents, does Wynne-Edwards' scholarship deserve the contempt with which it is treated in the literature.

[48]For example, Boyd and Richerson [99], Grafen [212], Nunney [393], Wade [540], Wilson [556], and Wilson and Sober [557]. Even Kropotkin's [311] attempt to justify anarchy against supposedly Darwinian criticism deserves to be treated with some respect.

generates knee-jerk hostility from those who insist that the term can be meaningful only if the known facts of genetics are denied. But what words are we then to use for processes that select among groups *without* contravening the laws of genetics? How are we even to discuss selection among *individuals*—for what is our genetic inheritance but a group of genes?

Fortunately, the nomenclature problem is easily solved for the purposes of this book. I replace Singer's [494] generalized references to group selection by the much more specific notion of *equilibrium selection* in a game. Such a usage has the advantage that "selfish gene" criticisms of the notion are obviously irrelevant. Restricting the domain of section to equilibria has the effect of excluding group organizations that are not internally stable, and therefore would not survive even if they were not competing with other groups. It may perhaps also serve to remind biologists that the games played by animals in real life typically have immense numbers of equilibria that occur in connected components. A population may drift through such a component of equilibria until it reaches an unstable equilibrium, whereupon selection can begin anew if the population is shocked slightly (Section 3.3.8). Such games pose the equilibrium selection problem in a particularly acute form.

2.5.1   Reciprocity

Before turning to the expanding circles of equilibrium selection, it is necessary to say something about the ground on which such circles are drawn— the sets of equilibria in the Game of Life from which a selection must be made. The ideas of reciprocal altruism (Trivers [524, 525]) and kin selection (Hamilton [216, 217]) are vital to an understanding of the range and variety of the menu of available equilibria, but very little will be said about reciprocal altruism in this chapter since it is the main topic of the next chapter.

As Section 2.2.6 of Volume I explains, it is possible for highly cooperative outcomes to be sustained by equilibrium strategies using I'll-scratch-your-back-if-you'll-scratch-mine principles in indefinitely repeated games. Each player continues to cooperate so long as his fellows reciprocate, but plans to switch to a punishment strategy should anyone deviate. The simplest such strategy is perhaps TIT-FOR-TAT in the indefinitely repeated Prisoners' Dilemma, whose pleasant properties have been widely advertised by Axelrod [31] in his celebrated book *The Evolution of Cooperation* (Section 3.2.5 of Volume I). However, the folk theorem of repeated game theory shows that TIT-FOR-TAT is just one of an enormous number of strategies that can be used to sustain cooperation among players whose motivations are entirely selfish.

But what have the equilibria of repeated games played by selfish agents to do with Trivers' [524, 525] notion of *reciprocal altruism?* Altruism should surely involve some element of self-sacrifice. However, no self-sacrifice is involved when Adam and Eve both play TIT-FOR-TAT in the indefinitely repeated Prisoners' Dilemma. They continue to cooperate by playing *dove* because they know that the victim of an exploitation attempt will retaliate by playing *hawk* until the deviant shows his contrition by cooperating again. But when the cooperative arrangement is working well, the darker side of their relationship will remain invisible to an observer. He will simply see Adam and Eve cooperating at every opportunity. A kibitzer may therefore be tempted to attribute their behavior to altruism. We are much moved, for example, by the mutual affection exhibited by pairs of lovebirds, but they arguably stick close to each other only because their partner is likely to be unfaithful if not watched continuously.

Reciprocal altruism is therefore something of a misnomer. One should rather say that the reciprocity mechanism makes some of the benefits of altruism available without the need for anyone to love his neighbor. As Section 1.2.2 of Volume I observes, Mr. Hyde is just as capable of getting on with his fellow men in repeated situations as Dr. Jekyll.

The possibility of reciprocal altruism assures us that Adam and Eve have a wide variety of cooperative outcomes that can be sustained as equilibria in an indefinitely repeated Game of Life. However, the question of how feasible outcomes are *sustained* is being downplayed in this chapter so that attention can be concentrated on how outcomes are *selected* from the set of all feasible possibilities. It seems to me that the origins of the equilibrium selection devices that we discuss under the heading of fairness or justice must be sought by looking first at how animals related by blood play the Game of Life that Nature deals them.

2.5.2 Kinship

Aristotle [16] tells us that it is within the household that we must first seek the "sources and springs of friendship, political organization, and justice". But only since genetics became a science has it been possible to do more than nod wisely at this good advice.

Hamilton's [216, 217] notion of *kin selection* is the foundation stone for the study of evolution and the family. People related by blood share genes. A gene that modifies some piece of behavior will therefore be replicated more often if it takes into account, not only the extra reproductive opportunities that the modified behavior confers on a host who carries the gene, but also the extra reproductive opportunities it confers on those among the host's relatives who also carry the same gene. The point was famously made in a semi-serious joke of J. B. S. Haldane. When asked whether he would give his life for another, he replied that the sacrifice would only be worthwhile if it saved two brothers or eight cousins!

Books on behavioral ecology like those of Barash [42], Hughes [263], Krebs and Davies [310], or Ridley [425] are ultimately convincing because biologists are able to appeal to thousands of case studies. Cooperative behavior in animals attributable to kin selection is very widespread indeed. African hunting dogs, for example, regurgitate food to help out a hungry pack brother. The evidence is particularly telling when special circumstances are revealed that explain why an apparent counterexample to the basic theory is in fact just one more supportive case. The explanation for the spectacular level of social organization among *hymenoptera* like bees and ants is particularly compelling. However, fascinating though it is, I plan to make no attempt to offer any kind of introduction to this huge literature. In particular, as everywhere else in the book, the games of sibling rivalry to be discussed are not intended to be realistic. Their purpose is simply to indicate why it would be a mistake to proceed on the assumption that blood relatives will act as though selfishly seeking to maximize their own reproductive potential at the expense of the prospects of other members of the family.

Little birds in their nest agree. Haldane's aphorism was based on the fact that we share half our genes with a brother and one eighth of our genes with a cousin. A gene that programs us to save a cousin therefore has one eighth of a chance of saving a copy of itself. Hamilton's [217] famous rule is an attempt to quantify the extent to which such considerations should be expected to result in one player making sacrifices on behalf of a relative. However, we shall have to wait some time before encountering this rule, and even then it will not be quoted in the standard form,[49] since it seems more useful for the purposes of this book to follow Bergstrom [55] or Myerson *et al* [375] in stating the results in the more abstract language of game theory.

In order to make a start on the kinship problem, imagine that children are always born in pairs. Two siblings, Adam and Eve, will then occupy the same nest in their infancy. In the nest they play a sibling rivalry game, the outcome of which affects their eventual ability to pass their genes to the next generation. To keep things simple, it will be assumed that the sibling rivalry game is symmetric, with only two pure strategies *dove* and *hawk*.

---

[49]Which says that altruistic behavior should be anticipated when $Br > C$, where B is the benefit to the recipient of the altruistic act, C is the cost of the act, and r is the degree of relationship between the benefactor and the beneficiary (who is assumed to be unique).

Some possible candidates for the sibling rivalry game are the Prisoners' Dilemma, Chicken[50] or the Stag Hunt Game of Volume I.

As in Section 2.2.1 of Volume I, the payoff $\pi(x, y)$ to Adam if he uses pure strategy x and Eve uses pure strategy y is to be interpreted as his biological fitness—the average number of extra offspring above some fixed background level π that he will produce as a result of the strategy pair (x, y) being played in the game.[51] Since the sibling rivalry game is symmetric, Eve's fitness is then $\pi(y, x)$.

When considering the evolutionary stability of a population (Section 3.2.3 of Volume I), it is important to bear in mind that the fact that Adam has fitness $\pi(x, y)$ and Eve has fitness $\pi(y, x)$ is a secondary consideration. The central issue is the rate at which the genes they carry are replicated. Matters are simplest in the case of a unisex species in which all siblings have the same genes as their mother and hence are clones. If the behavior of children in the nest is entirely genetically determined, all the children of the same mother will necessarily choose the same strategy. If normal families choose *dove* and mutant families choose *hawk*, the rates at which genes are replicated in normal and mutant families increase by $\pi(dove, dove)$ and $\pi(hawk, hawk)$ respectively. If the former exceeds the latter, a population of normals will be evolutionarily stable. Any bridgehead of mutants will then eventually disappear because mutant genes replicate more slowly than normal genes.

It is, of course, no accident that this reasoning parallels the argument behind the Paradox of the Twins given in Section 3.4 of Volume I. After all, Adam and Eve are indeed twins in this context. Evolution should therefore be expected to generate cooperation in the one-shot Prisoners' Dilemma, even though such behavior is not in equilibrium when Adam and Eve choose independently.

In more general games, one can summarize the behavior that a successful gene will instill in a population of clones using a crude version of Kant's categorical imperative: Maximize your payoff on the assumption that your siblings will choose the same strategy as yourself. In game-theoretic terms, the gene that ends up controlling the population will program Adam and Eve to optimize in the *one-player* game whose strategies are the same as in

---

[50]Biologists refer to Chicken as the Hawk-Dove Game. Volume I notes that Maynard Smith's [347] loose definition of the Hawk-Dove Game encompasses both Chicken and the Prisoners' Dilemma. If we stick by Maynard Smith's definition, neither the original nomenclature of game theory nor the new nomenclature of evolutionary biology need therefore be regarded as redundant.

[51]The fact that the number of children in a family is fixed at two does not contradict this interpretation of $\pi(x, y)$. Adam may raise several families in the breeding season should he survive so long.

Adam and Eve's sibling rivalry game, but in which the payoff that results from choosing x is $\pi(x, x)$. It would therefore be a mistake to use the *two-player* sibling rivalry game as the basis of a game-theoretic analysis of the behavior to be expected from Adam and Eve. Identical twins whose behavior is genetically determined are not independent players at all. The only true player is the single gene package that pulls their strings.

A more delicate analysis is necessary when studying sexual species. In the simple haploid case, the behavioral trait to be studied is determined by a single gene inherited with equal probability from the mother or the father. The population size will be assumed to be very large[52] and all individuals who survive to breeding age will be assumed to pair at random. To test a population for evolutionary stability, assume that a mutant gene has taken control of a very small fraction ϵ of a population that is currently pairing to raise families. The same fraction of their children will also be mutants, and so we must turn our attention to their grandchildren. The question is whether the fraction of mutants in this generation is greater or smaller than ϵ. If the latter, then the normal population is evolutionarily stable.

Since the number of mutant parents is small, a normal child will almost certainly come from a marriage in which both parents are normal, and a mutant child will almost certainly come from a mixed marriage with one normal parent and one mutant parent. As Bergstrom [55] observes, the question of evolutionary stability therefore reduces to comparing the fitnesses of normal children from normal marriages with that of mutant children from mixed marriages.[53] Since the sibling of a mutant child in a

---

[52]The population needs to be effectively infinite to justify the informal appeals to the law of large numbers that are standard in this context. See Binmore *et al* [87].

[53]To check the validity of criterion (2.1), we can begin by estimating the number N of grandchildren of all types and the number M of mutant grandchildren. If (2.1) is correct, it must be equivalent to the requirement that $M/N < \epsilon$ for sufficiently small values of $\epsilon > 0$. To this end, let F be the number of families and let $p(x, y) = \pi + \pi(x, y)$ be the total number of children that an individual will have on average if he uses strategy x and his sibling uses strategy y.

The fraction of marriages in which both parents are normal is $(1 - \epsilon)^2$, the fraction in which one parent is normal and the other mutant is $2\epsilon(1 - \epsilon)$, and the fraction in which both parents are mutants is ϵ^2. If we neglect terms of order ϵ or higher, it follows that we can proceed as though all marriages are normal in estimating the total number N of grandchildren of all types. Thus N is approximately $2Fp(dove, dove)$. In estimating M, we need to retain terms of order ϵ, but neglect all higher order terms. We can therefore proceed as though all mutant children come from mixed marriages, of which there are approximately $2\epsilon F$. For the same reason, the number of mutant grandchildren—who will nearly always be born into a mixed marriage—is approximately the same as the number of mutant children. One quarter of the time, both children from a mixed marriage will be normal and so we have nothing to count. One quarter of the time, both children will be mutants. The number of mutant grandchildren deriving from such a family is $4\epsilon Fp(hawk, hawk)$. One half of the time, one child will be normal and the other a mutant.

mixed family has half a chance of being a fellow mutant and half a chance of being normal, a criterion for a normal population to be evolutionarily stable is therefore

$$\pi(dove, dove) > \tfrac{1}{2}\pi(hawk, hawk) + \tfrac{1}{2}\pi(hawk, dove) . \qquad (2.1)$$

Criteria of this kind seem to have been first obtained by Grafen [211] and are used by behavioral ecologists to modify Hamilton's [217] "inclusive fitness" to a new notion of "personal fitness" (Hines and Maynard Smith [248, p.20]). However, both of these modifications to the classical notion of biological fitness are mentioned here only so that I can disclaim any intention of referring to them again. In what follows, fitness always refers to the reproductive success of a particular individual.

Bergstrom [55] refers to (2.1) as a semi-Kantian criterion, since the rule of behavior that a successful gene will instill in a haploid population is a hybrid of Kant's categorical imperative and the rule that was lightheartedly said in Section 2.4.3 of Volume I to be Nash's categorical imperative. The hybrid rule instructs you to maximize your payoff on the assumption that half the time your siblings will choose the same action as yourself and half the time they will not react to your behavior. Bergstrom [55] describes the outcome as a symmetric, strict[54] Nash equilibrium of the two-player game whose strategies are the same as in Adam and Eve's sibling rivalry game, but in which the payoff that Adam gets when he chooses x and Eve chooses y is

$$V(x,y) = \tfrac{1}{2}\pi(x,x) + \tfrac{1}{2}\pi(x,y) . \qquad (2.2)$$

However, I think one does better to conceive of the situation as a leader-follower game[55] in which the leader first chooses y with the aim of maximizing $\pi(x,x)$ and the follower then responds by choosing x with the aim of maximizing $\tfrac{1}{2}\pi(x,x) + \tfrac{1}{2}\pi(x,y)$. After the choice of y, the follower is then involved in a *one-player* game.

The Kantian nature of the intrafamilial ethics generated by considerations of this kind is apparent when the sibling rivalry game is the one-shot Prisoners' Dilemma. If Adam and Eve were really playing this game in

---

Since only one of the children in such a family will be producing mutant progeny, the number of mutant grandchildren is then $2\epsilon Fp(hawk, dove)$. The total number M of mutant grandchildren is therefore approximately

$$\tfrac{1}{4}\times 4\epsilon Fp(hawk, hawk) + \tfrac{1}{2}\times 2\epsilon Fp(hawk, dove) .$$

It follows that $M/N < \epsilon$ for small values of $\epsilon > 0$ if and only if (2.1) holds.

[54]In a Nash equilibrium, everybody is playing a best reply to the strategies chosen by their opponents. In a strict Nash equilibrium, nobody has any alternative best replies.

[55]To defend the leader stage of the game, one needs to appeal to some sort of equilibrium selection argument of the type discussed in Section 2.5.3.

the nest, then they would both choose *hawk*, since this strongly dominates *dove*. But the real players are the genes that control their behavior and, as we have seen, they have a different game to play. When Adam and Eve are clones, the single gene package that controls them will program them to cooperate by playing *dove*, for the reasons that Kant gave to motivate his categorical imperative. When they are siblings in a haploid population, they will be programmed to employ the semi-Kantian rule. If $2y > x + z$ and $y > z$ in the general one-shot Prisoner's Dilemma of Figure 2.2(a) of Volume I, the result will be that both cooperate by playing *dove*. They will similarly play like doves when the sibling rivalry game is Chicken or the Stag Hunt Game.

Of course, the situation analyzed above has been absurdly oversimplified. Sibling rivalry games have an indefinite number of players and need not be symmetric. Nor is the degree of relationship between the players likely to be sharply defined, if only because of the possible presence of cuckoos. Animals do not usually mate at random, nor are populations effectively infinite. Even the notion of a gene as an atomic entity becomes suspect when the molecular realities are closely studied. Above all, the manner in which genes control behavior must surely be much more complicated than the manner in which they control eye color. The human species is not even haploid.[56] However, we need to turn next to an even more fundamental difficulty that kicks in when the human species is under discussion.

Learned or instinctive behavior? The preceding discussion was set against a background of baby birds sharing a nest in order to emphasize that Adam and Eve's behavior was assumed to be entirely under the control of their genes. My reading of the literature suggests that sociobiologists sometimes forget their fundamental commitment to coevolution when reacting against the *tabula rasa* theory popular among social scientists by taking the same assumption for granted when extrapolating sociobiological conclusions to humans. In the language of Section 2.4.2 of Volume I, they then treat *homo sapiens* as *homo behavioralis*—a stimulus-response machine programmed directly with behavior like a chocolate-dispensing machine. The temptation is to follow Kant [297, p.63] in thinking that

---

[56]We are a diploid species which carries *two* genes at each locus. Even in a single-locus model, it is therefore necessary to take account of two genes. If the mutant gene is dominant, the necessary considerations are essentially the same as in the haploid case, but matters become more complicated when the mutant gene is recessive. Bergstrom [55] shows that $V(x,y)$ must then be replaced by $W(x,y) = \frac{1}{5}\pi(x,x) + \frac{3}{5}\pi(x,y) + \frac{1}{5}\pi(y,x)$. Bees and ants are haplodiploid—unfertilized eggs produce haploid males and fertilized eggs produce diploid females, which goes a long way toward explaining why such species can reach such high levels of sociality (Hamilton [217], Alexander *et al* [12]).

any mundane purpose humans might exist to pursue, such as ensuring that the genes which pull their strings are passed on to the next generation, would be best served by hominids who do not reason at all, but are simply hardwired with the optimal response to each relevant stimulus.[57]

But whatever sociobiologists may or may not believe about the extent to which human behavior is instinctive, my own view is that the Game of Life is too complicated for us to be hardwired with optimal strategies for all its subsidiary games. For most purposes, our genes therefore do not program us directly with behavior like *homo behavioralis*. It seems more realistic to proceed on the assumption taken for granted by folk psychology that our cognitive processes really do involve some use of the preferences and beliefs of *homo economicus* that revealed preference theory treats as convenient fictions.

It is important to recognize that, in postulating the actual existence of preferences and beliefs, I am abandoning the agnosticism of revealed preference theory maintained up to this point. Instead of just saying that Adam and Eve behave *as though* they were maximizing a utility function subject to their beliefs, I shall now be claiming that Adam and Eve sometimes *really do* maximize a utility function subject to their beliefs—albeit seldom in a calculated manner or with conscious intent.

Many of our personal preferences are doubtless genetically determined, like hunger, thirst and the sexual urge. Perhaps some of our beliefs are also hardwired—making the world not only stranger than we imagine, but stranger than we can imagine.[58] But some preferences and most beliefs must surely be acquired. That is to say, our genes do not always insist that we prefer or believe specific things; in some contexts they insist only that we organize our cognitive processes in terms of preferences and beliefs. On this view, we come equipped with algorithms that not only interpret the behavior patterns that we observe in ourselves and others in terms of preference-belief models, but actively build such models into our own operating systems. As Hume [267, p.422] remarks, "Nothing has a greater effect both to increase and diminish our passions, to convert pleasure into pain, and pain into pleasure, than custom and repetition."

---

[57] As Section 2.4.2 of Volume I records, Kant [297, p.64] thought that reason was therefore superfluous to our mundane needs and hence must exist to serve the transcendental purpose of creating a "will which is good". Such a will, so he argued, would necessarily operate his categorical imperative. As usual, trying to follow Kant's thoughts is like entering a Looking-Glass World. For example, we have just seen that a defense can be mounted for the type of intuition which led him to the categorical imperative in the case when the players belong to the species *homo behavioralis*. But this is precisely the case his own argument excludes.

[58] This is a paraphrase of an aphorism of J. B. S. Haldane.

It seems likely that the psychological mechanisms involved in learning new behavior or acquiring new preferences or beliefs are many and varied. However, I plan to speculate only about the minimal set of psychological mechanisms that seem consistent with the story I have to tell about human kin selection.

It is probably uncontroversial to suggest that we are natural imitators. Like proverbial monkeys, we tend to copy what we see others doing, whether the behavior makes much sense or not. But neither humans nor monkeys are totally uncritical. We test our newly acquired behaviors against our preferences, as expressed through our emotional responses. In short, we ask ourselves whether we like the consequences of our new behavior. If the behavior is found wanting, we seek to refine it, or else return to a tried-and-true alternative.

But where do we get the preferences used for this purpose when these are not wired in at birth? My guess is that we come equipped with algorithms that operate *as though* they were employing the principles of revealed preference theory to deduce preference-belief models from consistent sets of behavior. The process being proposed is recursive rather than circular. A well-established set of behaviors employed in a particular context is first encoded as a preference-belief model. This preference-belief model is then used to test or refine new behaviors[59] before they are admitted into an individual's repertoire of habituated responses. An adjustment period follows in which behaviors are refined until full consistency is achieved, whereupon a revised preference-belief model is constructed.

The evolutionary advantage of such an inductive process[60] is that new behaviors are tested against past experience in an internal laboratory before being put to use in the gladiatorial arena of life. If the environment is sufficiently stable that our past experience is relevant to present challenges, it then becomes possible to assimilate new behaviors quickly without taking large risks. Since a successful innovation by one individual can swiftly spread through a whole culture by imitation, the mechanism therefore makes us an unusually flexible animal.

But a price has to be paid for our flexibility. As Section 2.4.2 of Volume I observes, the fact that we must learn how to behave makes us a second-

---

[59]Such new behaviors may be constructed *de novo* from the preference-belief model, but calculating behavior of this kind must surely be comparatively rare compared with behavior acquired through imitation or serendipity.

[60]The idea that some such process is part of our biological heritage has become known as evolutionary Kantianism. However, Hume [267] seems to have a better grasp of reality when he tells us that, although we have no rational grounds for the principle of scientific induction, we nevertheless have no choice but to behave as though it were true. Stories like mine perhaps begin to explain why.

best species, in that Nature loses fine control over the way we play most games. In particular, for the reasons given in Section 3.4 of Volume I, human identical twins cannot emulate the behaviorally hardwired species for which the Paradox of the Twins can be made to work. If the preferences which mediate human behavior adequately reflect success or failure in the evolutionary race, then I shall shortly be arguing that we are condemned to choose *best replies* in the games we play—or else be displaced by rivals who do. But if everyone chooses a best reply, then the result will be a Nash equilibrium of the game.

Since our nature forces us to play Nash equilibria in games, our species must live with the cost of being unable to sustain first-best outcomes in games like the one-shot Prisoners' Dilemma. On the other hand, we are able to avoid the fate of a behaviorally hardwired species when playing games to which its members are not adapted. Rather than playing whatever third-best strategy might happen to be triggered by the available stimuli, humans enjoy the benefit of having the potential to learn a second-best strategy in any game whatsoever, whether it figured large in our evolutionary history or not.

Although a man who operates on the psychological principles I have been proposing will eventually respond to a new challenge as though he were *homo economicus*, it is important to insist yet again that it does not follow that he is more than dimly aware of what is going on in his head— even when the stories he tells himself and others about his inner life are entirely consistent with his observed behavior.

Introspection seems particularly problematic when trying to assess the extent to which we sympathize with the fate of our relatives. How much do I love my brother? Is "love" even the right word to describe my feelings? Personally, I find myself unable to give adequate answers to such questions—and I suspect that the more definite views offered by others owe more to romantic fiction than genuine self-knowledge. In accordance with the model I have been describing, it seems to me that we unconsciously learn how we feel toward our relatives by experimenting with different behaviors and observing the emotional responses of our bodies. At the end of the process, I may have no better way of articulating what I have learned to feel than saying that I love my brother so much that I am even prepared to lend him my blue suede shoes to impress his date tonight. But the fact that introspection seldom allows us to quantify our emotions adequately in a more direct way does not imply that they do not control our behavior very closely. On this subject at least, folk psychology surely hits the nail right on the head.

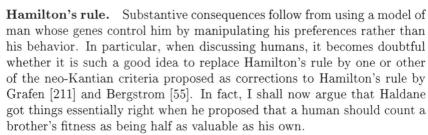

Hamilton's rule. Substantive consequences follow from using a model of man whose genes control him by manipulating his preferences rather than his behavior. In particular, when discussing humans, it becomes doubtful whether it is such a good idea to replace Hamilton's rule by one or other of the neo-Kantian criteria proposed as corrections to Hamilton's rule by Grafen [211] and Bergstrom [55]. In fact, I shall now argue that Haldane got things essentially right when he proposed that a human should count a brother's fitness as being half as valuable as his own.

The first step is to justify the claim made above that humans are condemned to play Nash equilibria in the Game of Life. I think that this is most easily seen by comparing what is involved when a human deviates from a socially determined norm with what happens when a mutant baby bird deviates from a genetically determined norm.

Imagine that Adam and Eve are equipped with preferences that are either genetically determined or else have been distilled from habituated behavior acquired in the past. In either case, it is important that the preferences remain relatively stable as new behavior is assimilated through a combination of individual trial-and-error adjustment and social imitation. I follow Aumann [25] in seeing the essence of such interactive learning as a two-stage process in which we first receive a social signal that tells us how to behave, and then test the behavior against our preferences to see whether we wish to follow its recommendation. Section 3.4.1 of Volume I discusses the notion of a correlated equilibrium that Aumann [25] built on this foundation. But matters are simpler here, since a player in a prehistoric family game is unlikely to have been able to receive signals from society of which his relatives were unaware. Operating a correlated equilibrium under such circumstances just reduces to specifying how a Nash equilibrium is to be selected. For example, in the Battle of the Sexes of Figure 2.1(c) in Volume I, the social signal might simply specify that the Nash equilibrium (*dove, hawk*) be played. But it could require that a coin be tossed, with the equilibrium (*dove, hawk*) being played if it falls heads and the equilibrium (*hawk, dove*) if it falls tails.

It is important that the social norm in use finally advocates the use of a Nash equilibrium, because the players are assumed to test whatever recommendation is made to them against their preferences. In practice, this means that they will occasionally experiment, either hypothetically or actually, with strategies that have not been recommended, in order to discover whether they can thereby gain a greater payoff. As discussed at great length in Section 3.4.1 of Volume I, deviations by different players must be expected to be *uncorrelated*. Even if Adam and Eve are identical twins, the fact that Adam happens to try out a deviation from the social norm just before bedtime on Tuesday provides no reason for supposing that Eve will simultaneously select precisely the same moment to deviate. The condition for the social norm to be stable or self-policing is therefore that it recommend that each player make a best reply to the behavior it recommends to the other players. In other words, the social norm must coordinate the players' behavior on a Nash equilibrium. This situation contrasts sharply with the case of behaviorally hardwired identical twins. The presence of a mutant gene in one player then guarantees its presence in the other. A deviation from the norm induced by the mutant gene in one player will then be matched exactly by a precisely similar deviation in the other.

Given the model proposed above for human behavior, what preferences will Nature write into the Game of Life played by relatives whose degree of relationship is r? If Adam and Eve are identical twins, $r = 1$ because they then share all their genes. If they are nonidentical twins, $r = \frac{1}{2}$ because they then share half their genes. If they are supposedly brother and sister, then r is something less than a half because of the risk of some unfaithfulness on the part of their mother. If they are known to have the same mother but different fathers, then $r = \frac{1}{4}$. If they are known to be full cousins, then $r = \frac{1}{8}$. If they are members of a wider kingroup, then r is some smaller but positive number.

In all these cases, it seems to me almost tautological that evolution will eventually program Adam and Eve with personal utility functions u_A and u_E that are computed from their respective fitnesses using Hamilton's rule in the form:

$$
\begin{aligned}
u_A(x,y) &= \pi(x,y) + r\pi(y,x)\,, \\
u_E(x,y) &= \pi(y,x) + r\pi(x,y)\,.
\end{aligned}
\tag{2.3}
$$

We will then be dealing with a case in which an individual's personal preferences are not narrowly selfish. When Adam and Eve are relatives, they will *sympathize* with each other's reproductive aspirations. Each player explicitly includes his relative's biological fitness as an argument in his personal utility function, which then determines his payoffs in the Game of Life. As a result, one is likely to see siblings cooperating even when their sibling rivalry game is the one-shot Prisoners' Dilemma. One can argue that they are nevertheless not behaving altruistically, since each is in fact optimizing in the Game of Life that they actually are playing. However, we shall get no closer to true altruism than in the kin-selection examples of this section.

The reasoning that leads to (2.3) is much simpler than that which led to the semi-Kantian rule embodied in (2.2). In the case of baby birds in their nest, the very fact that a mutant gene is planning to cause a player to deviate provides the gene with information. The gene learns that there is a significant probability that the other player will deviate simultaneously.

Equally importantly, it knows that if the other player does match the deviation, then both players will be carrying the mutant gene and so its payoff will be doubled.[61] On the other hand, if the other player doesn't match the deviation, then he isn't carrying the mutant gene and hence his fitness does not contribute to the gene's payoff.

Such complexities do not apply in the human case—nor presumably, to primates and other animals that transmit their culture from one generation to the next. After Adam deviates, a gene that modifies his preferences knows no more about its presence in Eve than it did before. If the degree of relationship between Adam and Eve in a human family is r, then one of Adam's genes will continue to believe that it is present in Eve with probability r even after Adam has deviated from the social norm. Assuming that Adam is genetically programmed to maximize something, the propagation rate of a controlling gene will therefore be optimized if the formula used in calculating whatever is to be maximized is (2.3). Similarly, given that the propagation rate of Adam's genes is optimized by maximizing some particular function, a gene that controls Adam's learning algorithm will do best if the algorithm eventually teaches him to maximize that function.

As usual, the benchmark question is how things go when the sibling rivalry game is the one-shot Prisoners' Dilemma. This is not a new issue, since the Game of Life obtained by transforming the Prisoners' Dilemma using Hamilton's rule in the form (2.3) appears in Figure 2.5(c) of Volume I as one of the various games that are sometimes substituted for the Prisoners' Dilemma by circle-squarers.[62] Such critics are right to argue that it is rational to cooperate in the transformed game when $r \geq \frac{1}{2}$, since *dove* then dominates *hawk*. They only go wrong in supposing that this conclusion is relevant to cases in which the players do not sympathize with each other, but really are playing the one-shot Prisoners' Dilemma.

The units in which biological fitnesses are expressed are expected numbers of offspring. However, nothing says that Adam and Eve must value each other's children equally. Indeed, Hamilton's rule requires that each player value his sibling's children at half the rate at which he values his own. When they are siblings, Adam therefore regards Eve's units of fitness as being worth half his own units of fitness. Similarly, Eve regards his units of fitness as being half as valuable as hers. One might then say that they act as semi-utilitarians. Just as a semi-Kantian chooses x to maximize $V(x, y) = \frac{1}{2}\pi(x, x) + \frac{1}{2}\pi(x, y)$, so a semi-utilitarian chooses x to maximize

---

[61]The principle is the same as that employed in the notion of evolutionary stability described in Section 3.2.3 of Volume I, except that the probability that the mutant is playing itself becomes much larger within the family context.

[62]In Figure 2.5(c) of Volume I, the degree of relationship is y rather than r.

$V(x,y) = \frac{1}{2}W_H(x,y) + \frac{1}{2}\pi(x,y)$, where $W_H(x,y) = \pi(x,y) + \pi(y,x)$ is the classical utilitarian social welfare function.

Adam and Eve share a common standard for making interpersonal comparisons of their fitnesses only in the extreme case when they are identical twins, so that $r = 1$. They then become classical utilitarians, whose personal payoffs in their Game of Life are found simply by adding their biological fitnesses. The Game of Life then becomes a common interest game, of which the Dodo Games of Figures 4.4(b) and 4.4(c) of Volume I provide examples. In a common interest game, Adam and Eve receive equal payoffs whatever the outcome of the game may be. No scope for conflict therefore exists. Their joint interest lies in maximizing the sum of their fitnesses, even if this means that Adam must lay down his life for Eve.

Unlike Kantian identical twins, whose behavior is genetically determined, a human pair of identical twins continue to play a *two-person* Game of Life. Regan [420] finds a problem for naive formulations of utilitarianism in the fact that such a common interest game may have Pareto-inferior equilibria (Figure 2.8 of Volume I). I don't see that classical utilitarians need lose any sleep over this problem, since the "greatest good" is clearly achieved at a Pareto-efficient equilibrium of the game. However, evolution does not care about the "greatest good" and so some accident of history may trap a human pair of identical twins into coordinating on a Pareto-inferior equilibrium of the Game of Life. Their family line will then be vulnerable to elimination by equilibrium selection as families who succeed in coordinating on a Pareto-superior equilibrium forge ahead in the reproductive race.

Our study of kin selection began with Kant's categorical imperative and has now reached classical utilitarianism. Does this fact have anything to say about the origins of the moral intuitions that philosophers debate so fiercely? I think it does. Indeed, as we shall see, I think that nearly all the supposedly universal moral principles in which we vainly instruct our children have their origins in the social contracts maintained by extended families or small tribes.

Extending the family. Returning to the problem of sibling rivalry, how does Adam know that Eve is his sister? More generally, how do people know the precise value of r to assign to all the different people within their family circle?

There are two possible answers. The first is that we receive a genetic signal that provides the necessary information. It may be, for example, that members of Adam's family always have green beards and that he can judge the extent of Eve's relationship to himself by comparing the shade of her beard with his own. The extreme case would require each player's genetic code to be printed on his forehead as envisaged by Howard [261], or for players to have transparent skulls as proposed in Section 3.2.2 of Volume I by Gauthier [196].

Some species can apparently sort out their siblings from strangers in some such way. Perhaps they can taste or smell the necessary genetic differences. However, it seems unlikely that humans operate in this manner. The evidence from the sexual preferences of unrelated children brought up together seems rather to suggest that we identify as relatives whomever we happen to encounter within the family circle.[63] The second answer is therefore that we deduce our degree of relationship to others from the extent to which we find ourselves thrown into their company. Thus Eve deduces that Adam is her brother from the fact that he is always hanging around at mealtimes competing for attention with her mother.

If this conjecture is correct, it has important consequences for the evolution of ethics. It means that the mechanisms that promote altruism within the family may also be triggered within a sufficiently close-knit group of unrelated individuals. The usual examples given are of an army platoon under combat conditions or a teenage street gang. In the *clan games* played by such groups, my guess is that the brotherhood of man becomes more than an empty metaphor and actually describes quite closely how members of small insider-groups relate to each other. Indeed, the formal brotherhood rituals used to adopt strangers into both primitive tribes and teenage gangs probably capture quite closely the biological mechanism being tapped. Even when the bonding rituals seem barely perceptible, one also feels curiously obligated even to old school friends or office colleagues, with whom one may otherwise have little else in common. Even establishing eye contact with a beggar in the street somehow creates enough inner discomfort at neglecting a fellow human being that we are sometimes moved to hand over our small change with no prospect of any recompense.

The comradeship and mutual trust to be found between friends and neighbors creates a danger for liberals. We are tempted to assume that the good fellowship that operates within the various small cliques to which we belong can be extrapolated to society in general. However, we are not programmed to treat strangers like members of the family and only disaster can be expected from utopian projects that proceed as though matters were otherwise. Dunbar [164, 165] offers evidence which suggests that the maximum size for a surrogate family of the type under discussion is of the order of a hundred individuals or so. He speculates that further growth in

---

[63]Children brought up together in communal nurseries on old-style kibbutzim very seldom showed a sexual interest in each other at a later stage.

the size of such clans is inhibited by the inability of our neocortex to handle the enormous volume of information necessary to keep track of the social relationships operating between large numbers of individuals.

Such a view is incompatible with Elster's [171] claim that love and duty are the cement of society. The unwelcome truth is that love and duty can be relied upon only within the family or in small groups whose atmosphere adequately mimics a prehistoric family environment. It is for this reason that Section 1.2.2 of Volume I rejects *homo ethicus* as a model of man outside these settings. To understand the social contracts that operate in large modern states, we need to look for mechanisms that will keep specimens of *homo economicus* cooperating without the need to appeal to an instinctive drive toward altruism.

Intergenerational transfers. In discussing kin selection, I have concentrated on sibling rivalry games, since it is Nature's solutions to such games that presumably provided the springboard for the later evolution of our more general notions of fairness and justice. But the relationship between parents and their descendants also has a role to play, since it provides a bridge between those alive today and the unborn generations of the future. With so much else to discuss, I have chosen to sideline this important issue, and so little more is offered here than a general indication of the approach I have in mind to intergenerational questions. Readers who omit the material altogether by skipping forward to Section 2.5.3 will not therefore feel the lack at a later stage.

A child shares half its genes with a parent and hence the degree of relationship between them is $r = \frac{1}{2}$. However, adulterous affairs in supposedly monogamous species turn out to be surprisingly common,[64] and so fathers presumably estimate their degree of relationship to their ostensible children at a significantly lower figure than their mothers. Matters are also complicated by the fact that an individual's fitness varies considerably with age. An aging mother may have little chance of reproducing herself again, while a newborn baby must first make it to puberty before it can breed. If money or food intake or some other variable is being used as a surrogate for fitness, such age considerations may make it necessary to multiply r by a discount factor that differs significantly from one. Nor does Nature seem very ready to relinquish instinctive control of the behavior of a mother and her babies in the game of life that they play together.

Applying Hamilton's rule therefore seems very much less safe than in games of sibling rivalry. However, perhaps we will not go too far astray

---

[64]Using genetic testing, Baker and Bellis [38] found that less than 80% of the children in a Liverpool high-rise were actually the issue of their ostensible fathers.

if we confine its application to the relationship between parents and their adolescent or adult children, and are not too specific about how degrees of relationship and discount factors are to be calculated. Macroeconomists routinely follow Barro [45, 47, 46] in taking this line. That is to say, they treat problems of intergenerational transfer by assuming that the current generation seeks to maximize a discounted sum of its own utility and that of all future generations.

In a sufficiently simple economy, such intergenerational altruism turns out to have the effect of neutralizing attempts by the government to alter society by transferring resources from one generation to another, since the result will simply be that people alter their wills to restore the *status quo* that previously held sway.[65] If there is sufficient interbreeding between our descendants that some of my ancestors will necessarily marry some of yours, one can even show that attempts by the government to transfer resources between members of the current generation will be nullified by a similar mechanism (Bernheim [57], Bernheim and Bagwell [58]). If taken at face-value, I guess that such results would remove all interest from the problem of justice in a free society. However, the models on which the results are based are so simplistic that they do no more than warn utopians that people will not sit idly by while their wealth is redistributed. They will respond instead in whatever way they can to keep things as they like them. In brief, reforms are pointless if they do not coordinate behavior on an equilibrium.

In any case, I do not believe that these results from macroeconomics defuse the problem of intergenerational equity raised by Rawls [417, p.284]. He proposes to solve the problem by requiring that we bind ourselves to the savings principle that would be agreed if all citizens from the current generation and all future generations were to bargain in the original position without knowledge of the generation to which they belong. Section 1.2.8 of Volume I outlines my difficulties with such an approach. In short, nobody could be more powerless than someone who has yet to be born. Only people alive right now have power to exercise, and so it is only their preferences that are relevant when computing the equilibria in the Game of Life that serve as candidates for a social contract. But it does not follow that a social contract will be chosen that takes no account of the needs of future generations. We all sympathize with our children to a greater or lesser degree. Our personal preferences therefore include a provision for their welfare. The bargain reached behind the veil of ignorance by those with actual power to wield will therefore not neglect their interests. Indeed, as the macroeconomic models of intergenerational transfer show, their interests may sometimes override other considerations altogether.

---

[65]An important proviso is that attention be confined to *interior* equilibria.

How many people? As an example of an intergenerational problem, consider the utilitarian dilemma of whether one should maximize total utility or average utility (Blackorby and Donaldson [93], Blackorby *et al* [92], Dasgupta [147, 148], Roemer [435]). If we can choose how many people there should be, it makes a great deal of difference which of these two quantities is maximized. Maximizing average utility creates a society with fewer people than maximizing total utility.[66]

I do not plan to come down on one side or the other in the debate between total and average utilitarianism, because I think it misconceived. There is no point in indulging in abstract disputation about what brand of utilitarianism is truly kosher. Like everything else, this is something that should be settled in the original position.

The simplest case of any interest involves a society that contains only cloned copies of Adam and Eve, who serve as delegates for each clone grouping in the original position. Adam and Eve will not be assumed to be shortsightedly concerned only with maximizing the number of children born to the current generation, as in the sibling rivalry games discussed above. Nor do they care directly about how many copies of themselves may be in existence at any time. Since they are clones, they care about the number of Adams and Eves alive at any time only to the extent that this is relevant to their true aim, which is to maximize the expected length of time for which their genetic code survives. Too few copies of Adam alive at any given time will make his tribe vulnerable to natural disaster or aggression from Eve's tribe. Too many people overall will wastefully consume resources and pollute the environment.

Behind the veil of ignorance, Adam and Eve will therefore reach some compromise that includes a specification of how many copies of each type should be permitted to exist under all possible contingencies. This compromise will be Pareto-efficient. No other arrangement would therefore allow one tribe a longer expected survival time without reducing the expected survival time of the other. The share of the total population that each tribe gets will depend on its relative power and the assumptions made about the circumstances under which the tribes bargain behind the veil of ignorance. Under the various assumptions of this chapter, the outcome will be utilitarian. But, even in this absurdly simplified example, the outcome

---

[66]If the utility of a citizen in a society with N people is $N^a e^{-bN}$, then total utility is maximized when $N = a/b$. Average utility is maximized when $N = (a-1)/b$. If b is small, the difference can be immense. In extreme cases, one may be led to Parfit's [399] "repugnant conclusion", according to which an overcrowded world in which everybody is only marginally better off than being dead is regarded as superior to a world with fewer people in which everybody enjoys a comfortable standard of living.

will be calculated in a much more complicated way than either total or average utilitarians envisage.

2.5.3   Equilibrium Selection

In game-theoretic terms, a necessary condition for a group to be stable against deviations from within is that it operate an equilibrium of whatever Game of Life its members play with each other. However, games typically have many equilibria. Some of these equilibria will be better than others, in the sense that animals from a group using one equilibrium may do better on average than animals from a group using another equilibrium. This will certainly be the case if the equilibrium used by the first group is a Pareto-improvement on the equilibrium used by the second. Groups using a Pareto-superior equilibrium will therefore grow in size or number at the expense of groups using a Pareto-inferior equilibrium. Eventually, the inferior groups will disappear. Evolution will then have operated to favor one kind of group organization over another.

The length of time it takes to eliminate a Pareto-inferior equilibrium currently being operated by all groups will depend on how likely it is that any single group will spontaneously shift to a Pareto-superior equilibrium. If mutation is the only source of relevant variability, the expected waiting times may well exceed the expected lifetime of the species. For example, in the order *hymenoptera* which includes ants, bees and wasps, we find different species operating many levels of social contract, from the Hobbesian "war of all against all", in which solitary individuals seek to maximize their own fitness with no regard to the help that they may give or receive from others, to the Rousseavian ideal, in which all individuals in a colony seemingly subordinate their own self-interest to a "general will". Evolutionary biology therefore provides no guarantee that a species will learn to cooperate even when the conditions are seemingly favorable.

Fortunately, human societies do not have to wait for the slow and uncertain processes of biological evolution to shift them between equilibria. However, the tale I tell in Chapter 4 about how they manage this feat is closely mirrored by the tale that biologists tell about the evolution of eusociality among the *hymenoptera*. I therefore have some hopes that an aside on this fascinating story will serve to respectabilize my own speculations on the evolution of sociality in humans. Biologists should be warned that my use of game-theoretic language contributes nothing of any substance to the eusociality debate. They may therefore prefer to skip forward to where the discussion of equilibrium selection in human societies resumes.

Kinship and equilibrium selection In Monod's [362] famous phrase, how far along the social road equilibrium selection has taken a species depends on both chance and necessity. Although chance is the prime mover, one can nevertheless sometimes say a great deal about the likely evolutionary history of a species by looking closely at the physical constraints imposed on a species by its genetic structure and the ecological niche it inhabits. For a game theorist, the physical constraints and opportunities that characterize a species appear as rules in the Game of Life played by its genes.

E. O. Wilson [558] noted that eusociality[67] is rare except among the *hymenoptera*. Alexander *et al* [12] suggest that true eusociality has actually evolved independently at least twelve times in the *hymenoptera*, but only twice elsewhere—the exceptional cases being the termites of the order *isoptera* and the recently studied naked mole-rats of the order *rodentia*. Hamilton [217] proposed kinship as a possible explanation of this phenomenon.

Hamilton's argument is based on the fact that the *hymenoptera* are haplodiploid, which means that unfertilized eggs grow into haploid males and fertilized eggs grow into diploid females. In a haploid species, each locus on a chromosome hosts just one gene. Humans are diploid, with each locus hosting two genes, one from the mother and one from the father. As explained in Section 2.5.2, the degree of relationship between human sisters is therefore $r = \frac{1}{2}$. By contrast, the degree of relationship between sisters in the *hymenoptera* is $r = \frac{3}{4}$, because each locus on their chromosomes gets the *same* gene from their father and a randomly chosen gene from the pair carried at that locus by their mother. Hamilton [217] therefore argues that eusociality is more frequent among the *hymenoptera* than other orders because the benefits of cooperation are greatest in the *hymenoptera*. In particular, a female ant or wasp will get her genes replicated more often if her sister founds a new colony rather than her daughter (with whom her degree of relationship is only $r = \frac{1}{2}$).

The view that eusociality is to be attributed primarily to kin selection in the *hymenoptera* has been questioned by various authors (Dugnatov [163, p.142]). For example, Lin and Michener [326], propose what has been called the mutualistic or communal aggregation hypothesis. This idea is familiar from Volume I: for cooperation to be workable in a society, it is not necessary that its citizens be Dr. Jekylls, who treat each other

---

[67]A species is *eusocial* if it lives in colonies with overlapping generations in which one or a few individuals produce all the offspring and the rest serve as functionally sterile helpers.

like brothers. Even a society of Mr. Hydes can cooperate if getting along together serves to advance the selfish aims of each individual citizen. In particular, the probability that a solitary member of a species succeeds in raising children may be so small that it pays for individuals to sacrifice their own reproductive opportunities to join a monarchy whose purpose is to guarantee a high probability of survival for children of the queen—even though the degree of relationship between the queen and her subjects may be relatively small. In a fascinating study of tropical and subtropical wasps that form multiqueen societies,[68] Itô [282] endorses this rival explanation of the evolution of eusociality in the *hymenoptera*. Although the queens in the colonies studied by Itô are sisters and some adopt a subordinate role that minimizes their reproductive opportunities, it remains true that the degree of relationship between a randomly chosen pair of daughters may not be at all high. But eusociality nevertheless survives.

How do these and other considerations translate into game-theoretic terms? If game theory is to be used as a tool of analysis, the first problem is to decide how best to model the Game of Life played by the genes of the species under study. Usually biologists look only at the strategic form of the Game of Life (Section 2.5.1 of Volume I). But even operating at such a level of abstraction requires specifying the *strategies* available to the genes of the species and the *payoffs* that the players will receive after each has chosen a strategy. Payoffs in the game are simply identified with properly calculated biological fitnesses. What counts as a strategy will depend on the time span under consideration. Usually all the strategies available to a species through sexual recombination of gene packages will be counted in. Strategies available as a consequence of mutations that are highly likely to occur during the period under study also belong on the list.[69] The guiding principle is that the game should be designed so that chance can be relied upon to throw all strategies into the ring during the time span under consideration. The necessities of Darwinian evolution can then be guaranteed to take the population to a suitably defined equilibrium of the game.

---

[68]The hazards facing wasps in the tropics are apparently much greater than in temperate climates, where single-queen colonies predominate. It is argued that multiqueen colonies grow faster and hence are more likely to survive in a risky tropical environment.

[69]A more ambitious approach would also take account of unlikely mutations or catastrophic changes in the environment by modeling these as low probability chance moves. The Game of Life would then need to be treated as a repeated game in which chance may alter the stage-game between successive plays by expanding or contracting its set of available strategies or changing its payoffs. Although one cannot avoid contemplating this more general case when considering the evolutionary history of a real species, I proceed in the text as though all catastrophes were impossible and all relevant mutations were certain to arise in the time available.

As we saw in Section 2.5.2, kinship considerations enter the Game of Life primarily through their effect on its payoffs.[70] In the case of the *hymenoptera*, Hamilton's [217] observation that $r = \frac{3}{4}$ means that these effects will be much stronger than in species with lower values of r. But, as Alexander *et al* [10, 12] observe, close kinship between sisters is not the only factor that might favor the evolution of eusociality in the *hymenoptera*. The fact that some species in the order have the capacity to develop a sting sufficiently powerful for an individual to deter even a large vertebrate from raiding their nests is also important, as is their potential to develop chemical controls that allow a mother to enslave some of her children so that they spend their time assisting her to lay more eggs rather than laying eggs of their own (Alexander's [10] parental manipulation hypothesis). The potential for such genetically determined traits to appear in the time span available needs to be incorporated in the strategies of the Game of Life to be attributed to a species.

Properties of the ecological niche inhabited by the species will affect both the payoffs and the strategies of the game. For example, the fact that tropical wasps face a more hazardous environment than their temperate cousins will be reflected in the fitnesses that they derive from using the strategies available to them. Similarly, the relatively safe underground environments exploited by both termites and the naked mole-rat will figure large in their games of life.

Where does this leave the communal aggregation hypothesis? This seems to me best seen, not as a rival to the kinship or parental manipulation hypotheses, or any other factors that need to be built into the Game of Life, but as a statement of the fact that the evolution of eusociality necessarily requires that Nature have solved an equilibrium selection problem. In brief, I see the communal aggregation hypothesis as a way of expressing the fact that equilibrium selection is a necessary component of any story about the evolution of sociality.

The difficulty of the equilibrium selection problem that Nature has to solve depends on the particular game of life played by the species under study. The problem is exacerbated by the fact that Nature has to do more than shift a species from a noncooperative equilibrium to a cooperative equilibrium. In accordance with the Linnaean dictum *Natura non facit saltus*, she must find a sequence of intermediary equilibria to serve as a route for the species to follow on its way from the original equilibrium to the final equilibrium.

---

[70]Although Section 2.5.2 also notes the effect that kinship can have on the *structure* of the game. For example, a multiplayer sibling rivalry game may correspond to a one-player Game of Life in extreme cases.

The set of available cooperative equilibria depends on the game of life. For example, in Section 2.5.2, we saw that kinship considerations can transform a sibling rivalry game with only a noncooperative equilibrium into a game of life with only a cooperative equilibrium. But real games of life must be expected to have vast numbers of equilibria among which Nature must choose. The larger the set of cooperative equilibria and, perhaps more importantly, the larger the set of routes leading to such equilibria, the easier it will be for chance and necessity to lead Nature to a eusocial solution of the equilibrium selection problem. As we have seen, various hypotheses have been proposed to explain why the games of life played by species in the order *hymenoptera* should make it easier for the equilibrium selection mechanism to find its way to a cooperative equilibrium. But whether the mechanism succeeds in doing so in the time available will depend on chance. We must therefore expect that sometimes Nature will find her way to a cooperative equilibrium in games that are not favorable to such an outcome while failing to do so in games that are favorable.[71]

All the best stories end with a moral, and the tale of eusociality in the *hymenoptera* is no exception. Section 2.3.1 of Volume I faced up to the problem of moving a society from an initial state of nature to a Pareto-improving social contract when nobody trusts each other and no philosopher-king exists to enforce the promises that people may make. Without a philosopher-king, a viable social contract must be an equilibrium in the Game of Life. Moreover, in the absence of trust, the actions the players need to take in moving their society from one equilibrium to a more cooperative alternative must also be in equilibrium. The easiest way to accomplish this trick is to envisage society moving *gradually* through a sequence of intermediary equilibria from the state of nature to the final civilized state.

The underlying problem faced by a reformer who seeks to reorganize a society along rational grounds is therefore essentially the same as that faced by Nature when she solves the equilibrium selection problem in the Game of Life of a species. The difference is that equilibrium selection among humans doesn't work in the same way as equilibrium selection among those animals

---

[71]My guess, for what it's worth, is that kinship will ultimately turn out not to be crucial for the evolution of sociality. The spectacular nature of the cooperative equilibria available to the *hymenoptera* perhaps blinds us to Nature's achievement in moving species in other orders to the more modest cooperative equilibria that are the best they can do given their different games of life. For example, in spite of being *hymenoptera*, Itô's tropical wasps seem to be barred from the full glory of eusociality displayed by their temperate cousins, by the fact that their environment is too hazardous to admit slow-growing, single-queen colonies. I suspect that the crucial stumbling block for the evolution of cooperation more commonly lies at the *beginning* of the route away from a noncooperative equilibrium. If so, then the real puzzle is how cooperation first gets off the ground at all.

whose behavior is entirely determined by their genes.

Equilibrium selection in humans. Insect species have to wait for chance to shift a society from one equilibrium to a more cooperative alternative. Societies operating the less efficient equilibrium will then eventually disappear if they are competing for the same resources. Human societies are more resilient, since our capacity to imitate makes it possible for one society to borrow the cultural innovations of another. However, my guess is that Nature has made us even more flexible. I think it likely that we are genetically programmed with algorithms that help to immunize our societies against competition from innovative rivals. Such algorithms actively seek out Pareto-improving equilibria as these become available through changes in the environment in which we live.

For example, Section 3.2.4 of Volume I describes a "cheap talk" mechanism that will eventually shift players in Rousseau's Stag Hunt Game (Figure 3.9) from the Pareto-inferior equilibrium to the Pareto-superior equilibrium.[72] However, such devices for switching a society to a Pareto-superior equilibrium are too simplistic for general purposes, because they depend on there being no ambiguity about *which* Pareto-superior equilibrium is to be selected. But there will frequently be many such equilibria—as when a society has the opportunity to exploit a new source of surplus, and the issue is who is to get how much of the gravy. Any mechanism for generating a Pareto-improvement in such circumstances must necessarily incorporate an equilibrium selection criterion.

My guess is that the fairness norms that seem universal in human societies have evolved primarily for this purpose. Their principal role is to single out one of the many equilibria typically available as Pareto-improvements on the *status quo* without the necessity for damaging and potentially destabilizing internal conflict.

Chapter 1 of Volume I explains at length why I think that the social tools of this type that we successfully employ in deciding what is fair in small-scale situations have the potential to be harnessed for use in finding answers to large-scale problems for which our evolutionary history has provided us with no solution techniques. In terms of the expanding circle metaphor to

---

[72]In discussing the evolution of cooperation in the indefinitely repeated Prisoners' Dilemma, Axelrod and Hamilton [34, 31] explain how such mechanisms for moving a population to a Pareto-efficient equilibrium become more effective when players interact only with their neighbors. Binmore [70, p.432] and Maynard Smith [347] provide simple introductions to their idea in the context of a much simpler game first analyzed by Zeeman [575]. In terms of the language of this section, such models show how evolution can insert equilibrium selection algorithms into a society by acting first on localized communities.

be pursued below, this is to suggest that we consciously seek to increase the radius of the domain in which we operate as moral animals. However, as I have repeatedly argued, such a proposal has no hope of working unless we are realistic about the second-best nature of the moral systems that we currently operate. This is why I am so anxious to trace the natural origins of our intuitions about justice rather than attributing them to inspiration from the realms of metaphysics.

Expanding the circle. To summarize the story so far, the claim is that the evolution of human ethics began as a consequence of equilibrium selection operating in family games. It began in the family because Nature found the equilibrium selection problem easier to solve in games played by close relatives than in games played by strangers. The reason is that humans actively sympathize with their relatives by building a direct concern for their welfare into their personal utility functions. In extreme cases, we have seen that this effect may be enough to convert a sibling rivalry game that looks like the one-shot Prisoners' Dilemma into a Game of Life in which the only equilibrium is Pareto-efficient

The more likely the people we encounter frequently are to be close relatives, the more we tend to sympathize with their needs and concerns. When we get to distant kinfolk we seldom encounter, the degree to which we sympathize with them becomes small. But my guess is that we still retain more than enough capacity to sympathize even with absolute strangers to explain the "warm glow" feelings that lead us to leave tips in restaurants we never expect to visit again, or to make charitable donations that are small compared with our income.

Such vestigial warm-glow feelings provide an inadequate foundation on which to build a modern state. It is simply not in our nature to love strangers as we love our near and dear. This is not to deny the existence of rare saintly individuals. Nor that we are all capable of acts of great self-sacrifice on rare occasions. But a utopian state that relies on saintly behavior from its citizens on a day-by-day basis will just not work. When moral behavior expanded from the extended family to the world at large, it did so in a more subtle way than by training us to love all men as we love our brothers. To understand the mechanism, it is necessary to give up the idea that Nature might have changed the structure of the game we play with strangers in the same way that it changed the structure of sibling rivalry games. We have to focus attention instead on how evolution succeeded in shifting us from one equilibrium to another in a *fixed* game.

My guess is that the moral circle sometimes expands through players misreading signals from their environment and so mistakenly applying a piece of behavior or a way of thinking that has evolved for use within some inner circle to a larger set of people or to a new game. When such a mistake is made, the players attempt to play their part in sustaining an equilibrium in the game played by the inner circle without appreciating that the game played in the outer circle has different rules. For example, Adam might mistakenly treat Eve as a sibling even though she is a complete stranger. Or he might mistake a one-shot game for an indefinitely repeated game.

A strategy profile that is an equilibrium for an inner-circle game will not normally be an equilibrium for an outer-circle game. The use of the inner-circle equilibrium strategy in the outer-circle game will therefore usually be selected against. But playing the outer-circle game as though it were the inner-circle game will sometimes result in the players coordinating on an equilibrium of the outer-circle game. The group will then have stumbled upon an *equilibrium selection device* for the outer-circle game. They succeed in coordinating on an equilibrium in this game by behaving as though they were playing another game with a more restrictive set of rules.[73]

In summary, we need to turn our attention away from circles within which Adam and Eve sympathize with each other's plight to wider circles in which their sympathetic identification is too weak to be significant. Insofar as they can, *empathetic preferences* then have to substitute for the sympathetic preferences that operate within families. But note that the same internal algorithms that allow us to use sympathetic preferences also allow us to handle empathetic preferences. When one gets down to brass tacks, both sympathetic and empathetic utility functions are computed simply as a weighted sum of the payoffs in a game. In the case of sympathetic preferences, the game is some analogue of a sibling rivalry game. In the case of empathetic preferences, the game is some analogue of the type of food sharing game to be discussed in Section 2.5.4. Nature therefore did not need to invent some entirely new mechanism to bridge the gap separating altruism within the family from fairness between strangers. She merely needed to supply some ramshackle scaffolding while adapting mechanisms that evolved to meet one set of needs to new and different purposes.

---

[73]This view of things underlies the paradigm behind Chapter 4. Players act as though they were bound by the rules of an artificial Game of Morals. But in so doing, they succeed in selecting an equilibrium in the Game of Life that they actually are playing. When things go smoothly, no conflict then arises between moral behavior and enlightened self-interest, although this may be far from clear to the players themselves. However, to study this approach properly, it is necessary to juggle many balls simultaneously. In the current chapter, the number of balls is reduced considerably by maintaining the fiction of a philosopher-king, whose presence allows us to ignore the problems that arise when working with equilibria in repeated games. This makes it possible to focus on the fairness criteria that I believe are built into morality games with a minimum of technical distractions, since we can proceed as though the Game of Life were a one-shot game.

2.5.4   Empathy and Fairness

Alexander [10] and Humphrey [275] are credited with the idea that we have
large brains as a result of an arms race within our species aimed at building
bigger and better internal computing machines for the purpose of outwitting
each other. This seems a very plausible speculation to me, but then the
idea is one that would naturally attract a game theorist. But whether or
not our capacity to empathize with our fellow men is the primary reason
that we have bigger brains than other primates, it seems uncontroversial
that we are genetically equipped to put ourselves in the position of others
to see things from their point of view.[74] Section 3.4.2 of Volume I even
went so far as to argue that that our concept of the "I" derives from this
empathetic capacity, its chief function being to act as a mirror of others in
our own minds and to reflect how we are similarly mirrored in the minds
of others.

The importance of empathetic identification in helping us to predict
the behavior of others will be obvious. Its role in facilitating learning by
imitation may not be quite so apparent. Since the spread of social norms
through imitation and education is an important backdrop to what follows,
it may therefore be worthwhile to observe that Adam needs to understand
why Eve behaved in a certain way if he is to know *when* the time has come
to copy her behavior (Hoffman [255]). To use his capacity for empathy to
understand what triggered her behavior, he needs to imagine himself in her
shoes with her preferences and her beliefs. It is then not such a long step
to comparing his personal preferences with her personal preferences, just
as he compares his fitness with his sister's fitness in a sibling rivalry game.

However, before embarking on a discussion of empathetic preferences, it
is first necessary to say something about the type of equilibrium selection
question to which empathetic preferences are Nature's answer.

Insurance contracts and the original position. As Isaac [280] has
persuasively argued, the opportunity to share food was perhaps the major
advantage to the human species in adopting a social life-style. Damas [141],
Evans-Pritchard [176], Lovejoy [331], Wiesner [552], Cashden [116], Sahlins
[455], and Smith [502] are typical references documenting the seemingly
universal practice of reciprocal food sharing in modern hunter-gatherer societies.

The consensus is that such food sharing is a form of mutual insurance
against the infrequent times when resources become scarce. If Adam is
lucky enough to have an excess of food this week, it then makes sense

---

[74]Victims of autism are said to suffer from damaged empathetic wiring.

for him to share with Eve in the expectation that she will be similarly generous when she is lucky in the future. Things are similar in the case of the alliances that operate within chimpanzee societies. One chimp comes to the aid of an ally unlucky enough to incur the enmity of a powerful foe in the expectation that the service will be reciprocated when their roles are reversed.

If the players are relatives, such relationships will be easier to get off the ground, since each player will sympathize with the other to some degree. But the reciprocal arrangements built into such mutual insurance pacts can work perfectly well without any need to attribute altruistic motives to the players. Indeed, the folk theorem of repeated game theory tells us that we must expect there to be an embarrassingly large number of alternative equilibria among which a choice must be made. In deciding which equilibrium to operate, the players are therefore confronted with a classic bargaining problem.

The possible agreements are sharing rules that must be negotiated before the players know who will be lucky. In predicting how much each is likely to get from any particular rule, the players will use their common experience of how things have gone in the past to assign a probability p to the event that it will be Adam who is lucky. My guess is that such mutual insurance pacts are more likely to have operated successfully between players of roughly equal prowess, but we shall see that it is not essential that we take $p = \frac{1}{2}$. Whatever the value of p, each possible sharing rule determines an expected payoff to Adam and Eve. The set T of all such payoff pairs is the set of feasible agreements for their bargaining problem. The *status quo* τ is the payoff pair that results if they operate without an insurance pact.

I don't suppose anyone knows to what extent our primitive ancestors bargained like buyers and sellers in a modern bazaar. The tradition is doubtless very ancient, but we don't need to assume that the bargaining was formalized in this particular way—or even that the proposals and counterproposals made by the players were verbalized at all. It would come to much the same thing if the players simply acted out their proposals and counterproposals physically over a period of time when both were frequently being buffetted by the winds of fortune. But however the bargaining may have been done, the important point is that bargaining of some sort must have taken place under circumstances very close to those envisaged by Rawls [417] and Harsanyi [233] when they independently proposed the notion of the *original position*.

Veil of uncertainty. The difference between bargaining over a mutual insurance pact and bargaining in the original position mirrors the distincₒween Buchanan and Tullock's [107] approach to "constitutional ₒracts" and the social contract theory of Rawls [417]. Instead of Rawls' veil of ignorance, Buchanan and Tullock study a *veil of uncertainty*.[75] Such a veil of uncertainty represents a genuine lack of knowledge about which event is actually going to occur, whereas Rawls' veil of ignorance represents a pretended lack of knowledge about events that are already fixed. Dworkin's [167] distinction between "brute luck" and "opportunity luck" has a similar flavor.

A simple model will perhaps help to clarify how bargaining behind a veil of uncertainty works. Instead of players I and II bargaining in ignorance of whether they will turn out to be Adam or Eve, think instead of Adam and Eve bargaining before they find out who will be lucky tomorrow. Take Mr. Lucky to be the hunter who has made a kill, and Ms. Unlucky to be the hunter whose efforts were unrewarded. Adam and Eve's relative hunting skills are then reflected in the probability p that both assign to the event that Adam will turn out to be Mr. Lucky.[76]

In the circumstances under which human sociality evolved, a mutual insurance pact must have been self-policing, using the reciprocity mechanism studied in the theory of repeated games. But the complications that arise when time enters the picture are left to later chapters (Section 3.4.4 and elsewhere).

In the model studied here, a fictional philosopher-king enforces the bargain negotiated behind the veil of uncertainty that conceals the identities of Mr. Lucky and Ms. Unlucky. Attention will be restricted to insurance contracts that merely specify the fraction x of the kill to be retained by Mr. Lucky.[77] Matters are simplified even further by assuming that Adam and Eve have the same preferences, so that $u_A(y) = u_E(y) = u(y)$, where y is the fraction of the kill a player receives. These utility functions will be normalized so that $u(0) = 0$ and $u(1) = 1$.

The utilities that Mr. Lucky and Ms. Unlucky assign to an insurance contract x are then $u_L(x) = u(x)$ and $u_U(x) = u(1 - x)$. Behind the veil

---

[75]Buchanan and Tullock [107] argue that uncertainty at a preliminary constitutional stage about what legislative issues will eventually prove to be contentious is likely to make it easier to achieve consensus than if all matters requiring decision are left to the legislative stage. Rawls' [416] original work features a similar veil of uncertainty, rather than a veil of ignorance.

[76]If Adam and Eve both turn out to lucky or unlucky, there is nothing about which to bargain. The probability p is therefore conditional on there being some point to writing an insurance contract.

[77]Section 2.6 considers *contingent* social contracts. A similar refinement here would make x depend on whether Mr. Lucky turns out to be Adam or Eve.

of uncertainty, Adam and Eve therefore respectively seek to maximize

$$
\begin{aligned}
w_1(x) &= pu_L(x) + (1-p)u_U(x) = pu(x) + (1-p)u(1-x), \\
w_2(x) &= (1-p)u_L(x) + pu_U(x) = (1-p)u(x) + pu(1-x).
\end{aligned} \qquad (2.4)
$$

Notice the similarity between these objective functions and those which arise without uncertainty when Adam and Eve are relatives whose degree of relationship is r. A player who receives a fraction y of the kill will then care that the relative with whom he is playing is to receive only $1-y$. According to Hamilton's rule, each player will then seek to maximize $u(y) + ru(1-y)$.

If no agreement on how the kill should be divided is reached, Mr. Lucky will eat it himself. The *status quo* for the bargaining problem Adam and Eve face behind the veil of uncertainty is therefore obtained by writing $x = 1$ in equations (2.4). As explained in Section 1.3.1, the result of using the symmetric Nash bargaining solution can therefore be determined by maximizing the Nash product

$$
\begin{aligned}
\Pi &= \{w_1(x) - w_1(1)\}\{w_2(x) - w_2(1)\} \\
&= \{(1-p)u(1-x) - p(1-u(x))\}\{pu(1-x) - (1-p)(1-u(x))\} \quad (2.5)
\end{aligned}
$$

As we shall find in Section 2.6.1, such problems are better tackled using geometric methods. However, no calculations at all are necessary to resolve the two easiest cases.

In the first case, Adam and Eve are risk neutral, so that $u(y) = y$. But risk-neutral people see no point in paying a premium to avoid taking a risk. We therefore know in advance that neither Adam nor Eve will put their signatures on a meaningful insurance contract. The symmetric Nash bargaining solution must therefore implement the empty contract $x = 1$, in which the lucky hunter simply eats the kill without sharing at all.[78]

For an insurance contract to make sense for Adam and Eve, they need to be *strictly* risk averse. Section 4.6 of Volume I explains that the utility function u must then not only be increasing, but strictly concave. With this assumption, we turn to the case when $p = \frac{1}{2}$. The bargaining problem behind the veil of uncertainty then becomes symmetric. An appeal to the Symmetry Axiom of Section 1.4.2 then reveals that $x = \frac{1}{2}$.[79] The lucky hunter therefore shares fifty:fifty with his unlucky partner.

---

[78]Write $V = u(1-x)$ and $W = 1 - u(x)$ in (2.5). Then $\Pi = p(1-p)(V-W)^2 + \{4p(1-p) - 1\}VW$. When $u(y) = y$, $V = W$, and so $\Pi = \{4p(1-p) - 1\}(1-x)^2$. Since $4p(1-p) \le 1$, Π is maximized when $x = 1$. Notice that an exception to the general case quoted in the text arises when $p = \frac{1}{2}$. Any x is then a solution.

[79]Following on from the previous footnote, note that $p = \frac{1}{2}$ implies that $\Pi = \frac{1}{4}(V - W)^2$. Since $\Pi' - \frac{1}{2}(V-W)(V'-W')$, the maximum occurs when $V = W$ or $V' = W'$. But $V = W$ minimizes Π. The maximum therefore occurs when $u'(1-x) = u'(x)$. Since u is strictly concave, u' is strictly decreasing. Thus $1 - x = x$ and so $x = \frac{1}{2}$.

Symmetric focal points. To symmetrize the bargaining problem behind the veil of uncertainty, it is enough to require that $p = \frac{1}{2}$, without insisting that Adam and Eve have identical preferences. The players then split the surplus equally, because the information available to them behind the veil of uncertainty neutralizes the differences in their personal characteristics. Similar results are a commonplace of the bargaining literature.

The case when Adam and Eve have been randomly chosen as bargaining partners from the *same* population of disparate types is particularly popular (Binmore [64], Skyrms [499], Young [571], and numerous others). As explained in Chapter 1, if a risk-neutral Adam is matched with a very risk-averse Eve, he will be able to appropriate the lion's share of the surplus because he is less fearful of a breakdown in the negotiations—but only if the difference in their attitudes to risks is common knowledge.

When neither player has any information about the other beyond knowing that he or she comes from the same population, Adam has a problem in bringing his power to bear. As far as Eve can see, her opponent is as likely to be a weak type like Ichabod as a strong type like Adam. If Adam had a verbal argument that could convince her that he was strong, nothing would prevent Ichabod from using the same argument. No amount of bluster from her opponent will therefore have any affect on her beliefs. Unless Adam has some way to signal his strength that Ichabod cannot copy, it follows that Eve will use the same bargaining strategy whether her opponent is Adam or Ichabod.[80] No such signals are available when two players negotiate about the division of a dollar using the smoothed Nash Demand Game of Section 1.5.1. In consequence, the only Pareto-efficient equilibrium requires that the dollar be split equally between the two players (Binmore [64]).

Symmetric games always have a symmetric equilibrium, If a symmetric game has a unique equilibrium, the equilibrium must therefore be symmetric. Once evolution had learned to use symmetry in such games to assist the players in focusing on the unique equilibrium of the game, the way was open for the same coordinating convention to be employed to select the symmetric equilibrium in symmetric games with multiple equilibria. The symmetry convention could then be extended to cases of asymmetric games that nevertheless exhibit some symmetries that are retained by the equilibria of the game.

Schelling's [464] famous essay on focal points largely consists of a list of increasingly difficult coordination problems for the reader to solve. Some

---

[80] Adam might advertise his power by publicly burning more money than Ichabod can afford to spend on a signal. Or he might delay agreement longer than Ichabod can afford to wait. Game theorists distinguish such costly signals from verbal signals by referring to the latter as "cheap talk".

of the focusing devices we are implicitly invited to apply require using alphabetical order or other tricks of language that wouldn't help two people from very different cultures to coordinate. But the appeal of symmetric resolutions seems to be universal.[81] My guess is that our bias for symmetric focal points has an instinctive basis.

It is sometimes argued that fairness consists of nothing more than the use of symmetric focal points in a moral context. But fairness norms actually recommend splitting a surplus equally only when the asymmetries that are common knowledge between the players are peripheral to their circumstances. I therefore prefer to think of equal-splitting conventions as *prefairness* norms, which represent only one step on the evolutionary road to the fairness norms of today. Modern fairness norms take account of factors like need, effort, and ability in assessing the value to a player of his share of the surplus (Section 4.7). The equal-splitting convention is then applied to these valuations—not to the raw shares of the surplus that each receives.

The origins of this practice are already evident in the simple model we have been using to model bargaining behind a veil of uncertainty. The probability p that Adam will be the lucky hunter is a measure of his ability relative to Eve. If we ensure that an insurance contract is worthwhile by assigning Adam and Eve the same strictly concave utility function $u(y) = \sqrt{y}$, then it is not hard to show that the lucky hunter's share of the kill is about $\frac{2}{3}$ when the same is true of p.[82] Adam and Eve's expected payoffs when Mr. Lucky gets $\frac{2}{3}$ of the kill are approximately $\frac{5}{15}$ and $\frac{2}{15}$ respectively. The insurance contract can therefore be seen as the result of applying the equalizing paradigm after Adam's hunting skill has been recognized by weighting his utils so that they are worth about $2\frac{1}{2}$ times as much as Eve's.

But what standard of interpersonal comparison is being applied? Insofar as the question is meaningful, the answer is the zero-one rule of Section 2.3.1. Adam and Eve certainly make no attempt to empathize with the plight of the other. When asking himself how it feels to be Mr. Lucky or Ms. Unlucky, Adam needs only to imagine himself in shoes that he *himself* might be wearing tomorrow. Even when prefairness norms are modified to take relevant asymmetries into account, they therefore fall a long way short

---

[81]To take a trivial case, Adam and Eve agree to meet at the exit in the belief that there is only one way out. Having discovered three exits a considerable distance apart, Eve will have no difficulty in deciding to wait at the one in the middle.

[82]Continuing Footnote 85, one now needs to maximize Π subject to the constraint $\Phi = V^2 + (1 - W)^2 - 1 = 0$. On writing down the conditions for a stationary point for the Lagrangian $\Pi + \lambda\Phi$, we are led to the equation $W/V = 2\lambda/q$, where $q = 1 - 2p(1-p)$. If we insist that $x = 16/25$, it follows that $\lambda = q/6$. Substituting this result back in the Lagrangian equations, we find that $q = 15/28$, and so $p = \frac{1}{2}(1 \pm \sqrt{1/14})$.

of the sophisticated fairness norms to which we are accustomed.

From uncertainty to ignorance.   To move from prefairness to fairness, evolution had to find a way to allow Buchanan and Tullock's [107] veil of uncertainty to give way to the veil of ignorance of Harsanyi [233] and Rawls [417].

To nail down the similarity between bargaining over mutual insurance pacts and bargaining in the original position, we need to rename Adam and Eve as they take their places behind the veil of ignorance. Adam and Eve retain their identities, but are now relabeled as players I and II. Instead of Adam and Eve being uncertain about whether they will turn to be Mr. Lucky or Ms. Unlucky, the new setup requires that players I and II pretend to be ignorant about whether they will turn out to be Adam or Eve. It then becomes clear that a move to the device of the original position requires only that the players put themselves in the shoes of somebody else—either Adam or Eve—rather than in the shoes of one of their own possible future selves.

However, on the face of it, a substantial gap still separates Rawls' or Harsanyi's proposed use of the original position to judge the fairness of political constitutions, and its use by our prehistoric ancestors in settling disputes over how a carcass should be divided.

In a prehistoric insurance contract, the parties to the agreement do not have to *pretend* that they might end in somebody else's shoes. On the contrary, it is the reality of the prospect that they might turn out to be Mr. Lucky or Ms. Unlucky that motivates their writing a contract in the first place. But when the device of the original position is used to adjudicate fairness questions à la Rawls, then player I knows perfectly well that he is actually Adam and that it is physically impossible that he could become Eve. To use the device as recommended by Rawls and Harsanyi, he therefore has to indulge in a *counterfactual* act of imagination. He cannot become Eve, but he must pretend that he could.

How is this gap between reality and pretence to be bridged without violating the Linnaean dictum that Nature makes no jumps? In spite of the fuss about punctuated equilibria (Dennett [158]), orthodox opinion among evolutionary biologists holds fast to the principle that Nature is a tinkerer (Jacob [283]). She proceeds by making small modifications to existing structures rather than by throwing "hopeful monsters" into the ring to see whether they survive. To make a naturalistic origin for the device of the original position plausible, it is therefore necessary to give some account of the structures with which Nature presumably tinkered when she first brought fairness norms into the world.

Natura non facit saltus. I see the step from the use of the device
of the original position in negotiating mutual protection arrangements to
its use in adjudicating fairness disputes as an example of how morality
can expand from one circle to another. To reiterate the theory, people
take a technique used within one circle of social problems and unthinkingly
apply it to a wider domain of problems. In so doing, they continue to play
by the rules of the game for which the technique originally evolved, not
noticing—or pretending not to notice—that the rules of the game played
in the wider circle may be quite different. Usually the result will not even
be an equilibrium in the wider game, and evolution will briskly sweep the
experiment away. But sometimes the procedure will succeed in coordinating
behavior on a Pareto-improving equilibrium of the wider game, whereupon
equilibrium selection will have moved the evolutionary ratchet one further
notch forward.

I have argued that the device of the original position has its roots in the
need of early mankind to negotiate Pareto-improving insurance contracts.
Earlier, I argued that the origins of empathetic preferences are to be found
in the games played by kinfolk. I now propose to argue that evolution
somehow found a way to combine these two developments to create the
equilibrium selection device of which we are dimly aware when making
appeals to fairness or justice. My guess is that at least some of the necessary
internal plumbing that allows us to operate this equilibrium selection device
is genetically fixed—and hence the universal attachment across cultures to
the basic notions of fairness and justice. However, I think it likely that
the empathetic preferences that serve as inputs to the justice algorithm are
almost entirely socially determined—and hence the different outcomes that
result from using the justice algorithm in different societies.

In negotiating an insurance contract, to accept that I may be unlucky
seems a long way from contemplating the possibility that I might become
another person in another body. But is the difference really so great? After
all, there is a sense in which none of us are the same person when comfortable and well fed as when tired and hungry. In different circumstances,
we reveal different personalities and want different things. When planning
ahead under uncertain conditions, it would therefore not be surprising if we
estimated our payoffs using the same wiring that we use when estimating
payoffs in family games.

When planning ahead, a player computes his expected utility as a
weighted average of the payoffs of all the future people that he might turn
out to be after the dice has ceased to roll. When choosing a strategy in
a family game, a player takes his payoff to be a weighted average of the
fitnesses of everybody in the family. In order to convert our ability to nego-

tiate insurance contracts into a capacity for using fairness as a coordinating device in the Game of Life, all that is then needed is for us to hybridize these two processes by allowing a player to replace one of the future persons that a roll of the dice might reveal him to be, by a person in another body who is to be treated in much the same way that he treats his sisters, his cousins or his aunts.

Figure 2.2 illustrates the evolutionary history of empathetic preferences and the original position in the story told so far. But a problem still remains. The weights we use when discounting the fitnesses of our partners in a family game are somehow obtained by estimating our degree of relationship to our kinfolk from the general dynamics of the family and our place in it. But where do we get the weights to discount Adam and Eve's personal utils when constructing an empathetic utility function?

Imitation and empathy equilibrium. The bargaining position of a player negotiating a prehistoric insurance contract is defined by a set of state-contingent preferences. As a minimum, the player needs to be able to compare his loss in giving away some food when lucky at hunting with his gain in being given some food when unlucky. That is to say, he needs to be able to compare being lucky in one situation with being unlucky in another. An empathetic preference has precisely the same structure, except that it is now necessary to compare being Adam in one situation with being Eve in another. For example, one might express a preference for being Adam wearing a fig leaf over being Eve eating an apple. Just as one cannot negotiate an insurance contract without being equipped with suitable state-contingent preferences, so one cannot employ the device of the original position to adjudicate fairness issues without being equipped with suitable empathetic preferences.

Empathetic preferences are fully discussed in Section 4.3.1 of Volume I. If one assumes that such preferences satisfy the axioms of Von Neumann and Morgenstern (Section 4.2.2 of Volume I), then they can be represented by a Von Neumann and Morgenstern utility function. If I am also sufficiently successful at empathetic identification that, when I put myself in the shoes of a fellow citizen, I make comparisons between *his* eating an apple or *his* wearing a fig leaf according to *his own* personal preferences, then it follows that the utility function representing my *empathetic* preferences can be expressed as a weighted sum of the Von Neumann and Morgenstern utility functions that represent Adam and Eve's *personal* preferences. The weights in this sum then serve as a standard for comparing Adam and Eve's personal utils.

The requirement that Adam should suppose himself to have Eve's per-

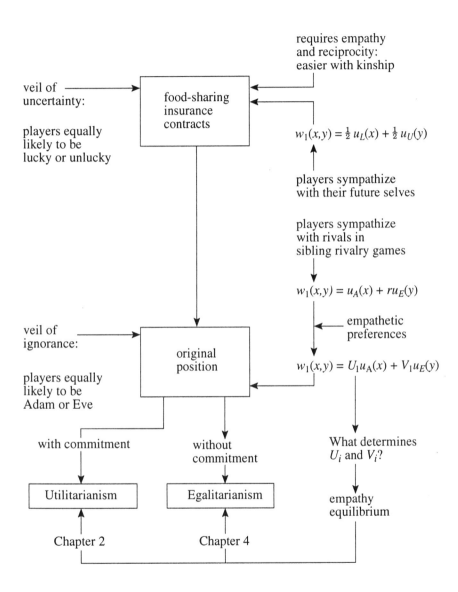

Figure 2.2: An evolutionary history of the original position?

sonal preferences when he imagines himself in her shoes is very troublesome for some critics. Dworkin [167], for example, makes no attempt to incorporate such a proviso into his theory of justice. But such an assumption is inevitable if my claims about the naturalistic origins of empathetic preferences are to be taken seriously. In negotiating an insurance contract, it would be absurd for Adam to assume that because he was unlucky at hunting and hence very hungry today, that he will still feel hungry after being lucky at hunting tomorrow. Similarly, in a sibling rivalry game, it would make no sense for Eve to assume that her fitness would be unchanged if she were her brother. Nor do I see any force in the objections sometimes made on intuitive moral grounds to assuming that Adam and Eve should empathize fully with each other.[83] If we are ever to escape the eternal round of strife and conflict that follows when moral disagreements are reduced to slanging matches between rival ipsedixists, it seems to me that we need to make a *genuine* attempt to see things from the point of view of others—no matter how repugnant their viewpoint may appear.

I see the empathetic preferences held by the individuals in a particular society as an artifact of their upbringing. As children mature, they are assimilated to the culture in which they grow up largely as a consequence of their natural disposition to imitate those around them. One of the social phenomena they will observe is the use of the device of the original position in achieving fair compromises. They are, of course, no more likely to recognize the device of the original position for what it is than we are when we use it in deciding such matters as who should wash how many dishes. Instead, they simply copy the behavior patterns of those they see using the device. An internal algorithm then distills this behavior into a preference-belief model against which they then test alternative patterns of behavior. The preferences in this model will be empathetic preferences—the inputs required when the device of the original position is employed.

I plan to short-circuit the complexities of the actual transmission mechanism by simply thinking of a set of empathetic preferences as being packaged in a social signal or meme—which is Dawkins' [151] name for the social equivalent of a gene. The imitative process is seen as a means of propagat-

---

[83]It is sometimes argued that it may be wrong for Adam to "legitimize" Eve's personal preferences by identifying so closely with her. Suppose, for example, that a child has been admitted to hospital after being injured in a road accident. Adam is a doctor and Eve is the child's mother. He wishes to save the child by giving a blood transfusion. She refuses permission on the grounds that the child's place in heaven will be lost if God's will is flouted. Adam may argue that if he were Eve, he would not be so blinded by prejudice, and therefore Eve's viewpoint should be disregarded—but how does such an attitude advance matters if Eve has some power to influence matters? (Perhaps the state is a theocracy in which hospitals are only barely tolerated.)

ing such memes in much the same way that the common cold virus finds its way from one head to another. As always, I keep things simple by assuming that all games to be played are games of complete information, which means that the rules of the game and the preferences of the players are taken to be common knowledge. In particular, it is assumed that the hypothetical bargaining game played behind the veil of ignorance has complete information. The empathetic preferences with which the players evaluate their predicament in the original position are therefore taken to be common knowledge.[84] Along with a set of empathetic preferences, a meme will also carry a recommendation about which bargaining strategy to use in the original position. Only when the stability of the system is threatened by the appearance of a "mutant" meme will they have reason to deviate from this normal bargaining strategy.

To explore the issue of evolutionary stability, imagine that all players are currently controlled by a normal meme N. A mutant meme M now appears. Will it be able to expand from its initial foothold in the population? If so, then the normal population is not evolutionarily stable.

Section 3.2.3 of Volume I describes two simple conditions for evolutionary stability, but only one of these is needed here—namely, the condition that (N, N) should be a *symmetric* Nash equilibrium of the underlying game in which M and N are strategies (Binmore [70, p.414]). In brief, N must be a best reply to itself.

To interpret this condition, imagine that player I has been infected with the mutant meme M while player II remains in thrall to the normal meme N.[85] Both players will test their recommended bargaining strategy against the empathetic preferences they find themselves holding, and adjust their behavior until they reach a Nash equilibrium of their bargaining game. I shall be assuming that this Nash equilibrium implements a suitable version of the Nash bargaining solution of the game (Section 1.3.1). As a consequence, player I and player II will each now receive some share of the benefits and burdens in dispute. The imitation mechanism that determines

---

[84]Without this heroic assumption, one would have to appeal to Harsanyi's [232] theory of games of incomplete information, and everything would become more complicated. In defense of the assumption in this particular case, one may observe that the players' empathetic preferences are assumed to be stable compared to their bargaining behavior. They would therefore have the opportunity to learn about each other's preferences in repeated play.

[85]If the fraction of the population controlled by M is ϵ, then the number of such mixed matches is $2\epsilon(1-\epsilon)$. The number of matches in which both players are controlled by M is ϵ^2, which is an order of magnitude smaller. What happens in mixed matches is therefore primary. Only when M and N do equally well in mixed matches does it become necessary to study what happens when both players are controlled by M. One is then led to the second condition for evolutionary stability given in Section 3.2.3 of Volume I.

224 CHAPTER 2. EVOLUTION IN EDEN

when it is appropriate to copy the memes we observe others using will take account of who gets what. Onlookers will almost all currently be subject to the normal meme N and so will evaluate the shares they see player I and II receiving in terms of the empathetic preferences embedded in N. If player I's payoff exceeds player II's, then I assume that onlookers who are vulnerable to infection are more likely to be taken over by the meme M controlling player I than by the meme N controlling player II. But then M will be a better reply to N than N is to itself.

A necessary condition for the evolutionary stability of a normal population is therefore that the empathetic preferences held by players I and II constitute what Section 1.2.7 of Volume I called a symmetric *empathy equilibrium*. In order to test whether a pair of empathetic preferences constitutes an empathy equilibrium each player should be asked the following question:

> Suppose that you could deceive everybody into believing that your empathetic preferences are whatever you find it expedient to claim them to be. Would such an act of deceit seem worthwhile to you *in the original position* relative to the empathetic preferences that *you actually hold?*

The right answer for an empathy equilibrium is *no*.

The tautological appearance of the argument leading to this criterion is misleading, since two substantive assumptions are concealed in the woodwork. In the simple two-person case with which we commonly work, player I is actually Adam and player II is actually Eve. However, players I and II are envisaged as being ignorant of their identities. Each acts as though he or she may turn out to be Adam or Eve with probability $\frac{1}{2}$. In this state of imagined uncertainty, they evaluate their prospects using their *empathetic* preferences. Someone who imitates player II in the preceding argument therefore joins player II in pretending that he doesn't know that player II is Eve. The imitator therefore doesn't evaluate player II's payoff in terms of Eve's *personal* utils, as he might well do if he were looking on from outside the players' society without empathizing with what was passing through Eve's mind. He joins Eve in adopting player II's point of view, and so interprets payoffs as player II interprets them. In the definition of an empathy equilibrium, this requirement appears as the condition that a player only considers whether an act of deceit would be worthwhile to him *in the original position*.

This assumption is entirely natural in the case of prehistoric insurance contracts. It would then be stupid to wait until after the uncertainty was resolved to evaluate the decisions made before it was known who would be the lucky hunter. I know that people very commonly do make such stupid evaluations—arguing that it was a mistake to bet on *Punter's Folly* because *Gambler's Ruin* actually won the race.[86] The equivalent prehistoric mistake would be to judge that the lucky hunter went wrong in negotiating an insurance contract with his unlucky colleague, because he then had to part with some of his food without receiving anything in return. But tribes that implicitly internalized such reasoning would eventually lose the evolutionary race.[87]

If evolution simply borrowed the mechanisms for operating the device of the original position from those already being used to operate primitive insurance contracts, then it would therefore be no surprise to find memes being evaluated from the point of view of players I and II rather than from the point of view of Adam and Eve. More generally, whenever I internally evaluate a meme being used by a neighbor with a view to incorporating it into my own repetoire, I presumably cannot avoid the need to empathize with my neighbor. Otherwise, I would be inadequately informed about the *context* in which it is appropriate to express the behavior packaged in the meme.[88] If my neighbor Eve models the context in which the meme is triggered in terms of a preference-belief model, then I must enter into that preference-belief model as best I can. That is to say, I must seek to appreciate *what* she was trying to achieve and *why* she acted as though this was the appropriate behavior to achieve her ends. But what preference-belief model occupies Eve's mind when she employs the device of the original position? It is the model of the world *as seen by player II*. It is true that this world is a fiction. When Eve adopts the viewpoint of player II, she is acting as though bound by the rules of a morality game rather than by the rules of the Game of Life. But if Immanuel Kant found it hard to distinguish between these games, why should we expect more from the intellectual powers of whatever algorithm our brains use when empathizing with others?

The second substantive assumption hidden in the woodwork concerns

---

[86]Politicians are typically evaluated in this way. Winston Churchill is a great statesman because things eventually worked out well after he chose to fight on at the beginning of the Second World War. But how would he be regarded if the future had followed a path that looked more probable at the time?

[87]Notice that I do not argue that prehistoric hunters would learn Bayesian decision theory as such. As Cosmides and Tooby [136] document, our behavior in a familiar social context often looks like it had been reasoned out very carefully, but we have great difficulty in applying the same principles in simpler but unfamiliar contexts. Those who have taught freshman calculus will not need to look elsewhere for evidence that Nature did not make us natural logicians.

[88]Autistic children have difficulty in understanding that something they know might not be known to others. If an autistic Adam observes Eve looking on while a toy is put away in a drawer, and then observes the toy being hidden elsewhere in Eve's absence, he is surprised that she looks for the toy in the drawer when she returns.

the *symmetry* of the empathy equilibria to which attention will be confined. For the argument supporting symmetry to make sense, it is essential that everyone learns from everyone else. If men only imitate men and women only imitate women, there is no reason why Adam and Eve's empathetic preferences should be the same. An essential precondition for confining attention to symmetric empathy equilibria is therefore that all the citizens of a society share in a common culture.[89]

It is at this level that I believe the community of spirit emphasized by modern communitarians and ancient moralists alike actually operates. As observed in Section 2.3.6 of Volume I, we are not bound together by iron shackles of duty and obligation, but by common understandings and conventions that propagate through societies like fungus through a rotten log. Each single filament in the system is so fragile that it can hardly survive the light of day. But, like Gulliver in Lilliput, we are no less tightly bound by the totality of such gossamer threads than if Nature had really provided us with the shackles that traditionalists invent.

However, although the subject of how a culture sustains itself is of enormous interest (Boyd and Richerson [98], Cavelli-Sforza and Feldman [118]), I hope my enthusiasm for my armchair speculations on the details of the underlying psychology of the process will be treated with the extreme caution that it deserves—especially since the formal moral theories developed in this book do not depend at all strongly on these details. For example, in this chapter, the notion of a symmetric empathy equilibrium will be employed only in arguing that social evolution will tend to equip everyone with the same empathetic preferences from which all moral content will eventually be eroded. But such conclusions could be derived from a variety of assumptions other than those I think most plausible.

2.5.5   The Long and the Short and the Medium

In an attempt to impose some order on the different dynamic processes studied in this book, Section 1.2.7 of Volume I proposed a crude classification of the time spans needed for various processes to converge to an equilibrium. As a minimum, we need to distinguish:

* **Short-run phenomena:** The short run corresponds to economic time—the time it takes for a market to adjust to an unanticipated piece of

---

[89]Sometimes, this requirement is expressed by saying that the norms respected by Adam and Eve are *universals* in their culture. Moral progress is then identified with successive widenings of the scope of universality—slaves are recognized as human or women as citizens. My theory attributes such expansions of the moral circle to the discovery that there is something to be gained by adopting the practices of outsiders or strangers.

news. I think of the short run being measured in minutes or hours.

• **Medium-run phenomena:** The medium run corresponds to social time—the time it takes for cultural norms or social conventions to adjust to a change in the underlying environment. I think of the medium run being measured in months or years.

• **Long-run phenomena:** One may regard the long run primarily as biological time—the time one has to wait for a gene pool to adjust to a new challenge. I think of it being measured in generations.

The modeling technique to be employed in adopting these distinctions is borrowed from the economic theory of the firm. Economic, social and biological processes actually proceed simultaneously, but models which reflect this reality would be prohibitively difficult to handle. One therefore attempts to approximate the way the world actually works by assuming that short-run processes are infinitely faster than medium-run processes, and medium-run processes are, in turn, infinitely faster than long-run processes. When studying short-run dynamic processes, one treats medium-run and long-run variables as fixed. The values of these medium-run and long-run variables then serve as parameters that determine the equilibrium to which the short-run process converges. In studying medium-run dynamic processes, one treats the long-run variables as fixed, but sets the short-run variables to their values at the economic equilibrium calculated in the short-run analysis. The long-run variables then serve as parameters that determine the social equilibrium to which the medium-run process converges.

Fortunately, it will not be necessary to ask how the long-run variables are determined in this chapter. However, in Section 4.8 a long-run dynamic process is studied by setting the medium-run variables to their values at the social equilibrium calculated in the medium-run analysis. The biological equilibrium to which the long-run process converges then determines the long-run variables. These then determine the medium-run variables, which then tie down the short-run variables.

My use of this modeling technique requires some heroic simplifying assumptions. I begin by taking Adam and Eve's personal preferences to be the long-run variables for the study. Our liking for clean water or healthy sex is therefore taken to be typical of all personal preferences. As observed above, treating personal preferences as being biologically determined has the advantage that they will be fixed in the current chapter, since time will only be allowed to vary in the short or medium run.

I know that some personal preferences actually do change from year to year.[90] Indeed, the argument of Section 2.5.2 concerning how we incorporate our degree of relationship with others into our personal preferences depends on some of our personal preferences being malleable over even shorter periods. But I shall ignore such complicating examples, just as I shall ignore the fact that some socially determined behaviors have endured without apparent change for millennia.

Adam and Eve's empathetic preferences are treated as the medium-run variables. A meme that packages a new set of empathetic preferences spreads through a population by imitation, and this cannot happen overnight. Empathetic preferences therefore remain stable in the short run. However, the players' strategy choices may vary in the short run as they experiment with deviations from the norm to test whether it remains optimal relative to their preferences. Recall that it is by appealing to such continuous jockeying for advantage that I justify confining attention to the equilibria of the game being played.

Having identified strategy choices as short-run variables, the first step is to appeal to Chapter 1 in order to characterize the economic equilibrium that determines Adam and Eve's strategy choices in the bargaining game they envisage playing in the original position. Behind the veil of ignorance, Adam and Eve evaluate social contracts using their empathetic preferences. The deal they reach in the original position will therefore depend on these medium-run variables, which remain fixed in the short-run. The next step is then to find out how Adam and Eve's empathetic preferences are determined in the medium run by their personal preferences. But this step will be delayed until Section 2.6.2. The final step of determining Adam and Eve's personal preferences in the long run will only be attempted in a different context in Chapter 4.

2.6   Nonteleological Utilitarianism

This chapter opened by proposing a division of ethics into theories of the Good, the Right and the Seemly. The leading theory of the Good is utilitarianism, which comes in various flavors, I am concerned only with the variety of that Goodin [209] refers to as "public" and Hardin [224] as "institutional". Institutional utilitarianism can in turn be split into ipsedixist, ideal observer, and philosopher-king versions. In Section 2.2, I argued that the modern consensus on political legitimacy is most closely matched by a

---

[90]Although the examples commonly quoted are not always apt. For example, people are not really dedicated followers of fashion because their preferences for high and low hemlines change over time. They follow fashion primarily because they want to be regarded as fashionable.

philosopher-king theory. However, in order to generate sound foundations for a philosopher-king version of utilitarianism, it is necessary to adopt a nonteleological approach, which recognizes that the things people say are good are good only because people say so. This amounts, of course, to reducing the Good to the Seemly.

In preparation for this enterprise, the previous section offered some speculations about the evolutionary origins of empathetic preferences and their role in operating the device of the original position. The current section brings these ideas together in order to illustrate the shape that a full-fledged theory of a utilitarian philosopher-king might possibly take.

The techniques employed in developing the theory are borrowed from Harsanyi's [233] nonteleological defense of utilitarianism. However, as explained in the preliminary discussion of his work offered in Sections 4.3 and 4.4 of Volume I, the cold waters of Kantian constructivism in which Harsanyi chooses to swim have no attraction for a Humean like myself. The foundations of the theory presented here are therefore very different.

Unfortunately, even Harsanyi's Kantian version of the theory has been largely neglected by moral philosophers. In consequence, it is necessary to discuss the nuts and bolts of his theory at length. This, in turn, entails mastering several pages of simple algebra. Fortunately, the algebra of his argument was described in detail in Sections 4.3 and 4.4 of Volume I. In this section, it will therefore be possible to move through the technical preliminaries at a brisk pace. Indeed, many readers may wish to skip the coming review of the underlying utility theory altogether.

2.6.1 Commitment in Eden

We begin with the short-run story, in which all preferences are fixed. It will be important later to turn to the medium run when the time comes to ask why Adam and Eve have one set of empathetic preferences rather than another, but for the moment, these are taken as given.

A marriage contract. In the fable to be told, an impartial philosopher-king represents a benign and enlightened government that exists solely to enforce the laws the people make for themselves. The state is simplified to the Garden of Eden, where the only citizens are Adam and Eve, whom we assume to be totally unrelated. Their problem is to negotiate a suitable marriage contract.

It is important to keep in mind that this is to be an old-style marriage, in that the philosopher-king will ensure that Adam and Eve remain bound by their marriage vows even though one or the other may beg for release once the honeymoon is over. He does not, of course, bind them against their will at the time of their marriage. On their wedding day, they *want* their vows to bind, because they know that they will be tempted to go astray in the future. Nor does the philosopher-king impose the form of the contract upon them. He accepts that they will determine their marriage contract using the device of the original position. He may hope that the marriage contract will not make Eve into Adam's chattel, but he agrees with Rawls that Adam and Eve should be bound by its terms whatever they may be.[91]

Harsanyi [233] and Rawls [417] assume that Adam and Eve are able and willing to commit *themselves* to the hypothetical deal they would reach in the original position.[92] Section 1.2.4 of Volume I explains the insuperable difficulties that such an assumption entails in the absence of some external mechanism that holds Adam and Eve to their commitments. But in the story to be told here, the philosopher-king provides just such an external mechanism. His presence allows us to put aside our doubts about the wisdom of attributing full commitment powers to Adam and Eve. In consequence, the viable social contracts need no longer be restricted to equilibria in the Game of Life as argued in Section 1.2.3 of Volume I. Instead, the set X of payoff pairs corresponding to feasible social contracts coincides with the entire cooperative payoff region of the Game of Life (Section 1.2.1).

The folk theorem mentioned in Section 2.2.6 of Volume I ensures that such an expansion of the set X makes no appreciable difference when the Game of Life is modeled as a *repeated* game. On the other hand, the fact that Adam and Eve have access to the entire cooperative payoff region makes an enormous difference when the Game of Life is modeled as a one-shot game like the Prisoners' Dilemma. However, whatever the character of the underlying Game of Life, its cooperative payoff region X is necessarily *convex*, for the reasons given in Section 1.2.2, which also discusses the reasons for assuming that X is closed, bounded above and comprehensive. In this chapter, it is also convenient to assume that its boundary is smooth. The set X then looks like Figure 2.3(a).

The asymmetries of the set X register the ineradicable inequalities between Adam and Eve for which the device of the original position provides redress (Section 1.2.5 of Volume I). To operate this social tool, Adam and Eve must be able to envisage the social contract on which they would have agreed after bargaining behind the veil of ignorance. To analyze their sit-

---

[91]See the slaveholder's argument of Section 1.2.4 of Volume I.

[92]Of course, both Harsanyi and Rawls qualify their commitment requirements. For example, Harsanyi [233] tells us that Adam and Eve are *morally* bound to honor the hypothetical deal that would be made in the original position. But such a statement has only rhetorical force. An author who argues that people should behave as though committed to a particular course of action because he chooses to label those who deviate as immoral, is simply saying that one ought to do something because one ought to do it.

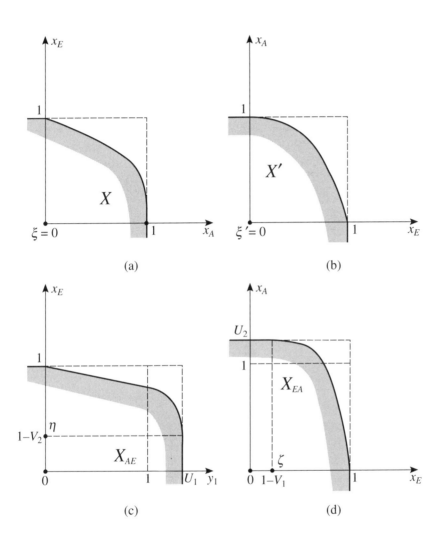

Figure 2.3: Various transformations of the set X.

uation during such a negotiation, they need names to refer to each other while ignorant of their true identities. As previously, Adam will be called player I when in the original position, and Eve will be called player II.

Behind the veil of ignorance, the players face an uncertain situation. They do not know which of two events, AE or EA, will be revealed after their negotiations are concluded. Because we chose the notation, we know that they will actually observe event AE, in which player I is Adam and player II is Eve. But the protagonists themselves must also take account of the event EA in which player I turns out to be Eve and player II to be Adam. Unlike Harsanyi [233], I allow Adam and Eve to come to an arrangement that makes the social contract contingent on who turns out to occupy which role. That is to say, they are assumed free to agree to operate one social contract C if event AE occurs and another social contract D if event EA occurs. This contingent social contract will be denoted by (C, D).

Utility theory. A contingent contract (C, D) is an uncertain prospect like that illustrated in Figure 4.6(a) of Volume I. In order to use Bayesian decision theory to assess (C, D), we need to know the players' *preferences* over the possible consequences C and D, and their *beliefs* about the events AE and EA (Section 4.5 of Volume I).

Player I's preferences are given by his empathetic Von Neumann and Morgenstern utility function v_1. His beliefs are represented by a subjective probability distribution, which reduces in this simple case to the probability p_1 he attaches to the event AE. Player I's expected utility for the contingent social contract (C, D) is then

$$w_1(C, D) = p_1 v_1(C, A) + (1 - p_1) v_1(D, E) \,. \qquad (2.6)$$

In this expression, $v_1(C, A)$ is the utility player I derives if the social contract C is operated with him in Adam's role. Similarly, $v_1(D, E)$ is the utility he derives if the social contract D is operated him in Eve's role.

The *empathetic* utility function v_1 should not be confused with Adam's *personal* utility function u_A. The former quantifies the preferences he might express when comparing being Adam wearing a fig leaf with being Eve eating an apple. The latter quantifies his own preferences about the consumption of fig leaves or apples. But the two types of preference are not independent of each other. For example, Adam's own personal preferences over fig leaves and apples may be the reverse of Eve's, but to empathize fully with Eve he must accept that, if he were in Eve's shoes, then he would have her personal preferences.

As Section 4.3.1 of Volume I explains, the assumption that a person empathizes fully with both Adam and Eve implies that his empathetic utility function can be expressed in terms of their personal utility functions u_A and u_E. In fact, once we know Adam and Eve's personal utility functions, we need only know the values of two positive constants U_1 and V_1 to be able to write down player I's empathetic utility function. From equation (4.8) of Section 4.3.1 in Volume I, we have that

$$v_1(\mathcal{X}, A) = U_1 u_A(\mathcal{X}), \qquad (2.7)$$

$$v_1(\mathcal{X}, E) = 1 - V_1(1 - u_E(\mathcal{X})), \qquad (2.8)$$

provided that all the utility scales are calibrated appropriately.[93] These expressions can be interpreted as saying that player I has an *intra*personal standard of utility comparison that equates V_1 of Adam's utils with U_1 of Eve's. Writing $w_1(\mathcal{C}, \mathcal{D})$ of (2.6) and the corresponding formula for $w_2(\mathcal{C}, \mathcal{D})$ in terms of Adam and Eve's *personal* preferences, we obtain that

$$w_1(\mathcal{C}, \mathcal{D}) = p_1 U_1 u_A(\mathcal{C}) + (1 - p_1)\{1 - V_1(1 - u_E(\mathcal{D}))\}, \qquad (2.9)$$

$$w_2(\mathcal{C}, \mathcal{D}) = p_2\{1 - V_2(1 - u_E(\mathcal{C}))\} + (1 - p_2)U_2 u_A(\mathcal{D}). \qquad (2.10)$$

Section 4.7 of Volume I argues that the putative manner in which the device of the original position adjudicates fairness questions in real life requires that $p_1 = p_2 = \frac{1}{2}$. Since practice is to be our guide, this assumption should therefore be part of the *definition* of the original position. That is to say, the act of imagination each player employs when passing behind the veil of ignorance should include behaving as though he believes that it is equally likely he will turn out to be Adam or Eve.

When the requirement $p_1 = p_2 = \frac{1}{2}$ is incorporated into the expressions (2.9) and (2.10), the utilities that players I and II respectively expect from an agreement to implement $\mathcal{C}$ if AE occurs and $\mathcal{D}$ if EA occurs become:

$$w_1(\mathcal{C}, \mathcal{D}) = \tfrac{1}{2}U_1 u_A(\mathcal{C}) + \tfrac{1}{2}\{1 - V_1(1 - u_E(\mathcal{D}))\}, \qquad (2.11)$$

---

[93] The calibration employed in Section 4.3.1 of Volume I requires anchoring the scales to two ideal outcomes $\mathcal{W}$ and $\mathcal{L}$, which one can think of as heaven and hell. Both personal utility scales are then chosen to have their zero at $\mathcal{L}$ and their unit at $\mathcal{W}$. No significance should be attached to this entirely arbitrary method of fixing Adam and Eve's utility scales. It is adopted only because their personal utility scales must be anchored somehow before one can begin to discuss how their utils are to be compared. Otherwise, to say that Adam's utils are worth more than Eve's would be like saying that a cubit is longer than a rood without having any idea of how these obsolete measures of length were defined. Unfortunately, the existence of the zero-one rule leads some people to believe that the mere act of anchoring Adam and Eve's utility scales *necessarily* implies that the utils so defined are to be traded one for one. But such a standard of interpersonal comparison would be just as arbitrary as the choice of Adam and Eve's personal utility scales. Similar confusion over the anchoring of player I and player II's empathetic utility scales is unlikely, since this is more obviously arbitrary. Their empathetic scales are fixed so that $v_i(\mathcal{L}, A) = 0$ and $v_i(\mathcal{W}, E) = 1$. The constants U_i and V_i are then given by $U_i = v_i(\mathcal{W}, A)$ and $1 - V_i = v_i(\mathcal{L}, E)$.

$$w_2(\mathcal{C}, \mathcal{D}) \;=\; \tfrac{1}{2}\{1 - V_2(1 - u_E(\mathcal{C}))\} + \tfrac{1}{2}U_2 u_A(\mathcal{D}). \qquad (2.12)$$

One may ask why the assumption $p_1 = p_2 = \tfrac{1}{2}$ is so readily taken for granted when the device of the original position is assessed as a fairness criterion. One possible reason is that the values of p_1 and p_2 do not actually matter in the medium run. To pursue this point, define $U_1^* = 2p_1 U_1$, $V_1^* = 2(1 - p_1)V_1$, $U_2^* = 2(1 - p_2)U_2$ and $V_2^* = 2p_2 V_2$. If the irrelevant constant terms are discarded, equations (2.9) and (2.10) then reduce to (2.11) and (2.12), provided that the coefficients in the latter equations are all starred. But the forces of social evolution modeled in Section 2.6.2 do not care whether these constants are starred or not.

□ **The bargaining problem behind the veil of ignorance.** Section 1.2 tells us that the simplest type of bargaining problem can be formulated as a pair (T, τ), where T consists of all payoff pairs on which the two players can agree, and τ is the payoff pair that will result if there is a disagreement. Our first problem is therefore to determine the set T of feasible payoff pairs for players I and II in the original position.

Figure 2.3(a) shows the set X of feasible personal payoffs pairs that Adam and Eve can achieve by coordinating on a suitable social contract. Behind the veil of ignorance, players I and II attach probability $\tfrac{1}{2}$ to both events AE and EA. Since they evaluate an uncertain prospect by calculating its expected utility, they regard a contingent social contract that leads to the payoff pair y when AE occurs and z when EA occurs as equivalent to the pair

$$t = \tfrac{1}{2}y + \tfrac{1}{2}z, \qquad (2.13)$$

which is simply a compressed version of (2.11) and (2.12).[94]

The payoff pair t is shown in Figure 2.4(a). It lies halfway along the line segment joining the payoff pairs y and z. The pair y lies in the set X_{AE} consisting of all payoff pairs that players I and II regard as attainable should AE occur. The set X_{AE} has a similar shape to X, but needs to be rescaled in order to reflect the relative worth that player I places on Adam's utils and player II places on Eve's utils. From (2.7), we know that player I regards a payoff of x_A to Adam as being worth $y_1 = U_1 x_A$. From the equivalent of (2.8) for player II, we know that she regards a payoff of x_E to Eve as being worth $y_2 = 1 - V_2(1 - x_E)$. To obtain X_{AE} from the set X, we must therefore replace each payoff pair $x = (x_A, x_E)$ in X by the rescaled pair $y = (y_1, y_2) = (U_1 x_A, 1 - V_2(1 - x_E))$. The resulting set X_{AE} is shown in Figure 2.3(c).

---

[94]Recall that $t = \tfrac{1}{2}y + \tfrac{1}{2}z$ simply means that $t_1 = \tfrac{1}{2}y_1 + \tfrac{1}{2}z_1$ and $t_2 = \tfrac{1}{2}y_2 + \tfrac{1}{2}z_2$.

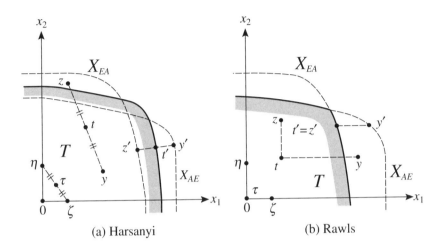

(a) Harsanyi   (b) Rawls

Figure 2.4: Constructing the set T.

To obtain X_{EA} from the set X is slightly more complicated because player I will become Eve if EA occurs. However, his payoffs are measured on the horizontal axis while hers are measured on the vertical axis. As shown in Figure 2.3(b), it is therefore necessary to begin by swapping over Adam and Eve's axes to obtain the set X' (which is simply the reflection of X in the line $x_A = x_E$). After this transformation, player I's payoffs and Eve's payoffs are both measured on the horizontal axis, and so we can proceed as before. The set X_{EA} has a similar shape to X', but needs to be rescaled in order to reflect the relative worth that player I places on Eve's utils and player II places on Adam's utils. To obtain X_{EA} from X', replace each payoff pair $x = (x_E, x_A)$ in X' by the rescaled pair $z = (z_1, z_2) = (1 - V_1(1 - x_E), U_2 x_A)$. The set X_{EA} is shown in Figure 2.3(d).

The preceding discussion of how X_{AE} and X_{EA} are constructed from X is a necessary preliminary to drawing the set T of all payoff pairs that are feasible for players I and II in the original position. As Figure 2.4(a) illustrates, T is the set of all $t = \frac{1}{2}y + \frac{1}{2}z$, with y in X_{AE} and z in X_{EA}. (The Pareto-frontier of T can be characterized as the set of all $t = \frac{1}{2}y + \frac{1}{2}z$ with the property that the tangent to the Pareto-frontier of X_{AE} at y has the same slope as the tangent to the Pareto-frontier of X_{EA} at z.)

Having tied down the set T, the next step in specifying the bargaining problem (T, τ) is to determine the *status quo* τ. This is not a problem in Harsanyi's own theory, which requires no knowledge of the properties of the Game of Life beyond those embodied in the set X. The same is not true of my own approach to the social contract problem to be described in Chapter 4. Nor is it true of my attempt in the current chapter to reconstruct Harsanyi's nonteleological defense of utilitarianism.

Like Wolff [565, 566], it seems to me that a social contract theory which makes no reference to a state of nature cannot be complete. We therefore cannot neglect the *strategic* structure of the Game of Life altogether. However, in a defense of utilitarianism, it would be pedantic to dwell at length on how the strategic structure of the Game of Life determines the state of nature, since one certainly does not need this information to locate the utilitarian outcome in the set X. The state-of-nature issue will therefore be pushed into the background in this chapter by using a particularly simple model of the Game of Life in which the question of what happens if Adam and Eve are unable to agree on a social contract is trivialized. But before I trivialize the state-of-nature problem in this way, it may be as well to outline the principles that determine τ in the general case.

The state of nature.   Hobbes' [251] gloomy social contract theory defends authoritarianism as the only viable alternative to his state of nature, which he famously characterizes as "solitary, poor, nasty, brutish and short".

Section 1.2.1 of Volume I accepts that a theory of the state based on an implicit bargain among its citizens makes no sense without some understanding on what the consequences of a failure to reach agreement would have been. However, I do not see that anything is to be gained by attempts to base the organization of a modern state on speculations about the political arrangements of prehistoric societies. The alternative to instituting a reform is not that we return to some forgotten or imaginary primeval state, but that we continue to operate without the reform. The state of nature in my own theory to be presented in Chapter 4 therefore consists of society's current social contract—as represented by the payoff pair ξ in Figure 1.3(b) of Volume I.

However, my own theory differs sharply from the paternalistic theory with which we are currently concerned in having no philosopher-king to enforce Adam and Eve's commitments. With a truly dispassionate philosopher-king in place, the alternative to an agreement in the original position will be that he enforces the threats the players make behind the veil of ignorance about what would happen if they were to disagree.

Nash's variable-threat theory of Section 1.5.2 describes the principles that players I and II would use in choosing their optimal threats. Matters are somewhat more complicated here, because the actions that a player threatens to make behind the veil of ignorance will need to be contingent on whether he turns out to be Adam or Eve. Suppose that implementing these threats will result in the payoff pair η if event AE occurs and ζ if event EA occurs. Then the *status quo* for the bargaining problem faced by players I and II in the original position is $\tau = \frac{1}{2}\eta + \frac{1}{2}\zeta$ as illustrated in Figure 2.4(a).

As noted in Section 1.5.2, if Adam and Eve use Nash's variable-threat theory to bargain over the one-shot Prisoners' Dilemma in full knowledge of their identities, then each will commit himself to play *hawk* should the negotiations fall through. The state of nature that players I and II create for themselves in the original position is therefore likely to be sufficiently miserable to give pause even to a misanthrope like Hobbes. Precisely how miserable τ will be depends on the strategic structure of the Game of Life— about which we have assumed nothing so far.

Specializing the Game of Life. Rather than face the bookkeeping problems involved in dealing with state-of-nature problems in the general case, I shall now reduce the predicament faced by Adam and Eve in negotiating a marriage contract to the crudest of monetary questions: how to split the dowry. As in Section 1.2.4, it can be assumed without loss of generality that the amount of money involved is one dollar. A social contract $\mathcal{C}$ is Pareto-efficient in such a divide-the-dollar problem when no money is wasted. A Pareto-efficient social contract $\mathcal{C}$ that assigns a fraction s of the dollar to Adam will therefore assign a fraction $1 - s$ of the dollar to Eve.

Rather than continuing to denote Adam's personal utility for the social contract $\mathcal{C}$ by $u_A(\mathcal{C})$, I shall redefine $u_A(s)$ for the divide-the-dollar case to be Adam's utility for the sum of money s that he receives when $\mathcal{C}$ is implemented. The notation will then be consistent with that of Section 1.2.4. Similarly, Eve's payoff when $\mathcal{C}$ is implemented is redefined to be $u_E(1 - s)$, which is her utility for the sum of money $1 - s$.

The Von Neumann and Morgenstern utility functions u_A and u_E will be assumed to be twice differentiable, strictly increasing, and concave. Among other things, this implies that Adam and Eve are risk averse and always prefer more money to less (Section 4.6 of Volume I). As for anchoring the utility scales, take $\mathcal{L}$ to be the state in which both Adam and Eve get nothing, so that $u_A(0) = u_E(0) = 0$. Similarly, take $\mathcal{W}$ to be the (infeasible) state in which both get a dollar, so that $u_A(1) = u_E(1) = 1$.

The Pareto-frontier of the set X of payoff pairs corresponding to feasible social contracts consists of the pairs $x = (u_A(s), u_E(1 - s))$ with $0 \leq s \leq 1$.

With our assumptions on u_A and u_E, the set X retains all the properties assigned to it hitherto.[95] In particular, although treating Adam and Eve's Game of Life as a divide-the-dollar problem symmetrizes their problem in terms of money, it does not symmetrize their problem in terms of their personal payoffs, unless we assume that they have precisely the same attitude to taking risks by taking $u_A = u_E$. Even in the divide-the-dollar problem, inequalities requiring redress will therefore remain.

It will be obvious to mathematicians that one gives up very little by way of generality in treating Adam and Eve's Game of Life as a divide-the-dollar problem.[96] Nor is much lost by going further and taking the strategic structure of their Game of Life to be that of the Nash Demand Game described in Section 1.5.1. In this game, Adam and Eve simultaneously announce payoff demands x_A and x_E. If the pair $x = (x_A, x_E)$ of demands lies in the feasible set X, each receives his demand. Otherwise each gets nothing. That is to say, the *status quo* outcome $\xi = (u_A(0), u_E(0)) = (0,0)$ shown in Figure 2.3(a) is implemented. When the utilities in Figure 2.3(a) are rescaled to transform X into the set X_{AE} of Figure 2.3(c), the *status quo* point $\xi = (0,0)$ transforms into $\eta = (0, 1 - V_2)$ of Figure 2.3(c). When the utilities in Figure 2.3(b) are rescaled to transform X' into the set X_{EA} of Figure 2.3(d), the point $\xi' = (0,0)$ transforms into $\zeta = (1 - V_1, 0)$ of Figure 2.3(d).

Taking the Game of Life to be the Nash Demand Game only seriously restricts the generality of conclusions that relate to the location of the state of nature. Indeed, my chief reason for choosing to specialize the Game of Life to the Nash Demand Game is to trivialize the state-of-nature issue. With this Game of Life, the worst threat that a player can make in the original position is to commit himself to make demands on emerging from behind the veil of ignorance that cannot be met whatever demand the other player may make. The *status quo* of the Nash Demand Game will then be implemented.

Having specialized the Game of Life to the Nash Demand Game, we can therefore simply assume that players I and II evaluate the result of a failure to agree in the original position as being equivalent to the payoff pair $\tau = \frac{1}{2}\eta + \frac{1}{2}\zeta$.

Solving the bargaining problem. Having specialized the Game of Life to trivialize the determination of τ, we can now forget the details of the

---

[95]Indeed, the fact that Adam and Eve are risk averse assures that X is convex without the need to postulate social contracts involving any randomization (other than the hypothetical random move that assigns identities to players I and II).

[96]Money simply serves as a parametrization of the curve that bounds X.

Game of Life until Section 2.6.2.

Recall that the bargaining problem faced by players I and II in the original position has been formulated as the pair (T, τ) shown in Figure 2.4(a). It is easy to describe the solution to this bargaining problem in geometric terms. As illustrated in Figure 2.5(a), Nash's theory of bargaining with commitment predicts that the bargaining outcome will be the symmetric Nash bargaining solution σ for the bargaining problem (T, τ).

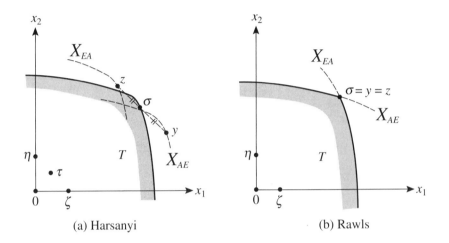

(a) Harsanyi	(b) Rawls

Figure 2.5: Applying the Nash bargaining solution to (T, τ).

Before discussing what this solution implies for Adam and Eve's personal payoffs, it is as well to emphasize the strong informational assumptions required by the argument leading to the payoff pair σ in Figure 2.5(a). In the original position, players I and II forget whether they are Adam or Eve. Since their own empathetic preferences are common knowledge, it follows that they must also forget which empathetic preference derives from Adam and which from Eve. But everything else is assumed to be common knowledge between players I and II. This assumption is essential in the case of *all* the data used to construct the bargaining problem (T, τ). Players I and II therefore know the rules of the Game of Life, and hence which potential social contracts are feasible. Each also knows Adam and Eve's personal preferences, and his own empathetic preferences together with those of his bargaining partner.

All this is a great deal to know and so it is worth observing that a more general approach is possible that does not depend on making such strong epistomological assumptions. However, without strong informational assumptions, it is necessary to appeal to Harsanyi's [232] theory of games of incomplete information, with the result that the technical problems would become much harder. But there are no conceptual obstacles to an advance in this direction.[97]

Returning to the question of how an agreement on σ in the original position translates into personal payoffs to Adam and Eve, it is necessary to recall that for σ to be admissible as a member of the set T, it must be of the form $\sigma = \frac{1}{2}y + \frac{1}{2}z$, where y is in X_{AE} and z is in X_{EA}. One must also remember that the bargaining that supposedly takes place behind the veil of ignorance is only hypothetical. Adam and Eve only pretend to forget their identities when using the device of the original position to compute a fair social contract. In fact, player I is actually Adam and player II is actually Eve. Of the two events AE and EA, it is therefore the former that actually obtains.

It follows that the social contract $\mathcal{C}$ which will actually be operated corresponds to the payoff pair $y = (y_1, y_2)$ in X_{AE} illustrated in Figure 2.5(a). In terms of Adam and Eve's original personal utility scales, the social contract $\mathcal{C}$ yields the payoff pair $h = (h_A, h_E)$ defined by $y_1 = U_1 h_A$ and $y_2 = 1 - V_2(1 - h_E)$. As far as I know, there is no neat way to summarize the payoff pair h in terms of the set X and the underlying Game of Life. However, matters become more promising in the symmetric case illustrated in Figure 2.6(a).

When $U_1 = U_2 = U$ and $V_1 = V_2 = V$, Figure 2.5(a) translates into the symmetric Figure 2.6(a). In particular, the bargaining problem (T, τ) becomes symmetric, and so the outcome

$$N = \tfrac{1}{2}H + \tfrac{1}{2}K \,, \tag{2.14}$$

obtained by using the symmetric Nash bargaining solution is symmetric as well.

The symmetry ensures that the payoff pair N can be achieved using the same social contract $\mathcal{C}$ whether AE or EA occurs. However, the event that actually obtains is AE and so the personal payoff pair $h = (h_A, h_E)$ that Adam and Eve actually experience when $\mathcal{C}$ is implemented is given by $H_1 = U h_A$ and $H_2 = 1 - V(1 - h_E)$.

In the asymmetric case, it proved difficult to characterize the payoff pair h as a point of X. But here it is easily identified as the point x in X at

---

[97] The Nash bargaining solution must be generalized as in Harsanyi and Selten [236]. A corresponding generalization of Nash's threat theory appears in Binmore [66].

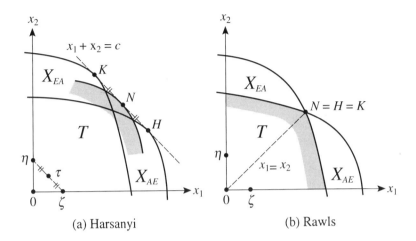

(a) Harsanyi   (b) Rawls

Figure 2.6: The symmetric case: $U_1 = U_2 = U$ and $V_1 = V_2 = V$.

which the weighted utilitarian social welfare function

$$W_h(x) = U x_A + V x_E$$

is maximized. To see this, observe that H and K in Figure 2.6(a) lie on a common tangent $x_1 + x_2 = c$ to the Pareto-frontiers of X_{AE} and X_{EA}. It follows that H is the point in X_{AE} at which the social welfare function $W_H(x) = x_1 + x_2$ is maximized. But the function defined by $x_1 = U x_A$ and $x_2 = 1 - V(1 - x_E)$ that maps X_{AE} to X, transforms $x_1 + x_2$ into $U x_A + 1 - V(1 - x_E)$. The constant $1 - V$ is irrelevant to the maximization, and so to maximize W_H on X_{AE} is the same as maximizing W_h on X.

Figure 2.7(a) duplicates Figure 1.3(a) of Volume I in showing the location of h as the point x in X at which W_h is maximized. Recall that such utilitarian bargaining solutions are independent of the *status quo*. This fact signals that, although one can never forget what would happen if there were a disagreement in a bargaining discussion, the precise location of τ proves irrelevant to the final utilitarian conclusion when U and V are given.

Provided that one is willing to swallow the assumption that $U_1 = U_2 = U$ and $V_1 = V_2 = V$, the argument leading to h in Figure 2.7(a) extends Harsanyi's nonteleological defense of utilitarianism to the case of contingent social contracts. As will be clear from the treatment of this topic in

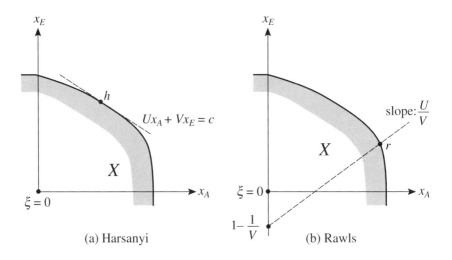

Figure 2.7: Utilitarian and Rawlsian solutions.

Appendix B, such an achievement would not be worth the effort if we did not need the apparatus developed along the way to tackle the question of how the parameters U_1, U_2, V_1 and V_2 are determined in the medium run.

2.6.2   Interpersonal Comparison in the Medium Run

The discussion of Section 2.6.1 is a short-run analysis in which all preferences are fixed. We now turn to the medium run, in which personal preferences are fixed but social evolution has the chance to mold empathetic preferences. The aim is to study the effect of the forces of social evolution on the values of U_1, U_2, V_1 and V_2 that determine the shape of the sets X_{AE} and X_{EA} in Figure 2.5(a).

Section 2.5.4 introduces the notion of a symmetric empathy equilibrium as the likely end product when evolutionary pressures based on imitation act on empathetic preferences in the medium run. The symmetry requirement was not emphasized in Section 2.5, but it is a standard requirement when the evolutionary stability concept of Maynard Smith and Price [348] is applied in a symmetric game.[98] To impose such a requirement in the current case is to assume that Adam and Eve share a common culture in which

---

[98]If mixed strategies are allowed, all finite symmetric games have a symmetric Nash equilibrium. They may not admit an evolutionarily stable strategy, but we shall not be using the full force of the definition of evolutionary stability.

everyone learns from everyone else. To argue for symmetry is then to claim that people who copy whatever behavior currently seems most successful are likely to end up behaving similarly.[99]

As Section 2.6.1 recalls, Harsanyi [233] makes a Kantian appeal to the Harsanyi doctrine in order to argue that $U_1 = U_2 = U$ and $V_1 = V_2 = V$. The same result is achieved equally cheaply here by restricting attention to symmetric empathy equilibria. The argument of Section 2.6.1 is then adequate to demonstrate that Adam and Eve will call upon the philosopher-king to maximize the utilitarian welfare function

$$W_h = U x_A + V x_E .$$

It is therefore entirely painless to translate Harsanyi's [233] nonteleological defense of utilitarianism into my naturalistic terms.

However, Harsanyi's appeal to his doctrine leaves us with no clue as to how the all-important ratio U/V is determined. If we were unconcerned about why the players think it appropriate to treat V of Adam's utils as being worth the same as U of Eve's, then Harsanyi's silence on this subject would create no problem. But the question of how interpersonal comparisons are made seems to me too important to be shrugged aside. On the contrary, the fact that insight into this question can be obtained using the concept of an empathy equilibrium is one of the major advantages of my approach.

Even for asymmetric empathy equilibria, we shall see that the personal payoffs Adam and Eve receive as a result of bargaining as though behind the veil of ignorance are precisely the same as the payoffs they would have gotten if they had solved the bargaining problem (X, ξ) directly using the symmetric Nash bargaining solution. When allowed to operate in the medium run, the effect of social evolution is therefore to leach out all moral content from the device of the original position. Reactions to this claim vary quite a lot. Those who think the result obvious are invited to skip forward to Section 2.7, where its implications are explored at some length. Those who are more skeptical will have to endure some mathematical huffing and puffing.

To keep the mathematics simple, the Game of Life is again specialized to be a divide-the-dollar problem. A Pareto-efficient, contingent social contract is therefore specified by the amount r that Adam is paid if AE occurs and the amount s that he is paid if EA occurs. Adam's personal payoffs in the two cases are $u_A(r)$ and $u_A(s)$. The corresponding personal payoffs for Eve are $u_E(1-r)$ and $u_E(1-s)$. If no agreement is reached on dividing the

---

[99]Teenage fashions are a notable example. It is amazing how uniformly adolescents can dress when left to choose for themselves.

dollar, then Adam and Eve each get nothing. This state of nature yields a
payoff of $u_A(0) = 0$ to Adam and $u_E(0) = 0$ to Eve.

The first step is to find an algebraic characterization of the symmetric
Nash bargaining solution σ of Figure 2.5(a). As explained in Section 1.3.1,
one of the ways of characterizing σ is as the point t in T at which the Nash
product

$$\Pi = (t_1 - \tau_1)(t_2 - \tau_2)$$

is maximized. In considering this product, we can restrict attention to
Pareto-efficient pairs t in T. Each such pair t corresponds to a contingent
social contract (r, s). It follows that

$$
\begin{aligned}
t_1 &= \tfrac{1}{2} U_1 u_A(r) + \tfrac{1}{2}\{1 - V_1(1 - u_E(1 - s))\}, \\
t_2 &= \tfrac{1}{2} U_2 u_A(s) + \tfrac{1}{2}\{1 - V_2(1 - u_E(1 - r))\}.
\end{aligned}
$$

Since $u_A(0) = u_E(0) = 0$, $\tau_1 = \tfrac{1}{2}(1 - V_1)$ and $\tau_2 = \tfrac{1}{2}(1 - V_2)$. Thus

$$\Pi = \tfrac{1}{4}\{U_1 u_A(r) + V_1 u_E(1 - s)\}\{U_2 u_A(s) + V_2 u_E(1 - r)\}.$$

This expression looks less formidable if we throw away the irrelevant factor
of $\tfrac{1}{4}$ and change the notation so that $f(x) = u_A(x)$, $g(x) = u_E(1 - x)$,
$\phi = U_1/V_1$ and $\psi = U_2/V_2$. The problem than reduces to maximizing

$$\Pi = \{\phi f(r) + g(s)\}\{\psi f(s) + g(r)\} \tag{2.15}$$

subject to the constraints $0 \le r \le 1$ and $0 \le s \le 1$.

The possibility of a corner solution is neglected.[100] A necessary condition for (r, s) to be an interior solution to the problem of maximizing Π is
that both partial derivatives $\partial\Pi/\partial r$ and $\partial\Pi/\partial s$ are zero. Thus

$$
\begin{aligned}
\phi f'(r)\{\psi f(s) + g(r)\} + g'(s)\{\phi f(r) + g(s)\} &= 0, & (2.16) \\
g'(r)\{\psi f(s) + g(r)\} + \psi f'(s)\{\phi f(r) + g(s)\} &= 0. & (2.17)
\end{aligned}
$$

Note that (2.16) and (2.17) constitute a system of two linear equations
in the unknowns $X = \psi f(s) + g(r)$ and $Y = \phi f(r) + g(s)$. A condition for
this system to have a solution other than $X = Y = 0$ is that its determinant
vanish. Thus,

$$\phi\psi f'(r)f'(s) = g'(r)g'(s). \tag{2.18}$$

Equation (2.18) has a geometric interpretation in Figure 2.5(a). It says that
the slope to the Pareto-frontier of X_{AE} at y is the same as the slope to the

---

[100]The product Π is maximized subject to the constraints $0 \le r \le 1$ and $0 \le s \le 1$.
A corner solution occurs if the maximum is achieved where one of these constraints is
active. An interior solution occurs when the maximum occurs at a point (r, s) with
$0 < r < 1$ and $0 < s < 1$.

Pareto-frontier of X_{EA} at z.[101] As noted in Section 2.6.1 when constructing the set T, this is actually always true of y and z whenever $t = \frac{1}{2}y + \frac{1}{2}z$ is a Pareto-efficient point of T.

Player I and II's empathetic preferences are determined by the parameters ϕ and ψ respectively. Once these parameters are fixed, the contingent social contract (r, s) that maximizes the Nash product Π is found by solving the equations (2.16) and (2.17). The solutions r and s of these equations are functions of ϕ and ψ. That is to say, $r = r(\phi, \psi)$ and $s = s(\phi, \psi)$.

Suppose that an empathy equilibrium occurs when $\phi = \Phi$ and $\psi = \Psi$. To use the definition of an empathy equilibrium given in Section 2.5.4, begin by fixing player I and II's empathetic preferences so that $\Phi = U_1/V_1$ and $\Psi = U_2/V_2$ respectively. Having fixed player I and II's preferences in the original position, we next ask whether player I could gain relative to his empathetic preferences if these were taken to be governed by ϕ instead of Φ. Similarly, we must ask whether player II could gain relative to her actual empathetic preferences if these were taken to be governed by ψ instead of Ψ.

If player I's empathetic preferences are determined by Φ and those of player II by Ψ, but Φ is misreported as ϕ while Ψ is reported accurately, then player I's expected payoff is proportional to $\Phi f(r(\phi, \Psi)) + g(s(\phi, \Psi))$. A necessary condition for the pair (Φ, Ψ) to determine an internal empathy equilibrium is that this payoff be maximized when $\phi = \Phi$. Differentiate player I's payoff with respect to ϕ and then differentiate the corresponding payoff for player II with respect to ψ. We are then led to the following pair of necessary conditions for (Φ, Ψ) to determine an internal empathy equilibrium:

$$\Phi f'(r)r_\phi + g'(s)s_\phi = 0, \tag{2.19}$$

$$\Psi f'(s)s_\psi + g'(r)r_\psi = 0. \tag{2.20}$$

In these and later expressions, the functions r and s, together with their partial derivatives,[102] are evaluated where $\phi = \Phi$ and $\psi = \Psi$.

To develop (2.19) and (2.20), it is necessary to know something about the partial derivatives of r and s. To this end, we differentiate (2.16) and (2.17) partially with respect to ϕ and ψ. After various algebraic

---

[101]Since $y = (U_1 u_A(r), 1 - V_2\{1 - u_E(1 - r)\})$, the slope of the Pareto-frontier of X_{AE} at y is $-V_2 u'_E(1 - r)/U_1 u'_A(r)$. Similarly, $z = (1 - V_1\{1 - u_E(1 - s)\}, U_2 u_A(s))$ and so the slope of the Pareto-frontier of X_{EA} at z is $U_2 u'_A(s)/ - V_1 u'_E(1 - s)$.

[102]It is standard to simplify the clumsy notation $\partial y/\partial x$ to y_x. For example, in (2.19) s_ϕ denotes the partial derivative $\partial s/\partial \phi$ evaluated where $\phi = \Phi$ and $\psi = \Psi$.

manipulations,[103] we are then led to the simple necessary condition that

$$\Phi\Psi f(r)f(s) = g(r)g(s) \,. \tag{2.29}$$

Like the functionally similar (2.18), this equation has a geometric interpretation in Figure 2.5(a). It says that the slope of the line joining η and y is the same as the slope of the line joining ζ and z.

Rather than pursuing the geometry, it is simpler to divide equations (2.18) and (2.29). The term $\Phi\Psi$ then cancels, leaving the equation

$$\frac{f'(r)f'(s)}{f(r)f(s)} = \frac{g'(r)g'(s)}{g(r)g(s)} \,. \tag{2.30}$$

This will typically have only one solution[104] in which $r = s$. It follows that,

---

[103]First differentiate (2.16) and (2.17) partially with respect to ϕ and ψ. With the help of (2.16) and (2.17), one may then reduce the resulting expressions to the form:

$$\Phi\Pi_{\phi r} = -g'(r)g(s) + \Phi Ar_\phi = 0 \,, \tag{2.21}$$

$$\Pi_{\psi r} = \Phi f'(r)f(s) + Ar_\psi = 0 \,, \tag{2.22}$$

$$\Pi_{\phi s} = \Psi f'(s)f(r) + Bs_\phi = 0 \,, \tag{2.23}$$

$$\Psi\Pi_{\psi s} = -g'(s)g(r) + \Psi Bs_\psi = 0 \,, \tag{2.24}$$

where

$$A = \Phi\Psi f''(r)f(s) + \Phi f''(r)g(r) + \Phi g''(r)f(r) + g''(r)g(s) \,, \tag{2.25}$$

$$B = \Psi\Phi f''(s)f(r) + \Psi f''(s)g(s) + \Psi g''(s)f(s) + g''(s)g(r) \,. \tag{2.26}$$

To illustrate how this is done, consider (2.21). From (2.16),

$$\Phi\Pi_{\phi r} = \Phi\{(fg)'(r) + \Psi f'(r)f(s)\} + \Phi r_\phi\{\Phi\Psi f''(r)f(s) + \Phi f''(r)g(r)$$
$$+2\Phi f'(r)g'(s) + \Phi g''(r)f(r) + g''(r)g(s)\} + \Phi s_\phi\{\Phi\Psi f'(r)f'(s) + g'(r)g'(s)\} \,.$$

The terms in the first bracket are equal to $-g'(r)g(s)$ by (2.16). The terms in the final bracket cancel with $2\Phi f'(r)g'(r)$ in the middle bracket as a consequence of (2.18) and (2.19).

The elimination of A and B from (2.21)–(2.24) yields that

$$g'(r)g(s)s_\psi = -\Phi^2 f'(r)f(s)r_\phi \,, \tag{2.27}$$

$$g'(s)g(r)r_\phi = -\Psi^2 f'(s)f(r)s_\psi \,, \tag{2.28}$$

Together with (2.19) and (2.20), equations (2.27) and (2.28) constitute a system of four homogeneous linear equations in the four unknowns $\Phi f'(r)s_\phi$, $g'(s)s_\phi$, $g'(r)r_\psi$ and $\Psi f'(s)s_\psi$. As the system has a nontrivial solution, equation (2.29) follows, since it says that the determinant of the system vanishes. (Equation (2.18) was obtained in an analogous fashion.)

[104]Provided that we take it as typical that

$$\frac{f'(x)}{f(x)} \bigg/ \frac{g'(x)}{g(x)}$$

is a strictly monotone function of x. This is true, for example, if $f(x) = x^\alpha$ and $g(x) = (1-x)^\beta$. Even in atypical cases, there is always a solution with $r = s$.

in equilibrium, the deal reached behind the veil of ignorance will not be contingent on who turns out to be who. It will specify that Adam gets a fraction r of the dollar and Eve gets a fraction $1 - r$ of the dollar whether player I turns out to be Adam or Eve.

Substituting r for s in (2.30) yields the equation

$$\frac{g(r)}{f(r)} = -\frac{g'(r)}{f'(r)} = \sqrt{\Phi\Psi}. \tag{2.31}$$

This equation says that the line joining $\xi = 0$ and $\nu = (f(r), g(r))$ makes the same angle to the vertical as the tangent to X at ν. By Section 1.3.1, ν is therefore the Nash bargaining solution for the problem (X, ξ).[105]

The preceding analysis does not require that the empathy equilibrium under discussion be symmetric[106]. In fact, since we have seen that the outcome is entirely determined by the ratios $\phi = U_1/V_1$ and $\psi = U_2/V_2$, it shows that the full symmetry requirement that $U_1 = U_2 = U$ and $V_1 = V_2 = V$ is overly strong—evolutionary pressures cannot be expected to shape the values of parameters that do not affect anything. This is not a particularly important point, since no harm is done by assigning irrelevant parameters any values that might be convenient, but I shall tie up the loose end anyway.

If the symmetry requirement on an empathy equilibrium is simply $\Phi = \Psi$, then (2.18) and the fact that $r = s$ imply that $\Phi = g(r)/f(r) = \nu_E/\nu_A$. Since ν is the symmetric Nash bargaining solution for the problem $(X, 0)$, it follows from Section 1.3.1 that the slope to the Pareto-frontier of X at ν is $-\Phi$. If $U/V = \Phi$, it follows that the utilitarian welfare function $W_h = Ux_A + Vx_E$ takes its maximum value on X at the point $x = \nu$. That is to say, we still arrive at Harsanyi's conclusion even if the irrelevant parameters are allowed to drift.

2.6.3 Retelling the Rawlsian Story

In his *Theory of Justice*, Rawls [417, p.115] attributes the same commitment powers to players I and II in the original position as Harsanyi [233]. As

---

[105]Alternatively, one can observe that the equation

$$\frac{f'(r)}{f(r)} + \frac{g'(r)}{g(r)} = 0. \tag{2.32}$$

characterizes ν as the point x in X at which the Nash product $\Pi = (x_A - \xi_A)(x_E - \xi_E)$ is maximized. Restricting our attention to Pareto-efficient values of x, we have that $x = (u_A(r), u_E(1 - r))$ and $\xi = (u_A(0), u_E(0)) = (0, 0)$. But $u_A(r) = f(r)$ and $u_E(1 - r) = g(r)$. The condition that $d\Pi/dr = 0$ therefore reduces to (2.32).

[106]If the equilibrium is asymmetric, then the outcome is the same as in the symmetric case with $U = \sqrt{U_1 U_2}$ and $V = \sqrt{V_1 V_2}$.

he puts it, we have a "fundamental natural duty... to comply with just institutions."

Chapter 4 argues that such commitment assumptions need to be abandoned if one is to mount a successful defense of Rawls' intuitions about fairness. However, completeness demands that we investigate what happens to Rawls' own theory when this is reconstructed along the same lines as Harsanyi's.[107] Since Rawls differs from Harsanyi in denying that Bayesian decision theory can be employed behind behind the veil of ignorance, the account that follows is not fully coherent. Purists are therefore urged to skip forward to Section 2.7.

Rawls [417] substitutes the maximin principle for Bayesian decision theory. Since the skepticism expressed in Section 4.6 of Volume I about the reasons Rawls [417] gives for making such a substitution is to be put on hold, we therefore assume that a player in the original position proceeds as though whichever of the two events AE and EA he dislikes more were certain to occur. Matters then proceed as in the reconstruction of Harsanyi's theory until equation (2.13) is reached. At this point, it is necessary to diverge from Harsanyi's argument because we are no longer to make the Bayesian assumption that player i seeks to maximize $t_i = \frac{1}{2}y_i + \frac{1}{2}z_i$. Instead, we must take the most pessimistic of all possible views, and assume that player i seeks to maximize

$$t_i = \min\{y_i, z_i\}.\tag{2.33}$$

The payoff pair $t = (t_1, t_2)$ defined by (2.33) is shown in Figure 2.4(b). To find t, draw a rectangle which has y and z at opposite corners. The point t then lies at its southwest corner. The set T in the bargaining (T, τ) faced by players I and II is the set of all such t.

The set T is easier to describe than in Harsanyi's case since it is simply the set of all payoff pairs that lie in both X_{AE} and X_{EA}. That is to say, $T = X_{AE} \cap X_{EA}$. Figure 2.5(b) shows how to compute the symmetric Nash bargaining solution σ in this new case. When the situation is not too far

---

[107]Readers unfamiliar with Rawls' [417] *Theory of Justice* should be warned that the model to be described bowdlerizes Rawls' position unmercifully, so that attention can be concentrated on the underlying structure of his model. Among other things it proceeds as though only the distribution of goods and services were relevant to a well-ordered society, and replaces his notion of an index of primary goods by the same utilities that appear in Harsanyi's nonteleological defense of utilitarianism. To use the maximin criterion in the resulting model is to create a hybrid monster for which no respectable lineage can possibly be found. But this problem is intrinsic to Rawls' approach. It is certainly no less severe in his own formulation. As Section 1.2.5 and Section 4.3 of Volume I explain, it is impossible to evade the problem of interpersonal comparison by using some index of primary or other goods. Indeed, Rawls [418] himself admits that his proposed index will seem no more than an *ad hoc* patchwork to an economist.

from being symmetric, σ lies at the point where the Pareto-frontiers of X_{AE} and X_{EA} cross.[108] Figure 2.6(b) illustrates the fully symmetric case when $U_1 = U_2 = U$ and $V_1 = V_2 = V$.

For both Harsanyi and Rawls, assuming symmetry ensures that the payoff pair N can be achieved using the *same* social contract $\mathcal{C}$ however the imaginary coin tossed to decide between AE and EA may fall. But it does not follow that players I and II will necessarily be indifferent to the fall of the coin. On the contrary, in Harsanyi's case, they will await its verdict with trepidation, since $\mathcal{C}$ may well specify that Adam is to be enslaved by Eve.[109] However, in the Rawlsian case no such anxiety is necessary. Players I and II will receive the same utility no matter how the coin falls. This version of Rawls' theory is therefore highly egalitarian.

The personal payoff pair $r = (r_A, r_E)$ shown in Figure 2.7(b) tells us how Adam and Eve evaluate the social contract $\mathcal{C}$ after emerging from behind the veil of ignorance to discover that AE actually obtains. It is determined by the requirement that

$$Ur_A = 1 - V(1 - r_E).$$

One may characterize r as the proportional bargaining solution with weights U and V for the bargaining problem (X, α), in which the *status quo* α is $(0, 1 - 1/V)$.

Gauthier redux? The case $U = V = 1$ is of some special interest, since the outcome r of Figure 2.7(b) then reduces to the Kalai-Smorodinsky solution of the bargaining problem (X, ξ).

Gauthier [196] proposes the Kalai-Smorodinsky solution as the necessary result of rational bargaining between Adam and Eve over which social contract to implement. His own defense of this claim is criticized in Section 1.2.9 of Volume I and elsewhere in this book. However, we now see that an alternative defense is available. To obtain Gauthier's solution to the social contract problem, we begin by employing the zero-one rule of Section 2.3 to determine a standard for making interpersonal comparisons of utility.[110] We then appeal to the device of the original position on the Rawlsian as-

---

[108]When the situation is more asymmetric, σ coincides with the symmetric Nash bargaining solution of either (X_{AE}, τ) or (X_{EA}, τ).

[109]Recall the discussion of Rawls' [417, 9.167] slaveholder's argument from Section 1.2.4 of Volume I. Rawls explains that, although Adam and Eve will not agree to a slaveholding society in his version of the original position, they would be committed to such a society if they were to do so.

[110]To set $U = V = 1$ is to require that players I and II regard Adam and Eve as being equally well off in heaven ($\mathcal{W}$) and also in hell ($\mathcal{L}$). That is to say, $v_i(\mathcal{W}, A) = v_i(\mathcal{W}, E) = 1$ and $v_i(\mathcal{L}, A) = v_i(\mathcal{L}, E) = 0$.

sumption that players behind the veil of ignorance solve decision problems using the maximin criterion.

Such a defense of Gauthier's solution is vulnerable to the standard criticisms. Why use the maximin criterion rather than Bayesian decision theory? What justifies the use of the zero-one rule for making interpersonal comparisons of utility? One might perhaps hope to to justify the zero-one rule by showing that a symmetric empathy equilibrium occurs when $U = V = 1$, but this forlorn hope is to be dashed immediately.

Garbage in, garbage out.   Only the briefest of remarks on empathy equilibria in the Rawlsian case will be offered here, since the subject will be discussed at great length in Chapter 4 in a more satisfactory context. It is not hard to prove that a symmetric empathy equilibrium cannot occur when the Nash bargaining solution of (T, τ) occurs at a point σ where the Pareto-frontiers of X_{AE} and X_{EA} cross, as in Figure 2.6(b). For a symmetric empathy equilibrium, it is therefore necessary that the frontiers touch at σ, so that any deviation by player I or II moves the Nash bargaining solution of (T, τ) to the Nash bargaining solution of either (X_{AE}, τ) or (X_{EA}, τ). However, it then turns out that either player I or II can improve his position by making $v > V$. No symmetric empathy equilibrium can therefore exist.

Such a disappointing conclusion should come as no surprise. The meaningless location of the *status quo* in Figure 2.7(b) already signals that something is awry. One cannot expect meaningful conclusions if one grafts the maximin principle onto a utility theory for which only the maximization of expected utility makes sense as a decision principle.

2.7   Morality as a Short-Run Phenomenon

This section returns to the utilitarian story at the point it was broken off at the end of Section 2.6.2. It begins by bewailing the fact that time will eventually erode all moral content from a fairness norm used regularly for a standard purpose. Section 2.6.2 tells us that, after social evolution has operated in the medium run, Adam and Eve will get precisely the same personal payoffs if they play fair by using the device of the original position as they would if they were to bargain face-to-face with no holds barred. So what use is morality if talk of justice merely conceals the iron fist in a velvet glove?

The answer is that the type of morality with which we are concerned has its bite in the *short run*. But before discussing the mechanics of the relevant processes, it will be helpful to review the medium-run situation.

2.7.1   The Princess and the Pea

The metaphor of the iron fist in the velvet glove nicely captures the reality of operating the device of the original position in the medium run. Although the actual outcome is determined as though by an iron fist, Adam and Eve are likely to see only the velvet glove. Just as it seldom occurs to the bourgeoisie that it might be unfair that they should dine in fancy restaurants while their less fortunate brethren rummage through garbage cans in the back alley, so Adam and Eve will think the compromise they achieve by applying the Nash bargaining solution to their personal utility functions is entirely fair. Nor will they be wrong, since they do not reach this deal via a bruising battle of wills during which they confront each other with the naked realities of power. On the contrary, they apply the device of the original position in an entirely civilized manner. Each therefore fully empathizes with the plight of the other. But the empathetic preferences with which their culture has provided them attach more importance to the inconveniences of the powerful than to the sufferings of the helpless. As the fairy story explains, a real princess will suffer so much discomfort from a pea in her bed that even ten mattresses will not suffice to allow her a proper night's sleep. By implication, the discomfort her servants endure as a consequence of being kept awake all night to pander to her needs counts for zilch. Nor are matters noticeably better in societies like the Soviet Union that were explicitly founded to promote equality. As in Orwell's *Animal Farm*, some pigs always eventually turn out to be more equal than their comrades.

Multiple models describing the same behavior.   Two different models describe Adam and Eve's behavior in the medium run. An amoral economist can appeal to the theory of revealed preference to argue that Adam and Eve behave as though they were exercising whatever power they may have at their disposal to grab as much for themselves as they can. On the other hand, a moral philosopher can simultaneously claim that Adam and Eve behave as though they were operating a utilitarian fairness norm.

   After social evolution has operated in the medium run, both the philosopher's model and the economist's model do equally well in predicting Adam and Eve's behavior. If the criterion of correctness is simply whether the observed behavior is predicted by a model, both models are therefore equally "correct". But the economist may congratulate himself on having offered a model that "explains" more than the philosopher's. The utilitarian model involves the unexplained parameters U and V that represent the standard for making interpersonal comparisons of utility in the society under study, but the economist's model requires no such parameters.

However, just because the economist's model has more explanatory power than its philosophical rival, it does not follow that its internal workings are a better representation of the way the world is put together. We can only guess at how the behavior of real people is generated, but the mental processes of the Adam and Eve of our story were laid bare in Section 2.5. A cognitive psychologist whose model of their internal clockwork incorporated this information would certainly not accept the economist's claims. To him, it would seem that the moral philosopher's model was a better approximation to the way that Adam and Eve's minds actually work.

If the economist were to protest, as they frequently do, the psychologist could then propose an experiment. If X changes in the *short run* to Y, then the economist's model predicts that Adam and Eve will cease to operate the symmetric Nash bargaining solution ν of the bargaining problem (X, ξ), and instead operate the symmetric Nash bargaining solution ω of the bargaining problem (Y, ξ). The philosopher's model predicts instead that Adam and Eve will operate the weighted utilitarian bargaining solution of (Y, ξ), with the *same* weights U and V that they used for the problem (X, ξ). Since we are proceeding on the assumption that the psychologist's model accurately predicts how Adam and Eve's minds work and the parameters U and V are fixed in the short run, we then know that the economist's model will be "refuted" and the philosopher's model will be "confirmed".

On the other hand, a sociologist whose model incorporates some understanding, not only of Adam and Eve's psychology, but also of the workings of the culture they inhabit, may propose the alternative experiment of changing X to Y, and observing how Adam and Eve respond in the *medium run*. It will then be the philosopher's model that is "refuted" and the economist's that is "confirmed". In short, it is naive to ask whether the economist or the philosopher has the better model. Which is the better model depends on the question being asked.

Naive reductionism? This last point is very difficult for some social scientists, who cannot see that a model may be useful even though it may not accurately represent the workings of the phenomena it seeks to clarify. For such thinkers, to propose a model is to claim that nothing in the world is relevant but what has been modeled. The modeler is then condemned as a naive reductionist,[111] who has fallen prey to the "nothing but" fallacy.

Let me therefore insist again that the models of this book are just tools. As my defense of two entirely different models that describe the same phenomenon demonstrates, I do not believe that either captures everything

---

[111]As calumnies are always vile, so reductionists are always naive.

that matters. Neither the economist's model nor the philosopher's model is adequate for more than very restricted purposes—and the same would be true of the psychologist's model and the sociologist's model if Adam and Eve were real people living in a real society. While it may seem unsatisfactory to purists that we should be stuck with trying to make sense of the world using an overlapping patchwork of sometimes inconsistent models, I make no apology for trying to achieve some small measure of understanding in any way I can. Indeed, holists seem to me to have fallen into precisely the trap of which they accuse those they label as reductionists. What are they doing if not trying to work with a single informal model of everything? In doing so, they seem to differ from genuine reductionists only in not being explicit about either their assumptions or their conclusions.

2.7.2 How Justice Works

Recall from Section 4.7 of Volume I that it is necessary to wear two hats when discussing fairness norms from a naturalistic perspective: a descriptive hat and a prescriptive hat. For the moment, it is the descriptive hat that needs to be worn—the hat one wears when discussing how fairness norms evolve and are used in ordinary life.

To tell the descriptive story, one must begin by imagining that groups of people assembled in different places at different times for various purposes find themselves continually facing the need to coordinate on Pareto-efficient solutions to new problems. Such minisocieties are simplified in my treatment to pairs of men and women seeking some *modus vivendi* that I refer to as a marriage contract. In our Garden of Eden fable, Adam is therefore a representative man and Eve is a representative woman.

There is no particular reason why each couple should be confronted with the same state-of-nature point ξ or the same set X of feasible social contracts. Nor need couples be well informed about the characteristics of the pair (X, ξ) that they face. Aside from pretending that certain information is concealed from them behind Rawls' veil of ignorance, couples then also have to accept that other information is hidden from them whether they like it or not. After they emerge from behind the veil of ignorance, they will still find themselves behind a veil of uncertainty of the type studied in Section 2.5.4. But the veil of uncertainty that now needs to be considered conceals more than the outcome of the chance move that selects a lucky hunter. The new veil of uncertainty hides the outcome of *all* the unrealized chance moves in their Game of Life.

I distinguish two polar cases with respect to such a veil of uncertainty. In the first case, couples can make their social contract contingent on how the uncertainty is resolved. The social contract they choose for the pair (X, ξ)

will then be the same as if they knew (X, ξ) in advance. The chief interest of this case lies simply in observing that the action of social evolution in the medium run may lead couples facing different problems to different standards for making interpersonal comparisons. As Elster [173] and Young [573] emphasize, the manner in which fairness judgments are made seems to vary markedly with the type of problem under consideration. My theory exhibits such relativism in an extreme form. Even the same people using the same fairness algorithm may be led to make different fairness judgments in different contexts as a consequence of finding themselves holding standards for making interpersonal comparisons of utility that are highly context-dependent. As Epicurus [174] observes:

> There is no such thing as justice in the abstract, it is merely a compact among men ... [it] is the same for all, a thing found useful by men in their relations with each other; but it does not follow that it is the same for all in each individual place and circumstance.

In the second of the two polar cases to be considered, couples are unable to make their social contract contingent on what they find to be true when they emerge from behind the veil of uncertainty. This case creates no extra mathematical difficulty, since a social contract $\mathcal{C}$ is then simply evaluated by taking the expectation of the payoffs it generates for Adam and Eve over all possible resolutions of the uncertainty they face. Roughly speaking, one simply replaces the pair (X, ξ) in the analysis of Section 2.6 by a weighted average of all the pairs that seem possible behind the veil of uncertainty. Whatever the true realization of (X, ξ) turns out to be, Adam and Eve will then operate the same standard for making interpersonal comparisons of utility. Even after social evolution has had a chance to operate in the medium run, nobody will then observe Adam and Eve operating the symmetric Nash bargaining solution of the pair (X, ξ) that is actually realized. They will operate a utilitarian bargaining solution for (X, ξ) with weights U and V chosen as described below—but with the symmetric Nash bargaining solution applied not to (X, ξ), but to a weighted average taken over all the pairs that (X, ξ) might have turned out to be.

However, in what follows, all such complications are swept away. To keep things simple, it will be assumed that all pairs always face the *same* set X of feasible social contracts, and the same state of nature ξ. For the reasons given in Section 2.5, it will be assumed that each Adam and Eve choose a social contract using the device of the original position. It will also be assumed that social evolution operates in the medium run to shape the manner in which Adam and Eve make interpersonal comparisons of utility. Eventually, everybody will therefore use the same weights U and V

when comparing Adam's utils with Eve's. In the presence of a philosopher-king with the power and the inclination to enforce any social contract that Adam and Eve may write for themselves, the weights U and V can then be computed as shown in Figure 2.8(a). The rules for this computation are:

(1) Find the symmetric Nash bargaining solution ν of the bargaining problem (X, ξ).

(2) Choose the weights U and V to make the weighted utilitarian solution for (X, ξ) coincide with ν.

Note that the decision to retain the state-of-nature point ξ has finally paid off. Although its location is irrelevant to the location of the weighted utilitarian solution ν when U and V are given, it is *not* irrelevant to how U and V are determined in the medium run.

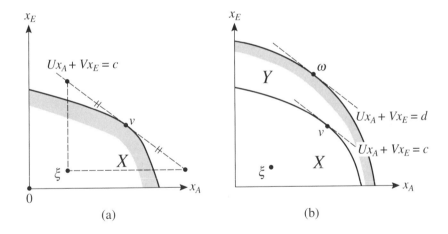

(a)   (b)

Figure 2.8: Justice in the short run.

If the symmetric Nash bargaining solution ν coincides with a weighted utilitarian bargaining solution as in Figures 2.8(a) and 2.8(b), then Section 1.3.1 shows that it necessarily also coincides with the proportional bargaining solution calculated with the same weights U and V. This observation should perhaps give pause to champions of the Good, since it shows that moral philosophers who champion the Right can equally well claim that Adam and Eve are operating a Rawlsian fairness norm. (As explained in

Section 1.2.5 of Volume I, one can see the proportional bargaining solution as the appropriate modification of the Rawlsian maximin principle when the necessity of comparing Adam and Eve's personal utils is taken into account.) I suspect this fact goes a long way to explaining why the egalitarian reforms supported by a whig like myself do not differ so very much from those that bourgeois moral philosophers of various other persuasions conjure from their different moral intuitions (Section 4.6.9).

However, although advocates of the Good, the Right, and the Seemly may well agree on what standard of interpersonal comparison of utility obtains after social evolution has operated in the medium run, their agreement is bought at a heavy price. As we have seen, nothing is gained by using a fairness norm in the medium run, since the compromise that results simply reflects the underlying balance of power.

But matters are very different in the short run. Suppose that some new innovation results in the set of available social contracts expanding from X to Y, as illustrated in Figure 2.8(b). The fairness norm now has a chance to fulfill the function for which it originally evolved—to shift its minisociety to a new Pareto-efficient social contract ω without damaging internal conflict. In the short run, U and V remain fixed, and so the new social contract ω is located as shown in Figure 2.8(b). In short:

(3) The new social contract is the weighted utilitarian solution ω for the bargaining problem (Y, ξ), computed with the weights U and V that evolved for the problem (X, ξ).

Of course, if the representative problem faced by Adam and Eve continues to be (Y, ξ) for long enough,[112] then the standard for making interpersonal comparisons will adjust to the new situation and so the moral content of the social contract will again be eroded away. But it would be wrong to deduce that morality has only a small and ephemeral role to play in regulating the conduct of our affairs. If matters seem otherwise, it is because we mislead ourselves by thinking of morality as something to be taken out of its glass case only when grand and difficult problems need to be addressed. The real truth is that we use our built-in sense of justice all the time in resolving the innumerable short-run coordination problems that

---

[112]Although the location of the *status quo* for the new bargaining problem does not affect the location of the weighted utilitarian solution ω, one may ask why I take the new problem to be (Y, ξ) rather than (Y, ν) as in Chapter 4. To answer the question, it is necessary to recall that the state-of-nature point in a philosopher-king story is determined by the threats available to players I and II behind the veil of ignorance. In taking the new problem to be (Y, ξ), I am therefore assuming that the threat situation is unchanged when X expands to Y. A problem with utilitarianism that I have ignored throughout is that nothing says that ω must be a Pareto-improvement on σ.

continually arise as we try to get along with those around us. Such coordinating problems are usually so mundane and we solve them so effortlessly, that we do not even think of them as problems—let alone moral problems. Like Molière's Monsieur Jourdain, who was delighted to discover that he had been speaking prose all his life, we are moral without knowing that we are moral.

Although I find few takers for the claim, I think it is of considerable significance that morality works so smoothly much of the time that we don't even notice it working. Just as we only take note of a thumb when it is sore, so moral philosophers tend to notice moral rules only when attempts are made to apply them in situations for which they are ill adapted. Section 2.3.3 of Volume I compares their situation with Konrad Lorenz's observations of a totally inexperienced baby jackdaw going through all the motions of taking a bath when placed on a marble-topped table. By triggering such instinctive behavior under pathological circumstances, Lorenz learned a great deal about what is instinctive and what is not when a bird takes a bath. But this vital information is gained only by avoiding the mistake of supposing that bath-taking behavior confers some evolutionary advantage on birds placed on marble-topped tables. Similarly, one can learn a great deal about the mechanics of moral algorithms by triggering them under pathological circumstances—but only if one does not make the mistake of supposing that the moral rules have been designed to cope with pathological problems.

Finally, I doff my shabby descriptive hat and optimistically don a prescriptive hat of the type worn by Harsanyi [233] or Rawls [417]. The argument will be familiar by now. We do not need to confine the device of the original position to the small-scale problems for which it evolved. We can deliberately seek to expand the circle by trying to apply this familiar social tool on a larger scale. That is to say, we can try to achieve Pareto-improving solutions to large-scale coordinating problems by appealing to the same fairness criteria we use to solve small-scale coordinating problems. But such an enterprise will not work unless we put aside the temptation to romanticize our fairness intuitions. In particular, people make interpersonal comparisons of utility like they do—not as we would wish them to.

If my theory is anywhere near correct, appeals to fairness that ignore the realities of power are doomed, because the underlying balance of power is what ultimately shapes the interpersonal comparisons necessary for fairness judgments to be meaningful. Philosophers with utopian ambitions for the human race tell me that such conclusions are unacceptable. But I think this is just another example of an argument being rejected because it has unwelcome implications. In particular, the fact that fairness norms do not work like utopian thinkers would like them to work should not discourage us from trying to use them in the manner in which they actually do work. Others are free to toy with grandiose plans to convert our planet into a New Jerusalem, but we whigs are content to aim at finding workable ways of making life just a little bit more bearable for everyone.

2.8   Why Not Utilitarianism?

William Godwin's [208] lively writings are best remembered for the story he told about a fire at the palace of Archbishop Fénelon. Only the cultivated and philanthropic archbishop or his obscure chambermaid can be rescued. Utilitarianism suggests that the archbishop ought to be saved, since this looks like maximizing the sum of utility in society as a whole. But what if the chambermaid is your mother? Godwin proposes this fable while arguing that utilitarian reasoning should take priority over conventional morality. But modern authors commonly reverse the logic by using similar stories to discredit utilitarianism. We are therefore told of sadists whose pleasure in torture outweighs the sufferings of their victims, or of an innocent bystander who must be executed to serve some overriding public good.

I don't fully understand why utilitarianism is singled out for such attention. One can just as easily invent stories that ridicule Christ's proposal that we love our enemies, or the Kantian claim that it is immoral to lie to a homicidal axman asking for the whereabouts of a neighbor. Nor do the moral intuitions to which the stories appeal seem to be at all strongly held. There is surely something hypocritical involved when they are told in living rooms sporting elegant oriental rugs that continue to be made to this day using the labor of enslaved children. If we are untroubled by such outrages in real life, why should we find them shocking in hypothetical examples? Even if we do, why should we allow our untutored gut feelings on moral questions to overturn conclusions reached using cool reason? After all, the Aztecs would doubtless have been similarly outraged at the immoral suggestion that the general good should suffer as a consequence of putting an end to the practice of human sacrifice. And if the Aztecs seem too savage an example, what of Aristotle on slavery, or Spinoza [504, p.386] on the subjugation of women?

The unwelcome truth is that practical morality—the morality by which we actually live—does in fact endorse the exploitation of those powerless to resist. We dismiss the homeless and the destitute as being an unfortunate consequence of the necessity that a productive society provide adequate incentives for its workers. Is this not to accept that an underclass must suffer in order that the rest of us can enjoy a higher standard of living? We do not, of course, say this openly. Instead, we square things with our consciences by dehumanizing those excluded from the feast. Since Mrs. Thatcher created a British underclass, I have lost count of the number of times I have heard people at dinner parties explain to each other how avoiding eye contact[113] makes it is easier to avoid parting with one's small change in the street. After all, who wants to spoil their day by interacting with a schizo?[114]

Attacks on utilitarianism based on intuitive appeals to fairness depend on persuading us not to dehumanize Adam if he falls on hard times, but to see him instead as "one of us". The mutual protection clauses of our implicit social contract then call for action to be taken on his behalf by those with the power to make a difference—just as Adam would be expected to exercise his power to intervene on our behalf if he found himself in a similarly pivotal position. But the more obviously Adam actually is a powerless outsider without the capacity or the knowledge to reciprocate any action we might take to protect him, the more easily we overlook the suffering he endures for the public good. Unborn generations are therefore particularly vulnerable, and so it is no surprise that we pollute their environment and consume their resources so recklessly. Nor is it any accident that we draft callow youths to fight our wars, or that the victims of libertarian gun laws largely live in ghettos. In short, the moral ideals to which opponents of utilitarianism appeal have a much more suspect origin than we like to admit.

This chapter shows that a coherent case can be made for the claim that utilitarianism would be the first-best choice for a static human society if people could commit themselves to the laws they would make for themselves under ideally fair conditions. But humans are not equipped with the same instinct for unthinking self-sacrifice that a bee exhibits when it commits suicide by stinging a wasp that invades its hive. We pay lip service to various supposedly transcendental moral duties that call for individuals to put the good of society before their own selfish concerns, but it is painfully obvious

---

[113]Why is eye contact so important? Because Adam cannot make eye contact with Eve without it becoming *common knowledge* between them that each is registering the other as an individual.

[114]It is instructive that the claim that street people are a bunch of schizos should frequently be advanced as an excuse for refusing to empathize with their plight. In the first place, the claim is transparently false. It is true that a substantial minority of people on the street are mentally ill, but the majority are just plain unlucky. In any case, why should the fact that a person is unable to help himself be thought to excuse someone else from offering assistance? The formal morality to which we pay lip service would rather seem to suggest that those of us who have benefited from lower tax bills as a result of the closure of public mental hospitals should do what we can to alleviate the misery of our victims. However, this is not the way that practical morality works. It is *because* schizophrenics are the objects of our callous disregard that it is possible to dehumanize others by tarring them with the same brush.

that we are incapable as a species of sustaining such ideals indefinitely when they run contrary to our own private incentives—unless, of course, our transgressions are monitored and punished by some outside authority. But the officers of authoritarian societies are notorious for their vulnerability to corruption.

Utopian attempts to organize a first-best society are therefore doomed from the outset. Authoritarian governments often have benevolent intentions when first founded, but some citizens soon turn out to be more equal than others. As Nozick [392] explains, it is therefore inevitable that societies like our own will evolve formal and informal mutual protection agencies by means of which citizens insure each other against victimization by a corrupt *nomenclatura*. The evolution of such agencies for mutual insurance provides individual citizens with some protection from the abuse of power by those in authority. But once such checks on the abuse of official power are in place, they make the establishment of a first-best society impossible, because they would also act as a check on the power of a benign philosopher-king.

In brief, a society that respects the freedom of the individual is necessarily a second-best society. This is not to argue against the institutions that protect our individual rights and liberties. First-best societies being incompatible with human nature, whigs like myself are delighted to live in a second-best society because we see that the alternative is to live in a third-best society or worse.[115]

Critics who reject utilitarianism because it is a second-best creed for a static society make the mistake that Thomas à Becket anticipates in Eliot's *Murder in the Cathedral:*

> The last temptation is the greatest treason:
> To do the right thing for the wrong reason.

They are right to dismiss utilitarianism—but not because the doctrine that some individuals should be required to make large sacrifices for the good of society as a whole is a second-best creed.

If eyes were transferable, the utilitarian expedient of running a fair lottery to decide who should be forced to give up an eye to a blind person would indeed seem to be first-best. However, we find such proposals repugnant, and so philosophers invent high-flown moral reasons why they should not be imposed. But the real reason for rejecting the proposal is that the prospect of having an eye surgically removed is so horrific that we are only too ready to join with our like-minded fellows to force a second-best solution on society—no matter what suffering we thereby inflict on the blind.

---

[115]We also see that the incentives provided by a utilitarian society are inimical to the creation of wealth, but such dynamic arguments have no place in the static model studied in this chapter (Section 4.7.7).

The reason for rejecting utilitarianism is not that it *isn't* first-best, but that it *is* first-best. Human nature puts first-best societies beyond our reach, and there is no point in yearning for a utopia that does not lie in our feasible set.

Enter a cherubim with flaming sword. In summary, the Garden of Eden serves as an ideal setting for a utilitarian society, because Adam and Eve then have God immediately at hand as the ultimate impartial philosopher-king who can be relied upon to enforce the promises they make at their marriage ceremony.[116] To express outrage at the idea that such a utilitarian society might compel some citizens to make sacrifices for the good of others is to blind ourselves to the realities under our noses. Consider, for example, the following much-quoted verse of John Ball, who was executed in 1381 for preaching social and economic equality:

> When Adam delf and Eve span,
> Who was thanne the gentilman?

The answer is supposedly *nobody*—although any feminist will be pleased to explain that Adam was actually the gentleman and Eve was his servant. But to the medieval mind, women were invisible when questions of equality were debated.

Utilitarians wear a completely different set of blinders from their egalitarian critics. Their failure lies in a lack of realism about the exercise of power. Without an incorruptible and omnipotent father figure to keep us on the strait and narrow path, a utilitarian society is bound to fall apart. To make progress, I therefore think it necessary to study the deal that Adam and Eve would make in the original position *without* a philosopher-king waiting on the other side of the veil of ignorance to enforce whatever social contract they may negotiate.

To pursue the biblical allegory, we need to allow Adam and Eve to eat of the tree of self-knowledge. They will then learn that God has left them to shape their lives for themselves without external help or supervision. Outside the Garden of Eden, the only policemen available are themselves. In exploring how Adam and Eve handle this harsh reality in the coming chapters, I think we shall also learn something about the real reasons that the gut feelings we acquire as children when being conditioned to operate our current social contract lead us to feel so strongly about the nonegalitarian nature of the utilitarian ideal.

---

[116]Voltaire might cynically observe that our conception of God has somehow evolved over the years to remain in tune with the changing moral laws that we pretend he enforces. The Archbishop of York was recently quoted as saying that not even living in sin is sinful any more.

The Day of Judgment

The merest hint of disapproval from those who determine our social status is often enough to keep us on the equilibrium path determined by a second-best social contract. If people could be persuaded to pay more than lip service to the idea that the Game of Life might continue after death, then the punishments envisaged by William Blake would make first-best social contracts available in the here and now.

Chapter 3

Rationalizing Reciprocity

> Keep me always at it, and I'll keep you always at it, you keep someone else always at it. There you are with the Whole Duty of Man in a commercial country.
>
> Mr. Pancks, from Dickens' *Little Dorrit*

3.1 Back-scratching

The previous chapter ended with the expulsion of Adam and Eve from the Garden of Eden. For the rest of the book, no philosopher-king or other external enforcement agency will be available to police any contract they make with each other. The only policemen available will be themselves.

If their Game of Life were the one-shot Prisoners' Dilemma of Figure 3.1(a), the results would be dire. With a philosopher-king to enforce agreements, any payoff pair in the cooperative payoff region shaded in Figure 3.1(b) would be available as a social contract (Section 1.2.1). Without a philosopher-king, Adam and Eve are restricted to operating equilibria in their Game of Life. But the one-shot Prisoners' Dilemma has only one equilibrium—that in which each player uses the strongly dominant strategy *hawk* and receives a payoff of zero.

However, as argued repeatedly in Volume I, it is a mistake to follow those political philosophers who seek to model our Game of Life as the Prisoners' Dilemma—or Chicken, or the Battle of the Sexes, or any other one-shot game. If our Game of Life really were the one-shot Prisoners' Dilemma, we should never have evolved as social animals. Although it

263

still involves a heroic batch of simplifications, it makes much more sense to model the Game of Life as an indefinitely repeated game. As Section 2.2.6 of Volume I explains, such a modeling approach sidelines the problem of finding the "right" game-theoretic paradigm for the human predicament. What matters is that some game or other gets repeated. Which game we choose to repeat is very much a secondary consideration.

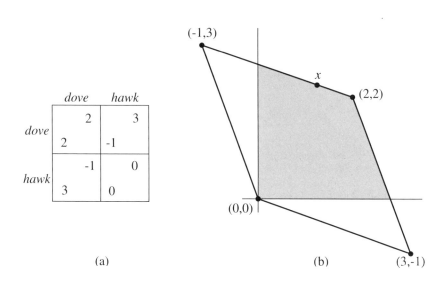

(a)   (b)

Figure 3.1: The Prisoners' Dilemma.

It remains an equilibrium in an indefinitely repeated game if the players continue to use a Nash equilibrium of the one-shot game. In the indefinitely repeated Prisoners' Dilemma, it is therefore an equilibrium if Adam and Eve always plan to play *hawk*. This equilibrium is commonly identified with the bleak state of nature that Hobbes [251] proposed as the only alternative to an authoritarian social contract. However, many more Nash equilibria exist. Every payoff pair in the shaded region of Figure 3.1(b) represents a Nash equilibrium outcome of the indefinitely repeated Prisoners' Dilemma (Section 2.2.6 of Volume I). In particular, Adam and Eve can choose as their social contract any Pareto-efficient payoff pair that is a Pareto-improvement on the Hobbesian state of nature.

As Section 2.2.6 of Volume I explains, the mechanism that supports such cooperative Nash equilibria in an indefinitely repeated game is *reciprocity*—as exemplified by the strategy TIT-FOR-TAT made famous by Axelrod's [31]

Evolution of Cooperation. This strategy requires that a player begin by playing *dove* and continue by choosing whatever action the opponent chose last time (Section 2.2.6 of Volume I). Two TIT-FOR-TAT strategies constitute a Nash equilibrium for the indefinitely repeated Prisoners' Dilemma. Since both defection and cooperation are reciprocated, an optimal response to TIT-FOR-TAT is to cooperate all the time (provided that the discount rate is not too high). Because a player using TIT-FOR-TAT will never be the first to defect, two players who have chosen the TIT-FOR-TAT strategy will always cooperate. Each will therefore be making a best reply to to the strategy choice of the other.

The realization that reciprocity is the mainspring of human sociality dates from at least as far back as the seventh century before Christ,[1] but a recognizably modern statement of its workings had to wait for David Hume. As Hume [267, p.521] explains:

> I learn to do service to another, without bearing him any real kindness, because I foresee, that he will return my service in expectation of another of the same kind, and in order to maintain the same correspondence of good offices with me and others. And accordingly, after I have serv'd him and he is in possession of the advantage arising from my action, he is induc'd to perform his part, as foreseeing the consequences of his refusal.

I suspect that the game theorists of the fifties who reinvented this idea were totally unaware of Hume's work. Since it was obvious to them that one-shot games have a limited range of application, they simply set to work to extend Nash's [381] newly minted equilibrium notion to repeated games. As is often the case when the time for an idea has come, a number of different researchers came up with versions of the basic result of the theory of repeated games more or less simultaneously (Aumann and Maschler [28]). The result is therefore now called the *folk theorem*, because nobody knows to whom it should properly be attributed. It says that all interesting outcomes in the cooperative payoff region of a one-shot game are also available as equilibria in an indefinitely repeated version of the game. We therefore do not need a philosopher-king to enforce contracts in repeated situations. There is nothing an external enforcement agency can do for us that we cannot do for ourselves.

I do not know to what extent game theorists of the time appreciated the wide-ranging significance of this discovery, but they certainly failed to spread the gospel to the biological community, with the result that the basic idea was reinvented again some fifteen years later by Trivers [524], who was

---

[1]Give to him who gives, and do not give to him who does not give—Hesiod [246].

a much more successful evangelist. In consequence, the mechanism behind the folk theorem has become known as *reciprocal altruism*, in spite of the fact that no genuine altruism is involved when the mechanism operates.

More than ten years after the publication of Maynard Smith's [347] widely admired *Evolution and the Theory of Games* and Axelrod's [31] even more successful *Evolution of Cooperation*, serious game theory still remains largely *terra incognita* for both evolutionary biologists and political scientists. For example, Dugnatov's [163, p.11] *Cooperation among Animals* tells us that Von Neumann and Morgenstern's contribution to game theory was to publicize the Prisoners' Dilemma! I therefore plan to discuss the folk theorem and its implications at some length in this chapter. But first an extended review of its place in a theory of the social contract is necessary.

Self-interest. Liberals tend to take for granted that the same bonds of mutual respect and fraternal feeling that hold together the various extended families to which they belong must also provide the glue that prevents larger societies from flying apart. Indeed, for many scholars, the idea that self-interest might provide sufficient centripetal force to keep a society going is downright abhorrent. I agree that altruists are more congenial companions than egoists, but I do not see how one can square the facts of human history with a belief that people need to be altruists to make a society work. Dostoyevsky's [162, p.35] autobiographical *House of the Dead*[2] makes this point very clearly when describing his experience as an inmate of a Czarist concentration camp:

> The majority of these men were depraved and hopelessly corrupt. The scandals and the gossip never ceased; this was a hell, a dark night of the soul. But no-one dared to rebel against the endogenous and accepted rules of the prison; everyone submitted to them. To the prison came men who had gone too far ... the terror of whole villages and towns ... [but] the new convict ... imperceptibly grew resigned and fitted in with the general tone.

In short, we do not need to follow Elster [172] in arguing that love and duty are the cement that holds society together. It is more realistic to join James Madison [215, p.285] in looking to David Hume for an appropriate

---

[2]It is commonplace for moral philosophers to offer quotes from great works of literature as evidence in support of whatever psychological theory they are defending. Homer and Henry James seem to be particularly popular sources of inspiration. But surely all that one learns from a work of fiction is the theory of human nature to which its author subscribes. However, Dostoyevsky's *House of the Dead* is not a work of fiction, but an eye-witness account of phenomena that he clearly found surprising at the time.

metaphor.[3] Stable societies need no cement to hold them together, because they are constructed on the same principle as the drystone walls that shepherds build to separate their holdings. Just as one baboon grooms another in anticipation of receiving a similar service in return, so each stone in a drystone wall is held in place by the weight of its neighbors and contributes its own weight to the task of keeping its neighbors in their place.

A society of Dr. Jekylls would doubtless be more genteel than a society of Mr. Hydes, but the folk theorem tells us that the latter are able to cooperate just as effectively as the former when the conditions are right. In fact, history would seem to offer substantial support to de Mandeville's [341] even more unpopular contention that the thrusting dynamism of a society of Hydes, each always looking out for ways to turn events to his own personal advantage, will create a larger social cake than a society of Jekylls. Even though Hydes divide their cake less equally than Jekylls, the poor in a Hyde society may still be materially better off than the poor in a Jekyll society. This is not to say that the poor in a Hyde society are likely to be more content than the poor in a society of Jekylls. One compares one's worldly wealth with what one sees others enjoying—not with might-have-beens in some world that never was. It is simply to observe that the poor in a commercially successful but inegalitarian society are likely to have more to eat than the poor in an unenterprising but egalitarian society.

What will the neighbors say?   To observe that a society of Hydes can cooperate effectively is not to argue that we ought to model real people as self-obsessed monsters. As argued in Section 2.5.3, the sympathy that we feel toward our family and friends must be expected to extend in some attentuated form even to complete strangers. Some admixture of Jekyll must therefore be stirred in with Hyde when preparing the human cocktail. Nor, as I argue next, should the notion of self-interest be interpreted naively.

Section 1.2.2 of Volume I introduces the notion of a *meme*, which is the word that Dawkins [151] uses for the social equivalent of a gene. A meme is a norm, an idea, a rule of thumb, a code of conduct—something that can be replicated from one head to another by imitation or education, and which determines some aspects of the behavior of the person in whose head it is lodged.

In this book I proceed as though most of our social behavior is dictated by the memes we carry around in our heads. Like everyone else who uses the

---

[3]The happiness and prosperity of mankind ... may be compared to the building of a vault, where each individual stone would, of itself, fall to the ground; nor is the whole fabric supported but by the mutual assistance and combination of its corresponding parts—Hume [266].

idea of a meme, I know perfectly well that to treat memes as atomic, self-replicating entities is vastly more problematic than in the case of genes. I also know that the concept of a mutation has to bear a much greater burden in a theory of memes than in a theory of genes. When talking about social evolution, one needs to imagine an equilibrium population of memes as under continual bombardment by streams of mutant memes deriving from a variety of sources. Not only must one take account of spontaneous errors that occur during the storage or transmission of a meme, one must also consider new memes which are deliberately constructed from other memes by creative thinking or which are generated by individual trial-and-error learning.[4]

Cavelli-Sforza and Feldmann [118], Boyd and Richerson [98] and Lumsden and Wilson [336] all offer models of social evolution in which imitation is seen as a central replication mechanism. Related models are studied in economics under the heading of "herd behavior"[5] (Banerjee [40], Bikhchandru *et al* [61], Lee [314], Scharfstein and Stein [463]). But in spite of these pioneering studies, nobody would want to claim that anything approaching a science of memetics yet exists. One adopts a conceptual scheme in which memes play a central role for want of anything that fits the bill better.

Section 2.2.1 of Volume I explains how Herbert Spencer's "survival of the fittest" translates into a tautology when game theory is used to model biological evolution. In a biological context, the payoffs in a game are the fitnesses of the players—the extra number of copies of themselves they will produce on average as a consequence of one strategy profile being employed rather than another. Players who maximize their expected payoff in such a game are therefore automatically ensuring that their progeny will increase in number at the expense of the progeny of players who play suboptimally. If strategy-choice breeds true, only maximizers of fitness will therefore survive in the long run.

To follow Dawkins [151] in treating the software carried in the human brain as a machine that memes construct to replicate themselves is therefore to commit oneself to an unorthodox view of the nature of self-interest. The payoffs in the Game of Life are not identified with the crude measures of wealth and well-being used by economists. They are interpreted as mea-

---

[4]Since acquired characteristics are thereby transmitted, social evolution driven by imitation is therefore Lamarckian to some extent. Notice that one does not need to attribute any autonomy to the individuals who are responsible for injecting new memes into the population. They can be assumed to be acting under the control of meme-generating memes.

[5]In herd behavior models, a small, endogenously determined fraction of agents take on themselves the cost of finding out what is currently optimal, while the rest of the population rides on their backs by simply imitating what they see others doing.

suring *social fitness*—the average number of times a meme gets replicated as a consequence of one strategy profile being used rather than another.

In the context of a theory of social evolution based on imitation, what matters in determining the payoffs in a game is therefore *whatever makes the bearer of a meme a locus for replication of the meme to other heads.* Social status is perhaps the nearest folk concept to this idea. As in the biological case, the identification is tautological once a sufficient level of abstraction has been adopted. Why does Adam imitate Eve? Because she has high status. Why does she have high status? Because people like Adam imitate her. To say that imitation is the sincerest form of flattery is therefore to hit the nail on the head. Of course, when one moves to a lower level of abstraction the tautology needs to be translated into something operational. Thus, just as biological fitness in spiders correlates with the number of eggs a spider lays, so social fitness correlates with wealth and power and all the other considerations that make some people in a community more respected than others.[6]

To what extent are we aware of being involved in a fierce competition for a place in the social pecking order? It is easy to find Americans who stoutly deny that the Land of the Free has any class structure at all, but my guess is that people are much more aware of what is going on than they readily admit. But however much we may consciously plot and scheme to claw our way one more rung up the social ladder, most of the jockeying for position must surely take place below the level of conscious awareness.

I think this last point is of considerable significance for a theory that models a social contract as an equilibrium in an indefinitely repeated game. Such equilibria are sustained by strategies that incorporate punishment algorithms whose purpose is to inhibit other players in the game from deviating from the implicitly agreed equilibrium path. As Abreu [1] has shown, one can often greatly simplify the analysis of equilibria in repeated games by studying supporting strategies whose built-in punishment schema require extracting the *maximum* credible penalty from any deviant. But it does not follow that we should expect to see such draconian remedies being operated in real life. On the contrary, my guess is that the punishments that deter us from cheating on the social contract are nearly always so mild that we hardly notice them at all. In my own country, I find it very hard to codify the subtle use of body language and shades of verbal expression by means of which my neighbors hint that my current behavior is putting my social

---

[6]It is no accident that three of Rawls' [417] primary goods are income and wealth, the powers and prerogatives of office, and the social basis of respect. (Rawls actually speaks of "self-respect", but I think that to respect oneself is simply to empathize with the respect one receives from those whose respect one reciprocates.)

status at risk. One is so habituated to responding appropriately, that such subliminal signals are automatically translated into behavior without any conscious control. Only when some minor social gaffe in a foreign country is greeted by an unfamiliar signal does the mechanism become apparent.

When discussing punishment in a repeated Game of Life, it is therefore necessary to get out of the habit of thinking only in terms of the formal remedies prescribed by the law. Electric chairs and houses of correction have only a relatively minor role in keeping our social contract on course. Much more important are the hints and cues that our peers use to convey their disapproval when we go astray.[7] Sometimes the hints are explicit, but a person's internal software more usually has to deduce the risk that he is taking with his social status from a barely perceptible coldness in the way he is treated by his friends and acquaintances.

Once the subtle nature of the web of reciprocal rewards and punishments that sustains a social contract has been appreciated, it becomes easier to understand why it is so hard to reform corrupt societies in which criminality has become socially acceptable. As the case of Prohibition in the United States amply demonstrates, one may impose the type of draconian penalty in which right-wingers delight on the criminals unlucky enough to be caught, but it is almost tautological that the resulting disincentives will be inadequate, since the probability of any individual being unlucky will be small if nearly everybody is guilty.

In the words of Edmund Burke [305, p.99]:

> A nation is not an idea only of local extent and individual momentary aggregation, but it is an idea of continuity which extends in time as well as in numbers and space ... it is made by the peculiar circumstances, occasions, tempers, dispositions, and moral, civil, and social habitudes of the people, which disclose themselves only in a long period of time.

The quick fixes of the right can therefore provide only temporary remedies. at best. Nor can anything more satisfactory be anticipated from adopting the left-wing expedient of appealing to people's better nature. One can only realistically hope to reform a society by convincing everybody who matters that there is something in it for him if he changes his attitude. In game-theoretic terms, to reform a society requires shifting it from one equilibrium to another in the Game of Life. Such a shift requires altering the culture of a society—its pool of common knowledge (Section 2.3.6 of

---

[7]Even those self-contained individuals who seem oblivious of the low esteem in which they are held by the world at large actually care desperately about the opinion held of them within whatever small clique or class they act out their role as a social animal.

Volume I). But people do not change the habits of a lifetime overnight. If our reforms are to have a lasting effect, we must therefore resign ourselves to changing people's hearts and minds a little at a time.

Quis custodiet ipsos custodes? Hume [267] referred to the rules which sustain an equilibrium that constitutes a society's social contract as *conventions*. A rational person honors such conventions because he believes that others will honor them, in which case it is in his own interest to honor them also.[8] To identify the social contract of a society with an equilibrium in a repeated game is therefore to face up to the fact that the internal stability of a society depends on nothing more than what people know or believe.

The gossamer threads of shared knowledge and experience may seem but flimsy bonds when compared with the iron shackles of duty and obligation that traditionalists imagine hold a society together. As Hume [271] puts it:

> Nothing appears more surprising to those who consider human affairs with a philosophical eye, than the ease with which the many are governed by the few, and the implicit submission with which men resign their own sentiments and passions to those of their rulers. When we inquire by what means this wonder is effected, we shall find that, as Force is always on the side of the governed, the governors have nothing to support them but opinion. It is therefore on opinion only that government is founded, and this maxim extends to the most despotic and most military governments as well as to the most free and most popular.

In short, the authority of popes, presidents, kings, judges, policemen and the like is just a matter of convention and habit. Adam obeys the king because such is the custom—and the custom survives because the king will order Eve to punish Adam if he fails to obey. But why does Eve obey the order to punish Adam? In brief, who guards the guardians?

Kant [296, p.417] absurdly thought that to answer this question is necessarily to initiate an infinite regress, but the proof of the folk theorem is explicit in *closing* the chains of responsibility. Eve obeys because she fears

---

[8]In adopting this viewpoint, I am rejecting Dworkin's [166] criticism of Hart's [237] version of conventionalism. Dworkin distinguishes two kinds of morality: conventional and concurrent. As an example of concurrent morality, he takes the Kantian case of lying. If I believe that I have a duty to tell the truth whatever may be the practice in my society, then I am expressing a *concurrent* moral principle. If I believe the Earth to be flat, then I am presumably expressing a concurrent physical principle. But why is a concurrent moral principle any more relevant to morality than a concurrent physical principle is to physics?

that the king will otherwise order Ichabod to punish her. Ichabod obeys because he fears that the king will otherwise order Adam to punish him. The game theory answer to *quis custodiet ipsos custodes?* is therefore that we must all guard each other by acting as official or unofficial policemen in keeping tabs on our neighbors. In the particularly simple case of a society with only two persons with which we shall mainly be concerned, Adam and Eve tread the strait and narrow path of rectitude because both fear incurring the wrath of the other.

It is a bad mistake to allow the paraphernalia of the modern state—its constitution, its legal system, its moral code, its pomp and ceremony—to blind us to this truth. In particular, laws and directives, even if written on tablets of stone, are simply a device to help us coordinate on an equilibrium in the Game of Life. The fact that a judge is supposedly an instrument of the Law no more excludes him from our social contract than the Great Seal of State that symbolizes supreme power excludes a king or a president.

In order that we coordinate efficiently, it is certainly necessary that our social contract assign leading roles to certain individuals, but Hume reminds us that the power we lend to the mighty remains in our collective hands. Since history warns us that the mighty cannot be trusted not to abuse their privileges, a liberal social contract makes effective provision for this collective power to be brought quickly to bear when corruption threatens the integrity of the state. As Hume [271] so aptly observes:

> In constraining any system of government and fixing the several checks and controls of the constitution, every man ought to be supposed a knave and to have no other end in all his actions than private interest.[9]

In the jargon of economics, the roles the social contract assigns to officers of the state must be compatible with their incentives. Or, to return to Juvenal's dictum, the bigger the guardian, the more he needs to be guarded.

Theories of the Good and the Right espoused by socialists and conservatives obscure the dangers of authoritarian political systems by their pretence that we can rely on metaphysical considerations to protect us from the abuse of power. Nothing, for example, in the ludicrous constitution of Plato's *Republic* constrains the philosopher-king and his guardians. We are asked instead to believe that their Rationality will suffice to ensure that they follow the Good.[10] Nowadays, nobody expresses his faith in the Good

---

[9]In this much misunderstood quotation, Hume is not saying that we ought always to treat our friends and neighbors as knaves. He is not even saying that politicians are always knaves. He is merely observing that the knaves among us will gravitate into positions of power if we make it worth their while.

[10]Plato found little to admire in the democracy of Athens when contrasted with the

and the Right in such naive terms, but we are equally afflicted with would-be philosopher-kings, who are just as sure of their own virtue as Plato was of his. But whatever else history may have to teach, surely we should have learned by now to trust Lord Acton's famous aphorism: "Power corrupts, and absolute power corrupts absolutely."

When whigs insist on this point, they are not just talking about taking money for services rendered. Corrupt officials are often utterly unconscious of their crimes against the social contract, but they undermine the social contract nevertheless. We are all only too ready to deceive ourselves with stories about why the insider groups to which we belong are entitled to regulate their affairs according to more relaxed versions of the rules than we think should bind outsiders. To whigs, a political philosophy that does not keep this fact about human nature in center stage is simply a distracting irrelevance. We see corruption as the major internal problem facing a modern state—and hence our attachment to theories of the Seemly.

3.2 Rights in a Theory of the Seemly

The Good, the Right and the Seemly are the respective themes of the final three chapters of this book. The previous chapter sought to trace our intuitions of the Good to their origins in prehistoric paternalistic societies. Rather than the unchanging beacon of popular legend, preordained by God or Reason to light the way to a New Jerusalem, we found instead a human artifact invented to describe the workings of a humble fairness norm that evolved from primitive food-sharing conventions. The absolute Good of teleological moral philosophers was thereby reduced to a relative good in a theory of the Seemly—a good that may not only differ between one society and another, but between different subsocieties of the same society, or even within the same subsociety at different times or in different circumstances.

A similar fate now awaits the various absolutist interpretations of the Right defended by deontological moral philosophers. As with the Good, the Right will be reduced to a system of relative rights and duties in a theory of the Seemly.

To say that it is Right to do something will be taken to mean the same as saying that one has a Duty to do it. The notion of an unconditional Duty that mysteriously binds all mankind will then be traced to attempts to universalize the duties that actually operate within particular social contracts. The English language invites confusion by allowing us to argue that

---

authoritarian constitution of Sparta. As he observes in the *Republic*, "So engorged with freedom are the citizens of a democracy that even their cattle swagger." But he overlooked the notorious venality of the officers of the Spartan state.

it is right that we should care for our own parents when they grow old, but that we have a right to leave elderly strangers to the care of others. Perhaps naively, I capture this distinction by saying that we have *a right* to take an action if and only if we have no duty to avoid it.

Adam Smith [500, p.160] put my view in a nutshell when he wrote:

> We do not originally approve or condemn particular actions because, upon examination, they appear to be agreeable or inconsistent with a certain general rule. The general rule, on the contrary, is formed by finding from experience, that all actions of a certain kind ... are approved of or disapproved of.

That is to say, rights and duties are embedded in a society's social contract. Since different societies operate different social contracts, citizens in different societies therefore face different limitations on the exercise of their personal freedom. Even in the same society, the rights and duties a citizen enjoys will vary as the social contract evolves over time in response to the vagaries of fortune and the efforts of reformers.

Talk of inalienable or imprescriptible natural rights in the style of 1789 is just so much whistling in the wind. We whigs value our personal liberty no less than the authors of the American *Bill of Rights* or the French *Declaration of the Rights of Man and the Citizen*[11]. But we do not see that anything but ephemeral advantage is to be gained by asserting the obviously false proposition that we *cannot* be stripped of our liberties.

Do we not imprison or execute criminals and therefore deprive them of their supposedly inalienable rights? The illusion that matters are otherwise is only maintained by pretending that those whose rights are alienated are not properly to be counted as citizens. As Thomas Moore wrote of Jefferson's slaves:

> The patriot, fresh from Freedom's councils come,
> Now pleas'd retires to lash his slaves at home;
> Or woo, perhaps, some black Aspasia's charms,
> And dream of freedom in his bondsmaid's arms.

It is an unwelcome truth, but a truth nevertheless, that we possess only those rights we are actually able to exercise—and history shows that we can be stripped of these very easily indeed.

We retain what rights we have only because enough of us keep sufficient power in our collective hands that authoritarians are unable to take them away. Any propaganda that conceals this harsh reality is a danger to those of us who do not wish to live under oppressive regimes. Whigs believe that

---

[11]The end in view of every political association is the preservation of the natural and imprescriptible rights of man—French National Assembly, 27 August, 1789.

we would do better to abandon all the rhetoric about inalienable natural rights—however effective it may be in the short run—lest we succeed in convincing our children that the price of freedom is not eternal vigilance.

3.2.1 Rights as Strategies?

The moral philosophy literature discusses rights and duties at enormous length, but surprisingly few attempts have been made to embed the notion of a right or a duty into a formal model of the type studied by economists.

Sen's [476] proposed definition is the best known of the various attempts to formalize the notion of a right. His definition of a minimal right requires that each citizen be a dictator over at least one pair of social alternatives. For example, if Adam has a right to decide how he dresses, then he is a dictator over at least one pair of alternatives, a and b, which are identical except that Adam wears his fig leaf in a but goes naked in b. This definition leads immediately to Sen's Paradox, which says that a society cannot simultaneously allow even the most minimal of rights to each citizen without violating the Pareto principle (Section 2.3.3 of Volume I).

Nozick [392, p.164] sees Sen's Paradox as reinforcing his view that a right can sensibly be interpreted as a "side constraint" to be satisfied *before* social choice considerations are invoked. Dworkin [166] expresses the same thought when he categorizes rights as "trumps" that take priority over other considerations (Roemer [435]). Such an attitude to the nature of rights is consistent with the criticism that Gaertner *et al* [190] and Sugden [511, 512] direct at Sen for neglecting to take account of the fact that people should be able to exercise their rights *independently* of each other.[12]

I agree with Sen's critics that the language of social choice theory is too thin to capture the idea of a right adequately. Sugden [511, 512] seeks to enrich the available language by defining rights in terms of the strategies of a game. Spinoza's notion of a "natural right" is easily expressed in this manner. Like Hobbes[13], Spinoza [505, p.200-204] holds that we have a natural right to take any action within our power. In particular:

> Everyone has by nature a right to act deceitfully, and to break his compacts, unless he be restrained by the hope of some greater good, or the fear of some greater evil.

---

[12]See also Gibbard [204], Hammond [222], Pattanaik and Suzumura [400], Riley [427, 428], Suzumura [516], and Sen [482]. Hammond [222] summarizes the literature and provides further references. A recent symposium in *Analyse & Kritik* is devoted to related issues. My own article in the symposium is a much abbreviated version of the philosophical part of this chapter (Binmore [73]).

[13]Every man has a Right to every thing; even to one anothers body—Hobbes [251].

A natural right in Spinoza's sense can be identified with the requirement that a player be free to choose any strategy he likes in whatever game he is playing. However, the example of Rock-Scissors-Paper given in Section 2.3.3 of Volume I shows that the exercise of such natural rights is not even consistent with the existence of a transitive collective preference relation, let alone the Pareto principle. One therefore does not escape Sen's Paradox simply by adopting the language of game theory.

However, Spinoza invents the idea of natural rights only to provide a suitably stark contrast between the Hobbesian brutality of his hypothetical state of nature and the civilized society that replaces it after a social contract has been agreed. But our intuitions about the nature of rights are not derived from such gedanken experiments. They derive from seeing how rights operate within the social contract that we currently operate. If I am anywhere near the mark in modeling a social contract as an equilibrium[14] in a fixed Game of Life, an understanding of our intuitive idea of a right therefore has to be sought by studying how equilibria in the Game of Life are *sustained*.

Since social choice theory is concerned with how social alternatives are *selected*, it therefore provides a very poor framework for discussing rights. In particular, we should not follow Nozick [392] in thinking of rights as side constraints on the set of social alternatives that need to be satisfied when a selection is made. We need rather to follow Pattanaik and Suzumura [400] in seeing rights as part of the *description* of a social alternative, which is abstracted away when a social alternative is envisaged as a point in the feasible set from which a social choice theorist has to make a selection.

Similar problems arise when a right is seen as a pure strategy in a game. It seems to me that our intuitions about rights depend in an important way on the opportunities for reciprocity that become available when the Game of Life has a repeated structure. However, when the extensive form of a game is replaced by its strategic form (Section 0.4), we can no longer study how rights and duties manifest themselves during the play of the game through the interaction of the pure strategies chosen by Adam and Eve. Just as we need to look inside the black box that represents a social alternative to find where rights are hiding in a social choice framework, so we need to look inside the black boxes that represent the pure strategies in an equilibrium profile to find where rights are hiding in a game-theoretic framework.

---

[14]Like Hobbes before him and Hume after, Spinoza [505, p.204] sees the need for the social contract that replaces the state of nature to incorporate incentive-compatibility requirements: "It is foolish to ask a man to keep his faith with us forever, unless we also endeavor that the violation of the compact we enter into shall involve for the violator more harm than good."

In summary, one can sensibly discuss how to *select* a social contract within the abstract frameworks that I am dismissing as inadequate for the purpose of talking about rights. However, the idea of a right belongs—not in a discussion of how social contracts are selected—but in a discussion of how they are *sustained*.

3.2.2 Rules for Sustaining an Equilibrium

This section takes for granted that rights and duties are determined by the social contract operated by a particular society. In particular, a player's *duty* simply lies in never deviating from the equilibrium path specified by the social contract. The custom of doing one's duty then survives because those who evade their obligation to honor the social contract suffer sufficient disapproval or punishment to make the deviation unattractive. It is right to do one's duty, but one has *a* right to take an action if and only if one does not have a duty to refrain from it.[15]

In simple cases, to identify a right action with an equilibrium action trivializes the notion of a right, since the set of actions from which a player is free to choose after fulfilling his obligations may then contain only one element. In seeking to exercise his rights, a player will then be left with Hobson's choice.

Suppose, for example, that the Game of Life is the indefinitely repeated Prisoners' Dilemma, and the social contract being operated calls for Adam and Eve to use the GRIM strategy of Section 3.2.5 of Volume I. The GRIM strategy requires that a player always cooperate by playing *dove* as long as the opponent reciprocates, but that he switch permanently to *hawk* if

---

[15]It is important to remember that a social contract is an equilibrium in the Game of Life. It therefore respects the realities of power. It follows that there is no point in saying that a tree has a right not to be cut down, nor that the generations to come have a right to be left a fair share of the world's resources. A tree or an unborn human is powerless and hence cannot be a player in the Game of Life. Animals, babies, the senile and the mentally ill are only marginally less helpless. However, it does not follow that a social contract will neglect the interests of the powerless. For example, those players in the Game of Life who care about the environment have a love of trees built into their personal preferences. But even when no players at all care about the fate of the helpless, a social contract need not throw them on the scrap heap. Section 3.4.1 describes a simple example of a social contract with overlapping generations in which the elderly are helpless cripples unloved by anyone, but who nevertheless fare as well as their younger compatriots. They are not helped because they have a *right* to be helped. They are helped because those who are actually playing the game have a *duty* to care for their elders—a duty that is enforced by other players in the game. For those who wish to live in a caring society, I believe it to be a dangerous mistake to allow sentiment to blur this point. There is no point in talking about rights without simultaneously discussing their concurrent duties. One then cannot evade the nub of the matter: how are these duties enforced?

the opponent ever cheats on the implicit deal.[16] With the GRIM social contract, Adam has a duty to choose the action *dove* unless Eve has previously deviated by playing *hawk*. In the latter case, Adam has a duty to punish Eve by always choosing *hawk*. After doing his duty, there is then no room left for any discretion in the exercise of his rights.

The notion of a right only becomes significant in games of life with more structure. A case discussed extensively in the literature concerns the right to choose what clothes one wears. Should Adam have the right to decide whether he goes naked or wears a fig leaf? The Fashionwise Prisoners' Dilemma of Figure 3.2 shows how to graft this problem onto the regular Prisoners' Dilemma. A small positive number ϵ is added or subtracted to Eve' payoffs to indicate that she always prefers that Adam cover his nakedness in public. However, Adam's preferences vary. A chance move decides whether Adam prefers to wear a fig leaf or go naked today by adding or subtracting ϵ to his payoffs, depending on what he wears. After observing the outcome of the chance move, Adam decides whether to wear a fig leaf or go naked. Then the Prisoners' Dilemma is played.

It is easy to write down two social contracts for the indefinitely repeated Fashionwise Prisoners' Dilemma. In the first, both players use the GRIM strategy without any reference to the manner in which Adam is dressed. In the second social contract, Eve plans to administer the same grim punishment if Adam leaves off his fig leaf as she would apply if he were ever to play *hawk*. He reciprocates by planning to switch to *hawk* should he ever forget himself by appearing naked in public.

In the first social contract, Adam has a duty never to play *hawk* unless Eve does so first, but he has a right to dress as he chooses. In the second social contract, Adam has a duty never to play *hawk* nor to appear naked. When it come to the exercise of his rights in this second case, his choice therefore reduces to wearing a fig leaf or wearing a fig leaf.

3.2.3 Moral Responsibility

Traditional theories of distributional justice differ as to the criteria to be applied in determining who gets what. We are variously told that Adam's share of the social cake should be determined by his need, his worth, his merit or his work.[17] For example, with respect to the last of these alternatives, the Bible tells us that a laborer is worthy of his hire. Aesop has the story of the ant and the grasshopper. Plato's *Republic* even tells us that

---

[16]Neither player therefore has an incentive to deviate unless the other does so first. As Section 3.3.4 explains at greater length, the pair (GRIM, GRIM) is therefore a Nash equilibrium.

[17]Vlastos [536] discusses how such intangibles as worth or merit are to be distinguished.

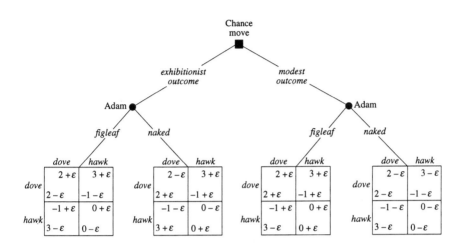

Figure 3.2: The Fashionwise Prisoners' Dilemma.

someone who can no longer work loses the right to live.

It seems to me that all such traditional theories have something valuable to say about how rewards and punishments are determined by our current social contract. Each theory has its own domain of application. For example, social benefits are supposedly determined by need, and Nobel prizes by merit. But none of the theories is adequate as a universal explanation of how we currently assign blame or desert. Nor is the claim that a utopian society would settle on one of the theories to the exclusion of the others often strongly pressed. I take this fact as a tacit acceptance of the need to look for a radically different type of theory—one that is not anchored in the practices or prejudices of a particular society or subsociety at some particular period in its history, but that explicitly recognizes that blame and desert are relative concepts that often vary sharply as we move from one social contract to another.

I suggest that to attribute *blame* to a player in the Game of Life is operationally equivalent to identifying him as an appropriate target for punishment within a particular social contract. To attribute *desert* to a player is to identify him as a suitable target for reward.[18] When the GRIM social contract is employed in the indefinitely repeated Prisoners' Dilemma, a player is therefore deemed to be worthy of blame if he deviates from equilibrium play, and deserving of credit if he doesn't. His reward in the latter case is that his opponent continues to play *dove*. His punishment if he deviates consists in his opponent playing *hawk* at all later times.

Such an approach to blame and desert recalls a major fallacy that critics of modern utility theory seem unable to avoid. Modern theorists do not say that Eve chooses *a* rather than *b* because the utility of *a* exceeds that of *b*. The utility of *a* is chosen to be greater than the utility of *b* because it is known that Eve always chooses *a* rather than *b*. Similarly, according to my account of blame and desert, Eve is not punished *because* she is to blame. Eve is held to be to blame *because* the social contract being operated requires that she be punished.

In spite of the various aphorisms that recognize the need we feel to hate those whom we have injured,[19] traditionalists find it hard to believe that anyone might be serious in proposing that the causal chain taken for granted by folk psychology needs to be reversed. How can it be possible to talk about just punishment in the absence of a prior understanding of the nature of moral responsibility? Without such a prior understanding, is it not inevitable that we shall find ourselves blaming or rewarding people for things over which they have no control?

But our current social contract *already* institutionalizes the practice of blaming people for matters outside their control. This is not to say that the level of punishment we administer to a defaulter is *never* determined by the extent he was in control of the events leading up to his crime. On the contrary, such cases are clearly the norm, because an efficient social contract will not call for penalties to be inflicted when the circumstances under which the crime was committed are such that there is no need to deter others from imitating the deed. If it can be demonstrated that Abel's death at the hand of Cain was accidental by pointing to the immediate cause of the crime, we do not need to follow the practice of our Saxon predecessors

---

[18]Scanlon [462] distinguishes responsibility from accountability. My definition equates them. But such an identification only makes sense if accountability is not narrowly interpreted in some quasi-legal sense.

[19]Sometimes a parallel is drawn between the hate we direct at those we have injured, and the hate directed at us by those we have helped. But I think the latter phenomenon is one of the pathologies of our capacity for envy (Section 4.2).

by holding Cain to blame and demanding the payment of weregeld.[20] But when the causal chain is uncertain, our social contract needs to be ruthless. All of us, for example, are held to be guilty until proved innocent when it comes to paying tax.

Some of the legal examples are even more blatant. Consider the doctrine that ignorance of the law is no excuse. It is clear why this doctrine is necessary. If the prosecution had to prove that the accused knew each letter of the law he had broken, everybody would plead ignorance nearly all the time and the legal system would collapse. Even more difficult for the traditional view are the examples multiplied in Nagel's [379] *Moral Luck*.[21] These include the doctrine that the seriousness of a crime is determined by the consequences for the victim. But the extent of the damage suffered by a victim is often largely outside the criminal's control. For example, a mugger may hit two victims equally hard, but he will only be tried for murder if one of them happens to have an unusually thin skull. Nearer to home, how many of us have escaped imprisonment and ruin as a consequence of negligent driving only because we happen not to have hurt anybody?

At the social level, we routinely inflict the most exquisite psychological tortures on our fellows for being ugly, boring, clumsy, or lacking in charm. Those of us who think of ourselves as intellectuals are relentless in the humiliations we heap on the stupid and uneducated. In medieval times, those suffering from clinical depression were held to account for the sin of *accidie*. Within living memory, bastards were held in contempt. Skin color and sex still provide ammunition for those unable to come up with other reasons why they are entitled to inflict injury on others. I find it particularly revealing that people excuse themselves from taking an interest in street people by calling them schizos (Section 2.8).

Such examples would seem to refute the claim that the principle on which our social contract works is that punishments are determined according to some *a priori* notion of moral responsibility. Presumably, people hang on to this idea because they believe that taking a more realistic attitude to the attribution of blame and desert will lead to the collapse of our moral institutions. But our prisons are *already* full of youngsters from deprived backgrounds who everyone agrees "never had a chance". All that will change if we stop pretending they had the same "free choice" as youngsters from privileged backgrounds is that we will stop kidding ourselves. The conclusion will certainly not be that criminals from the ghettos will be let off because "they couldn't help it". They have to be punished for the

---

[20]A practice that perhaps had more to do with providing insurance for the victim's dependents than punishing the culprit.

[21]Originally written in reply to a work of the same name by Bernard Williams [554].

same reason that we quarantine those suffering from dangerous diseases. A social contract that does not discourage the spread of criminal or antisocial memes will not survive.

 ### 3.2.4 Free Will

Section 3.4.2 of Volume I reveals that I hold mechanistic views on the nature of free will. But how can it be possible to discuss blame and desert if we are not free to choose between right and wrong? If everything we do is predetermined by the laws of physics and the conditions prevailing in the universe at the time of the big bang,[22] it seems obvious to a traditionalist that nobody can be held responsible for his actions. And, if we are not responsible for what we do, what basis can there be for apportioning blame or desert?

 In wand'ring mazes lost? Authors who defend metaphysical notions of free will seem to me determined to make complicated things simple and simple things complicated. It is perhaps because of this intellectual perversity that John Milton numbers them among the fallen angels. As *Paradise Lost* records:

> Others apart sat on a Hill retir'd,
> In thoughts more elevate, and reason'd high
> Of Providence, Foreknowledge, Will, and Fate,
> Fixt Fate, free will, foreknowledge absolute,
> And found no end, in wand'ring mazes lost.

Or it may be their lack of scruple in debate that led him to set them alongside Beelzebub and Belial. Can they really believe that those of us who refuse to build a personal "unmoved mover" into our model of man are denying the *phenomena* that led our ancestors to invent the notion of free will? But refusing to treat free will as anything other than a convenient fiction no more denies the mental phenomena we all experience than refusing to answer questions about why Zeus is angry today would deny that thunder is to be heard in the hills!

Section 3.2.4 of Volume I attempts to answer some of the more telling criticisms of modeling the mind as a machine, but I do not plan to expand

---

[22]Although a fuss is often made over this point, the fact that the laws of quantum physics are intrinsically statistical in character seems to me irrelevant to the issue. It may be that Eve's behavior can only be predicted statistically, but if she cannot alter the probabilities with which she takes one action rather than another, there seems no more reason for traditionalists to blame her for what she does than to blame a deck of cards for failing to deal a straight flush.

on these thoughts here, since Churchland [123, 124], Dennett [155, 156, 157] and others have already said most of what needs to be said much better than I could hope to say it. Nor do I plan to discuss how and why recent findings on the complicated workings of parallel distributed systems, like the human brain and artificial neural networks, refute the simplistic folk psychology to which traditionalists cling so firmly.[23] Instead, I plan only to explain why I think it unnecessary to make a mountain out of the molehill of personal freedom.

I think the problem of free will was solved by Spinoza [506, p.134] in 1677 when he argued that:

> Experience teaches us no less clearly than reason, that men believe themselves to be free, simply because they are conscious of their actions, and unconscious of the causes whereby those actions are determined.

People reject such a resolution of the supposedly difficult problem of free will as being too simple to merit serious attention. But the history of thought is awash with similar refusals to see that the only obstacle to resolving a problem is the stubborn insistence that an obstacle exists. For example, nobody nowadays sees any difficulty in agreeing that a moving body is always where it is at any instant. But this truism of Newtonian physics is the key to the problem of continuous motion that bedeviled philosophy for more than two millenia. Only when Weierstrass taught us how to model the continuum in a formal way did it become acceptable to recognize that Zeno had all along been offering a coherent model when he took for granted that a body in motion is just as truly where it is at any time as a body at rest. To quote Bertrand Russell [453]:

> Motion consists merely in the fact that bodies are sometimes in one place and sometimes in another, and that they are at intermediate places at intermediate times. Only those who have waded through the quagmire of philosophical speculation on this subject can realise what a liberation from antique prejudice is involved in this simple and straightforward commonplace.

---

[23]I am particularly puzzled by the hostility provoked by my innocent remarks on the physiology of color vision in Section 3.4.2 of Volume I. Perhaps Churchland's [124] much more detailed account of the same phenomena will be better received. (One does not need to follow Churchland in inventing someone who cannot imagine the sensation of perceiving yellow. Sachs [454] reports the case of an artist who lost the very concept of color along with his color vision. He even denied seeing everything in shades of gray, since gray is also a color.)

Deus ex machina? Section 2.2 quotes both Berlin [56] and Hayek [240] on the enormous profusion of definitions of the word "free". I shall distinguish between only two types of individual freedom, based on the negative and positive freedoms of Isaiah Berlin [56]. The extent of Adam's negative freedom is determined by the size of the feasible set from which he can choose. He exercises a positive freedom when he actually makes a choice.

The phenomena associated with both positive and negative freedoms seem to me to be consistent with even the crudest mechanical models of the mind. In the case of negative freedoms, we do not even have to postulate the existence of unobserved chance moves to explain the data, since Spinoza's approach to free will explains why even a deterministic computer can behave in a manner that metaphysicians attribute to the existence of a divine spark in the human race.

Consider the process of solving an optimization problem. Decision theory teaches us to take optimization problems in two steps. The feasible set X is first identified. Then an optimal point x inside X is located. Leibnitz even envisages God as passing through these two stages when contemplating which of all possible worlds to create.

After X has been determined, but before the optimizing calculation has been carried through, nobody knows what x is—least of all the decision-maker himself. If he knew what x was going to be before he began to calculate, he wouldn't have included other points in his feasible set. Before the optimizing calculation takes place, it therefore makes perfectly good sense to say that he is free to choose among any of the alternatives in X. It may even be blasphemous to maintain the contrary, since God was presumably free to create a world other than this best of all possible worlds.[24]

Not only is it perverse to argue that the feasible set was wrongly specified because only the optimal point is genuinely feasible for an optimizing decision-maker, it is incoherent. The decision maker's optimizing algorithm takes the feasible set as an *input* in determining the optimum. That is to say, the final choice is only defined after we have specified the set from which the decision maker is free to choose.[25]

---

[24]Since the world in which we live is clearly not perfect, I suppose one could respond to Dr. Pangloss by arguing that it therefore cannot be the world that God created. All our pain and effort would then merely be a simulation in the mind of God of a world too imperfect to create!

[25]One way of resolving Kant's [297] difficulty about the necessary behavior of rational agents not being necessitated is therefore to take into account the information that the analyst has at different times. We have no prior knowledge which necessitates that a rational agent will choose any particular point in his feasible set X. If we did, this would already have been incorporated into the definition of X. Our knowledge of what the agent will actually choose must therefore await a calculation of the optimal point x.

In formulating the Game of Life, Adam is therefore modeled as having a choice only when we do not know in advance how his choice problem will be resolved. We then express our current ignorance of what will determine his choice at any decision node in the game by saying that he is free to choose among any of the actions associated with that node. This negative freedom is unaffected by modeling Adam himself as a member of the species *homo economicus*, even though the behavior of a hominid from this species is entirely determined once we know enough about his preferences and beliefs.[26] He optimizes, and hence we can predict what he will do—just as we can predict what a computer program will do when it operates the simplex algorithm for solving a linear programming problem. In particular, if the equilibrium serving as Adam and Eve's social contract is common knowledge, then Adam will do his duty and remain on the equilibrium path.

Deus in machina? Samuel Johnson was not talking about negative freedom when he notoriously remarked: "Sir, we *know* that our will is free, and *there's* an end on't." The fact that a self-monitoring robot might report similar internal phenomena as it strove to decide whether its interests were best served by moving to the left or right in seeking a power outlet would move him not an inch. Johnson would doubtless claim that *his* mental processes were not determined in advance, because neither he nor anybody else could predict what they were going to be.

However, as Spinoza [506, p.134] explains, it is very easy for Eve to reconcile her inability to predict Adam's behavior with his being modeled as a robotic member of the species *homo economicus*, even if she has full access to his optimizing algorithm. She simply needs to accept that she is not fully informed about the preferences and beliefs that serve as inputs to the algorithm.[27] For example, in the Fashionwise Prisoners' Dilemma of Section 3.2.2, a chance move determines whether Adam prefers to wear a fig leaf or go naked. If Eve does not observe this chance move, then she is unable to predict for sure what action Adam will take. He will seem to her to be exercising a positive freedom to choose from the options left to him after fulfilling his duties.

Similar considerations apply to Adam's understanding of his own mental processes. Introspection tells us very little about what goes on inside

---

After completing this calculation, we will then know that it is necessary that a rational agent choose x. However, this posterior knowledge is available only after the appropriate calculation is over.

[26] Assuming no tie breaking is necessary as a result of multiple optima.

[27] I hide in this footnote the ritual reference to chaotic dynamics and their capacity to amplify tiny uncertainties about initial conditions. In the usual parable, the movement of a butterfly's wings in Brazil causes a hurricane in Texas.

our own heads. Far from the higher activities of our brains being open for inspection, creative thinkers are notoriously unable to say where they get their ideas from. Mozart famously remarked that he just listened and wrote down the music composed inside his head by "someone else". Poincaré suddenly discovered he knew the answer to the problem of Fuchsian groups while leaving an airplane after a trip during which he had not consciously considered the problem at all. Penrose [401] and others offer such phenomena in support of the claim that the mind cannot be treated as a mechanism. However, far from Gödel's theorem proving that we cannot operate algorithmically as Penrose argues,[28] Binmore [63] uses the theorem to show that some failures in self-knowledge are a *necessary* property of a self-monitoring mechanism (Section 3.4.2 of Volume I).

The beliefs and preferences that serve as inputs to *homo economicus* when he reasons his way to an optimal outcome are no more open to introspection than the reasoning processes themselves. Even when we think we are able to explain our behavior, the reasons we give are often absurdly at variance with what we actually do. As Hume [265] observed:

> Hear the verbal protestation of all men: Nothing is as certain as their religious tenets. Examine their lives. You will scarcely find that they repose the slightest confidence in them.

The truth is that we acquire our behaviors through a complex interaction between the genetic program with which we were born and the physical and social environment we were born into. In particular, we are instinctive imitators. We tend to imitate whatever behavior breeds social success with little or no conscious awareness of what is going on. Later, we try to rationalize what we find ourselves doing in the same kind of way that some accident victims whose brains have lost the concept of a left side to their body invent absurd rationalizations for the appearance of what they perceive as extraneous arms and legs in their beds.

---

[28]Penrose's elaboration of the Lucas Fallacy in his recent *Shadows of the Mind* [402] would seem to add little to the debate, since he does not seem willing to contradict Gödel's assessment that "it remains possible that there may exist (and even be empirically discoverable) a theorem-proving machine which is in fact equivalent to mathematical intuition, but cannot be proved to be so, nor even proved to yield *correct* theorems of finitary number theory" (Penrose [402, p.128]). The question therefore reduces to whether one thinks Gödel's hypothesis is plausible or not. But this is a subject on which even the greatest of mathematicians differ, Gödel believing it to be false and Turing true. In spite of his denials, those of us who side with Turing think that Penrose's Platonic beliefs about our capacity to tap into some inner hotline to an absolute world of mathematical truth classify him as a mystic. We cannot prove that his brand of mysticism is wrong, but nor does all his huffing and puffing about metamathematics and quantum physics come anywhere near showing that he is right. The plausibility of Gödel's hypothesis remains a matter of opinion.

In brief, since introspection provides little or no insight into the workings of our minds, the reasons why we like or believe or choose one thing rather than another are often largely a mystery to us. As people always do when confronted with a mysterious natural process, our ancestors therefore invented a metaphysical explanatory agency called free will—an unmoved mover or unintended intender who somehow causes the phenomena we cannot otherwise explain[29] (Dennett [156]).

It is not the introduction of a black box that distresses me about this traditional method of evading the problem of free will. Indeed, the well-trodden approach I advocate for tackling the problem of positive freedom also requires black boxes, albeit of a much more humdrum variety. When modeling the workings of our own mind or that of others, the modeling strategy used by game theorists is to represent matters that we do not understand or cannot observe using the standard technology of chance moves and information sets as described in Section 3.5 of Volume I.[30]

Harsanyi's [232] approach to modeling problems of incomplete information provides the purest example of the use of this technology.[31] When seeking to apply game theory to real problems, a major stumbling block largely ignored in this book is the fact that matters of importance often fail to be common knowledge among the players (Section 2.3.6 of Volume I). For example, in analyzing a game, it is taken for granted that nobody has any doubts about his opponents' preferences over its outcomes or their beliefs about its chance moves. But, when a game like the Nash Demand Game of Section 1.5.1 is played in practice, Adam may well have only a vague idea about the extent to which Eve is averse to taking risks. When such important characteristics of a player are not common knowledge, Harsanyi's approach requires that we consider the set of all possible types of person a player might be. For example, at each stage of Fashionwise Prisoners'

---

[29]Owing to the phenomenon "thought", the ego is taken for granted; but up to the present everybody believed ... that there was something unconditionally certain in the notion "I think", and that by analogy with our understanding of all other causal reactions this "I" was the given *cause* of the thinking. However customary and indispensable this fiction may have become now, this fact proves nothing against the imaginary nature of its origin: it might be a life-preserving belief and *still* be *false*—Nietzsche [384].

[30]Suppes [514, 515] argues that such chance moves should be seen as the source of the spontaneous uncaused inputs required by a genuine theory of personal freedom. But I don't see why a slave whose master will be determined by the fall of a coin should be thought to be any more free in the traditional sense than a slave whose master has already been determined. Indeed, as Suppes argues himself, it will generally be unknowable whether the slave will be acting under the instructions of a predetermined master of whose identity we are currently ignorant, or whether the master will actually be chosen at random. Game theorists certainly do not distinguish between these two possibilities, the same model being used to represent both cases.

[31]Chapter 11 of Binmore [70] contains a full account, with numerous examples.

Dilemma of Section 3.2.3, Adam may turn out to be a shrinking violet or an exhibitionist. A new game is then constructed in which notional chance moves assign a type to each player. The probabilities assigned to the various choices that can be made at these chance moves reflect the knowledge that players have about their opponents. Their ignorance about the type of person Chance has decided each player will be is reflected in the configuration of the information sets with which the new game is endowed.

The major conceptual difference between the metaphysical black box with which traditionalists model positive freedom and the humdrum black boxes represented by the chance moves of game theory is that the humdrum approach does not close the door on the possibility of opening a black box and examining its contents. Inside will be found, not only a great deal of information about the physiology of the brain and the wisdom stored in our genes, but also much information about the social inputs that are so important in determining our preferences and beliefs. The inputs that derive from our culture are particularly important in a social contract discussion.

The chance moves used to model these cultural inputs will be located *outside* our heads in a manner totally unacceptable to a traditionalist. A traditionalist holds that whatever is inside the black box of free will is to blame for any wrongdoing it might initiate and therefore merits punishment. But society itself lies inside the black boxes I propose to use. To a traditionalist, this seems to imply the "wishy-washy liberal" doctrine that we are not justified in punishing a transgressor because the responsibility for his behavior lies with the society that shaped him. Let me therefore repeat that the approach being defended makes *punishment* the fundamental notion, rather than blame. People are not punished because they are deserving of blame—they are blamed because our social contract requires that they be punished. Defaulters will therefore be punished whatever may or may not be awaiting discovery inside one black box or another. Traditionalists therefore need not fear the fate of Pandora if they open their own particular black box. Although no unmoved mover will be found inside, society will not collapse. Everything will go on pretty much as before.

 Counterfactuals.   Although it is not usually discussed in such terms, the preceding discussion of positive freedom is relevant to the problem of defining an equilibrium in a game. Rational players remain on the equilibrium path because of what would happen if they were to deviate. But how should we model the freedom that a rational player would exercise if he were to behave irrationally by deviating?

The appearance of a sentence in which a rational player behaves irrationally signals that we are now making contact with the philosophical tradition that seeks to define freedom in terms of counterfactuals.[32] As Section 3.4.1 of Volume I attempts to explain, the problem of interpreting such counterfactuals represents a major difficulty for game theory. I share with many game theorists, notably Selten and Leopold [474], the view that the appropriate approach requires embellishing the model by adding unobserved chance moves that can actually realize the various "possible worlds" in which the counterfactuals to be considered make sense. These are the same chance moves that I advocate placing inside the black box of positive freedom. By introducing such chance moves, one can make the counterfactual events that need to be contemplated merely unlikely rather than downright impossible. It is true that the methodology forces one to confront the problem of quantifying the probabilities that an observed irrationality was due to one cause rather than another, but this seems to me an advantage rather than a defect of the approach.

Since a social contract is to be modeled as an equilibrium in a repeated game, such problems about counterfactuals can be expressed by asking where primary responsibility lies for a deviation from the social contract in a Game of Life. Rather than pretending to answer this question by airy references to free will, I think we need to ask how the players in the game came to behave like *homo economicus,* that is, how they found their way to an equilibrium in the first place. The major mechanism proposed in this book is imitation. Unsuccessful players learn to play better by switching to memes or strategies that are seen to generate higher payoffs for their more successful rivals. However, before the learning process is over, one cannot count on an opponent being under the control of an optimizing meme. Moreover, new memes will continually be created by a small fraction of the population who think out clever new ideas or stumble on them by trial-and-error experimentation. Even after a society has got as close to an equilibrium in its Game of Life as it is ever likely to reach, its citizens will therefore still be vulnerable to infection by mutant memes.

In summary, the black box that explains why algorithmic players remain free to deviate from optimal play is located largely *outside* the players's heads in this story. A deviant player is explained by saying that he is infected by an outdated or mutant meme generated elsewhere. To an onlooker who does not know what meme is pulling a player's strings, it will seem that the player is acting freely, whether he deviates or not. After all, whatever choice he makes, it will seem to a kibitzer that he could have done something else. Since the player himself is unlikely to be much better

---

[32]G. E. Moore [363] defines "Eve could have done otherwise" to mean "Eve would have done otherwise, if she had so willed", thereby reducing something simple to something complicated.

informed about the relevant mental processes than an onlooker, he will also
see himself as acting freely.

Mixing metaphors. When game theory is used in modeling an evolutionary process, we treat genes as players with preferences and beliefs. We
attribute a purpose to a gene—that of maximizing its fitness. We even freely
discuss what the gene may or may not know when it is deciding what action
to take, as in Section 2.5.2. But a gene is just a self-replicating molecule
that nobody imagines has any more volition than a billiard ball or a cog
wheel. Nevertheless, game-theoretic models in which genes are personified
in this way have turned out to be a very successful tool in evolutionary
biology.

As in Section 2.7.1, we therefore find it necessary to confront two quite
different models that are capable of describing precisely the same phenomena. One is the mechanical model of molecular physics, in which a molecule
is a helpless victim of circumstance. The other is the model of evolutionary
biology, in which self-replicating molecules are treated as players in a game.
The fantasies of the second model are introduced because they provide a
useful shorthand for describing what happens after the forces of evolution
acting on a chaotic population of mindless molecules have reached an equilibrium. If such an equilibrium always had to be described in the longhand
of the first model, progress in evolutionary biology would be virtually impossible.

Dennett [154, 155, 156] argues that one can similarly describe human
behavior either with a mechanical model or with an intentional model, and
that much of the confusion in the free will debate lies in a fruitless attempt
to reconcile the internal structures of these irreconcilable models. Insofar as
he is talking about negative freedom, I agree that it is sometimes useful to
model man as *homo behavioralis* and sometimes as *homo economicus* (Section 2.4.2 of Volume I). The former is mechanical in the traditional sense,
being programmed to respond directly to stimuli like a chocolate-dispensing
machine. The behavior attributed to the latter is no less algorithmic, but
is expressed intentionally in terms of preferences and beliefs.

In as much as I part company with Dennett, it is because I am unsure
that the distinction between the mechanical and the intentional marks the
true battlefield in the free-will controversy. When positive freedoms enter
the picture, one also needs to distinguish between the models embodied by
homo sociologicus and what Kant [300, p.226] called *homo noumenon*—man
as a moral being.[33]

---

[33] One could write *homo ethicus*, or recognize the ghost in the machine by writing *homo
spiritus*. But Kant's *homo noumenon*, which he contrasts with *homo phaenomenon*—

We met *homo sociologicus* in Section 0.4.2. This fictional hominid is governed by a particular type of meme called a social norm. He differs from *homo behavioralis* in not being totally subservient to his genes. His behavior can change over time as the norms that govern his behavior adapt to the environment in which he finds himself. Given long enough, social evolution will sometimes succeed in making his behavior indistinguishable from that of a specimen of *homo economicus* equipped with suitable preferences and beliefs. Indeed, some of the social norms that control his behavior will explicitly take the form of preferences and beliefs, so that he not only comes to act like *homo economicus* in certain situations, he actually thinks like *homo economicus*. Unlike *homo economicus*, whose capacity for negative freedom lies in our inability to predict the output of his optimizing algorithm before it has been run, *homo sociologicus* has a capacity for positive freedom that lies in in our inability to predict the output of whatever algorithm he is using because we do not know its inputs.

Homo noumenon is harder to pin down, but he provides a model of man that is adequate for most workaday moral purposes—just as the principle "Nature abhors a vacuum" provides adequate guidance for dealing with most practical problems in pneumatics. The nebulous nature of *homo noumenon* only becomes apparent when one tries to evaluate the various claims made on his behalf. For example, undergraduate philosophers are apparently still taught that the dictum "ought implies can" entails that *homo noumenon* must have free will, since he would otherwise be unable to heed the call of a categorical imperative.[34]

I don't know whether an internally consistent model of *homo noumenon* could be built in which such results would be valid. But the important point is that one must avoid confusing the meaning of dicta like "Ought implies can" when used with a model of *homo noumenon* with their meaning when used in association with a model of *homo economicus* or *homo sociologicus*. In the case of *homo economicus*, "Ought implies can" means that his optimal outcome lies in his feasible set. In the case of *homo sociologicus*, it means that a social contract will specify punishments only for actions that somebody has the power to commit. Problems only arise when these different senses are confused—as when the fact that our social contract sometimes requires Adam to blame Eve is offered as evidence that she is morally responsible for her actions, and hence must have free will.

Similar considerations arise with the equally famous dictum "Is cannot

---

man as a natural being—would seem better to convey the flavor of what I have in mind.

[34]Recall from Section 4.2.1 of Volume I that a hypothetical imperative tells you what you ought to do if you want to achieve some particular objective. A categorical imperative tells you what to do unconditionally.

imply ought", sometimes referred to as Hume's law[35] and used to berate Humeans like myself who espouse naturalistic theories of ethics. But what Hume meant is only that noncontingent moral imperatives cannot be deduced from contingent matters of fact. Those who suppose otherwise fail to notice that they have mixed the language of *homo noumenon* with that of *homo economicus* or *homo sociologicus*. Within each of these domains separately, there is no intrinsic difficulty with the claim that an "Is can imply an ought". For example, *homo economicus* ought to use the simplex algorithm if he wants to maximize a linear objective function subject to linear constraints. *Homo sociologicus* ought do his duty if he is to avoid being condemned by his peers. Even *homo noumenon* ought to obey the categorical imperative if he is to be ideally "rational" in the Kantian sense.

3.2.5 Nil Desperandum!

The naturalistic views expressed in this book are often attacked as dehumanizing or dispiriting. Such criticism comes to a head with the ideas on moral responsibility and free will expressed in the current section. People often refuse to believe that anyone could really hold such supposedly bleak views on the nature of human existence. What would be the point of going on with life if such things were true?

One answer is that Nature doesn't care whether we like her truths or not. But to reply in such a vein is like saying yes or no when asked if you have stopped beating your wife. The right response is to deny the premise. The things one has to believe if one takes a naturalistic viewpoint seriously are neither dehumanizing nor dispiriting. In particular, the idea that telling ourselves the truth about our inner life will somehow throw a wrench into the works seems to me quite ridiculous—rather like the claim I just read in a popular science book that our bodies would fly apart if the quantum theory were false. Of course they wouldn't! They would continue to operate exactly as before. All that would change is that we would need to find a better explanation of how the universe works.

Matters are no different in the case of traditional theories of moral responsibility and personal freedom. As noted in Section 3.4.2 of Volume I, society will not collapse if people recognize that they are essentially no different from apes or robots. If we are indeed apes or robots, then everything that humans currently think or do is something that apes or robots can think or do. In particular, the way that a human society operates is one of

---

[35] It is said that to assert the contrary is to fall prey to the Naturalistic Fallacy. But the latter is actually a rather different notion of G. E. Moore that was dismissed in a footnote to Section 2.3.5 of Volume I.

the ways that a society of apes or robots can operate.

The loss of religious faith provides a good example. While holding onto their belief in God, people typically think that life would be impossible for them without their faith. Without God at the helm, life would lose its point, society would fall apart, wickedness would prevail, and so on. But after an apostate has recovered from the trauma of losing his faith, he finds that daily life goes on just as before. Nor are irreligious people noticeably less caring or good hearted than their churchgoing brethren—they simply find it possible to get on with their lives without the need to invent simplistic stories that supposedly explain everything around them. And so it is with those of us who have given up the rusty machinery of folk psychology. Do we seem to have lost our zest for life? Are we more prone to suicide? Not as far as I can see.

As Sextus Empiricus taught, contentment is possible without the need to cling to comforting beliefs. Or to say it with Goethe:

> Ich hab' mein Sach auf Nichts gestellt ...
> Drum ist's so wohl mir in der Welt.[36]

3.3 Folk Theorem

Thinkers of the left emphasize the importance of trust and commitment in maintaining cooperative understandings in human societies. Those on the right prefer to rely on the apparatus of law to enforce contractual arrangements. However, for indefinitely repeated games, the *folk theorem* tells us that we do not need to rely on anything but the enlightened self-interest of sufficiently forward-looking players to maintain the full panoply of cooperative possibilities. More precisely:

> Let x be any payoff pair in the cooperative payoff region of a game G that assigns each player at least his minimax value. If Adam and Eve are sufficiently patient, then there is a Nash equilibrium in the infinitely repeated version of G that generates an outcome y which is arbitrarily close to x.

This section will explain and defend this and other more sophisticated versions of the folk theorem in a nontechnical style. A more careful discussion appears in Chapter 8 of my *Fun and Games* (Binmore [70]), but serious scholars will need to consult the comprehensive treatments offered by Fudenberg and Tirole [187] and Myerson [374].

---

[36]I've built my life on naught but sand—that's the reason I feel grand!

The section also attempts to limit the damage created by the popular misconception that it makes sense to model every game in which cooperation can occur as the indefinitely repeated Prisoners' Dilemma in which everything one needs to know about the folk theorem is encapsulated in the phrase "tit for tat".[37] Axelrod's [31] *Evolution of Cooperation* certainly succeeded in focusing public attention on issues of vital importance, but his enthusiasm for the specific strategy TIT-FOR-TAT far outpaced his evidence, even for two-player games (Section 3.2.5 of Volume I). As for games with more than two players, TIT-FOR-TAT is badly misleading as a guide to the qualitatively different mechanisms that can then be used to sustain cooperative social contracts.

3.3.1   Memes

The notion of a *meme* was introduced by Dawkins [151] as the social equivalent of a gene. A meme is whatever gets replicated when people imitate their more successful neighbors or learn from their teachers. A meme may be anything from a hair style or a snatch of music to a scientific principle or a philosophical system. Perhaps the most important memes are those which act on other memes and so provide the raw material for the feedback systems that we explain in terms of intentionality and the ego. However, this section will confine its attention to memes that incorporate the how-to-do-it information required to implement a strategy in a game. Such memes will be modeled as a particular type of algorithm.[38]

The most familiar kind of algorithm is a computer program. Unfortunately, the language in which strategy-implementing algorithms are usually discussed diverts attention from the software that runs a computer to the hardware on which it is run. For example, the most general representation of an algorithm is called a Turing *machine*.[39] In an equally misleading manner, the algorithms studied in this section are called finite *automata* (Section 3.2.5 of Volume I). But we must not think of an automaton as a model of a player. We must think of a player as the computer hardware, and the automaton as the program that determines a player's strategy when he plays a certain kind of game.

---

[37]Dugnatov [163, p.25] devotes four pages to an uncritical listing of biological papers that elaborate on this theme. Even if these papers were uniformly reliable, they would be only indirectly relevant to most of the case studies in his book. For example, the mating of hermaphroditic sea bass is better modeled in terms of the Centipede Game (Section 1.6.2).

[38]Although the algorithms to be considered fall into the category to which Churchland [124] would confine the use of the word, I don't think he is entitled to deny Dennett's [158] more free-wheeling usage.

[39]On the assumption that the Church-Turing thesis holds (Section 3.4.2 of Volume I).

A finite automaton should therefore be seen as a type of meme that can be transferred from one head to another as a consequence of imitation or education. Social evolution then has the chance of achieving an equilibrium in a game relatively quickly. If an automaton were identified with a player's genetic hardware, convergence could only proceed at biological rates.

3.3.2 Finite Automata

Mathematicians distinguish various types of finite automata. Those suitable for playing repeated games are called Moore machines. When stimulated by being told what happened in the preceding round of a repeated game, a Moore machine responds by announcing a recommendation for the next round. In most of the examples considered here, an input for the automaton that represents Adam's strategy will simply be the action that Eve took last time. The possible outputs are the actions available to Adam at the next opportunity for play. All information other than Eve's last move that is used in computing Adam's next move must be held in the automaton's memory. However, sometimes it is useful to consider automata that accept more complicated inputs. For example, in Section 3.3.5, we shall encounter automata that take note of their own last action along with that of their opponent. In other contexts, one might wish to transfer some of the complexity of the processes being considered from the interior of the automata to their environment by allowing them to consult clocks or calculators.

The defining property of a finite automaton is that it has an upper bound on how much it can remember. For this reason, a finite automaton cannot keep track of all possible histories in an indefinitely repeated game. Confining attention to finite automata is therefore a definite restriction on the strategies that Adam and Eve can choose. Fortunately, nothing of great importance for the folk theorem is lost in submitting to this restriction.

Each conceivable configuration of memories that a finite automaton can sustain is a possible *state* of the machine. A careful description of a finite automaton would therefore begin by listing all its possible states. For a finite automaton, this set is necessarily finite. One of the states of the automaton is designated as its *initial state:* the state in which the automaton begins the game. An *output function* specifies the action that the automaton is to recommend in each of its states. A *transition function* tells the automaton how to move from one state to another after receiving as input the action taken by the opponent in the round of the game just concluded.

The mathematical formalism required to describe an automaton can sometimes be quite fearsome, but we shall find it adequate to represent automata capable of playing the indefinitely repeated Prisoners' Dilemma

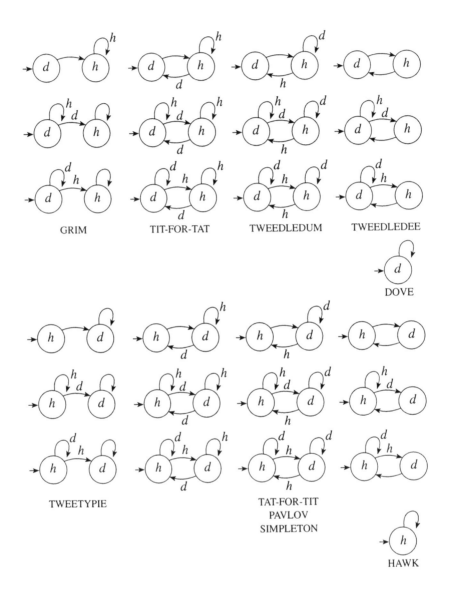

Figure 3.3: All one-state and two-state automata capable of playing the indefinitely repeated Prisoners' Dilemma.

with little pictures. Section 3.2.5 of Volume I gave some pictures of one-state and two-state automata in this class. The set of all twenty-six such automata appears as Figure 3.3.

The circles represent the possible states of an automaton, with the initial state indicated by a sourceless arrow. The letter inside a circle represents the automaton's output in that state. The letters d and h mean that *dove* or *hawk* are respectively to be played in the current round. Arrows leading from one circle to another show how the automaton transits from one state to another after emitting an output. Unlabeled arrows mean that the transition is made whatever the opponent just played. Arrows labeled with d or h indicate that the transition to be made is conditional on whether the opponent played *dove* or *hawk* in the round just completed.

The number of possible automata grows very rapidly with the number of states that are allowed. It is therefore necessary to be suspicious of the many computer simulations from which large conclusions are drawn on the basis of play between small numbers of automata said to be typical of this or that human characteristic. The complexity of the possible automata that evolution might throw into the ring defies all attempts at simple description.

Figure 3.4(a) shows what happens in the indefinitely repeated Prisoners' Dilemma when Adam uses the strategy TIT-FOR-TAT and Eve uses the strategy TAT-FOR-TIT from Figure 3.3. Play immediately gets into a cycle of length three, which is repeated over and over again.

When two automata play, they always end up in a cycle of some length. When TIT-FOR-TAT plays a copy of itself, the two automata immediately enter a cycle of length one, in which *dove* is always played. In the general case, some preliminary jostling may precede the cycle. For example, Figure 3.4(b) shows that two TAT-FOR-TITs first square off against each other by playing *hawk* before settling down into a cycle of length one, in which both always play *dove*. When TWEEDLEDUM of Figure 3.3 plays TAT-FOR-TAT, the initial jostling period is of length two.

3.3.3 Computing Payoffs

When Adam uses the strategy TIT-FOR-TAT against Eve's TAT-FOR-TIT in the repeated Prisoners' Dilemma of Figure 3.1(a), his income stream is $-1, 1, 3, -1, 1, 3, \ldots$. How does Adam evaluate such an income stream?

The standard method employed in economics is described in Section 1.2.4. If the time interval between successive rounds is τ, then Adam regards the income stream as being equivalent to the discounted sum:

$$s = -1 + \delta^\tau + 3\delta^{2\tau} + -\delta^{3\tau} + \delta^{4\tau} + 3\delta^{5\tau} \cdots,$$

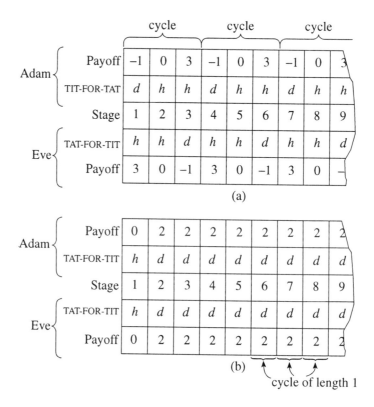

Figure 3.4: Computer wars.

where δ is Adam's discount factor. But game theorists usually prefer to assess an income stream by finding the constant income stream $x, x, x, x, \ldots$ that Adam likes equally well. In this book, x is called Adam's per-game payoff. As Section 1.6.1 explains, one finds Adam's per-game payoff x by multiplying the discounted sum s by $1 - \delta^\tau$.

The mathematics can be simplified either by assuming that Adam and Eve are infinitely patient, or by assuming that they play continuously. To study the case of infinite patience, we take the limit as $\delta \to 1$. To study the case of continuous play, we take the limit as $\tau \to 0$. In either case $\delta^\tau \to 1$, and so the analysis of Section 1.6.1 tells us that Adam's per-game payoff x is equal to his long-run average payoff.[40]

It is particularly easy to calculate long-run average payoffs when the

---

[40]In discussing the folk theorem, the phrase *long run* is used in its colloquial sense. The technical usage of Section 2.5.5 will be resumed in Chapter 4.

players use strategies that can be represented as finite automata. As we have seen, any two finite automata playing each other for long enough in a repeated game will eventually end up cycling through the same sequence of plays. A player's per-game payoff can therefore be calculated simply by taking his average payoff in a cycle. For example, Adam's per-game payoff when he uses TIT-FOR-TAT against Eve's TAT-FOR-TIT when playing the Prisoners' Dilemma continuously is

$$x = \tfrac{1}{3}(-1 + 1 + 3) = 1.$$

If Adam uses TWEEDLEDUM instead, we ignore the initial two rounds of jostling and take into account only the cycle of length one that follows. Once in the cycle, Adam always receives 2, and so this is his per-game payoff.

This book standardly assumes, not only that Adam and Eve's attitudes to time can be captured using a simple discount factor, but that play is continuous. However, it is important to appreciate that working with long-run averages is excusable only because it simplifies the mathematics. For example, in real life, we cannot afford to be infinitely patient about filling our bellies. As John Maynard Keynes famously put it, "In the long run we're all dead." Occasional asides will therefore be necessary to point out where more realistic assumptions about how people assess income streams are needed to justify the models employed.

3.3.4 Reciprocal Sharing

We are interested in the folk theorem of repeated game theory because it tells us what kind of sharing arrangements are sustainable using the reciprocity mechanism. Three ways in which costs or benefits can be shared will be considered. The first and second of these can be accommodated by expanding the set of strategies in the game under study. It is the third method that will therefore be the focus of attention in the folk theorem.

Casting lots. Section 1.2.1 describes how to construct the noncooperative and cooperative payoff regions of the game G of Figure 1.1(a). The noncooperative payoff region V is obtained by considering all the payoff pairs that can result when Adam and Eve randomize their strategy choices independently. However, we are more interested in the cooperative payoff region W obtained when Adam and Eve randomize jointly. As illustrated in Figure 3.5(a), this cooperative payoff region is the smallest convex set containing all the payoff pairs that can result when Adam and Eve are restricted to pure strategies in G,

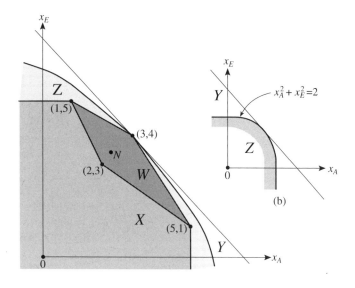

Figure 3.5: Payoff regions.

The cooperative payoff region of a game G is relevant when it is necessary to consider the kind of sharing that is accomplished by deciding who gets what by tossing a coin or spinning a roulette wheel. When such sharing is possible, it makes sense to replace G by a larger game H which is identical to G except that it is preceded by a publicly observed random event.[41] For example, if it matters how the team that kicks off at a soccer match is chosen, one would preface the soccer game G by allowing the captains to observe the fall of a fair coin. A pure strategy for one of the captains in the enlarged game H would then call for strategy s to be used in the game G if the coin falls heads and so Adam's team kicks off, and strategy t if it falls tails and Eve's team kicks off.

The point is that if we can prove a folk theorem about the use of pure strategies in a sufficiently large class of games, then we automatically include the case in which the players use a public randomizing device to coordinate their behavior.

---

[41]It is important in this context that the outcome of the random event be common knowledge. Mixed strategies are more commonly studied in the case where each player conditions his choice of pure strategy on independent, privately observed random events. If the players are restricted to such private randomizations, the resulting payoff region is V in Figure 1.1(b).

Physical transfers. Section 1.2.1 shows how the cooperative payoff region expands if one allows *free disposal* or *transferable utility.*

Wherever it makes a difference, this book assumes that free disposal is possible. This means that the players can always unilaterally throw utility away. Figure 3.5(a) illustrates how the cooperative payoff region W must then be replaced by the *comprehensive* payoff region X.

As in the case of public randomization, we do not need to allow the free disposal issue to complicate folk theorem questions, because we can expand the game G to a larger game H whose pure strategies allow the players to make wasteful actions. For example, one of the pure strategies in H might specify that Adam should use his second pure strategy in G *and* burn a ten dollar bill.

Before dismissing such wasteful strategies as a mathematical irrelevance, note that social contracts that insist on their use not only can exist— they do exist. Some of the inefficiencies in our own social contract can be traced to the punishments we inflict on our fellows when use their human capital to its best advantage and are therefore enviously categorized as getting above themselves. In some modern hunter-gatherer societies, this phenomenon can be so strong that children find it almost impossible to adopt the entrepreneurial spirit necessary to cope with the outside world.

Physical transfers from one person to another are equally unproblematic. For example, if Eve can offer Adam an apple as an inducement to play his first pure strategy rather than his second, then an expanded game H must be constructed with pure strategies that include Eve's transfering ownership of an apple to Adam.

In modeling such transfers, it is easiest if everything has a monetary value and the players are both risk neutral in money. One can then recalibrate their Von Neumann and Morgenstern utility scales so that each regards one util as being worth a dollar. Transfering an apple from Eve to Adam is then equivalent to taking some of Eve's utils and assigning them to Adam. If it were always possible for Adam or Eve simply to give each other some of their utils in this straightforward manner, then Section 1.2.1 explains that the payoff X of Figure 3.5(a) would need to be replaced by the payoff region Y.

In some branches of cooperative game theory, the idea that we can simply take a util from Eve and give it to Adam is taken for granted without any caveats about risk-neutrality. But Section 1.2.1 argues that this free-wheeling approach to transferable utility is deeply suspect. After all, if matters were so simple, all the fuss about interpersonal comparison of utility in Chapter 2 would be unnecessary. When the play of a game G can be supplemented by the transfer of goods or money as bribes or rewards, I shall therefore not normally assume that X can be expanded to the set Y. If .the economic environment is at all complicated, or the players are risk averse, the set Y of Figure 3.5(a) must be replaced by some smaller convex, comprehensive set Z, as in the following example.

Insurance deals.   Although the models of this book do not rule out the possibility that Adam and Eve might be risk neutral, I think it unwise to make this case the focus of attention. One reason is that one cannot systematically use dollars as the units on Adam's and Eve's utility scales without being thought to believe that Adam and Eve's utils should be counted the same in fairness calculations. But there is another important reason that arises from the argument of Section 2.5.4 that traces our use of fairness norms to primitive food-sharing understandings.

Such food-sharing arrangements are motivated by the need for the players to insure themselves against bad luck at hunting or gathering. In the simplest case, suppose that the players know that one of them will have two apples on Monday and that the other will have none, whereas the position will be reversed on Tuesday. Assume that the apples are perishable and cannot even be stored overnight. If the players are risk neutral over utils consumed at different times—as in the discounting paradigm—then neither will see any advantage in a deal that shares the apples so that each has one apple on both Monday and Tuesday. Each player would be equally happy with a lottery that left him hungry on a randomly chosen day and replete on the other.

To study physical transfers in situations where insurance contracts are meaningful, one therefore has to step outside the simplest mathematical model that comes to mind. However, it remains possible to accommodate the apple-sharing case without introducing anything very elaborate.

Rather than making Adam and Eve the ignorant folk behind a veil of uncertainty as in Section 2.5.4, it will be players I and II who are bereft of information. Behind their veil of uncertainty, they do not know who will turn out to be Adam or Eve. They then bargain in the belief that the two possibilities are equally likely. To evaluate this situation, they need to determine the cooperative payoff region of the game H played by Adam and Eve.

In the simple story told here, Adam finds two apples on Monday and none on Tuesday. Eve finds two apples on Tuesday but none on Monday. A pure strategy for Adam in the game H consists of the quantity a of apple he gives to Eve on Monday. A pure strategy for Eve consists of the quantity e of apple that she gives to Adam on Tuesday. Ignoring the possibility of free disposal, their choice of strategies will therefore leave Adam holding the bundle $(2 - a, e)$ and Eve holding the bundle $(a, 2 - e)$.

The simplest version of the discounting paradigm with infinitely patient players assigns utility $u(m, t) = m + t$ to a bundle (m, t) in which m is consumed on Monday and t on Tuesday. The players would then be equally happy with each of the bundles $(2, 0)$, $(1, 1)$ and $(0, 2)$, and we are led to the payoff region Y illustrated in Figure 3.5(b). But we have seen that the risk neutrality implied by such assumptions makes insurance deals pointless.

To model the insurance issue adequately, we can take

$$u(m, t) = \{(m + 1)(t + 1)\}^{\frac{1}{4}}.$$

Both players are then risk averse and would prefer to consume one apple on each day rather than two on a randomly chosen day. What is the payoff region Z in this situation? By the inequality of the arithmetic and geometric means,

$$
\begin{aligned}
x_A^2 + x_E^2 &= \{(3 - a)(e + 1)\}^{\frac{1}{2}} + \{(a + 1)(3 - e)\}^{\frac{1}{2}} \\
&\leq \tfrac{1}{2}(4 - a + e) + \tfrac{1}{2}(4 + a - e) = 4.
\end{aligned}
$$

Since equality is obtained when $a + e = 2$, it follows that the Pareto-frontier of the set Z is the quarter-circle of radius 2 illustrated in Figure 3.5(b).

Taking turns. Casting lots and transfering goods are ways of sharing that can be modeled successfully in one-shot games. Even the food-sharing insurance deal considered above can be squeezed into the one-shot format by lumping Monday and Tuesday together as a single period. But when sharing is achieved by taking turns, as when soccer teams alternate in defending the goal at the favorable end of the pitch, one cannot avoid studying the sharing mechanism in a repeated game alongside the punishment mechanism that sustains the exchange.

Figure 3.6(a) shows two automata, HUMPTY and DUMPTY, capable of playing the repeated Prisoners' Dilemma. If Adam chooses HUMPTY and Eve chooses DUMPTY, then play cycles through the sequence:

$$(dove, dove), (dove, dove), (dove, dove), (dove, hawk).$$

Adam and Eve therefore spend three quarters of the time at the payoff pair $(2, 2)$ in Figure 3.1(b), and one quarter of the time at the payoff pair $(-1, 3)$. Averaging over time leads to the pair of per-game payoffs $x = (1\frac{1}{4}, 2\frac{1}{4}) = \frac{3}{4}(2, 2) + \frac{1}{4}(-1, 3)$ shown in Figure 3.1(b). It is no accident that this is the same calculation that would be used to determine Adam and Eve's expected payoffs if they were to agree to play $(dove, dove)$ with probability $\frac{3}{4}$ and (*dove*, *hawk*) with probability $\frac{1}{4}$. After all, if we choose a time at random from the cycle, they will be found at (*dove*, *dove*) with probability $\frac{3}{4}$ and (*dove*, *hawk*) with probability $\frac{1}{4}$. For the same reason, we can approximate as closely as we like to any point in the payoff region of a game G by cycling through a suitable sequence of pure outcomes of the game.[42]

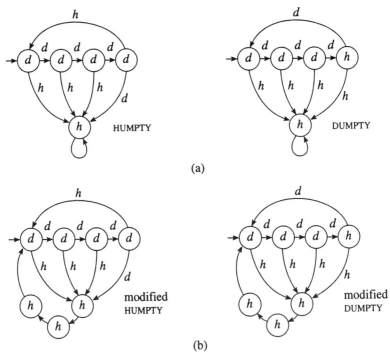

Figure 3.6: Humpty and Dumpty.

The essence of the folk theorem lies in seeing why the strategy pair (HUMPTY, DUMPTY) is a Nash equilibrium. Section 2.2.6 of Volume I used similar reasoning to show that (TIT-FOR-TAT, TIT-FOR-TAT) is a Nash equilibrium for the indefinitely repeated Prisoners' Dilemma when the players are not too impatient. The argument goes through most easily when applied to the pair (GRIM, GRIM), which Section 3.2.2 already used as the simplest example of an equilibrium that sustains a cooperative outcome in the indefinitely repeated Prisoners' Dilemma.

If both Adam and Eve use the strategy GRIM from Figure 3.3, then

---

[42] The identification is approximate because only rational weights can be achieved using finite automata.

they will always cooperate, thereby achieving a per-game payoff of 2. Of course, they could achieve the same result if they both use the strategy DOVE, which requires that a player always play *dove* irrespective of the other player's response. But (DOVE, DOVE) differs from (GRIM, GRIM) in not being a Nash equilibrium. If Eve sticks to DOVE, then Adam can push his per-game payoff from 2 to 3 by deviating to HAWK. However, if Eve sticks to GRIM, then Adam can do no better than to choose GRIM himself. If he deviates by making an unprovoked switch from *dove* to *hawk* at some stage, then the GRIM strategy requires that Eve move from her cooperative state to the punishment state in which she grimly plays *hawk* forever. Any deviation by Adam therefore results in his exchanging a per-game payoff of 2 for a per-game payoff of at most 0. It is therefore optimal for him not to deviate from the GRIM strategy. Since the same is true of Eve, (GRIM, GRIM) is a Nash equilibrium.

The same argument shows that (HUMPTY, DUMPTY) is a Nash equilibrium. If either player deviates from the prescribed equilibrium cycle, the opponent switches permanently to *hawk*, with the result that the deviant player gets a per-game payoff of at most 0. Since both players get more than 0 by playing their equilibrium strategies, neither has an incentive to deviate.

It will be evident that this argument can be made to work for any outcome in the shaded region of Figure 3.1(b). But what happens in less brutal games than the Prisoners' Dilemma, when the players do not have a strongly dominant strategy like *hawk* to use as a threat in keeping their opponent in line?

3.3.5 Crime and Punishment

To characterize the sharing arrangements that are sustainable as equilibria in the general case, we need to discuss a player's *minimax* value in a game.

Minimax. Adam's minimax value is found by looking for the worst punishment that Eve can inflict upon him in the long run. The long-run proviso matters because Adam will eventually figure out what punishment strategy Eve is using, and adapt his own strategy to minimize its impact on his payoff. Eve will predict that Adam will respond to her choice of punishment strategy t by choosing the strategy s that is a best reply to t. The worst that she can do to him is therefore to choose t so as to minimize the maximum payoff that Adam can achieve if he knows that she has chosen t.

If the players are restricted to using pure strategies, the first step in finding Adam's minimax value $\overline{m}_A$ in a game is to locate his maximum payoff in each column of its strategic form. In the game of Figure 3.7(a),

these maximum payoffs are enclosed in circles. The second and final step
is to locate the minimum of these maxima. In Figure 3.7(a), the value of
$\overline{m}_A$ found in this way is enclosed in a square. Similarly, $\overline{m}_E$ is the smallest
of the numbers found by locating Eve's maximum payoff in each row. The
minimax point $\overline{m}$ is simply $(\overline{m}_A, \overline{m}_E)$.

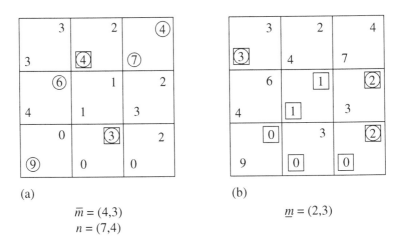

(a)

$\overline{m} = (4,3)$

$n = (7,4)$

(b)

$\underline{m} = (2,3)$

Figure 3.7: Minimax and maximin.

In the Prisoner's Dilemma of Figure 3.1(a), $\overline{m} = (0,0)$. Whatever punishment one player inflicts on the other, the opponent's best reply is to play
hawk. The first player can therefore inflict no worse punishment than by
playing *hawk* himself. In the game of Figure 3.7(a), $\overline{m} = (4,3)$ when the
players are restricted to using pure strategies.

 A more general folk theorem.   All the elements necessary to redefine
HUMPTY and DUMPTY of Figure 3.6(a) for the general case are now in place.

Pick any payoff pair $x \geq \overline{m}$ in the cooperative payoff region W of a
game. Because W is the convex hull of the set of payoff pairs of the game
that can be obtained using pure strategies, Adam and Eve can achieve
x by agreeing to play each pure-strategy pair (s, t) with an appropriate
probability p. To achieve the same effect by taking turns, they need to
cycle in a way that makes them spend the correct proportion p of the time
at each strategy pair (s, t).[43]

---

[43]If p is an irrational number, they will not be able to do this exactly, but they can
make the approximation as close as they choose.

The simplest way to defend such cycling against a deviation is to introduce a grim punishment scheme, as in HUMPTY and DUMPTY. If any player deviates from the cycle, the other switches permanently to a state in which he plays the strategy that holds the opponent to his minimax value. Since he gets at least his minimax value by sticking with the cycle, any deviation will therefore be unprofitable. A Nash equilibrium has then been constructed that sustains the outcome x.

Although the emphasis is usually on the fact that such reciprocity mechanisms can be used to sustain Pareto-efficient equilibria in indefinitely repeated games, it is important to recognize that the same mechanism can also be used to stabilize inefficient equilibria. When free disposal is permitted, the mechanism can even be used to sustain equilibria in which a player is required to burn enough money to reduce his payoff to his minimax value. Anthropological reports suggest that some modern hunter-gatherer societies do indeed operate social contracts in which this phenomenon is present to some degree. However, its chief relevance in this book is that it makes it possible to assume that the set X of feasible social contracts is not only *convex*, but *comprehensive* for all practical purposes.

Minimal punishments. The defense of the folk theorem offered above abstracts away many complicating features of the real world. In particular, the players are assumed to be ideally rational, so that they never deviate from the equilibrium path. The punishment that would await them if they were to lose their way therefore remains hypothetical. Provided that it is adequate to deter deviations, it follows that the severity of the punishment is irrelevant. Up to now, we have followed Abreu [1] by exploiting this fact to explore the range of outcomes that can be sustained as equilibria in indefinitely repeated games without allowing the issue to be confused by the level of punishments imposed. All punishments have been taken to be maximal. But there are sound reasons why the punishments built into real social contracts are seldom so draconian.

The reason for discounting the possibility that real social contracts will specify maximal punishments is that, unlike the ideally rational players of traditional game theory, real people cannot guarantee that they won't occasionally stray from the equilibrium path as a result of careless inattention or foolish miscalculation. No matter how good our intentions, we must therefore sometimes expect to be punished ourselves. Indeed, if we count all the subtle devices that people use to signal disapproval, we are all being punished in small ways all the time. What other way is there to interpret the subliminal withdrawals of polite courtesies, the half-turned shoulders, the irritating remarks passed off as badinage, the pursed lips, the awkward silences, the eyes that wander in search of a more socially acceptable partner, the failures to remember who we are or why we matter? How else do we learn to chart a course through the complexities of social life other than by using hints and cues from our neighbors as a control mechanism to bring us back to the equilibrium path when we stray?[44]

When specifying punishments that might apply to ourselves or those we love, we therefore all have a common interest in keeping their severity to the minimum level consistent with their deterrent function.[45] They may be barely perceptible, but it is these gossamer threads of convention that are most important in keeping a social contract on track.

How is it necessary to modify HUMPTY and DUMPTY so that the punishments they administer are minimal rather than maximal? To study this question adequately, it is necessary to suspend the assumption that time is continuous for the moment, since this assumption treats the short-term gains to be made by deviating as entirely negligible.

Instead of punishing any deviation by switching permanently to *hawk*, suppose that HUMPTY and DUMPTY are modified so that *hawk* is played only three times—ignoring anything that the opponent might do while being punished—after which the modified automaton switches back to the beginning of the equilibrium cycle. Figure 3.6(b) shows modified versions of HUMPTY and DUMPTY that operate in this way.

If Adam is contemplating cheating for the first time at the fifth round, and then responding optimally to the punishment from an Eve using the

---

[44]It is true that such signals are sometimes not directed at our behavior as individuals, but at the behavior of the social category we are perceived as representing. Traditional morality hopelessly condemns such behavior as "blaming the innocent". I don't like it either, but there is no point in pretending that it cannot be an effective technique for sustaining social contracts in multiperson societies (Section 3.4.4). This is why preachers commonly contrive to reinforce the xenophobia of their insider groups in the very act of paying lip service to the doctrine of universal brotherly love. A Neanderthal bigot like Paisley can even escape ridicule in Northern Ireland when he claims to be a reverend and founds a new Protestant sect.

[45]Of course, when the bourgoisie specify punishments for the underclasses, no such restraints apply. Victorian swindlers who cheated the poor out of their savings were therefore slapped on the wrist, while children who stole to assuage their hunger were imprisoned or transported. The current three-time loser doctrine is an unwelcome reminder that such Victorian thinking is far from dead. Even if we had two strikes against us, respectable folk like ourselves would never be sufficiently desperate or reckless to steal a pizza in the knowledge that we faced a mandatory life sentence if caught. Only the unworthy find themselves so embarrassed. Our standards of interpersonal comparison therefore allow us to maintain a sense of stern justice as we read of the suffering inflicted in our name on the socially halt and lame by a judicial system of which we are taught to be proud.

modified DUMPTY, then he needs to compare the income streams:

$$2, \quad 2, \quad 2, \quad -1, \quad 2, \quad 2, \quad 2, \quad -1, \quad 2, \quad 2, \quad 2, \quad -1, \quad \ldots;$$
$$2, \quad 2, \quad 2, \quad -1, \quad 3, \quad 0, \quad 0, \quad 0, \quad 2, \quad 2, \quad 2, \quad -1 \quad \ldots.$$

If the time interval τ between successive rounds of play is sufficiently small, then Adam's overall gain from deviating is approximately $1-2-2+1 = -2$. Since this is negative, Adam continues to be deterred from cheating even though the severity of Eve's punishments has been enormously reduced.

Mixed strategies and maximin. Confusion arises because people talk about Adam's minimax value $\overline{m}_A$ in a game when they really mean his maximin value $\underline{m}_A$. Recall from Section 0.4.1 that $\underline{m}_A$ is the largest payoff that Adam can guarantee no matter what strategy Eve may choose. For this reason $\underline{m}_A$ is called his *security level*. A strategy whose play ensures that Adam gets at least his security level is called a security strategy.

In Figure 3.7(b), Adam's maximin value is found by first enclosing his minimum payoff in each row in a square. His maximin value $\underline{m}_A$ is the maximum of these minima, which is enclosed in a circle. The maximin point for the game of Figure 3.7 is $\underline{m} = (\underline{m}_A, \underline{m}_E) = (3, 2)$.

The idea of a player's maximin value was discussed at length in Section 4.6 of Volume I while evaluating the claims that Rawls makes on behalf the maximin criterion as a means of solving the decision problem faced by players behind the veil of ignorance. Claims of this kind are usually based on a misapprehension arising from Von Neumann's [537] celebrated *minimax* theorem. This says that $\underline{m} = \overline{m}$ in any finite game—provided that we allow the players to use mixed strategies. The result is of fundamental importance in two-person, zero-sum games because it implies that the players' security levels sum to zero. As Section 4.6 of Volume I explains, it follows that a pair of mixed strategies for a two-person, zero-sum game is a Nash equilibrium if and only if each strategy in the pair is a security strategy for the player who uses it. In particular, if n is a Nash equilibrium outcome in a two-person, zero-sum game, then $\underline{m} = \overline{m} = n$.

If it is common knowledge that the players in a two-person, zero-sum game are rational, they will play a Nash equilibrium. In a two-person, zero-sum game, the minimax theorem therefore implies that each player will choose a strategy using the maximin criterion. But nothing says that the maximin criterion has any virtue for making decisions in other games. In the general case, all that can be said about the relationship between a Nash equilibrium outcome n and the maximin and minimax points in a game is that $\underline{m} \leq \overline{m} \leq n$. The game of Figure 3.6 demonstrates that both inequalities may be sharp when the players are restricted to pure strategies, but the minimax theorem continues to show that $\underline{m} = \overline{m}$ when we allow the players to use mixed strategies. However, almost every game in this book provides examples of Nash equilibria that yield payoffs to the players in excess of their security levels.

The chief reason for this digression on the minimax theorem is to observe that a version of the folk theorem remains valid even when the players are allowed to use mixed strategies. The theorem is no harder to prove if the players are allowed to see the random device that the opponent used to choose his pure strategy after each round, but without this unrealistic assumption, the players have to cope with the vexed problem of how to detect cheating that cannot be directly observed.

For example, it may be that Eve's equilibrium strategy requires that she always toss a coin to determine what action to take. If she doesn't behave in this way, Adam's duty is to punish her. However, he sees only the action that she actually takes, not the way that the coin really fell. Suppose, for example, that Eve always takes the action that favors her, but claims not to be cheating because her coin just happened to fall heads every time. How many times can she make this claim before Adam ceases to believe her and starts to administer punishment?

Rather than upset Auden's ghost by doing any statistics, the answer to this question is suppressed.[46] But the bottom line is that the folk theorem survives the introduction of mixed strategies virtually intact. The notions of maximin and minimax can therefore be interchanged when mixed strategies are permitted in a folk-theorem context. In particular, any outcome x in the cooperative payoff region of a game is sustainable as a social contract, provided that it assigns each player at least his security level.

3.3.6 Guardians Who Guard Each Other

As I have repeatedly stressed, a society that pretends its officers can always be relied upon to carry out their assigned duties when these conflict with their own private incentives is sowing the seeds of its own destruction. In consequence, we must learn to live with whatever second-best society can be patched together using incentive schemes designed to persuade officials that they advance their own interests by advancing the interests of the state. But who administers such schemes, and what are *their* incentives?

As Section 3.1 argues, the game theory answer to this classic question of who should guard the guardians is that *we must all guard each other*. One advantage in focusing on a society with only two citizens is that it becomes impossible to shut one's eyes to the need to close the chains of

---

[46]Thou shalt not sit with statisticians nor commit a social science—W. H. Auden.

responsibility in this way. Adam and Eve have to do what policing needs to be done because there is nobody else to do it.

However, so far the policing question has been evaded by confining attention to Nash equilibria. The story is then that Adam stays on the equilibrium path because he fears that Eve will punish him if he deviates. But the punishments available to Eve are frequently as painful for her as they are for him. So what incentive does she have to punish Adam should he stray from the strait and narrow path? If no incentive can be discerned, why should Adam believe that she will actually administer the punishment specified for his offense?

As Section 2.5.2 of Volume I explains, Reinhard Selten [471] introduced the notion of a *subgame-perfect* equilibrium to deal with such questions. To be subgame-perfect, a strategy profile must not only be a Nash equilibrium for the game as a whole, it must also specify Nash equilibrium play in each subgame—whether or not the subgame will be reached in equilibrium. When Adam contemplates making a deviant action that will lead to a subgame that is off the equilibrium path, he then knows that Eve's plans for playing this subgame are just as rational as her plans for playing the game as a whole. So if the subgame-perfect equilibrium prescribes that she punish him in this subgame, then punish him she will.

Fortunately, the folk theorem survives substantially intact if we demand that social contracts must be sustainable not only as Nash equilibria, but also as subgame-perfect equilibria. The methodology is not hard to explain, although all but one of the necessary technicalities will be suppressed.

The technicality to be addressed concerns the assumption that an automaton only inputs the last action of its opponent. With this assumption, it is not even clear what an automaton would do after it counterfactually contravened its own programming by deviating from the equilibrium path. In accordance with Selten's [471] trembling-hand resolution of such counterfactual scenarios (Section 3.4.1 of Volume I), the fact that the wrong action has somehow been taken is usually assumed to have no effect on an automaton's state. Neither of the strategy pairs (GRIM, GRIM) nor (TIT-FOR-TAT, TIT-FOR-TAT) is subgame-perfect with this understanding.[47]

The problem arises because we have not allowed a player's automaton to take account of the impact on future play caused by the player's failing to carry out the action prescribed by the automaton. However, the difficulty is easily fixed by redefining an automaton's input to be the pair of actions

---

[47]The alternative explanation of the counterfactual behavior is that the automaton somehow switched states before the last round and so took the wrong action because it was in the wrong state. With this understanding, (GRIM, GRIM) is subgame-perfect, but (TIT-FOR-TAT, TIT-FOR-TAT) continues to fail the test.

last taken by *both* players. This allows an automaton to take corrective action if it finds that the player it is supposedly controlling has erred by making an out-of-equilibrium move.

With this modification in the design of an automaton, it is easy to construct subgame-perfect equilibria by ensuring that players who fail to punish when the equilibrium being operated calls for punishment are punished themselves. As the following example shows, we can even force deviant players to join with their upright counterparts in punishing themselves!

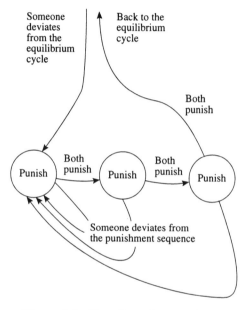

Figure 3.8: Guarding the guardians.

Figure 3.8 shows how the punishment schemes of the modified HUMPTY and DUMPTY automata of Figure 3.6(b) can be altered to make them a subgame-perfect pair. If someone decides to cease cooperating and so departs from the equilibrium path, both players join in punishing him for as many rounds as is necessary to render the deviation unprofitable. If the punishment is successfully administered, both players then return to the equilibrium path. But what if someone fails to punish when punishment is called for? Then *this* behavior is punished. And if someone fails to punish someone who has failed to punish when punishment is called for, then *this* behavior is punished. And so on.

The fact that people can be controlled in this way by setting them to spy on each other will come as no surprise to those who grew up in totalitarian states. But those of us who have lived more sheltered lives tend to be less willing to believe that our social contracts survive only because cheating is held in check by the ever-present threat of detection and punishment. We prefer to think that we keep off the grass in parks, not because we fear public disapproval, but because we don't want to be antisocial. But I think this is just another example of how folk psychology reverses causal chains. It isn't that we avoid public disapproval because we don't want to be antisocial—we have internalized socially responsible values because we thereby avoid disapproval. Those of us who lead a comfortable bourgeois existence manage to overlook this fact only because society has been so successful in conditioning our responses to the continual social prods and pushes with which are controlled that even our childhood memories of the conditioning process are inaccessible to us.

3.3.7 Tit for Tat?

In spite of its enormous importance for understanding the role of rights and duties in human social contracts, the folk theorem was largely ignored by scholars outside economics for more than twenty years. Only with Trivers' [524] coining of the term *reciprocal altruism* did its message begin to enjoy widespread currency. But it was the later publication of Axelrod's [31] *Evolution of Cooperation* that really put the folk theorem on the map. Nowadays, all social scientists know about TIT-FOR-TAT. Unfortunately, much of what has been said about the tit-for-tat paradigm is overblown or mistaken. Worse still, the popularity that the paradigm enjoys has obscured the important fact that there are many more ways of supporting a social contract than by naive pairwise reciprocation.

Axelrod's Olympiad. Section 3.2.5 of Volume I discusses the tit-for-tat bubble at length. To recapitulate, Axelrod invited various social scientists to submit computer programs for an Olympiad in which each entry would be matched against every other entry in the repeated Prisoners' Dilemma. After learning the outcome of a pilot round, contestants submitted computer programs that implemented sixty-three of the possible strategies of the game. For example, TIT-FOR-TAT and GRIM were submitted respectively by the psychologist Anatole Rapaport and the economist James Friedman.

In the Olympiad, TIT-FOR-TAT was the most successful strategy. To simulate the effect of evolution operating on his sixty-three strategies, Axelrod then used an updating rule which ensures that strategies achieving a high payoff in one generation are more numerous in the next. The fact that TIT-FOR-TAT was the most numerous of all the surviving programs at the end of the evolutionary simulation clinched the question for Axelrod, who then proceeded to propose tit-for-tat as a suitable paradigm for human cooperation in a very wide range of contexts. In describing its virtues, Axelrod [31, p.54] says:

> What accounts for TIT-FOR-TAT's robust success is its combination of being nice, retaliatory, forgiving and clear. Its niceness prevents it from getting into unnecessary trouble. Its retaliation discourages the other side from persisting whenever defection is tried. Its forgiveness helps restore mutual cooperation. And its clarity makes it intelligible to the other player, thereby eliciting long-term cooperation.

To what extent are these claims justified? On examination, it turns out that TIT-FOR-TAT is not so very successful in Axelrod's simulation. Nor is its limited success robust when the initial population of entries is varied. The unforgiving GRIM does extremely well when the initial population of entries consists of all the twenty-six automata of Figure 3.3. Nor should we expect evolution to generate nice machines that are never the first to play *hawk*, provided that some small fraction of suckers worth exploiting continually flows into the system. As for clarity, all that is necessary for cooperation to evolve is that a mutant be able to recognize a copy of itself. All that is then left on Axelrod's list is the requirement that a successful strategy be retaliatory. But this is a lesson that applies only in *pairwise* interactions. As Section 3.4 documents, in multiperson interactions, it need not be the injured party who punishes a cheater.

Before justifying these counterclaims, it is necessary to address Nachbar's [377] more radical criticism: that Axelrod mistakenly ran an evolutionary simulation of the *finitely* repeated Prisoners' Dilemma. Since the use of a Nash equilibrium in the finitely repeated Prisoners' Dilemma necessarily results in *hawk* always being played, we then wouldn't need a computer simulation to know what would survive if *every* strategy were present in the initial population of entries. The winning strategies would *never* cooperate.

As explained in Section 3.2.5 of Volume I, Nachbar is correct to claim that Axelrod inadvertently ran his evolutionary simulations with the finitely repeated Prisoners' Dilemma, but this fact does not necessarily invalidate Axelrod's conclusions, because none of the 63 entries were programmed to exploit the end effects of the finitely repeated game. In fact, Linster [327, 328] obtained results close to those reported by Axelrod when he replicated the simulations with the finitely repeated Prisoners' Dilemma replaced by the infinitely repeated version. However, contrary to popular opinion, the result obtained is not that only TIT-FOR-TAT survives. Actually a mixture of strategies survives, among which TIT-FOR-TAT is the most numerous, but nevertheless controls only a little more than $\frac{1}{6}$ of the population.

How significant is it that TIT-FOR-TAT commands a plurality among the survivors? Theory provides some help in answering this question. We know that Linster's simulation can only converge on one of the many Nash equilibria of the 63×63 strategic form whose pure strategies are the entries submitted to the Olympiad. If the population starts with each of these strategies controlling an equal share of the population, then Axelrod's work shows that the system converges on a mixed Nash equilibrium in which TIT-FOR-TAT is played with a probability of about $\frac{1}{6}$. However, we can make the system converge on a variety of the Nash equilibria of the 63×63 game by starting it off in the basin of attraction of whatever stable equilibrium takes our fancy. Axelrod tried six different initial conditions, and found that TIT-FOR-TAT was most numerous among the survivors five times out of six. Linster [327, 328] systematically explored all initial conditions, and found that TIT-FOR-TAT is played with greatest probability in the final mixture only about $\frac{1}{4}$ of the time.

But why restrict ourselves to Axelrod's sixty-three strategies? Why not follow Linster [327, 328] and begin with all automata having at most two states? The system then converges from a wide variety of initial conditions to a mixture in which the strategy GRIM is played with probability greater than $\frac{1}{2}$. But GRIM is not forgiving. On the contrary, it gets its name from its relentless punishment of any deviation for all eternity.

To evaluate the evolutionary claims for niceness that Axelrod makes on behalf of TIT-FOR-TAT, it is necessary to turn to simulations that mimic the noisy processes of mutation and sexual variation. In an innovative paper, Axelrod [32, 33] pioneered the use of Holland's [257, 258] genetic algorithm for this purpose.[48] Axelrod's pilot study considered only 40 simulations of 50 generations each, but Probst [411] later went the whole hog by running very large numbers of simulations for very long periods without imposing binding complexity constraints.[49] Axelrod [32] found that mean machines were thriving at the end of 11 of his 40 simulations, but Probst shows that Axelrod was wrong to dismiss this phenomenon as a transient blip preceding an ultimate takeover by naive reciprocators like TIT-FOR-TAT. On the contrary, it is the initial success of naive reciprocators like TIT-FOR-TAT that turns out to be transient. In the long run, mean machines triumph. Axelrod's [31] claim that evolution should be expected to generate nice machines in the indefinitely repeated Prisoners' Dilemma therefore turns out to be mistaken.

---

[48] Axelrod refers to the deterministic simulations of his earlier work as "ecological" to distinguish them from these "evolutionary" simulations.

[49] Section 3.2.5 of Volume I reports on the original simulations Probst ran for his master's thesis in Basel. The later studies of his doctoral thesis are shortly to be published (Pollack and Probst [411, 407]).

The strategy TAT-FOR-TIT of Figure 3.3 is the simplest example of the type of mean machine that emerges from Probst's simulations. A player using TAT-FOR-TIT begins by trying to exploit his opponent, and only starts to cooperate if he finds that she is trying to exploit him in the same way that he is trying to exploit her. Conclusions similar to Probst's are reached by the biologists Nowak and Sigmund [388, 389, 390, 391, 493], albeit in less decisive simulations. They refer to TAT-FOR-TIT as PAVLOV because it stays in the same state when it wins but shifts when it loses.[50]

Axelrod's [33, p.21] very recent sequel to the *Evolution of Cooperation* ignores the widespread criticism from game theorists surveyed in this section. It recognizes the unease registered by some biologists, but nevertheless reiterates his original claim that TIT-FOR-TAT embodies the essential features of a successful strategy for the indefinitely repeated Prisoners' Dilemma. But what of his own discovery that evolution sometimes favors mean machines? He argues that mean machines did well only because they can exploit suckers. If his simulations had been run for longer, he claims that the mean machines which initially did well would have fallen by the wayside after eliminating the suckers on which their success depends.

The intuition behind this argument is certainly valid for a population that initially consists of HAWKs, DOVEs and TIT-FOR-TATs, provided that no new strategies are allowed to intrude. If DOVEs predominate at the outset, then HAWKs will do well initially, but then fade away altogether as the DOVEs on which they prey become increasingly infrequent. But the intuition derived from this example doesn't extend to the case when HAWKs are replaced by TAT-FOR-TITs. If one starts the latter system off in the right basin of attraction, the final population will consist *only* of TAT-FOR-TITs.[51] Nor do the complexity considerations to which Axelrod [33] devotes his new book reverse these theoretical considerations. On the contrary, Binmore and Samuelson [84] find that introducing complexity considerations into

---

[50]In their early *Prisoner's Dilemma*, Rapaport and Chammah [415] prejudicially refer to TAT-FOR-TIT as SIMPLETON. (Recall that Anatole Rapaport figured large in Section 3.3 of Volume I as one of the circle-squarers seeking to prove that cooperation is rational in the one-shot Prisoners' Dilemma.) I prefer to stick with the terminology introduced in Binmore and Samuelson [41], because it recognizes the importance of the signalling role of the opening phase, during which two mean machines that will eventually cooperate with each other explore the possibility that the other machine might be exploitable. Abreu and Rubinstein [5] discuss this point in detail.

[51]Notwithstanding the equally correct fact that the final population will contain no TAT-FOR-TITs at all if the system is started off in the second of the two basins of attraction. Wu and Axelrod[568, 33] miss a similar point in their attempt to rebut the claims made by Nowak and Sigmund for PAVLOV. They add TAT-FOR-TIT and three other strategies to the initial population in Axelrod's original ecological simulation. This perturbation is not enough to shift the system into the basin of attraction of an equilibrium mixture containing TAT-FOR-TIT, but so what?

an evolutionary analysis destabilizes all equilibrium mixtures that fail to include a mean machine. The mixture consisting of the mean machine TAT-FOR-TIT and the two nice machines TWEEDLEDUM and TWEEDLEDEE proves to be particularly stable.

Why doesn't the tit-for-tat bubble burst? Why do science writers like Ridley [426] continue to use TIT-FOR-TAT as the paradigm for human cooperation? Do they not understand the criticisms of Axelrod's work surveyed in this section?

Sometimes TIT-FOR-TAT is endorsed under the misapprehension that it means no more than "tit for tat" in ordinary English. Further confusion then follows because the colloquial usage carries a fairness connotation. For example, a journalist recently told me that TIT-FOR-TAT is a scientific fact because badgers apparently split the time they spend grooming each other very equally. But why is this relevant to TIT-FOR-TAT? Even in ordinary English, the tit that follows a tat is a *punishment* that fits the crime.

To make this observation is not to deny that one needs to appeal to fairness when matching a tit to a tat. On the contrary, Section 3.7.3 argues that the natural response when Adam cheats on a fair deal is for Eve to take whatever action is necessary to reinstate the *status quo* that held good before the deal was reached. The loss that a cheating player sustains is then equal to the gain he anticipated receiving as a result of implementing the deal. Since these gains are calculated using the prevailing standards of fairness, so are the losses that result from Eve's punishment of Adam's deviation. The tit is therefore fairly determined by the tat. However, such fairness considerations are entirely absent in the type of Olympiad at which Axelrod [31] crowned TIT-FOR-TAT with the laurel wreath of victory.

Other popularizers are so seduced by the idea that evolution will necessarily make us nice that they see no need to examine the scientific evidence they quote in its support. When confronted with the shortcomings of the evidence, they give themselves away by changing their ground. Their enthusiasm for TIT-FOR-TAT then really turns out to be based on their experiences of being brought up in a comfortable middle-class household.[52] But the anecdotes about the social dynamics of the middle classes with

---

[52]Sometimes accounts of classroom role-playing exercises are also mentioned. It is true that unpaid students who know each other well behave pretty much as TIT-FOR-TAT prescribes in classroom exercises. But such students also typically play *dove* in the one-shot Prisoners' Dilemma! Section 0.4.2 explains why I believe that "stylized facts" derived from such classroom exercises are without scientific value. One certainly cannot extrapolate to real-world interactions between experienced strangers when a lot hangs on the outcome. Even properly controlled experiments run in the laboratory need to be interpreted with much caution.

which they defend TIT-FOR-TAT are irrelevant to the repeated Prisoners' Dilemma, which models the interaction between two strangers. To understand the social contracts that operate within middle-class insider groups, one must remember that the sons and daughters of bourgeois families enter a *multiplayer* game that *began long ago*.

The simplest game that seems to capture something of the intuition that popularizers have mistakenly learned to label with the TIT-FOR-TAT tag is an overlapping generations model in which three players are alive at any time. Occasionally, one of the players dies and is immediately replaced by a new player. In each period, two of the players are matched at random to play the Prisoners' Dilemma, while the third player looks on. Long ago, an equilibrium was somehow established that now requires each player always to play *dove*. A player who fails to do so will find that the opponent with whom he is next matched will punish him by playing *hawk*—whoever that opponent may be. Yesterday, the players were Adam, Eve and Ichabod. But Ichabod died overnight and has been replaced by Olive. She is now matched with Adam. Why does Adam treat her nicely by playing *dove?* After all, we know that there are many mean equilibria that might form the basis for a social contract between the minisociety consisting only of Adam and Olive. Some of these mean equilibria would allow Adam and Olive to explore the possibility that their new opponent is a sucker who can be exploited. But these equilibria are unavailable because of the presence of *Eve*. She enforces nice behavior from the word go by being ready to punish anyone who is nasty.

More generally, when children grow up within a middle-class insider group, they learn to treat other insiders with a consideration denied to outsiders. Insiders who don't conform soon find themselves treated as outsiders unless they mend their ways. However, this is as far as the analogy with TIT-FOR-TAT goes. Nature has not brought the same sweetness and light that operates within middle-class insider groups to the world at large. The outsiders who lurk in dark alleys with rape and mayhem in their hearts are neither nice nor forgiving. Nor do sharks only cruise in murky waters. They also swim in brightly lit boardrooms and patrol the corridors of power. Such upper-crust sharks show beautiful teeth as they prey upon our bank accounts and raid the pension funds of elderly widows. But we would be the fools they take us for if we returned the smiles with which they try to convince us that they are nice people like ourselves.

Political theorists cannot afford to ignore the nastiness in the world. It just isn't true that nastiness is irrational, or that evolution will eventually sweep it away. As Hume [271] warned, our constitutions therefore need to be armored against the modern methods that rogues and knaves posing as insiders have developed to subvert our social contract (Section 3.1). Even more urgent is the need to find ways of reducing conflict between mutually hostile groups. Can Serbs and Croats eventually be persuaded to stop treating each other as outsiders and start being nice to each other again? Is there any hope for Northern Ireland or the Middle East?

Axelrod [31] gives one striking example of the emergence of such cooperation. In the First World War, there were several fleeting outbreaks of implicit collusion between units of the British and German armies, in which each side ceased to shell the other. Axelrod [31] attributes this behavior to tit-for-tat reasoning, but such an explanation overlooks the obvious fact that the players didn't begin by being nice to each other. I agree that it is vital to understand how such cooperation between groups who treat each other as outsiders can get off the ground, but there seems no point at all in seeking to analyze the emergence of cooperation using a model that takes the conclusion for granted.

3.3.8   How Does Cooperation Evolve?

The folk theorem summarizes the many reciprocity mechanisms that can *sustain* Pareto-efficient outcomes in the indefinitely repeated Prisoners' Dilemma. The popular literature therefore does Axelrod [31] no favor when it credits him with having demonstrated the false proposition that TIT-FOR-TAT is the only such mechanism. His achievement was to pioneer the evolutionary approach to the problem of how equilibria are *selected*.

Even when the spotlight is turned from how equilibria are sustained to how they are selected, there remains room for misunderstanding Axelrod's work. Although he is a political scientist, his ideas on equilibrium selection are not immediately relevant to any social contract of recorded history. All known social contracts are obviously the product of long and fierce competition between rival forms of social organization dating back to prehistoric times.[53] As Section 2.5.3 explains, to understand this evolutionary struggle properly we need to study the fairness norms and other equilibrium selection devices that evolved for the purpose of giving one society an edge over its competitors in exploiting new sources of surplus. But Axelrod's [31] model of a *single* society operating in a vacuum lacks an arena within which such rival equilibrium selection devices can compete.

Rather than telling us anything about ancient or modern political sys-

---

[53]Which is far from saying that the evolutionary process is over. Indeed, the recent collapse of the Soviet empire must surely be one of the most dramatic events that the process has ever seen. Now that the mutant meme of communism has been expelled from the body politic, a billion new minds are available for colonization by memes from the west. Who knows what compromise will be eventually hammered out between the old and the new under all those recently acquired baseball hats?

tems, Axelrod's [31] simulations seem designed to offer insights into the *biological* problem of how evolution *first* got reciprocal altruism to work. Axelrod's work is therefore more obviously relevant to the sharing of blood among the vampire bats described by Wilkinson [553] than to *homo sapiens*. After all, we were social animals long before we were human. However, I think that Axelrod's work is also relevant to the much later period of retooling when *homo sapiens* abandoned the hierarchical forms of social organization of chimpanzees and baboons in favor of a *leaderless* style of social contract. This hypothesis is suggested by the anthropological studies of modern hunter-gather societies briefly reviewed in Section 4.5.

Modern foraging societies certainly have no bosses, but I differ from anthropologists who have written on the subject in thinking that the social contracts of these societies are nevertheless far too complex to serve as a pattern for the prehistoric bands of hunter-gatherers within which the human way of cooperating first took root. One might even argue that the social contracts of contemporary foraging societies are more sophisticated than our own insofar as order is maintained without resorting to such crudities as a police force and a judiciary. The small numbers in a modern hunter-gatherer society make it possible for the group *as a whole* to enforce the social contract—as in the three-person, overlapping generations model of Section 3.3.7. Section 4.5.2 therefore speculates that food sharing among primitive hunter-gatherers must *originally* have been more like the pairwise sharing of blood among unrelated vampire bats than the communal sharing of meat among chimpanzees, or the contractual sharing of a firm's residual revenues between the shareholders and the workers. Thus, when Adam and Olive shared food on a reciprocal basis, no third party like Eve would have been available to assist in curbing any cheating. Adam and Olive would necessarily have acted as their own policemen without outside assistance.

Evolutionary stability.   Having explained why I think that Axelrod's [31] attempt to study evolutionary equilibrium selection in *two-player* repeated games isolated from outside influences may be relevant to my enterprise, the next step is to ask to what extent his claims about the inevitability of the emergence of cooperation can be made to stick.

It is not hard to see the evolutionary advantages of cooperation. If we coordinate our efforts, we can all be better off. Section 2.5.3 spells out why equilibrium selection devices that Pareto-dominate their rivals should therefore be expected to evolve when different societies are in competition. In particular, failing societies will tend to imitate aspects of the social organization of their more successful rivals—just as Ford and General Motors have tried to copy the practices of the Japanese. But how did Nature get the cooperative rabbit out of the hat in the first place? What evolutionary trick threw the *very first* society that learned to exploit the benefits of cooperation into the melting pot? Was it an unlikely accident, or something that was bound to happen sooner or later?

Axelrod's [31] answer is based on the claims he makes for TIT-FOR-TAT, but we have seen that these claims are doubtful at best. Nor do I believe it wise to draw any broad conclusions from other computer simulations. As Sir Arthur Eddington said of astronomy, "Never believe a new piece of data until it has been confirmed by theory." So what does theory have to tell us about the evolution of cooperation in repeated games?

The obvious first question is to ask which strategies for the infinitely repeated Prisoners' Dilemma are evolutionarily stable in the sense of Maynard Smith and Price [348]. For the reasons given in Section 3.2.3 of Volume I, for a strategy s to be evolutionarily stable in a symmetric game, it must satisfy two conditions. First, it must be a best reply to itself, so that (s, s) is a Nash equilibrium of the game. Second, if t is alternative best reply to s, then s must do strictly better against t than t does against itself.

It follows immediately from this definition that *no* pure strategy s for the infinitely repeated Prisoners' Dilemma can be evolutionarily stable. The reason is that there are parts of the game tree that are not visited when s plays itself. For example, when TIT-FOR-TAT is matched against itself, the parts of the game tree reached when someone plays *hawk* remain unvisited. A mutation t can therefore appear that modifies s in ways that would only be revealed in unvisited parts of the game tree. The one-state automaton DOVE that always plays *dove* no matter what happens is an example of such a mutation of TIT-FOR-TAT. Since one cannot tell the difference between s and t from the way they play copies of themselves or each other, the mutation t is an alternative best reply to s that doesn't do *strictly* better than t does against itself. It follows that s is not evolutionarily stable. In particular, TIT-FOR-TAT is not evolutionarily stable. A mutant bridgehead established by DOVEs in a population of TIT-FOR-TATs will not be repelled.

Although this result has been proved over and over again by scholars from various disciplines starting with Selten [473], the myth that Axelrod [31] proved that TIT-FOR-TAT is evolutionarily stable continues to circulate. The myth is fueled by an argument in Maynard Smith's [347] *Evolution and the Theory of Games*, which shows that TIT-FOR-TAT is *weakly* evolutionarily stable.[54] To see this, simply observe that any alternative best reply to TIT-FOR-TAT must cooperate with TIT-FOR-TAT in the cycle into which play eventually settles. But when the alternative best reply plays itself,

---

[54]Weak evolutionary stability is obtained by deleting the word "strictly" from the definition of evolutionary stability.

it cannot get more than by cooperating all the time. Thus TIT-FOR-TAT gets at least as much against the alternative best reply as the latter gets against itself. For example, TIT-FOR-TAT gets exactly the same against the alternative best reply DOVE as DOVE gets against itself.

Is it worth fussing about the distinction between evolutionary stability and weak evolutionary stability? To see that the answer is *yes,* recall that no computer simulation of evolution in the infinitely repeated Prisoners' Dilemma ever comes up with a unique winner. The surviving population always contains mixtures of strategies. For example, the strategies GRIM, TIT-FOR-TAT, TWEEDLEDUM, and TWEEDLEDEE all survive in appreciable numbers in Linster's [327, 328] simulation using all twenty-six automata from Figure 3.3. Each of these strategies is evolutionarily stable in the weak sense, but so what? Even DOVE survives in small numbers in both Axelrod's and Linster's simulations, and DOVE has no stability properties at all.

I think at least two lessons can be learned from this elementary study of evolutionary stability. The first is that it is a mistake to focus attention on any single strategy, whether it is TIT-FOR-TAT, TAT-FOR-TIT or anything else. We have seen that a population of TIT-FOR-TATs can be invaded by DOVE and numerous other strategies. A population of TAT-FOR-TITs is not quite so easily subverted, but it can still be invaded by TWEEDLEDUM, TWEEDLEDEE, TWEETY-PIE, and other more complex strategies.

The second point is less obvious. Not only must we be prepared to think in terms of a surviving mixture of strategies coexisting in a quasi-symbiotic relationship, we must accept that their stability is likely to be precarious. The selection process is bound to be noisy, and so one must expect the proportions of surviving strategies to drift as the system adjusts to a continual bombardment of small shocks. In simpler situations, Binmore and Samuelson [84, 85, 458] find that the components of equilibria through which systems drift in this manner frequently contain unstable points. If random drift persists for long enough, the system is therefore likely to find its way to such an unstable equilibrium. A small further shock can then start the selection process up again, and so take the system to a completely different component of equilibria.

This scenario is to be compared with the commonly held misconception that equilibria are generically isolated objects. Evolution can then be envisaged in terms of a ball rolling down the sides of a pit to an equilibrium waiting at the bottom. But this picture can only be upheld in very simple games. In more realistic games, the pit becomes a valley. Each point on the flat floor of the valley corresponds to an equilibrium of the game under study. Once on the floor of the valley, the ball rolls at random under the influence of genetic drift. This is not in itself a very troubling phenomenon, but Binmore and Samuelson [86] find that *hanging* valleys are commonplace. If the ball finds its way to the flat floor of a hanging valley, it will drift as before. But it is now possible that the drift will take the ball to an unstable equilibrium on the cliffedge, where the hanging valley breaches the wall of a larger valley. The ball will then plunge into the deeper valley to a whole new set of equilibria waiting below. To an observer, it will seem as though natural selection had turned itself off after reaching an equilibrium, but then inexplicably started up again to whisk the system off to an entirely different equilibrium. In the repeated Prisoners' Dilemma at least, such dynamical high jinks are the rule.

Evolution and complexity. Our study of evolutionary stability has not taken us very far. Something else has to be added to the model if we are to be able to say anything useful about what kind of social contract evolution might eventually create. The consensus among modelers seems to be that the factor missing from orthodox formulations is the complexity of the equilibrium strategies. Abreu and Rubinstein [5], Anderlini and Sabourian [14], Ben-Porath [50], Binmore and Samuelson [84], David Cooper [133], Ben Cooper [132], Fudenberg and Maskin [186], Neyman [383], Nowak and Sigmund [388], Piccione [404], Probst [411], Rubinstein [446], and Schlag [465] are among those who have written relevant papers in which strategic complexity plays a crucial role. However, different modelers making different assumptions are led to different conclusions.

Perhaps the variety of results simply reflects the wide variety of social contracts that have been observed operating in isolated societies. However, the modeling judgments to be made are sufficiently subtle that it would be surprising if nobody at all has gone astray somewhere along the line. The questions that need to be answered in constructing a tractable model are endless. How patient should we make the players? What fraction of the population should we take to represent the size of a mutant bridgehead? It may not be unrealistic to make this fraction large in a particular locality. As Axelrod and Hamilton [34] show, a mutant invasion that establishes itself in such a locality may then easily spread to the rest of the population. How is complexity to be measured? How much does it cost to replace a simple automaton with a more complex alternative? Is there an upper bound on how complex an automaton can be, or is the complexity of automata determined only by cost? How does noise enter the system? Do the automata tremble in their outputs or their transitions or both? Even if one attempts to address all such questions simultaneously, one is still left with the problem of how much weight to attach to each consideration.

The following model of Binmore and Samuelson [84] represents one route through the jungle of alternatives to a possible world in which evolution inevitably generates cooperation. Fudenberg and Maskin [186] follow an alternative path to another cooperative world.[55] But no authors in this genre would think it appropriate to make sweeping generalizations about societies in general on the basis of their models.

Binmore and Samuelson [84] modify the definition of weak evolutionary stability to take complexity into account. This modification comes into play only when s fails the test for evolutionary stability, but is nevertheless weakly evolutionarily stable. The payoff that s gets when it plays an alternative best reply t is then equal to the payoff that t gets when it plays itself. Under these circumstances, we modify the definition by insisting on the extra requirement that s be no more complex than t. Just as an evolutionarily stable strategy is said to be an ESS, a modified evolutionarily stable strategy is a MESS.

One way of measuring the complexity of an automaton is by counting its states.[56] Thus TIT-FOR-TAT has two states, and so is more complex than DOVE. Taking $s = $ TIT-FOR-TAT and $t = $ DOVE in the definition, we find that TIT-FOR-TAT is not a MESS. But TAT-FOR-TIT is a MESS because neither DOVE nor HAWK satisfy the conditions necessary to bring the complexity requirement into play. Notice that TAT-FOR-TIT uses all its states when it plays itself, but TIT-FOR-TAT does not. As Abreu and Rubinstein [5] observe under more general circumstances, a MESS never has states that are unused when it plays itself. If s had an unused state, we could rewire it to exclude the unused state, and so construct a destabilizing automaton t.

Suppose a pure strategy for the infinitely repeated Prisoners' Dilemma were a MESS, but that play did not eventually settle into cooperation when the strategy was matched against itself. A normal population using the strategy could then be invaded by a certain type of mutant using Robson's [430] "secret handshake". This tactic allows the mutant to get the same payoff against a normal as a normal gets when playing another normal, but to get the cooperative per-game payoff of 2 when playing a copy of itself. Such a mutant does better than a normal, and hence will expand its bridgehead until it has taken over the whole population. Its existence therefore contradicts the hypothesis that a MESS can fail to cooperate with itself in the long run. That is to say, if a pure strategy is a MESS, then it eventually plays *dove* all the time when matched against a copy of itself.

---

[55]Fudenberg and Maskin [186] assume that trembles in a machine's output overpower other influences, whereas Binmore and Samuelson [84] treat such trembles as negligible.

[56]In a paper extending the technique to Rubinstein's [445] model of bargaining, Binmore et al [81] introduce a very much weaker complexity measure from which the same conclusions follow.

The secret handshake consists in choosing an abnormal action in the first round of play. This allows mutants to recognize each other. Once a mutant recognizes a twin soul, he plays *dove*. When a mutant plays a copy of itself, the two automata therefore obtain the cooperative per-game payoff of 2. However, unless players only interact with their neighbors, it will initially be rare for a mutant to be matched with another mutant. In deviating from normal play in the initial round, a mutant must therefore expect that he will be playing a normal player whom he has just provoked into entering his punishment phase. It is at this point that one needs to appeal to the fact that a normal automaton uses every state when playing against a copy of itself. Whatever state q the normal automaton enters when it begins punishing the mutant is also a state it uses at some time or another when playing another normal. The mutant can therefore now get the same long-run average payoff as a normal by switching to a state q of its own and behaving thereafter exactly like a normal. The mutant may lose a little by its deviant behavior at the outset of the game, but the loss becomes negligible when long-run averages are computed.

This argument suggests that TAT-FOR-TIT is the typical example of a MESS. Its secret handshake is to play *hawk*. If this signal is reciprocated, it decides that it is playing a copy of itself and switches to *dove*. However, it is always ready to punish any failure to reciprocate the switch to cooperation by reverting to the punishment schedule it advertised at the beginning of the game. Much more elaborate examples of MESSes can be constructed by inventing complicated secret handshakes that serve to exploit various types of suckers while simultaneously signaling to copies of themselves that they are indeed who they claim to be.

However, the preceding discussion of weak evolutionary stability warns us against pinning our hopes on any single pure strategy. We have to recognize that weakly stable strategies can be invaded. Although such invaders will not inevitably expand their bridgehead, a succession of such invasions followed by drift in the proportions of the strategies present will sully the purity of the original population beyond rescue.

Binmore and Samuelson's [84] definition of a polymorphic MESS takes into account that many different strategies must be expected to survive together in a population that can only be expected to be precariously stable. However, it turns out again that all the strategies from a polymorphic MESS cooperate with each other eventually. A particularly interesting polymorphic MESS consists of the three automata, TAT-FOR-TIT, TWEEDLEDUM, and TWEEDLEDEE. As long as each of these three automata controls some positive fraction of the population, they enjoy a symbiotic relationship that is able to repel any possible mutant invader—except for TWEETY-PIE. But if TWEETY-PIE appears, the system becomes vulnerable to more complex predators that will themselves be eliminated after wiping out all copies of the defenseless TWEETY-PIE. The transitory appearance of the predators will shift the proportions in which TAT-FOR-TIT, TWEEDLEDUM, and TWEEDLEDEE appear. If repeated often enough, such shocks will eventually lead to one or another of the normal automata drifting into oblivion. But then the remaining pair of automata become vulnerable to an assortment of mutants other than TWEETY-PIE.[57]

In summary, the question of whether cooperation is likely to evolve in a single society is much more complicated than is normally taken for granted. There are certainly grounds for optimism when the underlying mechanism is biological, but the arguments with which such optimism has been supported in the past have depended more on wishful thinking than serious analysis. Nor, as I argue next, are matters improved when we turn our attention to the special features of social evolution.

Nature a utilitarian? Sociobiological accounts of the evolution of man typically run out of steam by the time morality reaches the agenda. Insofar as speculations are offered about the type of morality with which Nature is likely to have endowed us, the consensus favors of some sketchily conceived version of utilitarianism. This view receives some support when the model of Binmore and Samuelson [84] is applied to general repeated games. When strategies from a polymorphic MESS play each other, the outcome is necessarily utilitarian. But hasty conclusions are inappropriate. Leaving aside the limitations of the model for biological purposes, we need to remember that human societies are the result of a process of *coevolution*, in which biological evolution and social evolution operate in tandem.

When the dynamics of biological and social evolution are better understood, my guess is that a major differentiating factor will be the modeling of mutations. For example, the notion of evolutionary stability proposed by Maynard Smith and Price [348] implicitly assumes that mutant challenges to a normal population come one at a time, so that the system has time to find its way back to an equilibrium before a new mutant challenge appears. But this will not do for social evolution. As Section 3.1 observes, one needs to imagine an equilibrium population of memes under continual bombardment by streams of mutant memes from a variety of sources—including creative and serendipitous thinking. It would be idle to pretend that anyone has a clear idea on how to model such a complicated problem, but it is clear enough that the stories that have been told so far about evolution in repeated games are hopelessly inadequate.

---

[57]Probst [411] has studied the stability properties of a group of five three-state automata with better stability properties, but the essential message remains the same.

In particular, current attempts to model evolutionary stability in the indefinitely repeated Prisoners' Dilemma not only presume that mutations come one at a time, but that they only occur between one generation and the next. In biology, this approach is sensible in the case of animals that have a well-defined breeding season. But in human societies, one must presumably think of the indefinitely repeated Prisoners' Dilemma as being played by generations that overlap—with new memes liable to appear at any time. A player who is controlled by one meme at the beginning of play may therefore be captured by another meme during the course of play. Thus players cannot be regarded as being *committed* to a strategy. Chapter 2 should therefore warn us not to be surprised if results that point in a utilitarian direction begin to evaporate.

However, rather than considering experienced players who may switch strategies, I want to focus on the possibility that new players may make copying errors when imitating successful veterans. It seems likely that a novice will frequently fail to appreciate the significance of rarely used aspects of the automaton controlling the veteran he chooses to copy, and hence put only a simplified version of the veteran's automaton into action. For example, a novice observing a population of veterans using TIT-FOR-TAT will normally only see *dove* being played and so might be led to adopt the strategy DOVE. Once enough DOVES are present, the population then becomes vulnerable to attack by HAWK or other mean machines.

Rubinstein [446] describes a model which can be thought of as exhibiting this phenomenon in an extreme form. The model differs from his later work with Abreu [5], in requiring that all states in an automaton representing an equilibrium strategy be used *infinitely often*. None of the properties of such a strategy will be hard to spot, and so it has a chance of being copied faithfully. Unfortunately, even the pair (TAT-FOR-TIT, TAT-FOR-TIT) fails to satisfy this equilibrium requirement, because the state in which *hawk* is played is used only once. In fact, Rubinstein finds that no symmetric Nash equilibrium satisfying his condition can be Pareto-efficient.

To despair at being led back to noncooperation after so long a journey would be to attach too much significance to Rubinstein's model. Analytic models are more reliable than computer simulations, but they are still no better than the assumptions on which they are based—and there is a great deal more going on when people interact than any of the models considered here admit. For example, Vanberg and Congleton [530] are among a number of authors who have pointed out that real Adams and Eves are seldom irrevocably paired together for life. Nor is the procedure by means of which they are matched in the first place totally outside their control.

The jury therefore remains out on the question of whether cooperation would inevitably evolve in a human society created by throwing a bunch of strangers together. My money is with Axelrod [31] on this question, but I hope that I have written enough to make it clear that nobody should be satisfied with the easy answers peddled in the popular literature. A lot of serious research needs to be done before we learn whether we have bet on the right horse.

3.4   Social Contracts in Big Societies

One unfortunate consequence of the star billing achieved by TIT-FOR-TAT has been to persuade a generation of evolutionary thinkers that no other forms of reciprocating behavior are worthy of attention. It is taken for granted that TIT-FOR-TAT is adequate to serve as a paradigm for the whole class of reciprocating mechanisms covered by the folk theorem. This mistake engenders the belief that cooperative understandings need to be supported by strategies that require a cheater to be punished by the party he cheats. But Hume [267, p.521] put his finger on the reason this view is inadequate more than two hundred years ago (Section 3.1). In a society with more than two players, a cheater need not be disciplined by the person he injures: he can be disciplined by a third party—as in the model of a middle-class insider group studied in Section 3.3.7.

3.4.1   Social Transfers

It is easy to give examples of transfers that are disciplined by players who do not actively gain from the transfer. In fact, we met such a case in Section 3.3.4 when discussing free disposal. We learned that Adam can be induced to throw money away under the threat of being punished by Eve. But what's in it for her? Isn't it irrational for her to insist that Adam waste a resource that could be shared between them?

Such questions confuse the problem of how equilibria are selected with the problem of how equilibria are sustained. It is true that Eve would have no reason to cast her vote in favor of a wasteful social contract. But nobody is offering her the option to change the social contract. The question she has to ask herself is whether it is rational for her to carry out the duty imposed on her by the *current* social contract. Since this social contract can be supported as a subgame-perfect equilibrium, the answer is *yes*. Her incentive to punish Adam is that a failure to do so will steer play into a subgame in which the equilibrium to be played is worse for her than the equilibrium to be played in the subgame that will be reached if she does her duty. One may object that this can only be possible if Adam plays his part in making subgames off the equilibrium path unattractive. But he isn't being asked his opinion on which social contract to operate any more than Eve. If subgames off the equilibrium path were reached, he too would find that his incentives always accord with doing his duty.

Following Hammond [218], another striking example can be obtained by adapting a model used by Allais [13] and Samuelson [459] to clarify Hume's [267] insight that money is only valuable by convention. The example will also serve to illustrate how readily the folk theorem generalizes to the case of overlapping generations (Kandori [291], Ellison [170]). The model of a middle-class insider group in Section 3.3.7 provides a related but less dramatic example showing that people do not have to believe that they might live forever to sustain a cooperative social contract.

As in Section 1.2.8 of Volume I, imagine an Amazonian society whose citizens are all female. Only two women are alive at any time, a mother and her daughter. Neither cares about anything but how much she gets to eat herself. In particular, daughters have no family feeling for their mothers. At the beginning of each period, the daughter of the previous period becomes old after having given birth to a daughter of her own. While young, she produces 200 loaves of bread. When old, she is a helpless cripple who is unable to produce anything at all. If she were able to store loaves from one period to the next, she would choose to eat 100 loaves in her youth and 100 loaves in her dotage.[58] However, bread perishes from one period to the next and so any pension the mother receives must come from the output of her daughter.

One possible equilibrium requires that each player eat the 200 loaves she produces while young and starve in her old age. On the face of it, this seems the end of the story, because mothers are totally powerless, and so are unable to bring any pressure to bear on their daughters. But the necessary threat of punishment need not come from a mother neglected by her daughter—it can come from her granddaughter! Using such a third-party punishment scheme, it easy to construct a subgame-perfect equilibrium that supports the Pareto-efficient outcome in which each player eats 100 loaves in her youth and gives the remaining 100 loaves to her helpless mother. The punishment threatened by the granddaughter is that she will not give the daughter 100 loaves in the next generation, should the daughter fail to do her duty in the current generation. More precisely, the sharing social contract requires that the players be split into conformists and nonconformists. It is conformist to give 100 loaves to your mother if she herself is a conformist, but to give nothing if she is a nonconformist. Notice that this definition ensures that daughters who do their duty by punishing a

---

[58]Preferences like those described in the insurance model of Section 3.3.4 suffice to make this her optimal consumption bundle.

nonconformist mother are not themselves punished. On the contrary, they would be punished for *failing* to withhold their mother's pension. Similarly, a granddaughter who fails to punish a daughter who failed to punish her nonconformist mother will be punished. And so on.

 Anything goes?   Section 2.2.4 of Volume I tries to explain why game theory is a morally neutral tool of analysis. To challenge the optimizing assumptions on which it is based because they are thought to entail morally abhorrent conclusions is particularly foolish in the context of the folk theorem. Far from the theorem implying that only selfish behavior can survive in equilibrium, we have just seen how it can be used to show that a caring attitude toward the weak and helpless can be a stable component of a social contract even between heartless Mr. Hydes. But it also works the other way. The same mechanisms can be used to sustain behavior that most readers of this book would prefer were absent from our social contract. Consider, for example, the exclusion of the Ainu from regular Japanese society. How come it is stable to treat the Ainu as outsiders? Because anyone who doesn't is treated as an outsider himself.

After listening to such conclusions, it is not uncommon for critics to react by saying that it is a waste of time to pay any attention at all to the folk theorem, because its conclusions simply echo the relativist folly that morality is a free-for-all, where anything goes. Let me therefore reiterate the reasons why this response is mistaken.

In the first place, the folk theorem doesn't say that anything goes. For an equilibrium to hold together, the system of duties it embodies must be finely tuned to provide all players with the right incentive to do just the right thing at the right time. It is true that the folk theorem nevertheless leaves the field open for a vast range of different social contracts. But if this were false, the folk theorem would necessarily have failed to take account of something important in human nature. After all, what do we find in the history books if not a vast range of diverse social contracts?

Moral relativism does indeed deny that it makes sense to say that one social contract is better than another in some absolute sense. But moral relativists nevertheless have their own opinions about the kinds of social contract they think desirable (Section 3.8.2). They differ in this respect from moral absolutists only in recognizing that their views on what is good or bad about a social contract are artifacts of the culture in which they grew to maturity. However, whether one thinks that moral values are absolute or relative, it is foolish to complain that they are absent from the statement of the folk theorem. The folk theorem is about *feasibility*. It tells us the class of social contracts that can be *sustained*. To say something about optimality, we need a theory of how social contracts are *selected* from the feasible set.

3.4.2 Friendship and Coalitions

Anthropologists have suggested that the advantages of food sharing among hunter-gatherers in times of scarcity may be the principal reason that sociality evolved in protohumans. But other forms of social exchange presumably became equally significant once social groups had become established. Shifting patterns of alliance for mutual attack and defense are a well-documented determinant of the social structure in chimpanzee societies. In human societies, the importance of such alliances is incalculable. Pretty much everything is transacted through the complex network of overlapping coalitions out of which all human societies are woven.

It is therefore ironic that even the concept of pairwise friendship should create problems for evolutionary psychology (Tooby and Cosmides [523]). Why should Adam continue to associate with the Eve, friend to whom he has sworn eternal fidelity when she is injured or falls sick? What good is she to him now? He needs someone powerful to stand by him when challenging or being challenged by an enemy.

The tit-for-tat paradigm offers no answers to such questions, because Eve's powerlessness after her injury makes it impossible for her to punish Adam should he cheat on their implicit agreement to look after each other. But the punishment does not need to be administered by Eve. It can be administered by Adam's other old friends, or the new friends that he may hope to make after abandoning Eve. After all, what use is Adam as a friend if he cannot be trusted to be faithful in times of need? In brief, just as most of us cannot afford to establish a reputation for being untruthful or dishonest, so Adam cannot afford to be labeled as faithless. If the social contract is sufficiently sophisticated, to be stuck with the label of faithless friend is to be condemned to being excluded from all future coalitions.

3.4.3 Police Forces

Section 4.5 discusses the circumstances under which some anthropologists believe that sociality evolved in protohuman species. The anarchic and egalitarian organization of modern hunter-gatherer bands is thought to provide a prototype for the structure of such early societies. The claim that protohuman societies were anarchic seems reasonable enough, since our species must have abandoned the hierarchical structure operated by ape societies at a relatively early stage. It therefore may well be that, like their modern counterparts, protohuman societies had no bosses to order people around.

But my guess is that the social contracts of protohuman societies were unlikely to be sufficiently sophisticated to maintain the egalitarian distribution arrangements of modern foragers. If the anthropologists are to be believed, food is gathered and distributed in such societies according to the Marxian principle that each contributes according to his ability and benefits according to his need (Erdal and Whiten [175]).

How can such a Marxian social contract survive? If only the tit-for-tat mechanism were available, nobody would ever offer food to powerless folk outside their immediate family. But the punishment for failing to share food need not be administered by the person who is left to go hungry. In modern foraging bands, it is administered by the whole group. Everyone is therefore a citizen within a social organization which resembles the Marxian utopia that supposedly follows the withering away of the state. No single individual can stand against the power of the rest of his group, and so almost any behavior can be sustained as a subgame-perfect equilibrium.

Detecting cheats. Someone brought up in a Western democracy would find the social ambience of a modern hunter-gatherer band unbearably stifling. Our nearest approach is perhaps the teenage clan, whose members discipline each other's behavior so severely one sometimes wonders whether they received their conformity training in the military. In both cases, the crucial feature that allows the dictatorship of the many to operate so successfully is the small size and closeness of the group, which ensure that a deviation by a member of the group is likely to become common knowledge very quickly. The coordination problem faced by the group in administering punishment is therefore easily solved.

Nevertheless, people living in hunter-gatherer societies still find ways to cheat on their social contracts. In some foraging societies, cheating is sufficiently close to the surface that food is shared via tolerated theft and aggressive begging. Cheating in such a society differs from cheating in ours only in that hunter-gatherers find it harder to get away with it than us, because they live under the continuous public scrutiny of neighbors who are shamelessly inquisitive. Their tight social contracts are therefore not viable for large societies in which it is necessary to work much harder at detecting and punishing cheaters. My guess is that we all devote a great deal more of our brain capacity to this activity than we realize. It is surely no accident that Cosmides and Tooby [134, 135, 136] find that people are very much better at solving logical problems when they arise in the course of tracking down cheaters in social contract problems than elsewhere.[59]

---

[59]Although I don't see that this fact implies that we necessarily have a special module for solving social contract problems in our brains.

In large societies, it is therefore efficient for some players to be rewarded for specializing in the detection role by becoming policemen. The problem of guarding the guardians then arises in its purest form. The folk theorem of Section 3.3 tells us that we can set the guardians to guard each other, but this version of the folk theorem assumes that information is perfect. It is obviously much harder to seek out and punish deviations when they may be committed in secret. It is still harder if the deviant has been given the power to punish those who might be moved to investigate his behavior.

In its current state, game theory can only suggest how such practical problems should be approached, but Hume [271] stated the necessary principle more than two hundred years ago. In designing a policing system, each officer should be treated as though he were a rogue and a knave.

3.4.4 Punishing the Innocent

A police force can serve to deter crimes that are traceable to a specific culprit, but what about the kind of antisocial act that it is very hard to pin on any particular individual? One might think that folk-theorem arguments would have no place in such a context, but Kandori [290] has shown that one can still sometimes sustain high levels of cooperation by using equilibria in which punishment is spread through a population by *contagion*.

As an extreme case, suppose that each person from a finite population is matched anonymously in each round to play the Prisoners' Dilemma with another person from the same population. Assume that the matching is renewed randomly every round, so that nobody is able to keep track of what any specific opponent is doing. How could a social contract be maintained in which everybody always plays *dove*?

The problem is that nobody ever knows who he is playing, and so cheaters can never be singled out for punishment. Cooperation can nevertheless be sustained as a Nash equilibrium if the victims of a deviation react by punishing whomever they happen to play with later, whether innocent or guilty of any offense.[60] For example, we can sustain cooperation by simply requiring that all players use the GRIM strategy, just as they would if their opponent were always the same. If Adam cheats by playing *hawk*, his opponent in that round will then start playing *hawk* thereafter. After the opponent is matched with another player in the following round, there will then be two players using *hawk*. Unless the same two players get

---

[60]Sir Walter Raleigh struck his son at a grand dinner for telling the story of a whore who refused to service the son on the grounds that she had lain with his father not an hour before. Some vestigial feelings of filial duty prevented his hitting his father back, and so he struck an innocent bystander instead, crying, "Box about, 'twill come to my father anon!"—Aubrey's [22] *Brief Lives*.

matched in the round after that, there will then be four players using *hawk*. Because the population is finite, the contagion will eventually spread with high probability to the whole population. Adam will then pay the price of cheating, since he could have secured a per-game payoff of 2 by honoring the social contract, but now is stuck with a per-game payoff of 0.

The effectiveness of such a policy of punishing the innocent depends on the size of the population. If Adam is not infinitely patient and the population is sufficiently large, then he won't be deterrred from deviating, because he will expect to get a sufficiently large payoff from exploiting the people he plays in the early part of the game to make up for the losses he will suffer once he starts meeting opponents who have been infected with the imperative to punish. This problem becomes more urgent when the problem of guarding the guardians is taken into account. Why should I start punishing the world because someone just cheated me? Perhaps my opponent was the first person ever to cheat and will never make the same mistake again. If so, then I can prevent myself from living in a society in which everybody eventually cheats by treating my experience of being cheated as though it never happened. On the other hand, perhaps contagion is already rife in my society, and all I shall succeed in doing by cooperating in the next round is to set myself up as a likely victim of a cheater. How these considerations balance against each other is a technical problem.[61]

Blaming groups for the crimes of individuals. Although the policy of punishing the world for the sins of an individual quickly ceases to make sense when the population gets large enough, equilibria in which the innocent are punished can still be viable when the cheater can be identified as belonging to a small enough group.[62]

Instead of punishing the world when cheated, a player then only punishes the cheater's group. The arrangement works better if all the members of the group being punished are related so that the cheater's personal utility function actually takes into account the pain suffered when others in his group are punished. It works better still when the fact that a member of one clan has been cheated by the member of another clan can be

---

[61]In the case of a population of size four, consider a player who is matched with someone who plays *hawk* in the first round. If his discount factor is $\delta = 1 - \Delta$, then his per-game payoff from punishing as prescribed by the equilibrium is approximately $1 + \Delta$ when Δ is sufficiently small. If he continues as though his opponent had played *dove*, his per-game payoff is approximately $1 - \frac{5}{2}\Delta$. A sufficiently patient player should therefore stick by the equilibrium.

[62]I don't suppose it would be allowed today, but I remember teachers sometimes keeping the whole class behind after school when nobody would own up to some misdeed. I dare say the tactic might even have been effective if pursued systematically.

communicated quickly by word of mouth so that the contagion spreads faster than the game can be played. Examples of cases when something goes wrong and the punishment schedule is implemented are depressingly familiar. One famous case is the hillbilly feud between the Hatfields and the McCoys. Another provides the setting for *Romeo and Juliet*.

But sweeping generalizations from such cases to the inevitability of xenophobia and war are inappropriate. The story explaining how such feuds get started offers no reason why we should be biologically hardwired to be hostile to strangers who have not injured us. Nor is the treatment received by explorers like Cook in such places as Hawaii supportive of such an hypothesis. As for war between nations, Serbs and Croats should note that it is just as irrational to punish everybody in a very large group for the misdeeds of its individuals as to punish the whole world. Insofar as the appetite for war is based on misapplying attitudes adapted for small-group interaction to large groups, we may therefore even have grounds for hoping that we will eventually learn to act more wisely.

3.4.5 Leadership and Authority

As Section 3.4.3 observes, modern foraging societies have no bosses. Moreover, their social contracts are equipped with mechanisms designed to inhibit the emergence of bosses. My guess is that these social mechanisms exist because such subsistence societies cannot afford to take the risk of allowing a reformer to persuade them to experiment with their traditional survival techhniques. But the immediate point is that the existence of such leaderless societies implies that humans do not need bosses to live in societies. So why do we have them?

One popular argument holds that leaders are necessary because, like Uncle Joe Stalin, they know what is good for us better than we know ourselves. But whether leaders know what they are doing better than their followers or not, they can be very useful to a society as a coordinating device for solving the equilibrium selection problem in games for which the traditional methods are too slow or uncertain. On a sailing ship in a storm or in a nation at war, one cannot afford to wait for due process to generate a compromise acceptable to all. Henry Ford told us that history is bunk, but at least it teaches us that the way to get a society moving together in a crisis is to delegate authority to a single leader.

The mention of authority may make it seem that one cannot discuss leadership without stepping outside the class of phenomena that can be explained by the folk theorem. But the authority of a leader does not need to be founded in some theory of the divine right of kings, or in a Hobbesian social contract theory where citizens trade their rights to self-determination for security, or in some metaphysical argument purporting to prove it rational to subordinate one's own desires to the general will as perceived by the head of state. If we see the leader's role as making a public choice of a subgame-perfect equilibrium from those available, then the fact that his authority is honored becomes simply a matter of convention like any other aspect of a social contract.

If Adam believes that everybody else will obey the leader by playing the equilibrium strategy he designates, then it is optimal for Adam to obey the leader as well. In extreme cases, when a madman is in power, the equilibrium he chooses may require both Adam and Eve to injure themselves under the threat that the other player will be ordered to punish them if they fail to comply. Why does Adam believe that Eve will carry out such a punishment order? Because she believes that he will obey the leader's order to punish her if she doesn't! When a leader has only two followers, the informational assumptions that need to be maintained to support such a subgame-perfect equilibrium admittedly stretch the imagination, but the same is not true in larger societies like those terrorized by Caligua or Adolf Hitler. In spite of their prominence in history books, such monsters are rare, and their regimes do not last very long. But even saintly leaders are human. Given long enough in power, they finally learn to tell themselves stories which allow them to respond to their incentives, while still remaining convinced of their dedication to the public interest.

To take a crude example, if Adam is the leader in the Battle of the Sexes (Figure 2.1(c) of Volume I), he could play fair by tossing a coin to decide which of the two pure Nash equilibria to nominate. If he nominates (*hawk, dove*) and Eve believes that he will therefore play *hawk*, it is optimal for her to play *dove*. Similarly, if he believes that his nomination of (*hawk, dove*) will induce her to play *dove*, then it is optimal for him to play *hawk*. A similar convergence of expectations applies if he nominates (*dove, hawk*). But experience strongly suggests that the opportunities for Adam to abuse his position of authority by cheating are too tempting to resist. Over time he will learn to weight the coin in his favor, so that the equilibrium (*hawk, dove*) is chosen more often than the equilibrium (*dove, hawk*). Eventually, he or his successors will convince themselves that they have a right to choose the equilibrium (*hawk, dove*) all the time.[63] Justice is therefore always a rare commodity in authoritarian states.

---

[63]Economists should note that it is only by accident that (*hawk, dove*) happens to be the so-called Stackelberg equilibrium of the Battle of the Sexes. But Adam is not a leader in the Stackelberg sense—he doesn't publicly make his move *before* Eve, leaving her with a take-it-or-leave-it problem. His initial choice of an equilibrium is merely a signal that commits nobody to anything.

Neofeudalism and whiggery. In Section 4.5, I follow the anthropological line that sees the emergence of hierarchical human societies as a social adaptation required by the need to turn to farming in order to cope with increasing population pressure. But such authoritarian forms of social organization have to operate within the framework of a biologically determined fairness norm that evolved as a coordinating tool among the leaderless foraging bands of prehistory. A tension therefore always exists between the conventional authority of leaders and the instinctive urge to coordinate using fairness criteria. Maryanski and Turner [343] use the metaphor of a social cage to capture the idea that authoritarian societies imprison us in a culture to which we are not biologically adapted.[64]

The instabilities created by our failure to be properly adapted to authoritarian social cages are not detected by the folk theorem, because the folk theorem depends on a notion of equilibrium that considers deviations by only one individual at a time. However, leaders who are too partial in choosing an equilibrium that favors the group that put them in power risk creating a *coalition* for mutual protection among those they treat unfairly. Such an alienated group will treat the *nomenclatura* as outsiders and coordinate on a rebellious equilibrium of its own, in which the first tenet is never to assist a boss in punishing one's own kind. Hamilton [215, p.3] explains how demagogues take over the leadership of such alienated groups by temporarily facilitating coordination on the fair equilibrium that serves as its focus. But if such demagogues are propelled into power, history shows that they soon become as corrupt as the tyrants they replace.

In Western democracies, we have institutionalized this incipient conflict between the rival coordination mechanisms of fairness and leadership in two ways. First, we constrain our leaders by a system of checks and balances built into the constitution that are intended to limit the extent to which leaders can act without some measure of popular support. Second, the constitution provides regular opportunities for switching leaders.

Although these measures do have the effect of preventing leaders taking a society to a markedly unfair equilibrium and keeping it there, they achieve this aim only haphazardly. Nor is it clear that matters are improving. American presidents are nowadays far more powerful than they were ever intended to be. In the United Kingdom, Mrs. Thatcher made it clear that British prime ministers have even greater opportunities to ride roughshod over constitutional arrangements that were originally intended to check the

---

[64]I agree with their assessment that social evolution is responsible for the bulk of the adaptation we have made to living in modern societies, but not that sociobiology therefore has little to teach us.

power of a monarch, who is now a mere figurehead.

In Section 2.2.6 and again in Section 4.10, I describe our current political systems as *neofeudal,* to reflect the extent to which we leave our equilibrium selection problem to be solved according to the prejudices of our leaders. Rather than seeing political philosophy in terms of a battle between a utilitarian left and a libertarian right, it seems to me that we need to think in terms of a battle between neofeudal forms of arbitrary authority and the due process of whiggish conceptions of justice. This is not to say that whigs believe that we can dispense with leaders altogether. Without entrepreneurs, we would never find the Pareto-frontier of the set of feasible social contracts. Nor is due process appropriate when quick decisions need to be made. But a fair society needs to hold its leaders in check—as the founding fathers of the American Republic knew only too well when they wrote its Constitution. One might therefore characterize the whiggish program by saying that it calls on us to think the thoughts the founding fathers would have thought if they had been faced with the more complex problems of the modern world.

3.5   The Role of the Emotions

Emotional responses are typically dismissed as irrational urges left over from our evolutionary history. The socially aroused emotions associated with pride, envy, and anger are even counted among the seven deadly sins. This bad press for the emotions is dangerously misleading. In this section, I therefore briefly make a case against the view that our emotions commonly lead us into inconsistent or self-destructive behavior. Gibbard [205] has much more to say on this subject.

3.5.1   Sore Thumbs

It seems to me that there are two reasons why misunderstandings arise about the role of the emotions. The first is that we attach undue prominence to those occasions when our emotions do indeed lead us into folly. Section 4.2 draws an analogy between this mistake and the proverbial sore thumb that always seems to be sticking out at the wrong angle. It would obviously be absurd to judge the usefulness of an opposable thumb by how well it works when it is sore. But when a thumb is operating as Nature intended, we tend not to notice it at all.

Consider, for example, the sin of pride. We say that pride goes before a fall when somebody misjudges his social status relative to those with whom he is interacting. Satan provides the most spectacular example. However, it doesn't follow that it isn't healthy for humans to be continually evaluating and reavaluating their social standing. How would Eve know not to imitate Adam if she were unaware that she outranked him in the social pecking order? It is true that only sick people devote large amounts of conscious thought to assessing their social status, but this fact only signals that we are hardwired with the facility to make such evaluations *unconsciously*. The sore thumb problem therefore arises in a particularly aggravated fashion. Only when we are not conscious of its operating at all is this particular human facility operating well.

3.5.2 Tunnel Vision

The second reason that emotional responses are misunderstood is that they tend to be evaluated within too narrow a context. O'Neill [395] argues that the prototypical scenario for the expression of anger occurs when Adam treats Eve unfairly. In her anger at this unjust treatment, she then inflicts some harm on Adam.[65] How are such emotional displays to be explained? Two examples of tunnel vision on this subject will be considered.

Transparent dispositions? The school of thought represented by Frank [185] or Ridley [426, p.133] sees anger as part of the operation of a commitment mechanism. To deter Adam from treating her unjustly, Eve advertises her determination to get back at him should he do so—whether or not such retaliation would be in her own best interests after the infringement has occurred.[66] The anger that follows an infringement is then seen as Nature's method of enforcing Eve's commitment.

Section 3.2 of Volume I criticizes this line of thought at some length while considering the Transparent Disposition Fallacy in the one-shot Prisoners' Dilemma. Skyrms [499] expresses similar doubts. The difficulty lies in the fact that people will learn to undermine the mechanism by sending the initial threat, but failing to carry it out should it damage them to do so. As game theorists put it, the threat then becomes "cheap talk", which one might as well disregard.

The response that such threats are commonly conveyed by *involuntary* body language lacks conviction. It is true that we cannot cry at will, but actors nevertheless produce tears on demand by internally simulating the kind of situation in which we cry involuntarily. In defending his transparent

---

[65] To this story, I would add that Eve needs to recognize that Adam has gained unfairly at her expense. If she makes such a judgment mistakenly, or fails to discharge her resulting anger adequately, we call her envious. But, as Section 4.2 argues, it does not follow that it is unhealthy to make fairness judgments.

[66] I recently followed a car bearing a sticker that read, "I don't get mad, I get even."

disposition theory, Frank [185] appeals to Darwin's [144] *Expression of the Emotions,* but this neglected masterpiece offers evidence only for the prosecution. The photographs in the book illustrating facial expressions typical of various emotional states actually show actors *simulating* the states in question.[67] Darwin [144, p.365] himself quotes *Hamlet* on this subject:

> Is it not monstrous that this player here,
> But in a fiction, in a dream of passion,
> Could form his soul so to his own conceit,
> That, from her working, all his visage wann'd;
> Tears in his eyes, distraction in's aspect,
> A broken voice, and his whole function suiting
> With forms to his conceit? And all for nothing!

Multiple personalities? The preceding summary of the evolutionary version of the Transparent Disposition Fallacy from Section 3.2 of Volume I is not intended to suggest that there are *no* phenomena that the theory does not adequately explain; only that the phenomena explained by the theory are the exception rather than the rule.[68] In the case of the second school of thought on the strategic function of the emotions, the issue is rather different, since I agree that the second theory is very successful in *describing* a wide range of behavior. But the question is whether the theory is at all successful in *explaining* the phenomena it describes.

The second theory models the human mind as a bundle of diverse personas, turned on and off by physiological triggers we perceive as emotions. Anger fits this paradigm particularly well. I have found nobody who fails to recognize the "out of body" experience commonly described as being "beside yourself with anger". One looks on in dismay while someone else seemingly stamps one's body up and down threatening mayhem.

However apt such a description of another persona taking over one's body may be, it is important to recognize that it provides no explanation of *why* it happens. Indeed, those scholars who insist that the paradigm must attribute conflicting aims and purposes to the various personas who inhabit our skulls would seem to rule out all hope of understanding the evolutionary origins of the phenomena the model describes. Why would Nature

---

[67] Darwin [144, p.191] observes that the muscles that depress the corners of the mouth when we grieve are particularly hard to control consciously. But old ladies in Germany nevertheless contrive to use these muscles in signaling their disapproval of unorthodox behavior in public. The effect is absurdly intimidating when several old ladies do this in unison!

[68] No such concession is appropriate for the versions of the Transparent Disposition Fallacy which argue that *rationality* demands the honoring of threats uttered in the past (Section 3.2 of Volume I). Skyrms [499] is equally firm on this point.

put our reproductive success into the hands of an unthinking mechanism that switches control between personas pursuing different goals, when she could leave control in the hands of a single persona whose aim is simply to maximize fitness?

The proper response to this question is to deny the premise. Insofar as it is useful to adopt a Cartesian model of the mind-body relationship, the data are consistent with the hypothesis that a single persona seeking to maximize fitness is running things. However, the rationality of this boss persona[69] is inadequate to cope with the enormous complexities of the Game of Life. Instead of optimizing, she therefore satisfices. In practical terms, this means that she monitors the cues she receives from her environment and uses the data to switch between one or other of the behavioral programs that she has inherited from the evolutionary past of her species.

Chess is not a bad analogy for the Game of Life in this context. Chess is too complicated to calculate an optimal strategy, but players do their best to win by sizing up the position, and then setting themselves subsidiary objectives like trying to take control of a particular file or denying room for an enemy bishop to maneuver. One might adequately describe the behavior of such players by saying that they love open files in some positions but develop an aversion to the movement of enemy bishops in others. However, such a description would be entirely superficial. To understand a Chess player, we need to know *why* he sometimes loves open files or hates enemy bishops. But such an understanding will always be denied us if we do not look beyond the current position to the strategic considerations faced by the player in the game *as a whole*.

Similarly, if we wish to understand why we switch between personas in passing from one social situation to another, it is necessary to see a player's menu of personas and the mechanism that switches between them as a grand strategy for the whole Game of Life. The evidence does not suggest that this grand strategy is fixed by our biology. My guess is that the mechanism which switches us from one emotional state to another adapts quite readily when an inappropriate state is repeatedly triggered in a newly encountered social environment. Indeed, most people are sufficiently good at acting to have a fair amount of conscious control over their current emotional state.

Finally, it should not be assumed that the grand strategy is optimal. The Game of Life has posed Nature with a problem that is far too hard for her to solve in the time available. In treating the grand strategy as though it were chosen by a boss persona seeking to maximize fitness, it is therefore important not to model this persona as being any less bounded

---

[69]Who is not to be confused with the Cartesian "I". Turning on her consciously cognitive mode is only one of way in which the boss persona may respond to stimuli.

in her rationality than the deans who boss professors around.

The strategic role of anger. Having criticized two possible explanations of the strategic role of the emotions, the time has now come for me to offer my own preferred explanation of the emotion of anger.

I never tire of arguing that it is pointless to model the Game of Life as anything simpler than an indefinitely repeated game. Only then is it possible to discuss reciprocity seriously. As our study of the folk theorem in Section 3.3 shows, maintaining high levels of cooperation as equilibria in indefinitely repeated games requires the use of strategies that punish deviations. But it would be absurd to suppose that people are often conscious of playing an equilibrium in a repeated game as they go about their daily lives. Chimpanzees are presumably not even capable of appreciating the concept of an equilibrium, but they nevertheless succeed in sustaining high levels of reciprocal altruism in their repeated Game of Life.

In spite of being a game theorist, I certainly make no Machiavellian calculations in deciding how much I can deviate from the local social contract without being punished by others, or how much I should punish others whom I catch cheating. Cosmides and Tooby [136] speculate that we are hardwired with the capacity to detect violations of the local social contract, without the need to process the data through the part of our minds that handles theorem proving and logic chopping.[70] Part of this process requires the use of the emotions. In particular, the emotion of anger is turned on when punishment is required. We then sometimes inflict injury on those we blame, at substantial cost to ourselves.

The temptation to dismiss such self-damaging behavior as being necessarily irrational is strong, but the folk theorem tells us that Pareto-efficient equilibria exist that make such behavior *optimal* for the players operating the equilibrium. If matters seem otherwise, it is because observers fail to notice that the behavior is not just a wild action in a one-shot game, but part of a grand strategy for playing an indefinitely repeated game.

The Ultimatum Game as a case study. Among economists, the case against the ideas on multiple personas just articulated focuses on the one-shot Ultimatum Game of Section 0.4. Recall that subjects do not play the subgame-perfect equilibrium of this game in the laboratory. Instead of demanding nearly everything, most people occupying the role of Adam

---

[70]To say that part of our mental hardware is set aside for calculating social relationships is not to assert that no socially acquired software is involved as well. Otherwise, those unfortunate children with normal brains who are deprived of human contact while growing up would not be socially dysfunctional.

offer Eve a substantial fraction of the money on the table. This behavior is wise, because most people occupying the role of Eve refuse proposals that assign them too small a share of the money.

Traditionally minded economists seek to explain this phenomenon by attributing exotic preferences to the players (Bolton [95], Levine [324]). In terms of the multiple personality theory, this amounts to assuming that Eve has at least two personas with different preferences. The first prefers more money to less, independently of Adam's welfare. The second is prepared to sacrifice money in order to ensure that Adam gets nothing. In the one-shot Ultimatum Game, so the story goes, Eve's spiteful persona is triggered by her angry reaction to an unfair offer from Adam.

Section 0.4.2 emphasizes that this explanation neglects the fact that the way subjects play games changes over time. Binmore *et al* [76] attribute this change in behavior to adaptive learning.[71] At first, subjects operate whatever social norm is triggered by the cues offered by the manner in which the experiment is framed. As they gather experience of the game, their behavior changes. If one waits long enough, a process of interactive trial-and-error learning may lead the subjects to an equilibrium.[72] As documented in Binmore *et al* [79, 91], they are then very ready to describe the equilibrium to which they have converged as *fair*—even though different experimental groups of subjects playing the same game may converge to different equilibria.

However, what is important here is the short-run behavior of subjects, since this is the focus of most commentaries on the Ultimatum Game. In evaluating this short-run behavior, it seems to me that one needs to recognize that the one-shot Ultimatum Game is a sore-thumb situation for the social norm that is initially triggered. One sometimes encounters ultimata in real life, but almost never in a one-shot situation where one's response to an ultimatum has no bearing on the future. On the contrary, in the Game of Life, it is frequently sensible to endure a substantial cost in the present in order to buy a reputation for being tough in the future. Equilibria calling for such behavior certainly exist in the indefinitely repeated Ultimatum Game. Indeed, it should come as no surprise that behavior appropriate to one of these equilibria is triggered when we plunge the players into a one-shot situation that only superficially resembles the repeated environment to which they have become adapted.

---

[71] The traditional economists' reply is that Eve has nothing to learn, since she already knows that one penny is more than nothing. But what she has to learn is something more difficult. She has to learn that the one-shot Ultimatum Game is not an environment in which it makes sense to get angry.

[72] Although not necessarily to a subgame-perfect equilibrium (Binmore *et al* [76]).

3.6   Due Process

The rights and duties with which we have been concerned hitherto form part of the story of how an equilibrium is sustained in a fixed Game of Life. But the Game of Life is not fixed. As it changes with the winds of fortune, a successful society will adjust its social contract to take account of the new opportunities and challenges that it now has to confront.

For example, after a technological innovation or the discovery of a new natural resource, the folk theorem guarantees that there will be many equilibria available for selection that are Pareto-improvements on society's current *status quo*. Since everybody can gain by moving to such a Pareto-improvement, the shift need not be accompanied by damaging internal strife, provided that the distributional questions raised by having a larger cake to split can be resolved satisfactorily. Section 2.5.3 speculates that fairness norms evolved precisely because they provide a relatively costless mechanism for resolving the equilibrium selection problem in such circumstances—societies that splintered into factional confrontation when extra gravy became available for division having been outperformed by societies that solved the resulting bargaining problem efficiently.

Figure 2.8 shows how the social contract operated by a society may change from ξ to ν and then to ω as the feasible set expands first to X and then to Y. When a philosopher-king is available to enforce the reforms, we need not concern ourselves with how such successive transitions are engineered. But, just as the absence of a philosopher-king forces us to focus on self-enforcing social contracts, so we are now forced to consider ways of shifting from one equilibrium in the Game of Life to another without destabilizing society along the way. After a society's opportunity set has expanded, we must therefore direct our attention to the dynamics of the *process* that a society employs when finding its way from the social contract that serves as its current state of nature to the Pareto-improving social contract it selects from those newly available.

The folk theorem arguments of the preceding section are not adequate for this purpose, since they only explain how *points* in the cooperative payoff region of a repeated Game of Life can be sustained as equilibrium outcomes. The study of such *constant-flow* equilibria needs to be supplemented by the study of *variable-flow* equilibria that move a society along a prespecified path through a cooperative payoff region.

This consideration is alien to the utopian tradition exemplified by Harsanyi [233] and the younger Rawls [417], for whom the historical experience of a society is irrelevant to the liberal constitutions they defend. Like the Jacobins of the eighteenth century or the Bolsheviks of the twentieth, modern utopians think it possible to make a bonfire of the baggage of the past, and to write a totally new social contract for tomorrow without any reference to the conventions that hold sway today. In the terms of Section 1.2.1 of Volume I, the utopias they defend are independent of society's current state of nature.

To find authors who think that history matters, one must turn to the very different conservative tradition exemplified by Buchanan [103, 106] and Nozick [392]. Such authors follow Edmund Burke in seeing reform as a continuous process in which the rights and duties of yesterday cannot be abolished at the stroke of a pen by some omnipotent philosopher-king, but instead evolve over time as a consequence of agreements freely entered into by agents who are motivated by considerations of individual self-interest rather than abstract utopian ideals.

3.6.1 Anarchy to Statehood

Nozick [392] makes due process the foundation stone of his theory of the Right (Roemer [435]). To know whether a particular situation is right, he argues that one must study its history to determine whether it came about as a consequence of applying right procedures to a previous situation that was also right. For example, if Adam holds a piece of property that was previously held by Eve, then he has a right to retain the property, provided that the same was true of Eve, and that she freely surrendered the right to him in return for some appropriate recompense.

The legitimacy of a right is thereby traced back to some primeval state of nature. But what counts as Right in such a state of nature? Here Nozick [392] falls back on Locke's [330] theory of natural rights, according to which: "No one ought to harm another in his life, health, liberty or possessions", and property rights in a good are acquired by mixing one's labor with it, provided that there be "enough and as good left in common for others".

It is easy to understand why conservatives find such a theory attractive, since it seems to legitimize the right of the rich to their wealth. I suspect this is why John Locke [330] continues to be treated as a supreme authority on rights, in spite of the widely acknowledged view that his ideas do not have a coherent philosophical foundation (Andrew [15, p.5]).[73] To modern authors, including Nozick [392], Locke's claim to have founded a theory of rights on the principles of Christian natural law seems merely a device to allow the newly emergent capitalists of his time the opportunity to eat their cake and have it too. That is to say, Locke invented a fairy story which

---

[73]To say nothing of Locke's failure to honor his own principles in his personal life. For example, Locke claims that our rights to our own bodies are inalienable unless we have committed a crime deserving of death, but nevertheless traded in slaves (Farr [179]).

provides a smoke-screen to conceal the fact that the moral rules built into the social contracts of modern commercial societies are not compatible with the moral rules built into the medieval social contracts which they replaced.

However, in spite of its popularity in bourgeois circles, Locke's theory would actually seem to make it impossible in practice to legitimize anyone's title to anything. As Hume [267] put it: "Philosophers may, if they please, extend their reasoning to the suppos'd *state of nature;* provided they allow it to be a mere philosophical fiction, which never had and never cou'd have any reality." But we must then not confuse such a philosophical invention with the way the world actually is. Property rights did not originate in any of the various states of nature that thinkers like Locke or Rousseau distilled from their romantic ideas about the New World awaiting settlement in America. Noble savages are few and far between. Still rarer are those who would willingly leave "enough and as good" for bums, schizos, and evil-smelling heretics when their own kind have yet to be sated. In short, Locke's state of nature never existed. In appealing to him as an authority, Nozick would therefore seem to be telling us that all property is theft, no matter how respectable its provenance may seem.

However, this would be to misunderstand Nozick [392]. In spite of conservative enthusiasm for his theory, Nozick's argument is not intended to absolve the rich from feeling uneasy when comparing their conspicuous consumption with the misery endured by the destitute. Nozick knows perfectly well that the Lockean state of nature is a fiction, and that scarcely any historical investigation of the provenance of a modern title deed would not uncover some hanky-panky along the way. His purpose in telling a mythical history of the evolution of an invented society from a Lockean state of limited anarchy to utopian statehood is to provide an ideal to serve as a benchmark in criticizing the shortcomings of our own society. But, as Spinoza [506] observes, comparing our own social contract with such a philosophical fiction is like comparing it with the social contract operated in some "golden age of the poets". If this is the name of the game, why not compare our way of doing things with the social organization of the angelic host in heaven above? So what if some counterfactual assumption about the human predicament leads to a utopia of sorts—so would the assumption that we are all little cherubs.

But although I question the value of telling just-so stories about how a society might have evolved under certain mythical circumstances, I think that Nozick's emphasis on the role of mutual protection agencies hits the nail on the head in many ways. But even here we differ on matters of importance. I think his optimism about the ease with which problems of coalition formation can be solved is misplaced. For example, I do not see that a monolithic mutual protection agency need prevail in each geographical neighborhood. On the contrary, the appearance of such a monster would seem to make it necessary for its subscribers to invent new mutual protection agencies to shield them from the abuses of power into which the officers of the monolith would soon fall. Indeed, it is because I do not know how to handle the system of overlapping coalitional institutions I believe would develop that I restrict attention in my own theory to societies containing only Adam and Eve.

3.6.2 Natural Equilibrium

Nozick's [392] emphasis on studying the dynamics of the evolution of sociopolitical institutions is very close to my heart, but my differences with him on coalitional questions are minor compared with my attitude to his telling his stories within a mythical world that somehow constrains citizens to honor an unexplained system of *a priori* natural rights. The historical record surely makes it plain enough that our concept of a right or duty evolved by fits and starts as the social contracts operating in our societies became increasingly complex with the passage of time.

Buchanan's [103] attempt to capture this phenomenon within a simple static model replaces the Lockean state of nature considered by Nozick with a Hobbesian state of nature in which distributional issues are decided by brute force. In each period, Adam and Eve are endowed with a given amount of each good, after which both seek to appropriate the endowment of the other, and to prevent their own endowment from being appropriated. If both play this appropriation game optimally, the result is what Buchanan [103] calls a "natural equilibrium". But there are Pareto-improving alternatives to playing such a "war of all against all". Adam and Eve would both benefit by respecting an agreed system of property rights that suitably apportions the flow of goods with which they are jointly endowed. Establishing such a "constitutional contract" allows the players to beat their swords into a plowshares, and so increase the available surplus by allowing their time to be used productively.[74]

Buchanan [103, p.74] supplements this version of a traditional social contract argument by pointing out that the contract needs to be continually renegotiated to take account of changing circumstances. At each point in time, Adam and Eve face a new bargaining problem for which Buchanan insists that the appropriate *status quo* is the current constitutional contract, rather than some idealized alternative. As he observes, to accept this point

---

[74]Once such a system of property rights is in place, Buchanan [103] sees the possibility for a further Pareto-improvement when there are many players and both public and private goods, the final outcome being a market-mediated "postconstitutional contract".

is not to register approval of our current sociopolitical system, but merely to accept that the political reality facing a society is how to get from *here* to *there*. This point is reinforced in his case because he postulates an external enforcement agency—a philosopher-king—which continues to maintain the old constitution while Adam and Eve are negotiating over the reforms to be built into its replacement.

Buchanan's [103] emphasis on realism in assessing the underlying balance of power appears in my own theory as the requirement that only equilibria in the Game of Life be considered as candidates for a social contract. I am also completely sold on his view that reform by mutual consent requires that Adam and Eve use their *current* social contract as the *status quo* when bargaining over a new social contract. But I then see no need to refer to some primeval state of nature at all—let alone the states of nature proposed by Hobbes, Locke, or Rousseau, which are totally uninformed by the scientific advances made by anthropologists in recent years.

Chapter 2 argues that behavioral ecology has things to tell us about the evolution and structure of fairness norms in human societies, but little would seem to be gained by asking what social contract our remote ancestors would have negotiated if they had been aware of the opportunities modern technology has made available to us. When speaking of the state of nature within my own social contract theory, I therefore refer neither to the utopian fictions of such authors as Nozick [392] or Gauthier [196] nor to the attempts of authors like Buchanan [103] to produce stylized models of how things might really have been. As explained in Section 1.2.1 of Volume I, the state of nature in my theory is simply the social contract currently in place—the *status quo* that serves as the raw material for a reformer.

When asked the way to Dublin, an Irish peasant famously replied that it would be better not to start from here. One might similarly wish for a state of nature into which so many accidents of history had not been frozen, but only grief can follow if we allow ourselves to be guided by wishful thinking.

3.7 Renegotiation

Following Buchanan [103], I see the task of a social contract theorist as going beyond the problem of describing our current state of nature and comparing it with other workable social contracts. A social contract theorist also needs to explain which social contracts can be attained from where we are now—and how to go about reaching them.

Section 1.6.2 argues at length that it is not enough for Adam and Eve to seal an agreement to move from their current *status quo* to a Pareto-improving equilibrium with the traditional handshake. If they were making a deal *within* the terms of their current social contract, such a publicly observed signal would normally be enough to ensure that each player would believe that the other planned to carry out his side of the bargain. Anyone caught cheating on the deal by trying to get more than his agreed share during the transaction would then suffer the punishment of losing his reputation for honest dealing. However, one cannot move from one social contract to another without suspending some of the rules for apportioning blame that held sway under the old social contract. In extreme cases, it may be necessary to seek a way to implement the agreement without either player having any grounds at all to trust the other.

Section 2.3.2 of Volume I uses the Stag Hunt Game of Figure 3.9(a) to illustrate the difficulties that arise when players do not trust each other. Suppose that Adam agrees to Eve's suggestion that they coordinate on the payoff-dominant equilibrium (*dove, dove*) instead of the risk-dominant equilibrium (*hawk, hawk*) that they have played in the past. Eve must now ask herself what Adam's word is worth. If he is planning to play *hawk* anyway, he stands to gain by pretending to agree—and hence perhaps persuading her to play *dove*—because he thereby gets a payoff of 4 instead of 2. But why should Adam perversely stick with *hawk* when he can get 5 by keeping his word? The reason that Eve has grounds for doubt is that *her* trustworthiness is also in question.

Adam has no more reason to trust Eve than she has to trust him. Agreeing to play *dove* while actually planning to play *hawk* allows him to insure himself against the possibility that she might cheat. He thereby gets a payoff 4 instead of 5 if she doesn't cheat, but a payoff of 2 instead of 0 if she does cheat. As Section 2.3.2 of Volume I explains, honoring his promise to play *dove* therefore only makes sense if Adam thinks Eve will honor her promise to play *dove* with probability $p \geq \frac{2}{3}$.

Section 1.6.2 explains how the problem of implementing agreements when there is minimal trust between the players in the indefinitely repeated case can be tackled by finding a continuous path from the *status quo* to the agreed outcome. In moving from one point on this curve to another close by, the players do not need to repose much trust in each other, since neither has much to lose if the transition goes awry. In the current section, this account is supplemented to take account of the special problems that arise in the original position.

3.7.1 Getting from Here to There

It is first necessary to explain how a shift from the state-of-nature point ξ to a Pareto-improving social contract ν can be achieved using strategies that are in equilibrium throughout the transition. But any enthusiasm this

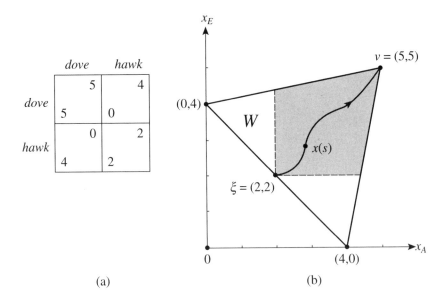

Figure 3.9: Stag Hunt Game.

news generates needs to be quelled immediately by observing that a set of players can only operate an equilibrium if the information needed for its implementation is common knowledge among the players. In particular, each player needs to have sound reasons for trusting that the other players will all carry out the parts allotted to them in implementing an equilibrium before the equilibrium becomes viable. But such mutual trust can only be built up a little at a time.

 Variable-flow equilibria. Wherever possible, this book works with the limiting version of the folk theorem obtained by passing to continuous time. In the equilibria studied in Section 3.3, the players then receive a payoff flow of x at each instant of time.[75] One might refer to such an equilibrium as static, but I prefer to speak of a *constant-flow* equilibrium in order to avoid confusion with other usages. Such equilibria are to be contrasted with *variable-flow* equilibria,[76] in which the players' behavior changes over

---

[75]The instantaneous payoff flow $x(t)$ at time t is obtained by first averaging the payoffs enjoyed in each round of the game played during a time interval of length $\tau > 0$ that contains t. The limit $\tau \to 0$ is then taken. With appropriate continuity assumptions, the precise location of t inside the interval of length τ is irrelevant.

[76]How are variable-flow equilibria possible in game played by finite automata? Suppose that we study an equilibrium played by finite automata in a repeated game, when rounds

time in such a way that their payoff flow $x(t)$ describes a curve through the cooperative payoff region of the game.[77]

The Stag Hunt Game first mentioned in Section 2.3.2 of Volume I will serve as an example. The payoff matrix of the one-shot version is shown in Figure 3.9(a). Figure 3.9(b) shows its cooperative payoff region W. The shaded region is the set of equilibrium outcomes in the continuously repeated version of the game. Since (*hawk, hawk*) is a Nash equilibrium of the one-shot game, it is a subgame-perfect equilibrium in the repeated game if everybody always plans to play *hawk*. The corresponding equilibrium outcome is then the payoff flow $\xi = (2,2)$. The strategy profile (*dove, dove*) is also a Nash equilibrium of the one-shot game. Another subgame-perfect equilibrium of the repeated game is therefore for everybody always to plan play *dove*. The corresponding equilibrium outcome is the flow $\nu = (5,5)$.

Figure 3.9(b) shows a continuous path from ξ to ν along which later points are always Pareto-improvements on points that come earlier. Since each point on such a path can be sustained as a constant-flow equilibrium outcome, it is easy to specify punishment strategies that are adequate to ensure that Adam and Eve have no incentive to deviate from the specified path. We only need to require that any deviation from equilibrium play be followed by the players switching to whatever constant-flow equilibrium supports the per-game payoff $x(s)$ current at the time s when the deviation took place. The benefit that accrues from a deviation will then be enjoyed only for a vanishingly small interval of time, and hence will be negligible when compared to the cost of the deviation—which is the loss

---

are separated by a time interval τ. Since the preliminary phase of jostling that may precede entry into an equilibrium cycle takes only a finite number of rounds, the payoffs received during this phase would seem to become irrelevant after taking the limit as $\tau \to 0$, and then computing the instantaneous flows at each time $t \geq 0$. A kibitzer watching the play of the continuous game would simply observe the players receiving their average payoff over the equilibrium cycle at each instant of time. But this argument is only valid if we refuse to allow the complexity of the automata to depend on τ. If we do, the number of jostling periods can grow without bound as $\tau \to 0$.

[77]Each such variable-flow equilibrium is equivalent in payoff terms to a constant-flow equilibrium whose per-game payoff pair is y. If Adam's discount rate is $\rho > 0$, then

$$\frac{y_A}{\rho} = \int_0^\infty y_A e^{-\rho t} dt = \int_0^\infty x_A(t) e^{-\rho t} dt \, .$$

To study the case when Adam is infinitely patient, one takes the limit as $\rho \to 0$. Given any fixed time T, Adam then cares only about what happens after T, because

$$\rho \int_0^T x_A(t) e^{-\rho t} dt \to 0 \text{ as } \rho \to 0 \, .$$

of the gains that the deviant player would have obtained by remaining on the equilibrium path and allowing $x(t)$ to climb through the awaiting set of Pareto-improvements on the current payoff flow of $x(s)$.

In accordance with the principle that punishments should be minimal, the players need not remain at the constant-flow equilibrium supporting $x(s)$ forever after a deviation at time s. It is only necessary to remain there long enough to make deviations unattractive. The players can then resume their passage along the path to ν. Or the punishment for a deviation might consist simply of temporarily slowing down the rate at which they move along the equilibrium path. The precise details are unimportant in the continuous-time limit, because any such crime-and-punishment flurry will be compressed into a vanishingly small interval.

Similar considerations apply to the length of time T that the players spend on the path from ξ to ν. They would like this time to be as short as possible in order to minimize the time they spend waiting to enjoy the full fruits of cooperation. But they are constrained in their choice of T by the need to allow trust to build up by degrees in a manner I make no attempt to model. However, in evaluating the worth of the type of variable-flow equilibrium considered here, it will be assumed that the length of time T that the players spend building up trust on their way to the Pareto-efficient equilibrium ν is negligible compared with the infinite period of time the players expect to spend at ν. This allows us to simplify the mathematics a little by treating a variable-flow equilibrium that results in the eventual implementation of ν as being worth the same as ν itself. One can justify this assumption by requiring that the players be infinitely patient, or by setting a fixed upper bound on the complexity of the finite automata allowed, but I prefer to think of the assumption merely as a simplifying approximation.[78]

3.7.2   Renegotiation in the Original Position

Section 2.5.4 suggests that fairness norms evolved as equilibrium selection devices. In particular, if ξ in Figure 3.10(b) is the state of nature, then the device of the original position provides a means of selecting a Pareto-efficient improvement ρ from the set of available equilibria. As we have

---

[78]Adam's per-game payoff up to time T is

$$\rho \int_0^T x_A(t) e^{-\rho t}\, dt \le \nu_A(1 - e^{-\rho T}) \le \rho T \nu_A \,.$$

His per-game payoff after time T is $\nu_A e^{-\rho T}$. The ratio of the former to the latter is negligible if $\rho e^{-\rho T} < \epsilon$, for a small enough value of $\epsilon > 0$. This inequality is satisfied when $T < 2\epsilon/\rho$. As ρ decreases, the requirement that the payoffs received on the way to ν are negligible therefore becomes less severe.

seen, there is no difficulty in finding a variable-flow equilibrium that will get Adam and Eve from ξ to ρ along a continuous connecting path of the type illustrated in Figure 3.10(b). But a new problem arises once an intermediary point ξ' has been reached.

While the players are temporarily operating the social contract ξ', it serves as their current state of nature. If the device of the original position is to be used consistently, the players should therefore pass behind the veil of ignorance again to *renegotiate* their ultimate destination. If the result of such a new appeal to the original position is ω, then a problem arises unless $\omega = \rho$. If Adam and Eve knew that the original agreement on ν would be renegotiated to ω on reaching ξ', there would be no point on their agreeing to implement the social contract ν in the first place.

To avoid such a renegotiation problem, matters have to be arranged so that the result of appealing to the device of the original position when the state of nature is ξ' is the same as when the state of nature is ξ. Otherwise, either Adam or Eve will have something to gain by demanding that a new appeal to justice be made on arrival at ξ'. In the presence of a philosopher-king, the player disadvantaged by such a renegotiation would be able to insist that the terms of the original agreement continue to be honored. But in the absence of any agency that enforces contracts, old agreements are water under the bridge.

The same difficulty was encountered in Section 1.7, when we studied the problem of bargaining without trust. In the special case when ρ is the regular Nash bargaining solution of the bargaining problem (X, ξ), the problem of getting safely from ξ to ρ without tempting either player to demand a renegotiation along the way was solved simply by requiring that they follow a *straight-line* path from ξ to ρ. If either player then insists on reopening the negotiations when an intermediary point ξ' has been reached, the result will just be a reconfirmation of ρ, since this is also the regular Nash bargaining solution of the problem (X, ξ').

Matters are only slightly more complicated when the device of the original position replaces the Nash bargaining solution as our equilibrium selection mechanism. In the next chapter, we shall return to Figure 2.5(b), which illustrates Rawls' version of the bargaining problem faced by the players behind the veil of ignorance. Although Rawls' use of the maximin criterion as a principle of individual choice is rejected, we shall nevertheless find good reasons to endorse his intuition that the feasible set T of social contracts as viewed from the original position should be identified with $X_{AE} \cap X_{EA}$ as in Figure 2.5(b). But the same reasons do not apply to the location of the *status quo* point τ. This must be located at $\tau = \frac{1}{2}\eta + \frac{1}{2}\zeta$, as in Harsanyi's Figure 2.5(a). The hybrid diagram that results is reproduced in Figure 3.10(a).

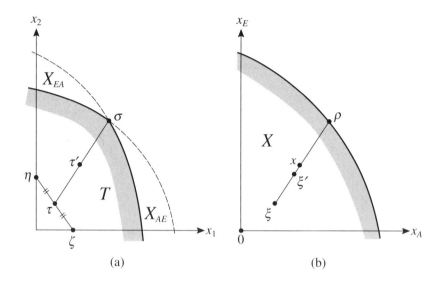

Figure 3.10: Renegotiation in the original position.

When the situation is not too asymmetric, the Nash bargaining solution σ for the bargaining problem (T, τ) faced by the players behind the veil of ignorance is located at the kink in the Pareto-frontier of T, as illustrated in Figure 3.10(a). But, wherever σ is located on the Pareto-frontier of X, the players in the original position can avoid the renegotiation problem studied in this section by agreeing to approach σ along the straight-line path from τ to σ. A new appeal to the original position once an intermediary point τ' on this straight line has been reached will simply reproduce the same agreement to coordinate on σ. Figure 3.10(b) shows the corresponding straight-line path from ξ to ρ that Adam and Eve's agreement in the original position calls upon them to follow once they have emerged from behind the veil of ignorance.

3.7.3   Making the Punishment Fit the Crime

In Gilbert and Sullivan's *Mikado,* the emperor of Japan is deeply interested in the administration of justice. As he puts it:

> My object all sublime
> I shall achieve in time—
> To let the punishment fit the crime—
> The punishment fit the crime.

Pool hustlers are condemned to play on a cloth untrue, with a twisted cue, and elliptic billiard balls. Teachers will be delighted to learn that chatterers whom lectures bore are sent to sermons by mystical Germans, who preach from ten till four.

Why do we take such satisfaction in seeing the biter bit? In hoisting him by his own petard? In returning a tit for a tat? In taking an eye for an eye and a tooth for a tooth? In brief, why is justice only poetic when the punishment fits the crime?

My guess is that poets are tapping an intuition about justice that is common to us all. It derives from the manner in which fair compromises between untrusting agents are policed. When Rawls is vindicated in Section 4.6.5, we shall find that the ray from the *status quo* ξ in Figure 3.10(b) to the fair outcome ρ has slope U/V, where the weights U and V are determined by the standard of interpersonal comparison relevant in the context (Section 2.7.2). It follows that ρ is the proportional bargaining solution for the problem (X, ξ) with weights U and V (Section 1.3.3). Psychological equity theory demands that each player's gain be proportional to his worthiness (Section 4.4). The use of the proportional bargaining solution is therefore equitable in the psychological sense if we identify Adam and Eve's worthiness with $1/U$ and $1/V$ respectively.

When shifting from ξ to ρ, the renegotiation considerations we have been studying suggest that Adam and Eve will actually make the transition by allowing the current social contract to move continuously along the ray that joins them in Figure 3.10(b). Eventually, they will get to ξ'. This point then serves as a temporary *status quo* that becomes operational if Adam or Eve demand a renegotiation of the deal. Having reached ξ', Adam and Eve must continue through the nearby point x on their way to ρ. The gain that each player will make in moving from ξ' to x is proportional to his worthiness, and hence will be perceived by the players as fair.

What if Adam now makes the mistake of cheating during the transition from ξ' to x? Although other punishment schemes would be equally effective, Section 3.7.1 suggests that Eve's natural response is to take whatever action is needed to restore the *status quo* ξ'. The losses that each player will suffer in surrendering x for ξ' are then equal to the gains they would have enjoyed in moving from ξ' to x. Since the gains were perceived as fair, so are the losses that result from Eve's punishment of Adam.

The natural punishment scheme for policing the transit from ξ to ρ therefore requires that poetic justice be done by choosing a punishment that fits the crime. In the language of Section 3.3.7, the *tit* that follows a *tat* is chosen fairly.

3.7.4   Renegotiation-Proofness

The renegotiation problems we have been discussing are closely related to the issues discussed by game theorists under the heading of *renegotiation-proofness*. However, it is important to note that the line I take on this question is not entirely orthodox. Section 3.3.5 endorses the arguments of the renegotiation literature that rational players will negotiate equilibria in which punishments are no more severe than necessary to deter deviations. I also agree that one must insist that a negotiated equilibrium be proof against demands for a renegotiation of its terms *on the equilibrium path*. But I differ from the literature in perceiving a difficulty with the idea that the same renegotiation criteria should also be applied after a deviation has taken the players *off* the equilibrium path.

The difficulty is related to the problem raised in Section 0.4.2 and elsewhere with the concept of a subgame-perfect equilibrium. Why should Eve continue to treat Adam as rational in a subgame that can only be reached if he plays irrationally? Should she not abandon her faith in his rationality altogether if he behaves irrationally often enough? Section 0.4.2 explains why I continue to rely on the notion of a subgame-perfect equilibrium in spite of such foundational difficulties. In brief, I believe that the distortions created by relying on the concept are outweighed by the discipline it imposes on the models that one permits oneself to contemplate. However, I believe the distortions become too severe to be neglected when one adds renegotiation-proofness off the equilibrium path to subgame-perfection.

The literature on renegotiation-proofness[79] has yet to reach a consensus on definitions, but all papers on the subject seem to me to neglect the vital question of how trust is established and maintained in reciprocal relationships. If Adam and Eve agree to coordinate on an equilibrium in a repeated game, but Adam breaks his word and deviates from the equilibrium strategy he agreed to play, Eve must presumably ask herself to what extent he can be trusted to honor any coordination agreement they may negotiate in the future. If he repeatedly breaks his word, the time must surely come when she wonders whether there is any point in negotiating with him in the future at all. But, just as the definition of a subgame-perfect equilibrium requires that we treat other players as rational no matter how often they may have behaved irrationally in the past, so the orthodox literature on renegotiation-proofness requires that we always trust other players no matter how often they have previously betrayed our trust.

The consequences of making such an assumption can be far-reaching.

---

[79]For example, Abreu and Pearce [3], Abreu *et al* [4] Asheim [21], Bergin and McCloud [54], Farrell and Maskin [180], Rubinstein and Wolinsky [452], and van Damme [528].

For example, when the approach of Abreu *et al* [4] is applied to the indefinitely repeated Stag Hunt Game, the only point in W of Figure 3.9(b) that can be supported as a renegotiation-proof equilibrium is $\nu = (5,5)$. In their theory, rational players who predict the renegotiation opportunities that might arise in the future are therefore restricted to agreements requiring that everyone always plan to play *dove*—no matter what may have happened hitherto in the game. The problem of how to get from ξ to ν therefore does not arise for Abreu *et al* [4]. According to their theory, rational players couldn't have been operating an equilibrium yielding the outcome ξ in the first place.

One might defend orthodox theories of renegotiation-proofness by arguing that the history of experience the players have shared before beginning to play the game under study is so extensive that no evidence of dishonesty a player might exhibit during the game is sufficient to outweigh his record of trustworthiness in the past. But this is one idealization too many for me. I believe that the problem of mutual trust is too close to the heart of a social contract theory to permit the adoption of a formalism that abstracts the whole issue away.

3.8 What about Moral Values?

Before passing from the study of how social contracts are sustained in this chapter to the study of how they are selected in the next, time must be found to discuss the language of moral values that folk wisdom employs when talking about sustaining or selecting social contracts.

Hobbes [250] characterizes human nature in terms of a person's strength of body, his experience, his reason, and his passion. In the jargon of economics, his strength of body is determined by the feasible set from which he is currently free to choose, his experience is summarized by his knowledge and his beliefs, and his passions are the fundamental preferences written into his personal utility function. A rational person uses his reason to choose an action from his feasible set that maximizes his expected utility relative to his beliefs.

This section argues that social and moral values need to be classified as part of a person's *experience* rather than his *passions*. We talk about the Good and the Right in much the same way that we talk about a good movie or the right way to skin a cat, but the folk psychology built into our language is no more reliable than the folk physics that makes the earth solid and the sun rise. The error of attributing behavior mediated by the use of a fairness norm to a player's having a taste for fairness is particularly misleading in the context of this book (Section 3.5.2).

3.8.1   Confusing Tastes and Values

Why do people drive on different sides of the road in London and Paris? The naive answer is that people like driving on the left in England but reverse their preference when they cross the channel to France. But this explanation certainly doesn't correspond with my own experience. Driving on the left in my native England offers me no delights that I miss when driving on the right in France.

My preferences about left and right are shared by Adam and Eve when playing the 2×2 Driving Game. In this game, both players assign a payoff of 1 to the outcomes (*left*, *left*) and (*right*, *right*), and a payoff of 0 to the outcomes (*left*, *right*) and (*right*, *left*). The players are therefore *indifferent* between the equilibrium in which everyone drives on the left, and the equilibrium in which everyone drives on the right.

When deciding whether to drive on the left or the right in the Driving Game, Eve needs to pay attention to more than her tastes. She needs to consult her beliefs about the coordinating conventions operating in her society. In Paris, the relevant convention selects the equilibrium in which everyone drives on the right. Eve uses her belief that this convention is common knowledge in Paris to predict that Adam will be driving on his right if she encounters him in an oncoming car. Her distaste for collisions then tells her that it is optimal for her to drive on her right as well. Adam reasons similarly, with the result that the two players succeed in using their common knowledge of French social values to coordinate on the equilibrium in which both drive on the right.

In the Driving Game, this analysis is so blindingly obvious that it hardly seems worth writing down. Tastes are held by individuals. Values are held in common by a society. *I* have tastes—*we* have values. How could the two concepts conceivably be confused?

Without the language of game theory to discipline our thoughts, the answer is that tastes and values are confused only too easily. Philosophers from Aristotle onwards have routinely confused tastes and values—and it matters. Unless the distinction is properly made, one is always at risk of falling prey to the fallacy that outsiders are intrinsically evil because their values differ from ours.

How are values confused with tastes? Two sources of error are mentioned here. The first occurs when people try to get special treatment for their personal tastes by calling them values. The second arises when appeals to folk versions of the principle of revealed preference result in values being mistaken for tastes.

Ipsedixists argue that the rest of us are held in thrall by low tastes and unsavory appetites, whereas they are motivated only by the highest of moral values. For example, Plato was a moral relativist to the extent that he recognized that the mores of Athens and Sparta were not the same. His own aristocratic tastes favored the values of Sparta. Given the power, he would have imposed a Spartan-like constitution on Athens. Other Athenian citizens had rival crackbrained schemes, and so Plato played the standard trick of claiming the status of values for his own personal tastes. However, a taste for one set of values over another is not itself a value. To confuse the two is to make it almost impossible to separate moral relativism from moral subjectivism (Section 3.8.4).

The second error is more insidious. Everybody readily agrees that the side of the road we drive on is a matter of social convention. Nobody therefore makes the mistake of deducing that Eve has a built-in preference for driving on the right from the fact that she is seen driving on the right. But people deny that our standards of fairness are no less matters of social convention than our traffic code. In consequence, they fall prey to the fallacy that someone who refuses unfair offers must have a built-in taste for fairness. However, if Eve is playing her part in operating an equilibrium in the Game of Life, her behavior depends on *all* of Hobbes' categories—not just on her passions.

As a preliminary to tracing the second kind of error to its source, this section reiterates the principles of revealed preference theory. Innocents may wonder how such simple and straightforward ideas could have resulted in so much dust being kicked in the air. One of these days, when the dust has settled, perhaps some intrepid scholar writing a history of decision theory will be able to tell us why.

By their fruits ye shall know them. The theory of revealed preference was invented to make economics independent of psychological speculations about the way our brains work. It therefore abandons any pretension to explain *why* Eve chooses one action rather than another. Gauthier [197, p.186] is typical of a widespread failure to understand this simple point. It isn't true that the received theory says that "only utilities provide reasons for acting". The pure theory doesn't claim to give any reasons for acting at all. It merely describes *what* actions Eve will choose if she acts consistently.

New definitions of consistency are being proposed all the time, but the orthodox definition permits Eve's behavior to be described by saying that she chooses actions as if maximizing the expected value of a utility function relative to a subjective probability distribution. Her utilities and subjective probabilities in this representation are *constructed* from data obtained either by observing her actions, or by making assumptions about what actions she would take under various hypothetical circumstances.[80]

To many critics, this statement of the aims of the theory sounds suspiciously like behaviorism. Am I really holding fast to Skinner's [497, 498] discredited doctrine that animals and people are simple stimulus-response machines? The answer is no. Skinner offered various hostages to fortune, but he didn't maintain the absurd hypothesis that human cognitive processes are irrelevant to the decisions we make—and neither do I. Revealed preference theory is *neutral* about what goes on inside our brains. In Section 2.5.2, I argue that preference-belief models probably do have an important role in describing our cognitive processes in some contexts, but to make such a speculation is to step outside the domain of revealed preference theory. One certainly would not wish to be committed to maintaining that trees model their environment using a preference-belief model when using the theory of revealed preference to describe their strategy choice in an evolutionary game!

When applying the theory of revealed preference, the first step is to determine Eve's set A of actions. As time goes by, we will observe what choices she makes when constrained to choose from a variety of different feasible subsets of A. For example, if we are observing her shopping behavior, we will eventually know what bundles of goods she buys when she has different amounts of money to spend and is faced by different price configurations. The original theory of revealed preference provided consistency conditions that allowed such shopping behavior to be described by saying that Eve chooses *as though* seeking to maximize a utility function defined over the set C of all bundles of goods.[81] However, the modern theory of revealed preference is much more ambitious. It recognizes that Eve's actions may depend on states of the world which she has not observed or which have yet to occur. For example, in considering how much to pay for an umbrella, Eve will take account of the prospects of future rain.

We then have three sets to consider: a set A of actions, a set B of possible states of the world, and a set C of final consequences. These are connected by a function $f : A \times B \to C$ that describes the consequence $c = f(a, b)$ of taking action a when the state of the world is b. Orthodox revealed preference theory then provides consistency conditions for Eve's behavior

---

[80]In the one-shot Prisoner's Dilemma, Eve's utility for the outcome (*dove, hawk*) is made larger than her utility for the outcome (*dove, dove*) because we are given that she would choose *hawk* if she knew that Adam were sure to choose *dove* (Section 2.2.4 of Volume I). Once this is understood, it becomes obvious that all the endless disputation over the standard game-theoretic analysis of the Prisoners' Dilemma is based on the simplest of misunderstandings (Chapter 3 of Volume I).

[81]One consistency requirement is called the *weak axiom of revealed preference*. Pigeons observe it quite well under laboratory conditions (Kagel *et al* [284]).

in A to be described by saying that she chooses *as though* maximizing the expected value of a utility function defined on C, relative to a subjective probability distribution defined on B.

One of the reasons that revealed preference theory is so badly misunderstood is that its practitioners feel that only the most formal of mathematics is suitable for discussing such an important topic. But the mathematics necessary to deduce expected utility theory from the principles of the theory is entirely trivial. Section 4.5 of Volume I manages the trick in a few pages of the most elementary algebra. As for the principles themselves, they can be expressed without any mathematics at all. They simply assert that a rational player will not allow what is going on in one of the domains A, B or C to affect his treatment of the other domains.

For example, the fox in Aesop's fable is irrational when he judges the grapes to be sour because he can't reach them. He thereby makes the mistake of allowing his beliefs in domain B to be influenced by what actions are feasible in domain A. If he were to decide that chickens must be available because they taste better than grapes, he would be guilty of the utopian mistake of allowing his assessment of what actions are available in domain A to be influenced by his preferences in domain C. The same kind of wishful thinking may lead him to judge that the grapes he can reach must be ripe because he likes ripe grapes better than sour grapes, or that he likes sour grapes better than ripe grapes because the only grapes that he can reach are probably sour. In both these cases, he fails to separate his beliefs in domain B from his preferences in domain C. More topical examples of such irrationalities can be found any day of the week in the speeches of great statesmen and political pundits.

One consequence of the emphasis on the mathematics of revealed preference theory at the expense of its foundational principles is that it distracts the attention of mathematically sophisticated scholars from the care that is necessary in choosing the underlying model. If there are linkages between the sets A, B and C that remain unexpressed in the consequence function $f : A{\times}B \to C$, then there is no reason why Eve's decisions should not reflect these unmodeled relationships.

For example, Sen [483] tells us that people never take the last apple from a bowl, and hence are inconsistent when they reveal a preference for no apples over one apple when offered a bowl containing only one apple but reverse this preference when offered a bowl containing two apples. The data supporting this claim were doubtless gathered in some last bastion of civilization where Miss Manners still reigns supreme—and this is relevant when modeling Eve's choice problem. Her belief space B must allow her to recognize that she is taking an apple from a bowl in a society that subscribes to the social values of Miss Manners rather than those of New

York City. Her consequence space C must register that the course of future play in her Game of Life is at least as important to her decision as whether she eats an apple right now. Otherwise, we shall not be able to model the punishment that her fellows will inflict on her for deviating from their social contract. Once we have set her problem in an appropriately wide context, Eve's apparent violation of the consistency postulates of revealed preference theory disappears. She likes apples enough to take one when no breach of etiquette is involved, but not otherwise.

The lesson to be learned from Sen's example is not that the consistency requirements of revealed preference theory are too strong to be useful, but that the first thing to do when they seem to fail is to ask whether the choice problem has been adequately modeled.

Tastes must be fixed. Imposing consistency requirements on Eve makes no sense unless the underlying determinants of her choice behavior are fixed. These determinants may not be fixed forever, but they must remain fixed while she makes choices from a sufficiently large set of different feasible sets to provide enough data to allow her behavior when choosing from other feasible sets to be predicted. This book refers to such a length of time as the short run. In the medium run, the knowledge of the social values built into her empathetic preferences may change, and so alter her beliefs over B. In the long run, her preferences over C may also change. It is therefore only in the short run that we are entitled to make *direct* use of the theory of revealed preference.

However, we can still use the theory of revealed preference *indirectly* if we are willing to make consistency assumptions about how Eve will update her beliefs over B in the medium run as she receives new pieces of information. The same rationality considerations which demand that beliefs and preferences be separated in the static case require that the preferences Eve reveals over C remain fixed as she updates her probability distribution over B.[82] For example, the fact that a fox tastes a grape and so learns that it is ripe should not lead him to change his mind about whether he prefers chicken to grapes.

Economists speak of a *direct* utility function when describing Eve's fixed preferences over consequences. The varying preferences she reveals over actions as her information about her environment changes are described by

---

[82]Much stronger consistency requirements are normally proposed, which imply that Eve's probability distribution will be updated according to Bayes' rule. I follow Savage [461] in thinking it "ridiculous" and "preposterous" to impose such consistency requirements outside a "small world" in which it is possible to predict how all possible future pieces of information might affect the underlying model summarized in Eve's prior probability distribution (Binmore [69]).

an *indirect* utility function. Such an indirect utility function can be interpreted as describing Eve's attitudes to the means available for achieving the ends described by her direct utility function. Final ends can then be distinguished from the means employed to obtain them by observing that the preferences Eve reveals over means are capable of varying in the medium run, whereas the preferences she reveals over ends remain fixed.

Unfortunately, the use of indirect utility functions creates yet more confusion, since authors insist on applying principles intended for use in the short run to medium-run problems. The Pareto principle is a leading example. If nothing that happens during or after the decision process will change anybody's beliefs, then it doesn't matter whether one applies the Pareto principle to a direct or an indirect utility function. But suppose the Hatfields and the McCoys are asked whether tax dollars should be spent on building a public library. Each clan believes the other to be illiterate and so votes *yes*. If we apply the Pareto principle to their actions in the voting booth, the library will be built—with disastrous results when the Hatfields and the McCoys discover that they voted for the same thing for diametrically opposed reasons.

To avoid such paradoxes, social welfare principles need to be applied to *direct* utility functions. Any study must therefore begin by providing a model of choice behavior that clearly distinguishes between means and ends. The language of game theory does this automatically. Eve's ends are summarized by the payoffs in the game she is playing. These payoffs are her direct utilities over the set C of consequences. The means she can employ to achieve these ends are represented by her strategies in the game. These strategies are the actions in the set A. The set B consists of the strategies that may be chosen by Chance and the other players.

Her indirect utilities over the set A of strategies reflect what she knows or believes about what is happening in B. In particular, they contain information about the social values operating in her society. If one social value requires using the Pareto principle in solving equilibrium selection problems, it becomes incoherent to apply the Pareto principle to the players' indirect utilities over strategies, because they already take for granted that the Pareto principle is in use. This may seem very obvious, but we shall find that the error involved in confusing means and ends is much harder to spot once it has been gift-wrapped in sufficiently obscure language.

Water or diamonds? In economics, the problem of separating personal tastes from social values appears in the distinction between use-value and exchange-value. Water is essential to our health and well-being, but diamonds are not. It follows that water has a high use-value and diamonds have a low use-value. Why should diamonds therefore be expensive and water cheap?[83]

This traditional question is answered in modern textbooks by saying that diamonds have a vastly higher exchange-value because they are in much shorter supply. But this reply evades the issue. As Alfred Marshall [342, p.820] explained, to discuss market prices only in terms of supply is like explaining how to cut paper with only half a pair of scissors. The interesting question is why diamonds are *in demand*.

It is common knowledge that displaying diamonds confers social status in some circles. In turn, social status allows access to benefits with high use-value—notably the opportunity to mate with good genetic stock. Eve's demand for diamonds is therefore derived from two sources—her personal tastes for genuinely useful commodities and her beliefs about the social values embedded in her cultural milieu. If the grubby social values that assign status to those who display diamonds were to go out of fashion, nobody would demand diamonds anymore, even though their personal tastes remain unchanged.

Aristotle. Aristotle [18] tells us that Adam may have three kinds of ends: the useful, the pleasurable, and the noble. Since he argues that a useful end is merely a means to a pleasurable or noble end, there is no difficulty in interpreting Aristotle's useful ends as actions to be assessed with an indirect utility function. Nor is there any problem in counting pleasurable ends among the consequences to be assessed with a direct utility function. However, he writes as though noble ends were to be counted on a par with pleasurable ends, thereby making the standard mistake of confusing social values and personal tastes. But this mistake does not lead him very far astray, since Aristotle is very clear that nobility is *socially* determined.

Adam is said to be noble if his behavior is *praiseworthy*. But what men praise depends on their culture. For example, Aristotle counts temperance among the noble virtues, but the Scythians of the hinterland of his home city doubtless regarded temperance as an ignoble vice in the manner of freebooting barbarians down the ages. Temperance is therefore a social value to be counted among Adam's beliefs about the social contract operated by the city in which he dwells. He is not temperate because he has a taste for being temperate. He is temperate because he will lose the respect of his fellow citizens if he fails to meet their standards—and he can gratify more

---

[83]Nothing is more useful than water; but it will purchase scarce anything; scarce anything can be had in exchange for it. A diamond, on the contrary, has scarce any value in use; but a very great quantity of goods may frequently be had in exchange for it—Adam Smith [501].

of his personal tastes with their respect than without it.

Temperance is a particularly telling example of a social value, because attempts to model it as a personal taste lead immediately to nonsense. Having ranked his feasible options according to his tastes, does Adam express a taste for temperance by choosing an option in the middle? Obviously not! Revealed preference theory ranks the top action at the top because Adam always chooses it when it is available. Critics of revealed preference theory say that the original ranking should represent Adam's "real" preferences rather than those he reveals. But the idea that we have real preferences which we prefer not to employ when making choices seems no less incoherent to me than saying that Adam will choose the middle option because it isn't the option he chooses.

Mill. Aristotle skirts a pit into which John Stuart Mill [358] falls headlong. In attempting to reconcile utilitarian paternalism with individual liberty, Mill argues that we have the social value of freedom built into our personal tastes. A social contract that maximizes the sum of our personal utilities would therefore guarantee some measure of liberty.

Freedom is measured by the size of the opportunity set from which we can choose without infringing the duties a social contract imposes on us (Section 3.2.2). We like to be a guaranteed a large opportunity set because we don't know what the future will bring. Securing a large opportunity set is therefore what Aristotle calls a useful end. If I don't know whether it will rain or shine tomorrow, having both an umbrella and a parasol on hand is useful because I will be able to maximize my direct utility function in either case. But if I were certain that it were going to rain for all eternity, how much would I be willing to pay to enlarge my opportunity set from an umbrella to an umbrella and a parasol? People think this question is ridiculous. They ask why anyone would ever buy something for which they have no conceivable need, thereby demonstrating their understanding of the irrationality of allowing their personal tastes over the set C of final consequences to be influenced by something going on in A or B.

Utilitarianism makes no more sense than the Pareto principle when applied to indirect utilities. It makes sense only when applied to the direct utilities that describe the players' personal tastes over the set C of final consequences. Thoughtful utilitarians sometimes express their perception of this observation by arguing that we should sum the preferences that the players would have if they were well-informed rather than the preferences they actually reveal.

As for the substance of Mill's attempt to reconcile liberty and utilitarianism, such an approach would lose all the advantages bought by postulating a

philosopher-king who *enforces* the laws that the people make for themselves
(Section 2.2.4). A culture that values freedom provides individuals with a
greater variety of means to achieve their individual ends, but one cannot
make people free in this sense without also offering them the opportunity to
act antisocially. If they are free not to buy tickets for the commuter train,
some of them will become free riders. A utilitarian philosopher-king makes
it possible to maximize the social cake available for division by cleansing
the system of such free riding. Since it has no philosopher-king, an egalitarian social contract is therefore second-best when compared with a first-best
utilitarian utopia (Section 2.8). But if we follow Mill in confusing the objects being evaluated with the manner in which we evaluate them, such a
comparison ceases to have any meaning.

Rawls.   Just as Mill wants to eat his cake and have it too by writing
deontological social values into our personal tastes, so Rawls [417] writes
consequentialist social values into our personal tastes by allowing them to
incorporate private conceptions of the good. I am not entirely clear how
these differing conceptions of the good are reflected in such primary goods
as the powers and prerogatives of office or the social basis of self-respect,
but perhaps it is enough to say that I think Aristotle would have classified
the former among his useful ends and the latter among his noble ends. In
short, neither should be counted as a personal taste at all.

Singer.   Peter Singer's [496] recent book sets out to answer Pliny's peren-
nial question: *How Are We to Live?* Since Singer's [495] earlier *Expanding
Circles* was the inspiration for the evolutionary approach to ethics offered
in Section 2.4, it is disappointing to find him still trying to persuade us to
adopt his greener than green personal preferences by representing them as
the ultimate in social values.

   Unlike most ipsedixits, Singer defends his propaganda with an argument. Since he is telling us what most of us would like to hear about
the suffering of animals and the destruction of the environment, the argument is not easy to evaluate dispassionately. However, when Singer's
eloquence looks like getting the better of me, I remind myself how uncon-
vincing Hegel's similar argument seems. But the fact that I like Singer's
vision of the ultimate social contract a lot better than Hegel's should be
irrelevant to my assessment of the quality their arguments.

   Singer [496, p.268] would prefer to live in a society in which we were all
committed to adopting the "point of view of the universe". He therefore
espouses a loose form of utilitarianism in which the universe serves as his

analogue of an ideal observer.[84] This ideal observer cares for all sentient beings, and therefore accords rights not only to human beings, but also to animals.[85] In representing this personal taste as a privileged social value, he observes that social contracts have tended to evolve so as to widen the range of those with whom we empathize. Singer [496, p.272] then parlays this evolutionary trend into an "escalator of reason" that will carry us onward and upward to his preferred social contract. No amount of reflection, so the story goes, can therefore make a commitment to his vision of an ethical life seem trivial or pointless (Singer [496, p.258]).

Unless my own ideas are hopelessly wrong, Singer is mistaken about what evolution is capable of doing with the human race, let alone with animals. However, my immediate complaint arises from his confusing an evolutionary process of equilibrium selection with the rules for sustaining an equilibrium. It is rational for me to obey the rules that sustain the social contract currently in place. To argue that reason demands instead that I obey the rules appropriate to some better social contract which would come into being if everybody were to do the same is to join the ranks of the circle-squarers who think it is rational to cooperate in the Prisoners' Dilemma (Chapter 3 of Volume I). In brief, a plant grown from a Humean seed cannot possibly produce a Kantian flower.

3.8.2   Das Adam Smith Problem

With his emphasis on the importance of empathetic identification, Adam Smith is perhaps the philosopher who brings us closest to understanding why social values are so easily confused with personal tastes. Since the fairness algorithm represented by the device of the original position is biologically determined, we do not think to question its fundamental validity. The social values necessary for its operation are therefore perceived as being carried entirely by the empathetic preferences we hold in common. But an empathetic preference is so close in structure to a personal preference that it is inevitable that the two should prove hard to separate.

According to Teichgraeber [518], Adam Smith [500, 501] saw his life's work as an interwoven braid with three strands—jurisprudence, moral philosophy, and economics. His views on the first of these topics are known only from the notes of his students and asides in his other works. But his

---

[84]Singer [496, p.206] appeals to the authority of Robert Hare on this front. Hare [225, 227] is sidelined in this book, because I feel that he manages to go only part of the way down a road that Harsanyi [231, 233] had fully explored some years before (Appendix B).

[85]Nothing is said about interpersonal comparison. Perhaps one counts heartbeats when trading the life of a lion against that of a lamb.

Theory of Moral Sentiments and *Wealth of Nations* bear eloquent testimony to his research in moral philosophy and economics.

Modern economists are often surprised to learn that Adam Smith saw no contradiction between the concern for social propriety that the first of these books attributes to *homo sapiens*, and his allegiance to *homo economicus* in the second. Except in Germany, the problem of bridging this gap seems to have attracted little attention until recently. The difficulty is illustrated by a famous passage from *The Wealth of Nations*: "It is not from the benevolence of the butcher, the brewer, or the baker that we expect our dinner, but from their regard to their own interest." But what has then become of their natural capacity for sympathizing with their fellow men, on which *The Theory of Moral Sentiments* relies?

Commentators on the Adam Smith problem seem largely agreed that we must look to the modern distinction between empathy and sympathy in order to achieve a reconciliation between his two books. For example, Raphael [414] takes critics to task for failing to note that Adam Smith *defines* sympathy to be something close to what I call empathy. However, modern apologists are too ready to forgive Adam Smith for failing to honor his own definition whenever he offers a serious argument. Instead, he repeatedly falls into the trap of appealing to sympathy in the sense that it is understood in this book. That is to say, he implicitly assumes that the welfare of others appears as an argument in our *personal* utility functions.

However, as with Rawls [417], I believe that Adam Smith's intuition was sound, even though his reasoning does not always bear close examination. His faith in the consistency of *Wealth of Nations* and *Moral Sentiments* can be vindicated by clearing away the confusion between the concepts of empathy and sympathy that he shared with David Hume.

Private tastes and public values.   Adam sympathizes with Eve when her well-being is an argument in his personal utility function. He then helps her because he wants to. The temptation to explain all moral behavior in such terms is very strong. But, as Section 3.8.1 explains, it is essential that we not confuse our tastes with our values in this kind of way. Tastes are held by individuals. Values are held in common by a whole society.

The Driving Game of Section 3.8.1 illustrates why Eve need only consult her personal preferences when considering whether to play her part in *sustaining* an equilibrium. If I am correct in modeling the intuitions that lie behind theories of the Right in terms of the rules we use to sustain a social contract, it is not surprising that conservatives who care only about maintaining our traditions should see no reason to look beyond the *homo economicus* model of Adam Smith's [501] *Wealth of Nations*. But when we consider how equilibria are *selected*, it is necessary to examine the social and moral values embedded in the pool of common knowledge that I identify with our culture. If it is correct that the intuitions which inspire theories of the Good are derived from observing how societies *select* equilibria, then it is equally unsurprising that socialists should construct variants of *homo ethicus* whose preferences reflect the values of a particular society. Although Adam Smith was certainly no socialist, the impartial spectator or ideal observer of his *Moral Sentiments* is just such a paragon of virtue.

When we turn from the social conventions that determine on which side of the road we drive to the moral conventions from which we derive our notions of justice, matters become less clear, because we actually do use preferences to encode some of the relevant values. But we do not write these values into our *personal* preferences. Society writes them into the commonly held *empathetic* preferences that serve as social inputs to the variant of the original position that Nature has sealed into our genes. Empathetic preferences are never revealed when Adam or Eve act alone in making decisions. They are revealed by the equilibrium selected when Adam and Eve *jointly* use the standard fairness norm to resolve a coordination problem.

In short, the solution to the Adam Smith problem requires reinterpreting many of his references to sympathy in *Moral Sentiments* as statements about empathetic preferences. Empathetic preferences are the bearers of the cultural *values* for which Adam Smith uses the notion of an impartial spectator as a metaphor. They are needed to select among equilibria. No contradiction arises with the traditional conception of *homo economicus* because economists have always been more concerned with how equilibria are sustained. For this purpose, nothing but the personal preferences of an agent are required.

3.8.3 Postwelfarism

Section 4.3.3. criticizes the reasons given by the postwelfarist movement for refusing to treat the personal preferences of individuals as paramount when planning reforms. For example, it is said that women living in subsistence societies have been conditioned to acquiesce in the subordinate role assigned to them. A reformer should therefore not ask such women what they want. Rather than paying attention to whether the recipients of his benevolence like it or not, the story is that he should seek instead to improve their lives according to his own objectively chosen standards of what is good for them.

The ipsedixism behind the approach is well meant, but it is based on the same philosophical muddle over personal tastes and moral values to which this section is devoted. As a matter of objective fact, I wonder whether a woman scratching a living in the jungle really feels much different about

matters that genuinely belong in the domain of personal tastes than a
Parisian socialite on a shopping expedition. Do they not both prefer to eat
rather than go hungry? To be warm rather than cold? To be safe from
violence? For their children to survive to have children of their own? A
primitive may take pleasure in eating grubs from which a sophisticate will
turn away in disgust, but the primitive would be equally horrified at the
prospect of dressing in the height of Parisian fashion. These differences arise
because people with the same fundamental preferences behave differently
when assigned different beliefs. Their beliefs about what is good to eat or
wear would probably change quickly if they found themselves sufficiently
hungry or cold, but the same is not true of their beliefs about the very
different social values built into their respective social contracts, because
these are based on long experience of the facts on the ground.

To miss this point is to confuse personal tastes with the social values
that we acquire as children through observing how the social contract operating in our community works. Postwelfarists are right to observe that the
Parisian social contract differs from those of the Brazilian jungle, but they
go wrong insofar as they think that the socially induced behavior caused
by the need for Eve to conform to the social contract of her society should
be attributed primarily to her personal tastes over fundamentals.

Tastes that are genuinely personal can be recognized by the fact that
they remain constant as the social contract varies in the medium run (Section 3.8.1). It seems to me foolish to think that we are better able to
assess the nature of such personal tastes than the person whose pleasure
centers actually get pinched or tickled when things go badly or well. What
better source of information could there be in seeking objective correlates
of a person's underlying personal tastes than the evidence revealed by her
own behavior? It is true that employing some version of the theory of revealed preference to this end is likely to be difficult, since one must first
carry out extensive sociological studies to separate the social values held
in common by a society from the fundamental tastes of its citizens. But
surely *all* attempts at reforming a society should routinely be prefixed by
a detailed scientific study of how its social contract currently works? Or
are humans not entitled to the same care and consideration that ecologists
think necessary when interfering with the habitats of plants and animals?

Nor does there seem any point in telling women in underdeveloped countries to alter their indirect utilities when these are properly adapted to their
current social contract. Why tell a woman in a society that treats her like
a slave that she ought to develop a taste for freedom? She already knows
that overwork makes her tired. She would therefore take more leisure if
she could. But the socially determined memes she carries in her head also
know that she will be punished by her friends and relatives if do-gooders

from outside persuade her to default on the duties written into the social contract of her village. It is therefore not surprising that she commonly rejects their advice.

Rather than telling other people what is good for them, postwelfarists would be more effectively occupied in seeking to reform the social contracts of primitive societies in ways the citizens of such societies themselves find acceptable. In particular, rather than lowering the self-esteem of women in primitive societies even further by treating their opinions on the way things ought to be as worthless, a reformer would do better to accept that their sense of their own worth is socially determined along the lines described in Section 4.7.4. The problem for the reformer is then not to convince Eve that her personal tastes fail to reflect her own best interests—but to persuade everybody in her society that it would be Pareto-improving to move to a social contract in which her perceived worthiness is greater.

3.8.4 What Moral Relativism Is Not

Section 0.3.1 rails against the stubborn tradition that confuses moral relativism with moral subjectivism. This section reverts to the same topic in the hope that my insistence on distinguishing between personal tastes and social values will help to clarify the issue.

Dead gods and supermen. Why has moral relativism become so entangled in people's minds with moral subjectivism that philosophers find it hard to write definitions clearly distinguishing the two? The dictionary has no such difficulty with the words *relative* and *subjective*.

My diagnosis makes Nietzsche [385, p.46] the archetypal author of the fallacious line of reasoning that creates the confusion. He was emphatic in his moral relativism. As Zarathustra spake it: "Every people speaketh its language of good and evil; this its neighbor understandeth not. Its language hath it devised for itself in laws and customs." Nietzsche correctly deduced that Church and State lie when they preach an Absolute Morality to bovine citizens conditioned not to think for themselves. But the outrage he felt at finding himself one of the herd led him badly astray.

Recall that Zarathustra came down from the mountain to proclaim the advent of Superman after learning of the death of God. He reveals what this means in terms of Nietzsche's [385, p.216] own personal development in the following passage:

> There is an old illusion—it is called Good and Evil. Around soothsayers and astrologers hath hitherto revolved the orbit of this illusion. Once did one *believe* in soothsayers and astrologers; and *therefore* did

one believe, "Everything is fate: thou shalt, for thou must!" Then
again did one distrust all soothsayers and astrologers, and *therefore*
did one believe, "Everything is freedom: thou canst, for thou willest!"

Nietzsche strove against the herd mentality he so despised, but he could
not break the habit of thinking in absolutes. Having lost faith in the public
pundits who invent Absolute Moralities for the herd to follow, he proposed
instead that each person should become his own private pundit to invent
an Absolute Morality for himself. Unlike Kant, he faced up to the hard fact
that abandoning conventional morality leaves no place to seek inspiration
for this purpose but in one's own personal tastes. He therefore fell afoul of
Hume's law: an Ought cannot be deduced from an is—not even a personal
Ought that applies only to oneself (Section 3.2.4).

But the classic error that Nietzsche makes on behalf of all the other
thinkers who contrive to muddle the distinction between relativism and subjectivism is to confuse moral values and personal tastes. Morality evolved
to regulate *interactive* human behavior. All the members of a particular
society need to be playing the *same* morality game for its rules to serve any
purpose. What would be the point of morality if it differed not only from
one *society* to another, but also from one *individual* to another?

Teaching children right and wrong. Nietzsche was used as a case
study to explain the uniformly bad press that moral relativism enjoys. I
think these negative attitudes are based on a simple logical error: people like
Nietzsche are moral relativists—therefore moral relativists must be people
like Nietzsche is thought to have been. We relativists are therefore told
we believe that the Will to Power of pedophiles justifies their disregarding
public disapproval in pursuing their claim that prepubic sex is good for
children. That we see nothing wrong in keeping slaves or beating our wives,
since both activities have been endorsed as morally sound by past societies.
More absurdly still, we are charged with believing that it makes no sense
to teach children the difference between right and wrong.

But scientific moral relativists do *not* claim that morality is just a matter
of subjective personal taste. On the contrary, a relativist argues that the
current moral values of a society are *objective* features of its culture. Like
anyone else, a relativist will also have his own personal views about how he
would like society to change, but this subjectivist hat is kept firmly locked
away in a cupboard when he argues that it is an objective fact that moral
codes are different in different societies. Like all other important scientific
facts, this truth can usefully be taught to children. But this is far from
saying that children should not simultaneously be taught the necessity of
observing the standards of right and wrong that hold *in their own society*.

On the contrary, the facts of history suggest that a society falls apart when it so loses faith in its own social contract that children cease to be given clear guidance on how it works.

Affirming that there have been societies in which it was not considered wrong to keep slaves or beat your wife does *not* entail teaching children the lie that we are therefore entitled to treat other human beings like property in a modern democracy. It is a matter of historical record that the moral absolutists of Plato's circle thought that making love to underage boys was entirely admirable. Presumably they thought it always would be. But moral relativists insist that it is also an objective fact that pedophilia is no longer acceptable in any of today's societies—and that we therefore have a duty to punish those who abuse children whenever we can catch them.

To echo the theme of Section 3.2.5, nothing dreadful will happen if we admit to ourselves that our notions of right and wrong evolve along with our societies. Our social contract certainly won't collapse if we stop pretending that the rules we actually follow in coordinating on an equilibrium are the absurdly impractical absolutist precepts to which we are trained to offer lip service. On the contrary, matching our words to our deeds can surely only have a *stabilizing* effect. Insofar as a danger exists, it lies in the possibility that an underclass may learn to use one or other of the absolutist fairy tales peddled by fundamentalists as a coordinating device in bringing its collective power to bear.

Nor does the truth obstruct any of the rational roads to reform. The enemies of progress are the absolutists, who cannot comprehend that our current standards of right and wrong might be improved. They are not perhaps so silly as the modern Sir Roger de Coverley[86] who recently assured me that the convenience of having British clocks show the same time as French and German clocks counts as nothing against the iniquity of abandoning the one true time. But the certainties of absolutists are nevertheless based on nothing more solid than Sir Roger's apparent belief that God drew the Prime Meridian through the former Royal Observatory in Greenwich at the same time that he fashioned the globe.[87]

---

[86]The Whig essayists, Addison and Steele, used this fictional country squire in their *Spectator* articles to poke fun at the antiquated attitudes of the Tory backwoodsmen of their day.

[87]History sometimes provides a fossil record of the evolution of memes. No classical author ever mentions "natural rights" or "free will". Along with "romantic love", these ideas were invented in medieval times (Macintyre [337]). Greenwich Mean Time dates from Maskelyne's period as fifth Astronomer Royal, when he used the Royal Greenwich Observatory as his reference location in preparing issues of the *Nautical Almanac* between 1765 and 1811.

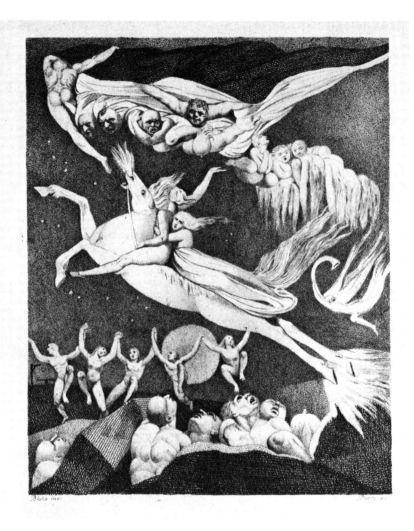

The second-best, bourgeois utopia defended in this chapter
would not appeal to William Blake. But neither does an
eternity of singing the praises of the Lord appeal to me.

Chapter 4

Yearning for Utopia

> Let the State suffer from coughing and short breath, the world from vice, let them be barbarous as they are, let them tyrannize, epicurize, oppress, luxuriate, consume themselves with factions, superstitions, law-suits, wars and contentions, live in riot, poverty, want, misery, rebel, wallow as so many swine in their own dung with Ulysses' companions, let them be fools, since that's their wish. I will yet make a Utopia of my own, a new Atlantis, a poetical Commonwealth ...
>
> Robert Burton's *Anatomy of Melancholy*

4.1 Introduction

Harsanyi's [233] utilitarian theory of the Good was reconstructed in Chapter 2 as a theory of the Seemly. Within such a reconstruction, the absolute Good of traditional moral philosophy is reduced to a relative good—a human artifact invented to describe the workings of a fairness norm that evolved from primitive food-sharing conventions. However, the resulting theory incorporates a major flaw since the players are assumed to be *committed* to the terms of the utilitarian social contract studied. But where is the omnipotent philosopher-king needed to enforce such commitments?

Abandoning the notion of a philosopher-king raises the questions treated in traditional theories of the Right. Chapter 3 reinterprets such theories in terms of the self-policing rules that players must honor to sustain an equilibrium in a repeated Game of Life. In such a decapitalized theory of the seemly, a player has a right to take any action not forbidden by duty, his duty being to remain on the equilibrium path specified by the social contract his society operates.

This final chapter offers a theory of the seemly within which no opposition is seen between what is right and what is good. The theory insists that all officers of the state are treated as being part of society so that no external enforcement agency is possible. Ideas of the right are therefore necessary to describe how equilibria are *sustained*. Ideas of the good are required to determine how equilibria are *selected*.

In accordance with Rawls' [417] undertaking to reduce the Good to the Right, rights are therefore logically prior to the good, for the mundane reason that one must always determine what is feasible before asking what is optimal. When Harsanyi's [233] methodology is applied within such a theory of the seemly, it no longer leads to a utilitarian conclusion. We are led instead to a second-best, bourgeois utopia characterized by a version of the Rawlsian maximin criterion. My theory can therefore be seen as a synthesis of the approaches of Rawls [417] and Harsanyi [233] that uses Harsanyi's methodology without his commitment assumptions to obtain a Rawlsian conclusion.

In terms of the philosophical ancestries illustrated with a diagram in the Reading Guide, theories of the Good are represented by the road to utilitarianism labeled Route 2 and associated with the names of Bentham, Mill and Harsanyi. Theories of the Right are represented by the road to egalitarianism labeled Route 1 and associated with the names of Locke, Rousseau, Kant, and Rawls. Route 3 diverts Route 1 to utilitarianism by replacing Rawls' analysis of rational bargaining in the original position by Harsanyi's. Chapter 2 diverts this road even further to Route 4 by relocating its origins in the ideas of Hobbes and Hume. It points the way to a utilitarian theory of the Seemly. This chapter presents an alternative theory of the seemly that dispenses with commitment assumptions. This theory is represented by the road to egalitarianism labeled Route 5.

4.2   Envy

Another source of inspiration needs to be recognized in addition to Harsanyi and Rawls. This is the work on envy-freeness that originates with Foley [183] and Kolm [309]. To appreciate this idea and to understand how my approach differs from the conventional line taken by economists, it is necessary to ask some questions about the nature of envy. In particular, what is the relation between envy and fairness?

Sore thumbs. Nobody thinks about the central heating when the room temperature is comfortable. People start looking for the thermostat only when things get too hot or cold. Matters are similar with the moral rules that sustain equilibria in the Game of Life. When they are operating smoothly, we tend not to notice them at all. Our moral program then runs below the level of consciousness, like our internal routines for driving cars or tying shoelaces. But such programs stick out like a sore thumb when things go wrong and they are somehow invoked in circumstances for which they are not adapted.

Section 2.3.3 of Volume I uses the baby jackdaw studied by Konrad Lorenz as an example of such sore thumb phenomena, but the mention of shoelaces brings to mind another example. The male bower bird has an impressive ability to tie elaborate knots when constructing a bower to impress a prospective mate. But the system of rules he uses for this purpose only becomes apparent when he somehow loses the piece of fiber being tied in the middle of the process. Although the activity no longer serves any useful purpose, the bird nevertheless continues to perform all the steps of the knot-tying program with an empty beak.

I believe that folk notions of vice and virtue are largely based on observing moral rules being invoked in similar sore thumb situations. It does not occur to us to congratulate our fellows on their virtue or to complain about their vices when the social contract is working well. Only when unusual circumstances occur for which the current social contract is not adapted does the opportunity arise to sanctify those rare individuals who nevertheless persist in operating its moral rules.

To argue that a sick society becomes dysfunctional *because* vice triumphs over virtue is therefore to reverse the causal chain. It is true that a public preoccupation with vice and virtue is symptomatic of an unplanned shift from one social contract to another, but it is not a flight from virtue that creates such a shift. What happens is that the shift creates a multitude of sore thumb situations in which conventional morality demands that people should hold fast to the old virtues—but mostly they don't. The inadequacy of traditional notions of vice and virtue then becomes manifest. Eventually, a new view of the nature of vice and virtue evolves, thereby reducing the number of sore thumb situations. After the change, these new virtues and vices are said to have been valid in perpetuity.

Only when the historical perspective is sufficiently long are such shifts from one social contract to another readily perceived. In commenting on the decline of Rome, MacMullen [339] observes that certain forms of extortion first became customary, and then enforceable as legitimate rights. In the interim, those rare folk who refused to offer bribes played the part of the

sore thumb. But eventually everybody learned to think of the bribes as
fees it was only right and proper to pay (Section 2.3.3 of Volume I).

The foule sinne of envye. The remedy for a sick society therefore does
not lie in demanding a return to the virtues of the past. When was such a
campaign ever successful? Nor am I convinced that the vices denounced in
such campaigns are always dysfunctional. Indeed, like de Mandeville [341],
I think there is a case to be made for referring to the seven deadly virtues
rather than the seven deadly sins. In particular, anger, pride, and envy
would seem to be manifestations of human emotional capacities that are
not only healthy when they are triggered in the circumstances for which
they evolved, but are essential ingredients for human sociality.

Envy gets a particularly bad press.[1] As Schroek [468] documents, envy
is universally condemned as an unconditional bad. To join Schroek in
questioning this view is not to deny that envy takes a pathological form
in some societies. According to anthropologists, innovation in societies like
the Navajo is almost impossible, because any attempt by an individual to
better himself is bitterly resented by his neighbors. In some particularly
squalid societies, individuals apparently eat alone and at night lest they be
forced to share the fruits of their labors with others. Nor is it in question
that the appearance of envy in sore thumb situations usually signals that
everybody is likely to end up worse off. But if envy in a society were entirely
dysfunctional, how could our capacity for envy have evolved?

I believe that Chaucer's[2] "foule sinne of envye" is a dysfunctional variant
of a human capacity that has a strongly positive side. When manifested in
its functional form, it is the engine that allows us to operate fairness norms.
As Schroek [468] puts it:

> Those feelings that are so vitally important to a political order—
> namely, the sense of equity, of justice and injustice—are inherent in
> man because of his capacity for envy.

For Adam to envy Eve, he must be able to compare his lot with hers. He
must have the capacity to feel that it is better to be Eve eating an apple
than Adam wearing a fig leaf. A precondition for envy is therefore the
ability to hold an empathetic preference. Section 2.5.4 argues that we are
endowed with such empathetic preferences because we need them to serve

---

[1] I consider that envy is an oral-sadistic and anal-sadistic expression of destructive
impulses, operating from the beginning of life, and that it has a constitutional basis—
Klein [306, p.176].

[2] After Pride, wol I speken of the foule sinne of Envye, which that is, as by the word
of the philosophre, sorwe of oother mennes prosperitee—*Parson's Tale.*

as inputs when using fairness norms, for which the device of the original position serves as a stylized representative. In turn, fairness norms serve as equilibrium selection devices that can take a society quickly to a Pareto-improving equilibrium when some new source of surplus becomes available. One might therefore say that we have the capacity for envy because it helps a healthy society to operate efficiently.

Compassion and pity. Envy is a degenerate form of the healthy impulse to seek redress when unjustly treated. But do we need to rely on the clamor of the downtrodden to ensure that deviations from a fair social contract are corrected? Why not give equal weight to the magnanimous impulses of those who find themselves unjustly favored? Is not compassion or pity as important as envy? In responding to such questions, it is worth beginning by pointing out that compassion, pity and magnanimity are not necessarily different names for the same thing.

Section 1.2.2 of Volume I discusses the pros and cons of replacing *homo economicus* as our model of man with some version of *homo ethicus*. It notes that one does not need to step outside the *homo economicus* paradigm in order to accommodate most of the evidence that *homo sapiens* shares some of the virtues of *homo ethicus*. In particular, one can build a capacity for *compassion* into *homo economicus* simply by assigning him a utility function whose arguments include the well-being of his fellows. Section 2.5.2 argues that we actually do sympathize in this way with our family and friends to a considerable extent. We also sympathize with strangers to some small degree, but the extent of our sympathy diminishes with distance and our ability to identify with their circumstances. It would therefore be most unwise to depend on such compassionate feelings to glue together a large modern state. Compassion makes social life more pleasant, but society would continue in its absence. Even Kant [298] agrees that the "problem of organizing a state, however hard it may seem, can be solved even for a race of devils if only they are intelligent."

Aristotle's [17] *Rhetoric* is usually translated as saying that three conditions are required for *pity* to arise in the human heart. The first is that the misfortune suffered by those pitied must be serious. The second is that their misfortune must be no fault of their own. The third is that we must be vulnerable to the same type of misfortune ourselves.

I agree with Nussbaum [394] that Aristotle could not have had in mind the notion of pity that is current in modern times, but his third condition seems to me also to rule out compassion as an alternative translation. Rather than describing the heartache we sometimes feel at the plight of others, he seems rather to be describing how we feel about someone to whom we owe a debt under the terms of a fair social contract. I think it more appropriate to refer to the empathetic feelings aroused under such circumstances as *magnanimity*. Notice that I therefore place magnanimity in the domain of *empathetic* preferences and compassion in the domain of *sympathetic* preferences. With my definitions, a person has an internal motive to act out of compassion, but needs external pressure from his peers to act magnanimously.

In its modern usage, pity differs from both compassion and magnanimity. We think it only our due to be dealt with generously by those who have unfairly benefited at our expense. It is bearable to be the object of compassion. But who can tolerate being pitied?

The reason is that pity is frequently used to *exclude* people from the class of those with whom we regard ourselves as sharing a social contract. We pity drunks and down-and-outs, schizos and dumb animals, without making much of an attempt to put ourselves in their position with a view to learning how to cope with the problems they have to face (Section 2.5.4). We see their problems as irrelevant to *our* class or creed. Instead of offering them an inclusive legup, we therefore offer them an exclusive handout.

Westerners who find out often express surprise that the thirdworld recipients of their charitable largesse are frequently bitter at being on the receiving end. But this only goes to emphasize the failure of Westerners to empathize with the objects of their pity. After all, how would you feel if you had to turn up each day at my back door to beg for some gruel?

Charity is therefore far from being the greatest of the virtues. It is true that our charitable donations do succeed in relieving physical suffering. But they also serve to separate human beings into distinct groups—those who give and those who have no choice but to receive. Faith and hope may be for all, but charity is exclusively for the gentry.

These sour remarks should not be taken to mean that I advocate ceasing to donate the small fraction of our incomes we currently earmark for charity. Personally, I sympathize with the sick and hungry in far-off places more than enough to make me willing to spend a small proportion of my income on their relief.[3] But we shall never succeed in expanding the circle within which justice operates if we allow our charitable impulses to distance ourselves from those we help.

As Section 2.5.4 argues, a just society needs to be underpinned by a common culture, in which everybody is prepared to learn from everybody else. Charitable impulses based on pity obstruct this aim. Compassion is more promising, but it is futile to suggest that human nature can somehow

---

[3]Only a small proportion because, as everybody but saints and unworldly teenagers knows, charity begins at home.

be adjusted so that we come to sympathize with strangers as we sympathize with our kin. On the other hand, it is not futile to teach our children that they have much to gain from learning to empathize with as wide a group as possible,[4] so that we can widen the domain within which both our envious and our magnanimous impulses can usefully operate to maintain a just social contract. But I fear that envy will always play more urgently on our heartstrings than magnanimity.

4.3 Equity in Economics

The idea that humans have a capacity for envy because it can promote efficiency in a healthy society is completely at variance with a line thought among economists which holds that equity and efficiency mix like oil and water. Economics textbooks taking this line also emphasize the supposed impossibility of making interpersonal comparisons of utility. They are then able to define the problem of equity away by identifying social optimality with Pareto-efficiency (Section 1.8.1).

It somehow seems to pass unnoticed that this line of thought is incompatible with the foundations of the equally respectable subject of welfare economics, within which the comparability of the utility functions used to measure individual well-being is often so taken for granted that authors see no reason to offer any account of how such comparisons are made. Their lack of interest in the foundations of their subject is presumably explained by the fact that ideas from welfare economics would become impossibly difficult to apply in practice if one tried to measure costs and benefits in units more sophisticated than the dollar. But, as observed in Section 2.3, to measure the social value of a reform in terms of how willing people are to pay for it is to ignore that it is patently unfair to take a dollar out of the pocket of a poor man in order to put it into the pocket of a rich man.

Fortunately, the schizophrenia implicit in the coexistence of these two lines of thought is not shared by all economic theorists. Both traditions have spawned numerous deviant offshoots that take the problem of equity more seriously than their parent lines. Two of these deviant approaches to equity are discussed here, one from each tradition. Aspects of both will be apparent when my own theory is unveiled in Section 4.6.

---

[4]Exclusive schooling is obviously counterproductive in this regard. Rousseau's [441] *Emile* must surely be one of the most foolish and dishonest philosophical works ever written. (Rousseau cynically abandoned all his children at birth to the tender mercies of orphanages.) But he was right about the enormous importance of allowing children to learn *at first hand* about the way those born to less fortunate families must live.

4.3.1   Envy-Freeness

Economists who hold fast to the view that interpersonal comparisons of utility are impossible or meaningless have sought to define equity in terms of what people possess. But how is one to compare the different commodity bundles owned by different people?

Rawls [417] postulates a commonly accepted index by means of which the primary goods that constitute his abstract commodities are to be measured on a common scale. As for concrete goods like apples and fig leaves, he argues that they have an "exchange value" and hence can be aggregated into a single primary good, which he calls wealth and income (Section 1.2.5 of Volume I). Economists are impatient with such an approach. They see no point in referring to intangibles such as "the powers and prerogatives of office" or "the social basis of self-respect" as goods. Their instinct is to treat what cannot be weighed or measured as though it does not exist. Nor do economic theorists see any reason why the rates at which tangible goods are exchanged in a market should be thought relevant in deciding what is equitable. In making fairness judgments, one cannot escape the reality that different people value different commodities differently.

Foley [183], Kolm [309], Baumol [49], Gardenfors [194], Varian [531, 532], and others have recognized the role that envy plays in the operation of fairness norms by proposing that an allocation of goods be deemed equitable if nobody would prefer to swap his assigned commodity bundle for a bundle held by someone else. More precisely, an allocation (c_A, c_E) that assigns the bundle c_A to Adam and the bundle c_E to Eve is said to be *envy-free* if $c_A \succeq_A c_E$ and $c_E \succeq_E c_A$.

Divide-and-choose.   As an example, consider how a cake containing an uneven distribution of nuts and raisins might be divided between two people who have different preferences about nuts and raisins. The classic divide-and-choose method is of very ancient vintage.[5] Adam divides the cake into two pieces and Eve chooses one of them.

One strategy open to Adam is to cut the cake so that he is indifferent between the two parts. Whichever piece of cake Eve chooses, he will then not envy her. If she chooses the piece that she prefers, neither will she envy him. An algorithm therefore exists which implements an envy-free outcome

---

[5]In one of Aesop's fables, an ass, a fox, and a lion have to divide the spoils of a hunt. The ass splits the kill into three equal piles and invites the lion to make the first choice. The lion responds by eating the ass. The fox then offers the lion a choice between a very large pile and a very small pile. However, Aesop is a comparative latecomer. According to Brams and Taylor [100], Hesiod tells a story in which Zeus is the divider and Prometheus the chooser!

independently of the players' preferences.

Of course, Adam can do better for himself if he knows Eve's preferences. He can then divide the cake into two pieces, A and B, between which *Eve* is indifferent. For example, if Adam cares only about nuts and Eve cares only about raisins, he can divide the cake into two pieces that each contain an equal number of raisins.[6] This can be done in many ways, but Adam should choose the method that maximizes the number of nuts in A. By transferring a tiny slice of A to B, Adam can then assure that Eve will choose B. He then ends up with A, which he prefers to B. Eve ends up with B, which she prefers to A. Although Adam is now better off then before, the result therefore remains envy-free.

The fascinating history of such divide-and-choose problems recorded in the recent book of Brams and Taylor [100] is long and not yet over. But economists have naturally been more interested in the possibility of implementing envy-free allocations using the market mechanism.

The market. Section 1.2.5 explains how a simple problem of pure exchange can be represented as an Edgeworth box $\mathcal{E}$. In the problem represented in Figure 4.1(a), Adam and Eve have a total of A apples and F fig leaves to divide between them. Recall that each point t in $\mathcal{E}$ represents a possible trade between Adam and Eve in which Adam receives (a, f) and Eve receives $(A - a, F - f)$. Assuming that Adam begins with all the apples and Eve begins with all the fig leaves, the no-trade or endowment point is located at e. The diagram shows two indifference curves, $u_A(a, f) = c$ and $u_E(A - a, F - f) = d$, drawn through the point t. The set C consists of all trades that Adam likes at least as much as t. The set D shows the set of trades that Eve likes at least as much as t.

The most important idea in economics is that of a *Walrasian equilibrium* in a perfectly competitive market. To introduce this idea, imagine Adam and Eve trade through a middleman or auctioneer. He begins by announcing prices p and q at which he undertakes to trade apples and fig leaves for money. Adam and Eve then exchange their endowments for the scrip that he issues as money.

After the exchange, Adam has pA units of scrip with which he can buy back any commodity bundle (a, f) satisfying $pa + qf \leq pA$. His budget set B_A therefore consists of all points in $\mathcal{E}$ that lie beneath his budget line $pa + qf = pA$. Adam's optimal choice r from his feasible set B_A is illustrated in Figure 4.1(b).

---

[6]If the cake could be deconstructed altogether, the problem could be resolved without difficulty by assigning Adam all the nuts and Eve all the raisins. But it is assumed that the normal courtesies of cake-cutting are to be observed.

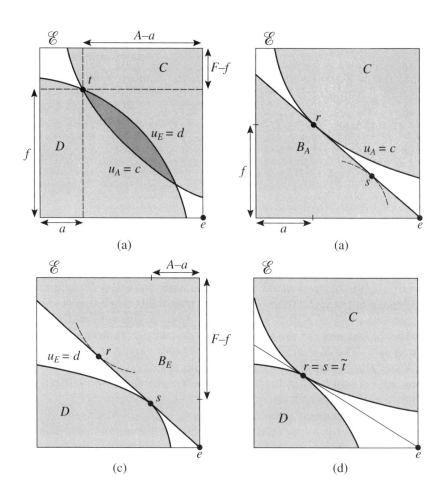

Figure 4.1: The Edgeworth box.

Similarly, Eve has qF units of scrip with which she can buy back any commodity bundle $(A - a, F - f)$ satisfying $p(A - a) + q(F - f) \leq qF$. Her budget set B_E therefore consists of all points in $\mathcal{E}$ that lie above her budget line—which coincides with Adam's budget line in the diagrams because $pa + qf = pA$ is the same equation as $p(A - a) + q(F - f) = qF$. Eve's optimal choice s from her feasible set B_E is illustrated in Figure 4.1(c).

This train of events may leave the auctioneer with a problem. He has undertaken to redeem the scrip he has issued with apples or fig leaves at prices p and q. But what is he to do if someone demands more apples or fig leaves than he has in stock? Such a shortfall will occur whenever $r \neq s$ in Figures 4.1(b) and (c). To avoid getting into a mess, the auctioneer therefore needs to choose prices $\tilde{p}$ and $\tilde{q}$ that *clear the market* by ensuring that $r = s = \tilde{t}$, as shown in Figure 4.1(d). The resulting allocation of commodities, together with the prices that support this allocation, are said to be a Walrasian equilibrium.

The invisible hand. The Walrasian auctioneer of the preceding story is a formal embodiment of the *invisible hand* that moves prices to their market-clearing values in Adam Smith's [501] seminal *Wealth of Nations*. Of course, neither Adam Smith [501] nor Walras [545] are asking us to believe that some official, whether visible or not, actually sets the prices at the Chicago wheat market. The auctioneer is a *deus ex machina* introduced to model the end-product of the dynamic process by means of which prices actually adjust to disparities of supply and demand in the real world. If the demand for apples exceeds the supply of apples at the price at which they are currently trading, then people selling apples will be able to raise the price. If the supply of fig leaves exceeds the demand for fig leaves, then people selling fig leaves will have to lower the price. Prices will therefore adjust until the supply and demand of both commodities is equalized. A perfectly competitive market therefore does not need an auctioneer to set prices to clear the market. Indeed, the great virtue of a market system is that markets *clear themselves* under perfect competition.

Price-takers What conditions need to be satisfied to make a market perfectly competitive? The most important requirement is that all traders act as *price-takers* in equilibrium. Each trader takes it as given that exchange is mediated through a price system and that his own actions cannot affect the prices at which goods are traded. A perfectly competitive market is therefore to be contrasted with a monopolized market in which the monopolist is a *price-maker*, because she can choose the price at which to trade her product. Our story of Adam and Eve therefore fits rather uncomfortably within the perfect competition paradigm, since Adam initially monopolizes the supply of apples and Eve the supply of fig leaves. Economists therefore refer to their plight as a problem in *bilateral monopoly*.

In a bilateral monopoly problem, there is no reason why the protagonists should use the price system to mediate their trade. The natural approach is to cease to frame their problem in commodity space and to transform it into a Nash bargaining problem (X, ξ), as described in Section 1.2.5.

Figure 4.2(a) is a version of Figure 1.4(c) which has been engineered to ensure that the resulting bargaining problem (X, ξ) is symmetric. The use of the symmetric Nash bargaining solution then yields a symmetric utility pair σ corresponding to the trade t in the Edgeworth box $\mathcal{E}$, where both players share the available apples and fig leaves equally. This is to be contrasted with the Walrasian allocation w, where the distribution of goods is highly asymmetric.

One way of obtaining the Walrasian allocation w as the equilibrium outcome of a bargaining game played between Adam and Eve, requires that the players be constrained to communicate in terms of *prices*. To be more precise, suppose that Adam can only announce the maximum quantity of apples he is willing to trade, and the minimum number of fig leaves he is willing to take for each apple he exchanges. Eve simultaneously announces the maximum quantity of fig leaves she is willing to trade, and the minimum number of apples she is willing to take for each fig leaf she exchanges. Trade then takes place at the maximum level consistent with the two announcements.

Adam's announcement confines the possible trades to the set C shown in Figure 4.2(b). In particular, it fixes the best price Eve can get for her fig leaves in terms of apples. Figure 4.2(b) shows that her optimal trade in the set C is t. She can secure t by ensuring that her own announcement confines the possible trades to the set D. Thus D is a best reply to C. However, for a Nash equilibrium, we also require that C be a best reply to D. Such a Nash equilibrium is shown in Figure 4.2(c). As the diagram shows, if the resulting trade w is Pareto-efficient, then it is Walrasian.

Replicating the players. How might Adam and Eve come to be price-takers? The standard defense requires that the market contain a very large number of economic agents, each of whom is small compared with all his fellows. Each agent can then realistically proceed on the assumption that his own actions have a negligible effect on the market aggregates that determine the rates at which goods are traded.

Edgeworth [169] ingeniously showed how to illustrate this idea without leaving the confines of the Edgeworth box. As Shubik [490] pointed out,

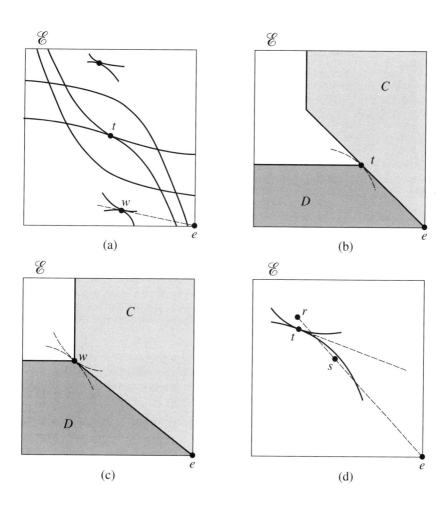

Figure 4.2: Bargaining in the Edgeworth box.

Edgeworth's approach amounts to computing the core (Section 0.5.1) of the large trading game obtained by replicating Adam and Eve many times. Any non-Walrasian allocation can be shown to lie outside the core of the game if a sufficiently large number of copies of Adam and Eve are available (Debreu and Scarf [153], Hildenbrand [247]). Aumann [23] obtains similar results with the core replaced by the Shapley value (Section 2.3.1).

Figure 4.2(d) illustrates the crucial case of a non-Walrasian allocation on the contract curve (Section 1.2.5) that assigns the same bundle $t = (a, f)$ to each Adam and the same bundle $(A - a, F - f)$ to each Eve. Observe that each Adam would rather trade at $r = (b, g)$ than t, and each Eve would rather trade at $s = (c, h)$ than t (because she prefers $(A - c, F - h)$ to $(A - a, F - f)$). The point s has been chosen so that integers M and N can be found satisfying $(M + N)s = Mr + Ne$, where e is the endowment point. The point s must then lie on the straight-line segment joining e and r, which is possible if and only if t is not Walrasian. (To see this, begin with $r = t$, and then move r slightly.)

The equation $(M + N)s = Mr + Ne$ ensures that a coalition of $M + N$ Adams and M Eves can redistribute their endowments so that each Adam in the coalition gets (b, g) and each Eve in the coalition gets $(A - c, F - h)$.[7] Each Adam in the coalition therefore trades at r and each Eve trades at s. It follows that they all prefer the result of the redistribution to trading at t. In the terminology of Section 0.5.1, they will therefore block[8] the allocation t, and so t lies outside the core.

Section 0.5.1 suggests that economists defending the core would be more critical of the myopic behavior they attribute to players if they were not so pleased with the conclusion to which the reasoning leads when employed in market games. Personally, I see no particular reason why N pairs of Adams and Eves should not coalesce into N separate coalitions, each of which trades at the Nash bargaining solution so that the final outcome in Figure 4.2(a) is that everybody gets $(\frac{1}{2}A, \frac{1}{2}F)$.

Equity via the market?   History provides little support for those who think that fully planned economies administered by armies of bureaucrats perform better than economies in which markets are used wherever they

---

[7]The vector equation $(M + N)s = Mr + Ne$ reduces to the two scalar equations: $(M + N)c = Mb + NA$ and $(M + N)h = Mg$. Rewritten as $(M + N)c + M(A - b) = (M + N)A$ and $(M + N)h + M(F - g) = MF$, these equations say that the coalition as a whole is endowed with just the right number of apples and fig leaves for the redistribution to be possible.

[8]To block an outcome t, a coalition S must be able to find an alternative outcome u that all members of the coalition prefer to t and that they can jointly enforce independently of what players excluded from the coalition may do.

work well. But the price paid for not intervening at all in the operation of the market is that any concern for equity must be surrendered. Figure 4.2(a) illustrates this point very effectively. Although the Edgeworth box is entirely symmetric, the Walrasian equilibrium assigns nearly all the fig leaves to Eve, while splitting the stock of apples fairly evenly.

However, we are never faced with the stark choice of either planning everything or nothing. We can and do plan some aspects of our economy, leaving others to be determined by the market. Even though no bureaucrat may have any idea what Adam and Eve's preferences are, it is still possible to implement an envy-free but Pareto-efficient outcome in a pure exchange economy by a judicious combination of planning and the market mechanism.[9] In an economy with a large number of Adams and Eves, all that is necessary is for the bureaucrats to redistribute the initial endowments until everybody has the same bundle $(\frac{1}{2}A, \frac{1}{2}F)$, and then allow trade to proceed to a Walrasian equilibrium. The progressive taxation operated in all modern states already goes some way down this road.

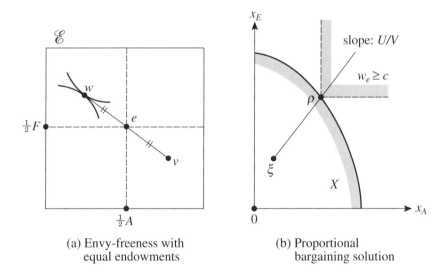

(a) Envy-freeness with
equal endowments

(b) Proportional
bargaining solution

Figure 4.3: Welfarist and postwelfarist concepts.

Figure 4.3(a) shows the new endowment point $e = (\frac{1}{2}A, \frac{1}{2}F)$ reached after the bureaucrats have redistributed the intitial endowments until ev-

---

[9]In certain markets, Varian [533] shows that all envy-free, Pareto-efficient allocations can be obtained in this way. Thomson [521] offers a more general characterization of the equal-income, Walrasian correspondence.

erybody has the same amount of every good. The Walrasian allocation obtained with e as the endowment point is located at $w = (a, f)$. Why is w envy-free? If an Adam and an Eve were to swap their Walrasian allocations, they would move from the trade w to the trade v, but Figure 4.3(a) shows that both prefer w to v.

What's wrong with envy-freeness? The standard example used in criticizing the concept of envy-freeness assigns Adam and Eve each a bottle of gin and a bottle of vermouth for which they have no use except to make martinis according to a formula from which they never vary. Adam is an unsophisticated soul who shakes together equal measures of gin and vermouth to make a martini. He can therefore make two bottles of martinis with his endowment. Eve is more discerning and tolerates martinis only if made by diluting straight gin with no more than one drop of vermouth. She can therefore manage to make only a little more than one bottle of martinis with her endowment.

The initial endowment is envy-free. It is also Pareto-efficient. But is it sensible to call it fair? Why are we measuring Adam and Eve's situation in terms of the number of bottles of gin and vermouth in their possession, when all they really care about are martinis? Perhaps we should assign $\frac{2}{3}$ of a bottle of gin to Adam and $\frac{4}{3}$ to Eve. Each will then be able to make the same number of martinis.

Arguments against such a resolution of the fairness problem are not hard to find. One might argue that Eve deserves to be penalized because her tastes are more expensive to satisfy, but presumably nobody would persist with this argument if she needed the martini for urgent medical purposes. Or one could argue that Adam should be penalized because his undiscriminating attitude to the mixing of martinis demonstrates that he doesn't appreciate them as much as Eve.

Such arguments show that it is no more satisfactory to assess equity in terms of bottles of martinis than to assess it in terms of bottles of gin and vermouth. When Eve considers whether to envy Adam, she doesn't only compare her worldly goods with his. She compares his whole situation with hers. Even if she is poor and he is rich, she will not envy him if he is suffering from incurable clinical depression. She literally wouldn't swap places with him for a million dollars. When Eve compares her lot with Adam, it is not enough for her to imagine how it would be to have his possessions and *her* preferences. She must imagine how it would be to have his possessions and *his* preferences.

In brief, although authors who emphasize envy-freeness have surely put their fingers on a vital insight in recognizing the close connection between our vulnerability to envy and our thirst for justice, I think their definition of envy is inadequate. To determine whether Eve has grounds for envying Adam, we need to take account of her empathetic preferences. The need to make interpersonal comparisons of utility therefore cannot be evaded.

4.3.2 Welfarism

According to Sen [481], welfarism is the approach to social choice that pays no attention to anything but the utilities that citizens assign to the available social alternatives. Welfarists therefore have no choice but to face up to the problem of interpersonal comparison of utility from the outset.

Unfortunately, the issue has been confused in recent years by a movement led by Scanlon [462], Sen [479, 481], Dworkin [167], Cohen [129] and others, who argue that welfarism is an inadequate foundation on which to build a theory of fairness. Sometimes it is argued that subjective criteria like individual preferences have no place at all in such a theory, and that only objective criteria—like who owns how much of what—should be considered. Moulin [367] refers to such theories as *postwelfarist*. As in the case of envy-freeness, postwelfarist theories usually involve some mixture of subjective and objective considerations. In Cohen's [129] notion of "midfare", the mixing of subjective and objective criteria is explicit.

For the reasons outlined in Section 3.8.1, I believe that the movement away from welfarism is retrograde. Apart from their incipient ipsedixism,[10] postwelfarists ignore the fact that modern utility theory was invented *because* attempts to encompass human aspirations in terms of *a priori* objective criteria have been universally unsuccessful.

I agree with postwelfarists that the Victorian ideas on utility held by Bentham and Mill (Section 2.3) are inadequate for making fairness judgments. If hedonistic dials could somehow be wired into Adam and Eve's pleasure centers, justice would not be achieved merely by ensuring that both dials gave the same reading. Among many other objections to such a naive approach to justice, there is the point made by Scanlon [462] that some people have "champagne tastes" which are costly to satisfy, and that such costs need to be taken into account along with the benefits people enjoy when considering how social decisions are made.[11]

Such criticisms miss the target when directed at versions of welfarism

---

[10]Bentham [53] defines an ipsedixist to be someone who proposes his own prejudices as moral imperatives for others to follow (Section 2.2.2). Such prejudices are all too apparent in the lists that moral philosophers compile when seeking to characterize the good life. But, as Bentham [53] observes, who are we to urge our preference for poetry on those who prefer push-pin?

[11]We are to be held responsible for our tastes and punished or rewarded accordingly.

based on modern theories of utility. Economists who understand the foundations of their subject really do mean what they say when they insist that their conception of utility merely provides a summary of consistent behavior (Section 3.8.1). It is therefore incoherent to attack modern welfarism on the grounds that the utilities attributed to the players in the game do not adequately reflect their motivations. According to the modern view, if their utilities did not adequately reflect their motivations, they would be the wrong utilities. In particular, the charge that utility theory necessarily pays attention only to the benefits that players receive while neglecting the costs that gratifying these benefits may impose on other players seems very strange to a game theorist. Modern utility theory was invented as a theoretical tool precisely to ensure that nothing that matters when decisions are made is left out of account.

One would have thought that a more appropriate criticism from an objectivist would be to complain that utility theory takes *too much* into account for it to be useful as a guide to practical decision making. But the appropriate response to the latter very reasonable criticism is not to invent a bad theory which pretends we know how to characterize all the fears and aspirations people nurse in their bosoms in terms of a simple set of universally valid, objective criteria. The appropriate response is to continue to work with a good theory, but to recognize that each application requires a new search for objective criteria that adequately distinguish people's underlying personal tastes from their beliefs about the social values of their society. Such objective criteria will usually depend very strongly on the *context* in which practical problems arise.

 Cooperative game theory. Traditional cooperative game theory (Section 0.5) is entirely welfarist in its assumptions. Its methodology consists of proposing axioms that the solution to a particular problem should satisfy, and then characterizing the set of solutions that satisfy these axioms.

The types of problems considered are usually split into two broad categories: those in which utility is assumed to be *transferable* between the players (TU problems), and those in which it is not (NTU problems). In Section 4.6 of Volume I and elsewhere, I make the uncontroversial point that Von Neumann and Morgenstern's [538] approach to transferable utility is fraught with difficulty. No attempt is therefore made in this book to provide any foundations for the concept. A more useful separation into categories is obtained by distinguishing between problems where utility is assumed to be *comparable* between players, and those where it is not.

Chapter 1 concentrated on the Nash!bargaining solution Nash and Kalai-Smorodinsky bargaining solutions, whose axiomatic characterizations specifically exclude the possibility of comparing Adam and Eve's utils. According to the Nash program (Section 0.6), one can assess the predictive power of such cooperative solution concepts by evaluating the realism of the noncooperative bargaining games whose equilibrium outcomes coincide with the concept under study. Chapter 1 found in favor of the Nash bargaining solution on the basis of this criterion.

Chapter 1 also briefly considered two cooperative solution concepts that do depend on the possibility of comparing utils across individuals: the utilitarian bargaining solution and the proportional bargaining solution (Section 1.3.3). Chapters 2 and 4 are attempts to apply the Nash program to such representatives from the class of cooperative solution concepts that depend on making interpersonal comparisons of utility. Just as noncooperative bargaining games were used in Chapter 1 to model the *process* by which a deal is reached, so Chapters 2 and 4 use the device of the original position to model the *process* by which fairness judgments are made.

A weighted version of the utilitarian bargaining solution was defended in this manner in Chapter 2, using the fiction of a philosopher-king. When this fiction is abandoned, the current chapter shows that the same argument leads to a weighted version of the proportional bargaining solution.

The proportional bargaining solution. Figure 4.3(b) illustrates the proportional bargaining solution ρ with weights U and V for the bargaining problem (X, ξ). It is sometimes called the egalitarian bargaining solution, on the grounds that all players receive an equal increment on their *status quo* payoff after their utils have been suitably weighted. As Section 1.3.3 explains, the idea is closely connected to the Rawlsian maximin criterion. If X is strictly comprehensive, the proportional bargaining solution ρ is the point x in X at which the Rawlsian social welfare function

$$W_\rho(x) = \min\left\{U(x_A - \xi_A), V(x_E - \xi_E)\right\} \tag{4.1}$$

is maximized subject to the constraint $x \geq \xi$. This social welfare function is said to be Rawlsian because it requires applying the maximin criterion to Adam and Eve's utilities after these have been normalized to be zero at the *status quo* ξ and then weighted by the constants U and V to reflect how the players compare each other's utils.

The proportional bargaining solution has been extensively studied by Raiffa [413], Isbell [281], Kalai [288], Myerson [373], Roth [438], Peters [403] and others. The reason that it has been the focus of so much attention is that incorporating the assumption of full interpersonal comparison of utility into the axioms that characterize a cooperative solution concept leaves little room for any rivals.[12]   Since a serious competitor is absent, it is not particularly surprising that the proportional bargaining solution should emerge from an attempt to apply an analogue of the Nash program to fairness problems. However, as in the case when the Nash program was applied in defending the Nash bargaining solution in Chapter 1, there is much to be learned about *how* the solution should be applied. In particular, what determines the location of ξ? How are U and V to be found?

The reason the assumption of full interpersonal comparison of utility leaves little room for maneuver in formulating cooperative bargaining solutions is easily explained. The assumption allows Adam and Eve to break the surplus to be divided into bite-size pieces, each of which seems worth the same as the last to both players. For example, if each piece is worth one util to Adam, then it is worth U/V utils to Eve. They can then treat their problem as a succession of bite-sized problems much more easily than was possible in Section 1.4.3. Assuming that each new bite-size piece is split in the same way independently of the growth in each player's utility level, then the sum of all the bite-size pieces that are split will be divided in the same proportion as each separate piece. A proportional bargaining solution will then have been implemented.

Kalai's [288] characterization of the proportional bargaining solution captures this intuition using a set of axioms. In addition to the standard requirement that ρ belong to the weak bargaining set[13] B_0 for the problem (X, ξ), he requires three further axioms:

• Independence of Utility Levels;

• Independence of Common Rescaling of Units;

• Decomposability.

The first and second of these axioms need to to be compared with the axiom of Independence of Utility Calibration used in Sections 1.4.1 and 1.4.2 in characterizing the Nash and Kalai-Smorodinsky bargaining solutions. In Section 1.4.1, the function $R : \mathbb{R}^2 \to \mathbb{R}^2$ defined by

$$R(x_A, x_E) = (ax_A + c, bx_E + d)$$

is used to recalibrate Adam and Eve's utility scales. Such a recalibration transforms the bargaining problem from (X, ξ) to $(Y, \eta) = (R(X), R(\xi))$.

---

[12]The utilitarian bargaining solution is not a rival, since one can alter the zeros on Adam and Eve's utility scales without changing its value. It therefore does not involve full interpersonal comparison.

[13]Recall from Section 1.3 that we cannot insist that ρ lies in the regular bargaining set B, as with the other bargaining solutions considered in this book, because ρ may be only *weakly* Pareto-efficient when X is not strictly comprehensive.

The axiom of Independence of Utility Calibration demands that the recalibration have no impact on the bargaining solution $\sigma = f(X, \xi)$ beyond requiring that it be rewritten in terms of the new utils like everything else, so that $f(Y, \eta) = R(\sigma)$.

The axiom of Independence of Utility Levels replaces the function R by the function $R_0 : \mathbb{R}^2 \to \mathbb{R}^2$ defined by $R_0(x_A, x_E) = (x_A + c, x_E + d)$. This shifts the zeros on Adam and Eve's utility scales but leaves the units in which their utils are measured fixed. To impose the requirement that $f(Y, \eta) = R_0(\sigma)$ is therefore to insist that Adam and Eve's wealth levels are irrelevant to the manner in which their bargaining problem is resolved.

The axiom of Independence of Common Rescaling of Units need only be applied to the bargaining problem $(X, 0)$. It then replaces the function R by the function $R_1 : \mathbb{R}^2 \to \mathbb{R}^2$ defined by $R_1(x_A, x_E) = (ax_A, ax_E)$. This function leaves the *status quo* point 0 fixed, but rescales the units on both Adam and Eve's utility scales by the *same* factor $a > 0$. To replace R by R_1 is to recognize that it matters whether each of Adam's utils is worth half or twice as much as each of Eve's utils. With R_1, we preserve such distinctions by always rescaling both player's utils simultaneously.

The Decomposability axiom requires that the bargaining can be broken down into steps in any way whatsoever without affecting the final outcome. Thus, if $\sigma = f(X, \xi)$ and $\tau = f(Y, \xi)$, then $\sigma = f(X, \tau)$ whenever Y is a subset of X.

To prove that Kalai's axioms characterize the proportional bargaining solution is very easy. Fix a half-space Z as in Figure 4.4(a), and let Y be any subset of Z. The heart of the proof is to confirm that $\eta = f(Y, 0)$ lies on the straight line joining 0 and $\zeta = f(Z, 0)$.[14] As illustrated in Figure 4.4(b), choose R_0 to map η onto 0. Then choose R_1 to map $R_0(Z)$ back to Z. Since R_1 maps 0 to itself, it follows that we have mapped the bargaining problem (Z, η) onto the bargaining problem $(Z, 0)$. Decomposability implies that $\zeta = f(Z, \eta)$. The other axioms therefore imply that $R_1(R_0(\zeta)) = f(Z, 0) = \zeta$. A constant a therefore exists for which $a(\zeta - \eta) = \zeta$. Hence η is on the line joining 0 and ζ.

Kalai's [288] axioms have been criticized as being overly strong. In particular, why should we assume Decomposability? However, Peters [403, p.65] has shown that we have little choice since, in the presence of the other assumptions to be made, Decomposability follows from the much weaker requirement that the bargaining solution never assigns a player less when the problem $(X, 0)$ is replaced by $(X + Y, 0)$.

---

[14]It is then easy to deduce that $f(X, 0)$ lies on the same straight line for all X containing 0. One then appeals to the axiom of Independence of Utility Levels to deduce that $f(X, \xi)$ lies on the line of the same slope passing through ξ.

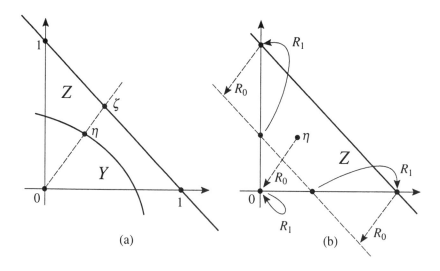

Figure 4.4: Characterizing the proportional solution.

Efficiency and equity. The theory that efficiency and equity are necessarily opposing principles is kept alive by the type of story that Nozick [392] tells about Wilt Chamberlain in Section 1.2.8 of Volume I. It is true that people sometimes so focus on the question of how a cake is to be divided that they forget that it may be worse for everybody to divide a small cake equally than a large cake unequally, but such inefficiencies are not somehow inevitable when fairness criteria are applied. On the contrary, I claim that fairness norms evolved precisely because it is important in intergroup competition that a group not allow internal dissension to obstruct the opportunity to exploit a new resource to the full (Section 2.5.4). When fairness criteria are employed in the manner for which they evolved, they are used to select among the set of *efficient* equilibria. Efficiency therefore takes priority over equity. In particular, the assumption of weak Pareto-efficiency is built into all the bargaining solutions of this book.

4.4   Equity in Psychology

In social psychology, equity theory is centered around an empirically based law which predicts that problems of social exchange will be resolved by equalizing the ratio of each person's gain to his worth. People who are deemed worthy therefore get more of the gravy than others. As in Wilson

[561], the theory is usually called "modern" equity theory, although it originates with Aristotle[15] [18], and has fallen somewhat out of favor since it was first introduced to social psychologists by Homans [260] and Adams [7, 8] more than thirty years ago.[16]

Written as an equation, the psychological theory of equity asserts that

$$\frac{g_A}{w_A} = \frac{g_E}{w_E}, \qquad (4.2)$$

where g_A and g_E are the respective gains to Adam and Eve, and w_A and w_E quantify how worthy they are. But how are gains to be measured? Where is the zero to be located on whatever scale is chosen? How is worthiness to be construed? Is it measured in terms of social status, merit, effort, need, or what?

Early psychological papers assume that effort is the prime determinant of worthiness. For example, Homans [260] argues that a fair division of the proceeds of a cooperative enterprise will make each partner's share proportional to his investment. However, critics of the theory are able to produce experimental evidence suggesting that social considerations other than relative effort are more relevant to fairness judgments in some contexts.[17] Need is thought to be particularly important.

The equity law has therefore been joined by a menu of other laws proposed as better descriptions of the behavior of subjects in social contexts with a less obvious economic focus than the original experiments. Other authors have simultaneously complicated the simple algebra of the original equity law in an attempt to get a better fit to the data in contexts where the law does work reasonably well. The result is a literature that begins to resemble traditional cooperative game theory (Section 0.5) in offering a plethora of descriptive criteria without any underlying principles that explain why experimental subjects should think it appropriate to apply different fairness norms in different contexts.

Defenders of equity theory, notably Walster *et al* [546, 547], have argued that the experimental results of their critics can be accommodated within the framework of the orthodox law of equity by broadening the sense in which worthiness is to be understood. For example, one can explain the

---

[15]What is just ... is what is proportional—*Nicemachean Ethics*.

[16]See, for example, Adams and Freedman [9] Austin and Hatfield [29], Austin and Walster [30], Baron [44], Cohen and Greenberg [131], Furby [188], Mellers [354], Mellers and Baron [355], Messick and Cook [356], Pritchard [410], Wagstaff *et al* [541, 544, 543], Walster *et al* [546, 547, 548]. Wagstaff [542] has a user-friendly book in draft that sets the philosophical scene, and reviews the history and current status of modern equity theory. Selten [472] provides an account of the theory which is easily accessible to economists.

[17]For example, Deutsch [159], Kayser *et al* [303], Lerner [320, 321], Reis [422], Sampson [456], Schwartz [469].

fact that experimental subjects sometimes take account of need by arguing that the worthiness coefficients in the equity law are determined by the whole social context and not just by one particular social dimension. But critics like Schwartz [469, p.132] have no time for what they see as a piece of "sleight of hand" that empties the equity law of any substantive content. However, equity theorists can reasonably respond that one does not gain much understanding of human fairness norms by inventing an entirely new descriptive law each time a context is found in which none of the existing laws seem to fit the data.

My own approach offers a refuge to both sides in the debate. As explained in Chapter 2, I believe that fairness evolved in the human species because the ability to share food efficiently has survival value. More speculatively, Section 2.5.4 argues that the mechanism we use for this purpose is written into our genes. If so, then the universal appeal of Rawls' original position can be attributed to its providing a stylized model of a mechanism that we unconsciously use in solving everyday social coordination problems. However, to use the original position, Adam and Eve need a common standard for making interpersonal comparisons of utility. Outside the family context, my guess is that this standard is not biologically hardwired, but is determined instead by social or cultural evolution. Since the worthiness coefficients w_A and w_E determine the relevant standard for interpersonal comparison in the law of equity, my approach therefore provides a possible means for determining how they may vary with the social context.

Chapter 2 offers a prototheory that describes how standards of interpersonal comparison should be expected to adjust in the medium run when a new source of surplus changes the underlying Game of Life. In principle, my theory provides a function that maps the social context onto the worthiness coefficients that apply once Adam and Eve have become familiar with the context. The criticism that equity theory becomes irrefutable if the worthiness coefficients do not have a fixed value then loses its sting. On the contrary, the simplistic predictions offered in Section 4.7 about how w_A and w_E should be expected to vary with need, effort, ability, and social standing will doubtless be refuted only too readily given the current primitive state of my theory.

I therefore agree with those equity theorists who think it is a mistake to attribute differences in fairness judgments in different social contexts to differences in the underlying social mechanism. On the contrary, I believe that the underlying social mechanism is universal, but that the inputs required to operate the mechanism are vulnerable to social evolution, and hence vary with the context. However, in my theory the underlying mechanism is not the psychological law of equity but the device of the original position. Nor is it guaranteed that the use of the original position will lead to the same result as the law of equity. Indeed, Section 2.6 shows that Adam and Eve would agree that the weighted *utilitarian* bargaining solution were fair if they knew that some external enforcement mechanism were available to police the deal that they hypothetically make behind the veil of ignorance.

Since utilitarians and egalitarians regard each other with mutual horror, further support for equity theory from my approach may therefore seem unlikely. But despair would be premature, since the assumption that everyday fairness transactions are monitored by a godlike philosopher-king is far from realistic. In reviewing the anthropological evidence, the next section observes that public opinion can sometimes serve as a substitute for an external enforcement agency, but that transactions between individuals in prehistoric hunter-gatherer communities must largely have been *self-policing.* Section 4.6 therefore considers the implications of using the device of the original position when the players have no opportunity whatever to make commitments. We are then no longer led to the weighted utilitarian bargaining solution, but to the egalitarian or proportional bargaining solution introduced in Section 1.3.3.

The appearance of the proportional bargaining solution is significant because it can be seen as a missing link between Rawls' [417] widely endorsed maximin criterion of justice and the psychological law of equity. Only the simplest of algebraic manipulations are necessary to recognize that these apparently diverse concepts are essentially the same.

The fact that one gets the proportional bargaining solution by maximizing a Rawlsian social welfare function has been noted several times already, most recently in Section 4.3. To see that the law of equity is equivalent to the proportional bargaining solution is even easier. One need only compare the form of the equity law given in (4.2) and the equation

$$U(x_A - \xi_A) = V(x_E - \xi_E)$$

describing the line on which the proportional bargaining solution lies in Figure 1.7(b). To identify the two equations, simply write $g_A = x_A - \xi_A$, $g_E = x_E - \xi_E$, $w_A = 1/U$ and $w_E = 1/V$. The worthiness coefficients w_A and w_E of social psychology are then inversely proportional to the weights U and V that economists use to quantify how Adam and Eve's utils are compared.

It is encouraging that the social contexts mimicked in the psychological experiments that support equity theory are broadly consistent with the assumptions used in deriving the proportional bargaining solution from the original position in Section 4.6. However, the theoretical support that my approach offers to equity theory is subject to various caveats, of which three are mentioned here:

- Not only may the worthiness coefficients vary with a number of social parameters in a manner that need not be easy to guess, they may also vary with time as the subjects adjust to a novel social situation.

- It should not be assumed that objective correlates of gain are easy to identify. Even when money can be used to measure individual welfare, it does not follow that it will serve as a correlate of personal utility in small group situations, especially when sex or kinship enter the picture explicitly (Section 4.5.4). Nor is it likely that the notional units of esteem that Homans [260] introduces when discussing social status will often find an easily measurable equivalent (Section 4.7.6).

- The location of a *status quo* will be problematic in many contexts. Particular difficulties are raised by attempts to apply the equity law with negative gains, or to determine fair punishments (Section 3.7.3).

Such caveats may mollify the critics of psychological equity theory to some extent, but my guess is that their hostility is more deeply rooted. If one is convinced that morality exists to *transcend* the struggle for power, it is hard to suspend disbelief when faced with a theory that makes the underlying balance of power fundamental to what a society counts as fair. But if the morality of everyday life required people to take actions that made them less fit, how could it have evolved? Philosophers can afford to laugh at such arguments because their discipline does not force them to confront the readiness with which experimental subjects adjust their perceptions of what is fair to their circumstances. But one of the achievements of modern social psychology has been to expose the gap that lies between how we like to think of ourselves and the reality of what we actually are.

In Milgram's [357] well-known experiments, an authority figure was able to induce subjects to inflict apparently serious electric shocks on a victim in order to punish him for performing a task badly. Later work in the same vein is directly relevant to equity theory.[18] In a typical example, Lerner and Simmons [323] required subjects to view supposedly live pictures of an innocent victim suffering painful electric shocks for some considerable time. When powerless to intervene, subjects placed in situations of this kind comfort themselves when debriefed by inventing reasons why the victim is unworthy in some respect or other. Lerner [319] argues that subjects denigrate the victim in order to sustain the false belief that we live in a "just world", but I think that the subjects are simply demonstrating that the justice of the "just world" in which we actually live isn't the same as the idealized justice of moral philosophy.

---

[18]For example, Berscheid *et al* [59], Glass [206], Legant and Mettee [316], Lerner *et al* [317, 318, 322, 323].

I hope that the unwelcome predictions offered in Section 4.7 on how need and effort are likely to influence the worthiness coefficients in the equity law will be received in the same spirit. Nobody denies that only a second-best utopia would allow power to affect distribution so meanly. Even egalitarians share the utilitarian yearning for a first-best utopia in which suckers are all given an even break (Section 4.6.9). But realism requires that we take the same attitude to how society actually works as Oliver Cromwell took to his appearance. He too would have liked to look better, but he nevertheless insisted on being painted "warts and all".

4.5 Equity in Anthropology

Naturalists think that traditional disputes over the Good or the Right lack any firm foundation. We believe that the ethical imperatives which philosophers claim to derive from universal metaphysical principles are actually intuitions about the workings of whatever conventions are regarded as seemly in the society with which they are familiar. It is not by accident that Locke's system justifies the commercial social contract of the England of his time. Nor that Hegel's system does the same for the authoritarian social contract of Prussia.

Examples of such relativism are more striking when they do not force us to confront our own cultural taboos. The anthropologist Fürer-Haimendorf [189] therefore compares the horror with which Don Carlo expresses his love for his stepmother in Verdi's opera with the matter of fact assumption in Tibet that the son of a widower might sleep with his father's new wife.[19] If universal moral principles are to be found, naturalists believe that the place to look for them is in our shared evolutionary history. In particular, if egalitarian social contracts are more stable than utilitarian alternatives, it is because the conditions under which biological and cultural coevolution operated in the formative years of our species were hostile to the survival of utilitarian fairness norms.

A naturalist comparing egalitarianism and utilitarianism therefore asks empirical questions. But there are good reasons why it is not so easy to tell whether a given social contract is egalitarian or utilitarian. Section 4.6.7 argues that the utilitarian and proportional bargaining solutions will be identical once a society's common standard for comparing utils has adjusted

---

[19]But it should be noted that anthropologists usually softpedal their moral relativism. Hobhouse [254, 252, 253] even had absolutist tendencies. Westermarck [549], Edal and Edal [168], Ladd [312], and Fürer-Haimendorf are among those who discuss morality explicitly, but the modern fashion seems to be to sideline the moral implications of anthropological studies.

to the underlying realities of power. To distinguish utilitarianism from egalitarianism in an operating society, it is necessary to look at its *short-run* response to a new coordination problem, since the two systems then reward worthiness in opposite ways (Section 4.6.9). But whatever else my theory may have to say, it certainly does not suggest that evaluating the data derived from setting people new coordination problems is at all trivial.

An alternative approach is to examine the evolutionary pressures that shaped the fairness norms of our prehistoric ancestors directly. Within the framework of my theory, the key issue is whether an implicit insurance deal between Adam and Eve could be enforced by the rest of their community, or whether the only viable deals were those they could enforce themselves. In the former case, Section 2.6 argues that any biological hardwiring that we use when operating a fairness norm will have a utilitarian bias. In the latter case, Section 4.6 argues that the bias will be egalitarian.

So how did our prehistoric ancestors live? Evolutionary psychologists have suggested that inspiration is best sought in the social contracts of the few remaining hunter-gatherer societies that have survived into modern times. I agree that anthropology has much to teach us about the variety of social contracts that are possible for the human species, but it will emerge that I have grave doubts about the extent to which we can usefully extrapolate from modern foraging societies to those of prehistory.

4.5.1   Sharing and Caring

The consensus is very strong among anthropologists that the remaining hunter-gatherers of the world operate societies that share and care for each other without bosses or social distinctions.[20]

I must confess to some suspicion of the more politically correct of such reports. As in Axelrod's [31, 33] work (Section 3.3.7), the need to believe that people are intrinsically nice seems very strong. For example, Megarry [352, p.230] counters the theory that men are inherently violent because they are descended from hunters by demanding that we "rigorously reject" the idea that hunting as practiced by humans represents a continuation of our prehuman way of life. He says, "Among primates, life is usually contained within a troop but feeding remains a solitary activity." But everybody has seen video footage of West African chimpanzees collectively

---

[20]Bailey [37], Damas [141], Erdal and Whiten [175], Evans-Pritchard [176], Fürer-Haimendorf [189], Gardner [195], Hawkes *at al* [239], Helm [244], Isaac [280], Kaplan and Hill [301], Knauft [308], Lee [315], Riches [424], Tanaka [517], Megarry [352], Meggitt [353], Rogers [436], Sahlins [455], and Turnbull [526]. Usually, modern foraging societies are said to be "egalitarian", but I prefer not to use this word in a sense that would include a utilitarian society.

hunting monkeys and then dividing and eating the carcass together. We have also read Darwin on the squalid lives of the natives he encountered in Tierra del Fuego. We know too much about infanticide, headhunting, cannibalism and the like to be taken in by the more utopian claims made for the noble savage. But even after making allowances for the rosy colored spectacles of some observers, one is left with an overwhelming mass of evidence suggesting that the sharing of food, especially meat, is universal in hunter-gatherer societies. Nor is there reason to doubt that foraging societies get along without vesting authority in chiefs or bosses.

Why share food? In Section 2.5.4, I followed the traditional line of attributing the evolutionary origins of the food-sharing phenomenon to the need for individuals to insure each other against privation. As the anthropologist Evans-Pritchard [176, p.85] explains:

> The habit of share and share alike is easily understandable in a community where everyone is likely to find himself in difficulties from time to time, for it is scarcity and not sufficiency that makes people generous, since everybody is then insured against hunger. He who is in need today receives help from him who may be in need tomorrow.

At least three criticisms of this explanation of food-sharing need to be mentioned. The first is that prehistoric hominids are unlikely to have been provided with the "Machiavellian intelligence" necessary to sustain such insurance contracts (Byrne and Whiten [112]). This piece of jargon expresses the familiar claim that to explain a piece of behavior in terms of rational self-interest is to assert that it was carefully planned in advance by a coldly calculating intellect. Economists commonly disclaim such straw men by pointing out that a person riding a motorcycle is implicitly solving a very difficult mathematical control problem, but nobody would think to deduce that Hells Angels must therefore be master mathematicians. As Evans-Pritchard explains, people in hunter-gatherer societies acquire the *habit* of sharing—and this habit survives because it coordinates behavior on an equilibrium of the Game of Life without anyone even needing to be aware that a game is being played.

The second criticism disputes the suggestion that hunter-gatherer societies commonly live on the edge of extinction. Sahlins [455] observes to the contrary that modern hunter-gatherer societies have a relatively affluent lifestyle if one compares the amount of leisure they enjoy after meeting their needs with that of a modern agricultural laborer or a university professor.

Although they are not foragers, it seems to me that there is much to be learned on this subject from the conflicting reports on the Ik of Uganda.

Turnbull's [527] notorious account of their supposedly ultra-selfish society was contradicted by the later observations of Heine [243]. Presumably, Heine observed the Ik in a fat year and Turnbull in an unusually lean year. To reconcile Evans-Pritchard and Sahlins, it is similarly only necessary to observe that every year cannot be a fat year, and it is in the lean years that the invisible hand of evolution strikes down unfit groups. Nor can foragers always rely on bringing home the bacon even in the fattest years, so that there will always be good reasons for sharing meat on a reciprocal basis. Nor should we assume that the natural methods of birth control with which modern hunter-gatherers help to adjust their populations to the food supply preceded the evolution of the food-sharing phenomenon. My guess is that population control is a relatively recent adaptation to the marginal territories currently occupied by hunter-gatherers. But, without controls of some kind, the iron law of Malthus [340] would soon turn plenty into scarcity. How else does one explain the spread of hunter-gatherer societies over the whole world, even to the most inhospitable of environments?

The third criticism challenges Evans-Pritchard's appeal to reciprocity as an explanatory factor in food sharing. For example, Erdal and Whiten [175] conclude that their survey of more than a hundred studies demonstrates that the sharing of food observed

> goes beyond the explanatory power of either kinship or reciprocation. Individuals do sometimes attempt to obtain a disproportionate share of resources or influence for themselves, but this is contained through vigilance and counter-dominant behaviour by their group members.

But what is the second sentence about if not a social contract in which everybody looks after everybody else because those who don't are punished by their fellows? It is true that the mechanism that supports the reciprocal arrangement is not one of the simple models of bilateral exchange that people usually have in mind when they refer to TIT-FOR-TAT. But, as we saw in Section 3.4, the punishment strategies that support efficient equilibria in repeated games do not necessarily require that the player injured by a deviant also be the person who punishes the deviation. In the case of modern hunter-gatherer societies, the whole band combines to punish anyone who fails to cooperate in operating the scheme of mutual insurance by which it succeeds in surviving when times are bad.

To a layman like myself, a disturbing feature of the anthropological literature lies in the differences between modern assessments of hunter-gatherer societies and those of fifty years ago. It is easy to accuse the earlier authors of prejudice, but it must surely also be important to recognize that they were able to study hunter-gatherer societies inhabiting much less hostile and constrictive environments. The opportunity for groups to resolve disagreements by breaking apart and expanding into new territory must have been very significant in shaping the prehistoric social contracts that existed before population pressures began to bite, and so reports on the foraging societies of the forties that still occupied unsaturated forest habitats are of particular importance from an evolutionary viewpoint.

The older anthropologists continue to describe sharing societies without bosses, but the sharp discontinuity that separates modern accounts of the loving kindness of hunter-gatherers like the bushmen of the Kalahari desert from the hysterical ferocity of the Yanomamö of the Brazilian jungle or the bleak inwardness of the Sironi indians of Eastern Bolivia is no longer so evident.[21] For example, Fürer-Haimendorf's [189] study of moral attitudes and social organization in South Asian communities presents a seamless continuum through the foraging Chenchu, the slash-and-burn Reddis and the headhunting Nagas to the conformity of the orthodox Hindu peasantry. Violence between individuals increases as property rights are extended with the advance of agriculture but then diminishes as authority is concentrated in the hands of individuals who are able to coordinate communal reprisals. At productive levels that render property largely irrelevant, individuals or groups who feel unfairly treated can deal with the situation by putting distance between themselves and their enemies, but it is significant that the administration of justice remains decentralized long after groups have begun to till the soil and keep animals. According to Fürer-Haimendorf [189], punishment of those who treat others unjustly is not seen the responsibility of everybody living in a village, even among the relatively sophisticated Daflas.

I think it important to consider the recent speculations of evolutionary psychologists on ancient and modern hunter-gatherer societies at length, if only to reassure authors like Boehm [94] or Erdal and Whiten [175] that the anthropological data is explicable without the need to postulate the evolution of true altruism in the human species. However, Section 4.5.3 returns to my contention that the social contracts of modern hunter-gathering societies are not necessarily a good guide to the social contracts of our prehistoric ancestors.

4.5.2 Enforcement in Foraging Societies

Both Hobbes and Hume repeatedly insist that the mighty are powerful only because the people over whom they rule believe in their power. Even the haughtiest of autocrats is therefore no less a player in the Game of Life

---

[21]Wiessner [552], Chagnon [119], Holmberg [259].

than the most abject of his subjects. If his regime does not operate an equilibrium in the Game of Life, it will not survive.[22]

Knauft [308] argues that the evolution of authority in human societies can be seen in terms of a U-shaped curve, in which dominance-structured prehuman societies give way to anarchic bands of human hunter-gatherers that are then replaced by the authoritarian herding and agricultural societies with which recorded history begins. As Erdal and Whiten [175] document, the evidence is strong that leadership in modern hunter-gatherer societies lies only in influencing the consensus: "But when a consensus has been reached, no-one has to follow it against their will—there is no enforcement mechanism."

My naturalistic theory of fairness depends heavily on how one models the enforcement of fairness norms. As Chapter 2 shows, if the hypothetical deal reached in the original position is enforced by a philosopher-king standing outside the system, then the outcome is utilitarian. However, this chapter argues instead that the fairness norms we actually use in our daily lives operate as though no enforcement at all were available. At first sight, the apparently anarchic structure of modern hunter-gatherer societies would seem to support the idea that enforcement structures were indeed absent when the fairness algorithms I believe to be biologically determined were evolving. However, one has to be careful not to put the cart before the horse. One cannot argue that food-sharing is the key to human sociality and simultaneously proceed as though humans were already living in organized communities in the style of modern hunter-gatherers after the fairness norms governing the sharing of food evolved.

Nor does the fact that modern hunter-gatherers operate social mechanisms which prevent potentially authoritarian leaders from getting established imply that their societies do not enforce norms. On the contrary, the evidence is that the social contract operated by a hunter-gatherer community is enforced with a rod of iron. No individual occupies the role of philosopher-king, but the relatively small size of a hunter-gatherer band makes it possible for *public opinion* to fulfill the same function. When Adam asks himself whether he should offer some of his meat to Eve, he

---

[22]To insist that all viable regimes should ultimately be modeled as equilibria in the Game of Life is not to deny that the manner in which authority is exercised in some societies is so stably established that it makes sense when studying a subsociety to treat the government as lying outside the system—as with the fiction of the philosopher-king used in Chapter 2 to model the modern consensus on political legitimacy. However, when the very ground of political legitimacy is in question, it is necessary to revert to the strict line that all regimes, whether anarchic or authoritarian, are merely examples of social contracts that survive because deviations from the conduct they prescribe are unprofitable. In the discussion of different types of society that follows, the question is therefore never *whether* an equilibrium is being operated, but *what kind* of equilibrium.

knows very well that he will be relentlessly mocked and ridiculed by the band as a whole should he fail to share in the customary fashion. Full-scale ostracism would follow if he nevertheless persisted in behaving unfairly.

Reports that modern hunter-gatherer communities share on a quasi-utilitarian basis are consistent with the view that public opinion serves as a substitute for a philosopher-king in such societies. But it is hard to share the enthusiasm expressed by some anthropologists for the oppressive social mechanisms by which discipline is maintained. Envy is endemic. For example, among the !Kung[23] of the Kalahari desert, nobody cares to keep a particularly fine tool for too long. It is passed along to someone else as a gift lest the owner be thought to be getting above himself. But such gifts do not come without strings. In due course, a fair return will be expected. As observed in Section 4.2, the close attention to the accountancy of envy in such a social contract makes progress almost impossible. According to Hayek's [240, p.153] definition (Section 2.2.3), the citizens of such a society are free because they are subject to no man's will, but it would be a bad mistake for libertarians to idolize such societies, which would better serve as a role model for the socialist utopia that Marx envisaged would emerge after the state had withered away.

I therefore diverge from evolutionary psychologists like Erdal and Whiten [175] who believe that the social contracts of prehistoric hunter-gatherers are preserved in fossilized form by the foraging bands of today. I don't doubt that prehistoric bands were equally free of bosses, but I think it un-likely they operated a form of social contract that seems to me at least as sophisticated as the authoritarian alternatives operated by ancient tillers of the soil. This claim is of considerable importance to my speculations about the circumstances under which fairness norms evolved, and so it will be necessary for me to defend it at some length.

Farming versus foraging. Cohen [130] attributes the origins of agricul-
ture to a food crisis in prehistory that arose when human hunter-gatherer bands had expanded until the locally available habitat was no longer able to support their economies. The response to this overpopulation problem was twofold. I am particularly interested in the adaptations that allowed foraging to continue in marginal habitats, but anthropologists naturally concentrate on what proved to be the mainstream cultural adaptation— the emergence of agriculture and herding as new modes of production.

The organization necessary both to exploit the increasing returns to scale available in these new modes of production and to prevent the surplus from being appropriated by outsiders made it necessary to abandon the

---

[23]I have no idea how to make the tongue click the exclamation mark denotes!

anarchic structure of prehistoric foraging bands. Instead, authority began to be vested in leaders. This readoption of the hierarchical organization typical of ape societies did not require a new set of biological adaptations. We did not lose our capacity to submit to leadership when we acquired the new program that permitted our protohuman ancestors the flexibility necessary to sustain the anarchic lifestyle of hunter-gatherers with a whole world into which to expand. Even in the uncompetitive ambience of a modern foraging society, our natural urge to dominate one another is not extinguished by our natural urge to be fair—otherwise social mechanisms that inhibit dominance behavior would not be necessary.

Anthropologists attribute the social retooling necessary for a return to the type of hierarchical social contract needed to maintain a communal farming society to *cultural* evolution. The time available seems too short for a further *biological* adaptation to have been responsible.

It has been argued that the human species paid a heavy price for the opportunity to become farmers. When social evolution erected an authoritarian superstructure on a biological foundation that had evolved to permit our ancestors to live a free-wheeling leaderless existence, a war began between part of our biological nature and our social conditioning. Commentators like Maryanski and Turner [343] believe that we are still fighting this war. In the language of game theory, their characterization of a modern industrial society as a *social cage* is expressed by saying that our habituated use of leadership as an equilibrium selection device conflicts with our natural instinct to employ fairness for this purpose.[24]

Speculative though it is, such a story about the origins of the farming communities from which our own industrial societies are descended seems relatively uncontroversial, but the same is not true of my belief that the social contracts of protohuman foraging bands were too unlike the complex social contracts of modern foragers to allow the analogy to be useful.

Recall Cohen's [130] suggestion that a prehistoric food crisis caused by overpopulation spelled the end of foraging as the normal productive mode among humans. But it did not wipe hunter-gatherers out altogether from those Old World territories where the problem arose. Foragers continue to survive in marginal habitats on the fringes of deserts or the polar ice-cap, where growing crops or herding animals is not feasible. Indeed, the fact that such habitats were colonized is one piece of evidence that favors the overpopulation theory.

---

[24]My guess is that we succeed in tolerating leaders by inventing the social fiction that they are responsible as individuals for the capabilities of the groups they coordinate. The worthiness that would be attributed to the group if it were a person is then conferred on its leader. His claim to more than his fair share is thereby rationalized away. But maintaining such a charade is endemically stressful.

To survive in such marginal habitats without the possibility of emigration, foragers had to develop a culture that was no less at variance with their natural instincts than those who took up the farming option. Three cultural adaptations universally observed in modern hunter-gatherer societies seem especially significant. The first is their use of "natural" methods of birth control—like the delayed weaning of children. This adaptation goes some way toward solving their population problem. The harshness of their environment when times are bad probably does the rest. The second cultural adaptation was the development of extremely effective social mechanisms that prevent the emergence of leaders or entrepreneurs—except perhaps temporarily in emergencies.

Why should mechanisms that inhibit leadership confer an evolutionary advantage? The reason is presumably that innovators are poison for foraging bands occupying marginal habitats. The survival of the memes that regulate the life of a hunter-gatherer society depends on the equilibrium on which its members coordinate in a bad year, when food is scarce. If the crisis is sufficiently severe, some members of the band die, and the rest seek refuge with neighboring bands. But such lean years are infrequent (Sahlins [455]). In the fat years that intervene, memories of the privations of the last lean year fade. The band is then at risk of being seduced by a charismatic entrepreneur into coordinating on a new equilibrium that does better at exploiting the surpluses available in fat years. Disaster then ensues if this new social contract is being operated when a lean year comes along.

In brief, the memes that inhibit the appearance of leaders, who are likely to tamper with a traditional social contract tailored to the conditions that prevail in lean years, serve as a kind of collective unconscious that preserves a folk memory of disasters narrowly avoided in the past. The stubborn conservatism of supposedly stupid peasants occupied in subsistence farming doubtless has a similar explanation.

The third cultural adaptation has already been mentioned. Public opinion can serve as a substitute for a philosopher-king in small close-knit communities. Section 2.5.4 then provides us with an explanation of how the use of the fairness algorithm I believe is written in our genes can result in quasi-utilitarian sharing of the type reported to be universal among modern hunter-gatherers.

Why *quasi*-utilitarian, rather than utilitarian? The reason is that the assumptions this book makes about the way players evaluate payoff flows ignores the precariousness of life in subsistence communities (Cashden [117]). If Adam and Eve live in such a society, they face the Gambler's Ruin problem that regularly shatters the pipe dreams of the inventers of surefire schemes for breaking the bank at Monte Carlo. The memes which control modern foragers cannot afford to see the gains from future fat years as providing adequate compensation for the losses endured in a lean year. To take such a long view is to ignore the very real possibility that the culture they embody may not survive to enjoy the bounty of the fat years to come.[25]

A meme that inhabits a hunter-gatherer head will therefore not compute its utility as a long-run arithmetic mean. For the kind of reasons given by Robson [431], it may perhaps evaluate payoff flows in terms of a suitably weighted long-run *geometric* mean, so that the impact of low payoffs in lean years has a much larger impact than high payoffs in fat years. The Adam and Eve of Section 2.6.1 would then cease to use a weighted utilitarian bargaining solution in the short run. Instead, they would use the weighted Nash bargaining solution with bargaining powers U and V in the short run!

4.5.3   Anarchy in Prehistory?

Figure 4.5 illustrates the speculations about the evolutionary history of modern social contracts offered in Section 4.5.2. Its most significant feature is that the sophisticated social contracts of contemporary hunter-gatherer societies appear on a twig on the branch of the tree that leads to societies like our own. To understand the origins of our instinct for justice, we therefore need to go back to the common ancestor of both types of social contract. But this does not mean that there is nothing to be learned from the social contracts of modern foraging bands. They show that human biological hardwiring allows us to operate social contracts without ape-like dominance hierarchies. Since modern hunter-gatherer societies manage without bosses, humans do not need bosses to survive as social animals.

Knauft's [308] U-shaped curve must therefore be correct to the extent that it embodies the claim that the imperative for authoritarianism was somehow cleansed from the genes of our prehuman ancestors, only to be revived in relatively recent history as a *cultural* adaptation to the need to domesticate plants and animals in response to population pressures. To understand the circumstances under which human fairness algorithms evolved, we must therefore return to a time after the biological imperative to organize in terms of dominance structures had weakened sufficiently to permit more flexible social systems to evolve, but before population

---

[25]In the case of subsistence-level farming communities, the point needs to be pressed even more firmly. Hunter-gatherers can ameliorate the risks they face in bad years to some extent by adapting their foraging techniques to the climate. But farmers have much less flexibility, since they have to choose what mix of crops to plant before they know how much rain will fall. A subsistence farming commnunity might therefore even use the maximin criterion in evaluating payoff flows. The analysis of Section 2.6.1 only begins to make sense when surpluses are sufficient to permit adequate reserves being held from year to year.

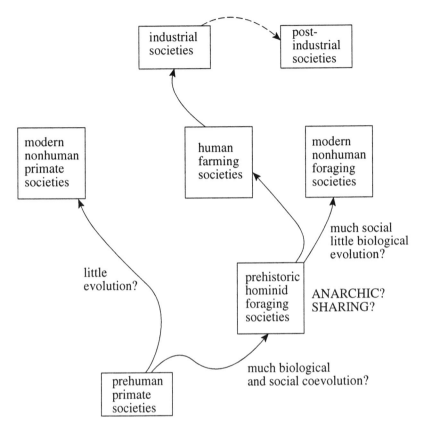

Figure 4.5: In what environment did fairness evolve?

pressures had led those societies that continued to forage to develop socially sophisticated methods of controlling both their population as a whole and their selfishly inclined entrepreneurs.

Such prehistoric foraging bands must have differed from their modern descendants in several important respects. In particular, barriers to emigration were absent when the whole world was available for colonization. Under such circumstances, their social organization must have been anarchic to an extent that would make modern hunter-gatherer societies look positively paternalistic. How could it have been otherwise when a dissident group always had the Lockean option of breaking away at relatively low cost to set up shop in pastures new? Neither public opinion nor personal authority can act as Big Brother when punishment can always be evaded simply by walking off.

I believe that the distress we all feel at the exercise of arbitrary authority derives from attitudes that were hardwired into our heads under such circumstances. Who has not felt the frustration that Nietzsche [385] so eloquently attributes to Zaruthustra?

> My brethren, will ye suffocate in the fumes of their maws and appetites? Better break the windows and jump into the open air!
> ... Empty still are many sites around which floateth the odor of tranquil seas. Open still remaineth a free life for great souls.

In 1888, there were indeed still wild and free places in America, but the tragedy of today is that nowhere is left to which we can escape. In this respect, our own overpopulation crisis is a replay of the prehistoric crisis that brought authoritarianism into the world.

With no external enforcement agency available to enforce food-sharing agreements, the Adams and Eves of prehistoric times presumably must have shared food much as Wilkinson [553] reports that vampire bats share blood.[26] Just as each bat in a reciprocating pair has to act as his own policeman in disciplining any tendency by his partner to cheat, so each Adam and Eve who learned to cooperate would have found it pointless to appeal to the rest of the band about any bad behavior by their partners. Cooperation was therefore originally based on *pairwise* interactions, as assumed by Axelrod [31] when propounding the virtues of TIT-FOR-TAT.[27]

Some support for the suggestion that pairwise interaction should form the basis of any attempt to understand how human cooperation evolved is to be found in anthropological studies from earlier in the century, when hunter-gatherer and other economically primitive communities still occupied relatively benign habitats. Fürer-Haimendorf's [189] account of Indian tribal societies in various developmental stages in the forties has been mentioned already. The groups he describes are perceptibly less well-behaved than the surviving hunter-gatherer communities cited by modern evolutionary psychologists. Information on how they punished infractions of their social contract is therefore considerably richer. It is particularly interesting to trace how the punishment function becomes concentrated in the hands

---

[26]Vampire bats starve if deprived of blood for more than 60 hours. Wilkinson [553, p.182] observes that close kinship is not necessary for a bat to regurtitate blood for a regular roostmate. Not only are bats able to recognize each other, the evidence shows that they are more likely to help out a neighbor who has helped them out in the past.

[27]But it must be remembered that TIT-FOR-TAT is only one of many reciprocating strategies that can support cooperation as an equilibrium in the indefinitely repeated Prisoners' Dilemma (Section 3.3.8). Nor must we forget that TIT-FOR-TAT is hopelessly inadequate as a description of the type of social interaction that becomes possible once a third player starts taking an interest in Adam and Eve's transactions (Section 3.4).

of increasingly less rustic philosopher-kings as societies become more economically productive. But political sophistication lags behind economic development. For example, although the hill-dwelling Daflas of the North East frontier had passed way beyond the subsistence stage and maintained an active trading economy, Fürer-Haimendorf [189, pp55-83] found that "justice by retaliation" was still the only resource, even for an Adam and Eve occupying immediately adjacent longhouses.

If I am right about the anarchic structure of prehuman hunter-gatherer societies, then the tale of Section 2.5.4 must be retold without a philosopher-king to keep everyone honest while the original position evolved. The evolutionary progression proposed in Figure 2.2 continues to apply, but recycling the arguments of Chapter 2 without the fiction of a philosopher-king leads to a society much more like our own than the quasi-utilitarian societies of modern hunter-gatherers (Section 4.6).

4.5.4 Kinship in Small Groups

In Section 2.5.2 and elsewhere, warnings are offered against proceeding as though the camaradie that enlivens the small groups within which we work and play together extends to the world at large. The daily life of such small groups awakens the memories buried in our genes of how to interact with others enjoying the same campfire or sleeping in the same longhouse. The brotherhood of man then manifests itself, because we are literally programmed to treat each other like brothers or sisters in such circumstances. But Dunbar [164] has plausibly argued that the size of a group within which it is possible for people to treat each other like family is limited by the capacity of the human neocortex to sustain a social model in which each person in the group and his relationships with others is modeled individually.

This theory about the social dynamics of small groups would seem to be undermined by Erdal and Whiten's [175] claim that kinship theories fail to explain how food is shared among modern hunter-gatherers. It is doubtless true that a player's share cannot be calculated from his family relationships using some simple formula. But it is hard to see how reports that food is shared according to *need* can be explained without assuming that the players actively sympathize with each other's plight, as predicted by Hamilton's rule (Section 2.5.2).

An example will be useful in illustrating the mechanism envisaged. The same example will then be used to compare the food-sharing norms of modern hunter-gatherers with the different food-sharing norms that my theory attributes to the more anarchic hunter-gatherers of prehistory.

Modern hunter-gatherers.   The apparatus that will be needed here was introduced in Section 2.4.5 to discuss bargaining between two players over insurance contracts behind a veil of uncertainty. But we cannot again keep things simple by assigning both players the same individual utility function, since it would then be impossible to study what happens when Adam is more in want than Eve. It will therefore be necessary to introduce some new individual utility functions for Adam and Eve, which are normalized so that $u_A(0) = u_E(0) = 0$ and $u_A(1) = u_E(1) = 1$. The new utility functions have a particularly simple form. Adam and Eve are assumed to be indifferent between obtaining all of the kill brought home by a hunter and receiving some smaller share σ of the kill. In Adam's case, $\sigma = \frac{2}{3}$. In Eve's case, $\sigma = \frac{5}{6}$. Finally, both Adam and Eve are assumed to be risk neutral about shares x that lie between 0 and σ.[28]

Need is a complex concept to be discussed at greater length in Section 4.7.3. Only one aspect of what it means to say that a person is in need will be relevant here—the extent to which he is in want. Adam and Eve's need to eat will be measured by the smallness of the largest share σ for which they are willing to take a risk of ending up with nothing. Their desperation is strengthened by the assumption that they are risk neutral over smaller shares (Section 4.8.2). Since $\frac{2}{3} + \frac{5}{6} > 1$, Adam and Eve's needs cannot be satisfied simultaneously, even though we shall assume that they are the only members of the band who are thought worthy of a share of this particular kill. To what extent will the norm they operate recognize that Adam's need is greater than Eve's?

If I am right about public opinion in a modern foraging band acting as a collective philosopher-king, then Adam and Eve's prowess at hunting or gathering will be irrelevant to how the surplus is divided. Unlike the case studied in Section 2.5.4, the social contract will make the lucky hunter's share contingent only on whether he is Adam or Eve. An able hunter may protest and perhaps try to keep the kill for himself, but he is only one against the combined might of the whole band. The Marxian principle that each should contribute according to his ability is therefore realized.

Distribution is not as easy to deal with as production. Following Section

---

[28]The graphs of their individual utility functions therefore consist of two line segments, one joining $(0,0)$ to $(\sigma,1)$ and another joining $(\sigma,1)$ to $(1,1)$. A better definition of need in this context would replace the first line segment by a line segment joining $(0,0)$ to $(\sigma,0)$. A player would then reject any share $x < \sigma$ in favor of any lottery with prizes $x = 0$ and $x = \sigma$. The shares in Table 4.1 would then be replaced by the probabilities that a player's need is met. However, such a model would have to be complicated by providing players with some form of compensation for settling for a lower probability of winning σ.

2.6.2, I will proceed as though the band operates a utilitarian norm in which the standard of interpersonal comparison has adjusted until pairs of players divide the surplus according to the Nash bargaining solution. The band as a whole is assumed to enforce its social contract by confiscating the kill should squabbling replace a dignified application of the relevant fairness norm. The relevant *status quo* for the Nash bargaining solution is therefore $(0,0)$. It is then easy to show that the kill will be split fifty-fifty between Adam and Eve—whatever values of σ we write into their individual utility functions. This split seems to be egalitarian, but it appears less so when translated into utility terms. The slope of the relevant part of the Pareto-frontier to their bargaining set X is $-\frac{4}{5}$. To implement the outcome in which Adam's share is $x = \frac{1}{2}$ using a weighted utilitarian solution, we can therefore take $U = 4$ and $V = 5$. Eve's lesser need is then reflected in her individual utils being counted as worth only $\frac{4}{5}$ of Adam's. But since the moral content of a fairness norm is eroded away in the medium run, this difference in their perceived worthiness does not result in a move in Adam's direction away from the fifty-fifty split.

Now alter the story by making Adam and Eve relatives. Their individual utility functions then have to be replaced by personal utility functions that incorporate Hamilton's inclusive fitness criterion (Section 2.5.2). If Adam's degree of relationship to Eve is r, his individual utility $u_A(x)$ for a split of the kill in which he gets x and Eve gets $1 - x$ must be replaced by his true personal utility:

$$v_A(x) = u_A(x) + r u_E(1 - x).$$

Similarly, if Eve's degree of relationship to Adam is s, then $u_E(x)$ must be replaced by $v_E(x) = u_E(x) + s u_A(1 - x)$.

The inbreeding that is inevitable in small isolated groups will be ignored in the first instance. When Adam and Eve are siblings, the gene-counting arguments of Section 2.5.2 then imply that $r = s = \frac{1}{2}$. When they are first cousins, $r = s = \frac{1}{8}$. The possibility that $r \neq s$ is included to take account of parent-child relationships. If a son has survived until puberty and his mother is no longer nubile, then the degree of their relationship has to be altered to take account of their different chances of reproducing their genes. In this example, I consider the extreme cases when $r = \frac{1}{2}$ and $s = 0$, and $r = 0$ and $s = \frac{1}{2}$.[29]

---

[29] People who find it shocking to suggest that mothers are hardwired with an algorithm that estimates the relevant values of r and s when deciding how food should be shared within a family should recall how mothers inured to famine are reported to feed only the stronger of their children, so that at least one will survive. Our algorithms wouldn't let us do the same thing because our data includes no firsthand experience of children dying of starvation—but they would if famine were endemic in our society too.

Adam	Eve	Adam's share	Eve's share	U	V	U^*	V^*
stranger	stranger	$\frac{1}{2}$	$\frac{1}{2}$	4	5	4	5
cousin	cousin	$\frac{6}{11}$	$\frac{5}{11}$	3	4	4	5
brother	sister	$\frac{2}{3}$	$\frac{1}{3}$	1	2	4	5
son	mother	$\frac{2}{3}$	$\frac{1}{3}$	3	10	4	5
father	daughter	$\frac{1}{6}$	$\frac{5}{6}$	4	3	4	5

Table 4.1: Families sharing in modern foraging bands.

Table 4.1 shows how the kill is divided when family relationships are taken into account. The constants U and V are weights whose use ensures that the agreed split maximizes the weighted utilitarian solution, provided that Adam and Eve's personal utilities are properly evaluated to show their sympathy. with each other. The constants U^* and V^* perform the same function in the case when Adam and Eve's personal utilities are mistakenly replaced by their individual utilities.

Except when Adam is Eve's father, the closer the relationship between Adam and Eve, the more his greater need is recognized. When he interacts with his mother or his sister, his needs are met in full. Such recognition of Adam's need is also evident in the standards of interpersonal comparison that operate in the different cases. For example, when Adam and Eve are siblings, one of Adam's utils is deemed to be worth only half of one of Eve's. When they are cousins, his utils are worth three quarters of hers. In the exceptional case when Adam is Eve's father, his unreciprocated concern for her welfare results in her needs taking total precedence over his. Her utils are then deemed to be worth three quarters of his.

Although no simple formula connects who gets what with how they are related, kinship clearly provides a good explanation of why the needy receive special treatment in modern hunter-gatherer societies. The phenomenon is strengthened if we take account of the fact that inbreeding will increase the degrees of relationship. A simple model that assigns probability ρ to the event that any married couple share a particular gene, attributes a degree of relationship $r = \frac{1}{2} + \frac{1}{2}\rho$ to siblings, and $r = \frac{1}{8} + \frac{7}{8}\rho$ to cousins. If $r = \frac{1}{7}$, then Adam's needs are met in full, even if he and Eve are only cousins. If $\rho = \frac{1}{4}$, his needs will be met in full, although Adam and Eve may have no obvious family connection at all.

An observer will then see all members of the same generation sharing food as though only need mattered. But one cannot deduce that kinship is irrelevant to the way food is shared. On the contrary, the needy are cared for *because* they are kin.

Prehistoric hunter-gatherers. Section 4.5.3 argues against assuming that modern foraging bands will serve as a model of the prehistoric foraging bands of our ancestors. If I am right, the social contracts of prehistoric hunter-gatherers were enforced neither by a powerful leader nor by the whole group acting in concert. The parties to a sharing agreement therefore had to police the deal themselves. A form of social organization in which each citizen produces according to his ability and consumes according to his need would have been beyond their comprehension.

In the simple example we have been studying, the difference between the two forms of social contract emerges in the location of the state of nature. In a prehistoric foraging band, the band as a whole would not have disciplined Adam and Eve by confiscating their product if they fought over its division instead of operating the conventional fairness norm. The analysis therefore needs to be modified so that the *status quo* used when applying the Nash bargaining solution becomes some analogue of Buchanan's [103] *natural equilibrium*.[30]

I assume that the probability the kill is left in Adam's hands after a failure to agree on an insurance contract is $p = \frac{1}{5}$. In the cleanest case, there is no fighting and a failure to agree simply leaves the carcass in the hands of the player who made the kill. The same assumption is made in Section 2.5.4, but here we allow deals that make the amount the hunter gets contingent on whether he is Adam or Eve. With no fighting, we can identify p with the probability that Adam is the successful hunter, and so p serves as a measure of ability.

The parameter choices in the model imply that we are to study the case in which Eve is more able and Adam is more needy. Table 4.2 compares the shares each now receives with the case of a modern hunter-gatherer society. Notice that the standards for making interpersonal comparisons have not changed, but the new power structure in their Game of Life dramatically alters Adam and Eve's share of the surplus. Only Eve's needs are satisfied when Adam and Eve are no more related than cousins. When Eve is Adam's mother, she still reserves more than half the surplus for herself. Even if a high level of inbreeding with $\rho = \frac{1}{3}$ is postulated, Adam still gets less than

---

[30]My theory predicts that prehistoric foragers used the proportional bargaining solution rather than the weighted utilitarian solution, but the medium-run implications are the same in both cases (Section 4.6.6).

$\frac{4}{7}$ of the surplus when Eve is as closely related as a sister.

Adam	Eve	Adam's share	Eve's share	U	V	U^*	V^*
stranger	stranger	$\frac{1}{6}$	$\frac{5}{6}$	4	5	4	5
cousin	cousin	$\frac{1}{6}$	$\frac{5}{6}$	3	4	4	5
brother	sister	$\frac{1}{3}$	$\frac{2}{3}$	1	2	4	5
son	mother	$\frac{7}{15}$	$\frac{8}{15}$	3	10	4	5
father	daughter	$\frac{1}{6}$	$\frac{5}{6}$	4	3	4	5

Table 4.2: Families sharing in prehistoric foraging bands.

Outbreeding and genetic diversity. Inbreeding increases the degree of relationship between members of a foraging group and therefore makes them more sympathetic to the needs of their companions. But it also has adverse effects. Genetic variation is important to the survival of the group if their Game of Life is subject to shocks to which they are not adapted. If their response to the shock is determined by a gene which they all carry, then they will all respond identically. But if a variety of genes is present in the group, different responses to the shock will compete against each other. The gene that programs the best response to the shock will then become more prevalent in the population, and so the group will adapt to the shock without needing to wait for an unlikely mutation to come along.

Three factors are usually quoted as being responsible for maintaining genetic variation. The first is genetic drift. If the gene at the locus on a chromosome which becomes relevant after the shock has not been selected for in the past, nothing prevents the gene that sits there changing its identity at random during sexual reproduction without any price in lost fitness being paid. The second is assortative mating. An example of such mating occurs when siblings fall in love with people outside their immediate family rather than with each other. (Our societies maintain incest taboos to preserve genetic variation in the group as a whole, rather than to avert the rather small risk of defective babies being born.) The third is outbreeding with members of other groups.

A toy model that incorporates all three phenomena will perhaps serve to illustrate the essential mechanisms. It is a model of genetic drift because selection pressures are absent. It is a model of assortative mating because siblings marry outside the family. It will be a model of outbreeding as well, but only inbreeding will be considered in the first instance.

Imagine a genetically haploid society that always contains precisely four citizens, two male and two female (Section 2.5.2). In each generation, every boy and every girl falls in love and marries. Each happy couple then produces exactly one boy and one girl before passing away. Siblings never marry, and so each child has exactly one available partner unless some outbreeding occurs.

Suppose, for example, that Adam marries Eve and Ichabod marries Olive. If Adam and Eve both carry the same gene a, but Ichabod and Olive carry two different genes b and c, what can be said about the genetic diversity of their offspring? Since each child gets the relevant gene with equal probability from each parent, there are two possible gene distributions among the marriages of the next generation. The first case occurs half the time. Either b or c disappears, so that each marriage in the next generation will have one partner carrying the gene a and the other carrying whichever of b and c survives. The second case occurs the other half of the time. One marriage will carry genes a and b, and the other genes a and c.[31]

There are seven such scenarios that need to be studied, each of which reproduces one or other of the same scenarios in the next generation with probabilities that can be computed fairly easily. We are then looking at a Markov chain with a 7×7 matrix A of transition probabilities. A limiting distribution for such a Markov chain is a vector p of probabilities that satisfies $p = Ap$. This equation ensures that, if one knows that the population in a given generation is distributed over each scenario i with probability p_i, then the same is also true of the next generation. Whatever the scenario in which the population started, genetic drift will operate until things are scrambled up as much as they will ever get scrambled. What you and I and evolution know about the distribution will then be summarized by one of the limiting distributions of the Markov chain.

Without outbreeding, no fancy calculations are necessary to see what will happen. Given long enough, all the genes except one will gradually disappear from the the population. With probability one, we will then be left with the scenario in which everybody in the population has the same gene. Genetic drift will therefore have eliminated genetic variation altogether.

To avoid such a dangerous outcome, human groups typically breed out-

---

[31] Adam and Eve's marriage can only produce children who carry the gene a. Olive and Ichabod's marriage can produce four equally likely outcomes, in which their girl and boy respectively carry the genes bb, bc, cb and cc. In the first and last cases, a gene has been lost from the population.

side their group of origin. This is presumably why mysterious strangers are so much more sexually attractive than the boy or girl next door. The mechanism to be studied here consists of groups swapping their members. To keep things simple, I assume that a group never swaps with another group carrying a gene already present in its own population. This assumption will obviously result in an overestimate of the genetic diversity that the mechanism generates.

Suppose that the group never swaps more than one of its members in any generation, and that the probability a swap takes place at all in any given generation is ϵ, where $0 < \epsilon < 1$. A new matrix B of transition probabilities can now be constructed on the assumption that the person swapped is chosen at random. We then have to study the Markov chain with transition matrix AB obtained when both processes are in operation together. We have seen that genetic drift is hostile to diversity even with assortative mating. Outbreeding pushes things in the opposite direction. Which of the two tendencies is victorious?

In fact, the two processes compromise, yielding a unique limiting distribution p that is independent of the initial scenario. Genetic diversity is therefore maintained, but at what level? To answer this question, it is necessary to compute the vector m that lists the average degree of relationship between married couples in each of the seven possible scenarios. In the limiting case, the average degree of relationship between married couples in the group is then given by $\rho = m^\top p = m_1 p_1 + m_2 p_2 + \cdots + m_7 p_7$.

The value of ρ obviously depends on the parameter ϵ, which measures how much outbreeding is going on. When there is no outbreeding at all so that $\epsilon = 0$, we already know that $\rho = 1$. When $\epsilon = \frac{1}{3}$, ρ is somewhat more than $\frac{1}{2}$. When $\epsilon = \frac{2}{3}$, ρ is approximately $\frac{1}{3}$. When someone is swapped in every generation so that $\epsilon = 1$, ρ is a little more than $\frac{1}{5}$.

Of course, the backgound level of relatedness in a real group of foragers will depend on all kinds of things that are excluded from the toy model studied. The size of the group will be particularly important, as will the diversity of the whole genetic pool from which the groups who swap members are drawn. My guess is that the figure for ρ written in our genes varies with the evolutionary history of the human species in different parts of the world, but that it is never small enough to be negligible.

This background degree of relationship perhaps rationalizes the "warm glow" feelings that lead us to part with our small change to down-and-out strangers in the street, or to contribute small fractions of our income to the charitable relief of people in famine-ridden countries across the ocean that we will never visit or understand.

Sharing between clans. Section 2.5.4 argues that the empathetic preferences we use as inputs when operating the device of the original position are only a short step away from the sympathetic preferences we use in coordinating our behavior within the family. However, speculation on how morality expanded its circle of application from interactions within the family to interactions with strangers was left until the example of hunter-gatherer societies was available.

We were shaped by evolution while living in family groups. It therefore seems likely that our genetic programming allows us no other way of dealing with other human beings than as relatives. Pet cats are presumably in the same boat. They have no choice but to treat us as oversized feline parents, just as we treat them as undersized human children. But we have a much more complex social life than cats. To say that biological evolution has found no way for us to relate to strangers except as relatives leaves open the question of how prehistoric hunter-gatherers first learned to slot strangers into their complicated kinship structures.

The opportunities for sharing food are said to be the primary reason that evolution taught humans to live in groups. But the inbreeding problem that arises in very isolated groups creates one of a number of other reasons for sharing. Different groups or clans need to share genes. This imperative made it necessary for groups to interact with other groups. Intergroup marriages had to be arranged—perhaps first by tolerated theft as in food-sharing among some modern foragers. Some way of assessing the worthiness of strangers adopted into the clan by this method was then necessary. How do I perceive my brother-in-law? Is he really to be treated as a brother? If not, how do I estimate my degree of relationship with him? Waifs and strays who were adopted into the clan after their own group had disbanded present a rather different problem. Doubtless, such recruits had a much harder struggle to prove themselves worthy to be counted as one of the inner family.

My guess is that the prowess of a stranger adopted into the clan was the chief determinant of his perceived degree of relationship among his new companions. Although the example is not entirely apt, consider the shares Adam and Eve receive in Table 4.2 when they are unrelated. Since Eve is a mighty hunter, her demands are met in full and Adam gets only what is left after she is sated. Her worthiness relative to Adam in this transaction is $U/V = \frac{4}{5}$. Although not based on any family relationship, this quantity differs only numerically from the relative worthiness of $U/V = \frac{3}{4}$ that would apply if Adam and Eve were cousins. Slotting a stranger into the clan structure as an honorary relative must therefore have been relatively painless.

Dealing amicably with strangers who were not assimilated into the group must have been a larger step. But the biological imperative to share genes provided an incentive for evolution to overcome the problem. The differential worthiness of the bartered brides or grooms must have created a need for the exchange to be accompanied by gifts of food or other goods. I don't mean to imply that prehistoric bands went in for the bizarre potlatching rituals[32] that still survive in some parts of the world—only that something must have been exchanged on the side for both clans to feel satisfied that the transaction was fair. Assessing the worthiness of people in other clans who were not well known to you cannot have been easy, but this is where I suspect the social conventions we still use for this purpose first saw the light of day.

4.6   The Game of Morals

The scene having been set, the time has come at last to unveil my own theory of the social contract. This was previewed extensively in Chapter 1 of Volume I, but the more difficult ideas all had to be taken on trust at that stage. However, it is now possible to use the evolutionary ideas introduced in Sections 2.5 and 2.6 to close the gaps.

4.6.1   Fair Social Contracts

Section 2.5.4 argues that fairness norms evolved because they allow groups who employ them to coordinate quickly on Pareto-improving equilibria as they become available, and hence to outperform groups that remain stuck at the old equilibrium. The metaphor of the expanding circle was borrowed from Singer [495] to describe the mechanism envisaged. The suggestion is that players sometimes learn how to coordinate on equilibria in a new circle of games by playing these games as though they were games from a familiar class. Occasionally, they thereby succeed in coordinating on an equilibrium in games from the wider circle. If this behavior becomes established, they

---

[32] A delightful paper by Carmichael and McCloud [115] uses the theory of repeated games to argue that the traditional practice of exchanging expensive gifts when a match is made serves to deter deviant behavior in the marriage. Such gift giving makes it difficult for a clan to rematch their half of a broken marriage. The most effective gifts for this purpose are items that are costly to the giver, but without value to the receiver after a breakup—like engagement rings or wedding feasts. Some such story perhaps explains why such large gifts are commonplace at tribal weddings. But I suspect that the arms race element involved when the gifts exchanged grow exponentially over time has more to do with the fact that high social standing can be demonstrated in some contexts by establishing low worthiness (Section 4.7.7).

will then have discovered an equilibrium selection device for some games that works by the players pretending to be bound by the rules of more restrictive games.

I refer to the latter class of games as *morality games* when the way they select equilibria in the Game of Life permits a fairness interpretation, but morally neutral equilibrium selection devices frequently operate in much the same way. For example, if all players proceed as though it were unthinkable to break the law by choosing *left* in the Driving Game, then they will succeed in coordinating on the equilibrium in which everyone chooses *right*. A less trivial example is described in Section 3.4.1. In an overlapping generations model, an equilibrium is described in which entirely selfish daughters are induced to share the product of their labor with their powerless mothers. The rules of the morality game acccompanying this social contract will specify that caring for elderly parents is a moral duty that children are "bound indispensibly to perform" (Burke [109]).

It is anything but original to observe that the rules of the Game of Life are physically binding while the rules of a society's morality game bind only by convention—like the rules of Chess. The *physis* versus *nomos* controversy dates from the fifth century before Christ! In the words of the ancient sophist Antiphon: "The rules of law are adventitious, while the rules of nature are inevitable; and the rules of law are created by covenant and not produced by nature" (Gough [210]). But this simple truth somehow became obscured during the transition from papyrus to paper. Modern scholars nowadays routinely confuse the issue by referring to the rules of morality games as "natural laws". However, if this misleading terminology is to be used, it seems to me necessary either to follow Hobbes [251] and Hume [267] in making it clear that natural laws are actually unnatural and do not compel obedience—or else to follow Spinoza [505] in arguing that we have a natural right to do anything that Nature has put within our power.

I consider only one morality game, which I call *the* Game of Morals. My guess is that the notion of a Game of Life and a morality game being played simultaneously has substantial descriptive validity for the way *homo sapiens* runs his societies, but it would be absurd to claim that any simply characterized game of morals could come close to encompassing the full richness of human moral interaction. The rules of the Game of Morals I propose come to grips only with fairness phenomena.

The *Game of Morals* is defined to be a twin to the Game of Life except that it offers the players additional moves that are not available in the real world. Between each and every round of the repeated Game of Life, the rules of the Game of Morals specify that any player has the opportunity to appeal to the device of the original position—whether or not such appeals to justice have been made in the past. When an appeal is made, the players disappear behind a veil of ignorance where they negotiate in ignorance of their current and future identities about what equilibrium in the Game of Morals should be operated in the future. A chance move then reshuffles their places in society so that each player has an equal chance of ending up either as Adam or Eve.[33]

A *fair social contract* is defined as an equilibrium in the Game of Life that calls for the use of strategies which, if used in the Game of Morals, would never leave a player with an incentive to exercise his right of appeal to the device of the original position. A fair social contract is therefore an equilibrium in the Game of Morals, but it must never be forgotten that it is also an equilibrium in the Game of Life—otherwise evolution will sweep it away. Indeed, the Game of Morals is nothing more than a coordination device for selecting one of the equilibria in the Game of Life. People *can* cheat in the Game of Morals, just as they can move a bishop like a knight when playing Chess. But they have no incentive to do so, because playing the Game of Morals as though its rules were binding leads to an equilibrium in the Game of Life. No player can therefore gain by deviating unless some other player acts against his own best interest by deviating first.

4.6.2 Paradise Lost

Section 2.5.4 argues that the device of the original position evolved from a primitive food-sharing norm that still remains deeply embedded in the human psyche. Section 4.2 traces our continuing capacity for envy to this source. If such speculations about the evolutionary history of the original position are correct, it is not surprising that the device should be so successful in capturing our intuitions about how we actually use fairness norms in our day-to-day affairs.

Chapter 2 is largely devoted to studying how and why we routinely use devices like the original position to solve small-scale coordination problems. However, in this chapter, I am more interested in the possibility of using the concept of fair play with which evolution has gifted the human species to engineer the kind of large-scale reforms that appeal to whigs like myself—reforms achievable by mutual consent because they make everybody better off. But, if appeals to our natural sense of justice are to succeed in garnering support for such large-scale reforms, they need to be based on a sound understanding of how we actually use fairness norms in our daily lives and why they evolved in the first place.

---

[33]As explained in Section 2.6.1, the requirement that players behind the veil of ignorance have an *equal* chance of becoming Adam or Eve is inessential to the results obtained in the medium run.

As Section 2.7.2 insists, the fact that fairness norms do not work as philosophers with utopian ambitions would like them to work should not discourage us from trying to use them in the manner in which they actually do work. In particular, we have to live with the unwelcome truth that the interpersonal comparisons of utility necessary to make fairness judgments meaningful are ultimately determined by the underlying balance of power. When moving from small-scale applications of fairness criteria to large-scale applications involving the whole of society, we also have to accept the equally unwelcome fact that no philosopher-king will be available to police the hypothetical agreements reached behind the veil of ignorance.

In a burst of poetic enthusiasm, Section 2.8 likens the need to dispense with a philosopher-king to the expulsion of Adam and Eve from the Garden of Eden. Outside the garden, they find that God has left them to shape their lives for themselves without outside supervision. On eating the apple, Milton's *Paradise Lost* tells us that the unfortunate pair discover:

> Both Good and Evil, Good lost, and Evil got,
> Bad Fruit of Knowledge, if this be to know,
> Which leaves us naked thus, of Honour void.

This chapter's Adam and Eve are void of honor in that neither will observe the terms of any marriage contract unless it is in his or her interest to do so. Viable contracts must therefore be self-policing.

Chapter 3 studied the self-policing mechanisms by which social contracts can be *sustained* as equilibria in a repeated Game of Life. This chapter studies the implications of using the Game of Morals as a device for *selecting* an equilibrium. But before embarking on this project, we need to clarify and defend the structure of the Game of Morals in more detail.

4.6.3 Modeling the Original Position

This section brings together the basic tools that will be needed in analyzing the Game of Morals. These ideas have already been discussed earlier in the book, sometimes more than once. Many readers will therefore wish to skim forward to Section 4.6.4.

The set of feasible social contracts. Section 1.2.5 of Volume I simplifies the problem faced by Adam and Eve in writing a marriage contract to that of selecting a payoff pair $x = (x_A, x_E)$ from a set X representing the set of feasible social contracts. Asymmetries built into the set X represent the "inequalities of birth and natural endowment" that Rawls [417, p.100] believes a just society should seek to redress.

Figure 2.1(a) illustrates a typical set X of feasible social contracts. How-ever, the interpretation of the payoff pairs in X offered in Section 2.6.1 needs to be revised very substantially. With the unlimited commitment assumptions of Section 2.6.1, it made sense to identify X with the cooper-ative payoff region (Section 1.2.1) of a Game of Life that could have been anything at all—even the one-shot Prisoners' Dilemma. But we must now restrict attention to social contracts that are no more than conventions for coordinating behavior on a particular equilibrium in the Game of Life. We therefore have to rely on the mechanism that biologists refer to as recipro-cal altruism if social contracts incorporating high levels of cooperation are to be achievable. This mechanism can only operate in situations in which the players interact repeatedly. It is therefore important to everything that follows that the Game of Life be modeled as an *indefinitely repeated game*.

Section 3.3 describes the folk theorem of repeated game theory. The theorem uses the reciprocity mechanism to show that all payoff pairs of any interest in the cooperative payoff region of a game remain available as *equilibrium* outcomes in an indefinitely repeated version of the game, provided that the outcomes in the latter are assessed on a per-game basis. That is to say, we must reinterpret Adam and Eve's payoffs in the set X of Figure 2.1(a) as *payoff flows* in the sense of Section 1.2.4. Among its other virtues, the folk theorem assures us of the important fact that the set X is both *convex* and *comprehensive* to the extent discussed in Section 3.3.4. In fact, we shall take X to be closed, bounded above, convex, and comprehensive, as standardly assumed in Chapter 1. Where necessary, its boundary will also be assumed to be smooth.

The state of nature. Reinterpreting payoffs as utility flows is particu-larly important in respect of the state-of-nature point ξ. As we shall find, Wolff's [565, 566] criticism of Rawls' [417] neglect of this issue will be amply justified.

Section 2.6.1 simplified the question of determining a state of nature in the presence of a philosopher-king by looking only at a special case in which the problem of appealing to Nash's [382] variable-threats theory was trivialized. But Nash's variable threat theory does not apply in the absence of a philosopher-king, because Nash's theory is based on the understanding that unilateral commitments can be made (Section 1.5.2). Rather than asking what dreadful actions Adam and Eve would commit themselves to perform if they were to fail to agree on moving to a new social contract, we now have a much simpler question to pose: What will Adam and Eve's payoff flows be while they are negotiating over a new social contract? When commitments cannot be made, it is this pair of payoff flows that is the appropriate *status quo* for Adam and Eve's bargaining problem—not the Hobbesian state of nature that would result from an application of Nash's variable-threat theory (Section 1.7.1).

My theory therefore locates the state-of-nature point ξ at the pair of payoff flows that Adam and Eve receive in whatever social contract is currently being operated. As Section 1.2.1 of Volume I observes, the theory is therefore not vulnerable to the telling criticisms of authors like Gellner[34] [200, p.249], who attack Rawls and other social contract theorists for proceeding as though a society were free to discard its history and plan its future as though newly born.

In Figure 2.1(a), the zeros and units on Adam and Eve's personal utility scales have been chosen so that the state-of-nature point ξ is located at 0 and the maximum feasible payoff for both players is 1. The same normalization will be maintained in this section along with as much of the notation of Section 2.6 as possible. However, let me insist that this normalization of Adam and Eve's personal utility scales is entirely arbitrary. In particular, it does *not* signal that their utils are to be compared using the zero-one rule of Section 2.3.

Bargaining. In Section 2.6.1, the use of the Nash bargaining solution to model bargaining outcomes is defended with an appeal to the Nash Demand Game. This is defended in turn in Section 1.5.1 as the bargaining model appropriate to the case when the bargainers are able to make unilateral, take-it-or-leave-it commitments. However, without a philosopher-king to enforce commitments, the Nash Demand Game ceases to have any attraction as a bargaining model.

For the reasons given in Section 1.7.1, it is necessary to turn instead to Rubinstein's [445] Alternating Offers Game. Recall from Sections 1.6.2 and 1.7.2 that one does not thereby escape the need to make commitment assumptions altogether. The players must still be assumed to trust their bargaining partners to honor their word with respect to sufficiently small amounts or over sufficiently small periods of time. Nor is Rubinstein without his critics on other grounds (Section 1.7.2). Nevertheless, the Alternating Offers Game seems to me to represent as realistic a model of the way bargaining actually takes place in the real world as one might reasonably hope to find—and such realism is important to a theory that argues that we already use the device of the original position in settling small-scale coordination problems in our daily lives. After all, if the bargaining hypothesized in the original position were outside Adam and Eve's experience, how could

---

[34]Ernest Gellner [201] recently died, but he left behind him the book *Reason and Culture* to remind us of his breadth of knowledge and clarity of thought.

they be using it at present, albeit only for such petty matters as deciding who should wash tonight's dishes?

Section 1.7.1 explains why the unique subgame-perfect equilibrium of the Alternating Offers Game implements a weighted version of the Nash bargaining solution when the interval between successive proposals becomes vanishingly small. The weights α and β are the bargaining powers that Adam or Eve enjoy by virtue of their differing attitudes to the unproductive passage of time. To be precise, $\alpha = 1/\rho_A$ and $\beta = 1/\rho_E$, where ρ_A and ρ_E are the respective rates at which Adam and Eve discount time.

In what follows, it will always be assumed that Adam and Eve discount time at the same rate.[35] In spite of being unable to follow Nash [380] in defending his bargaining solution with the Demand Game, we are therefore nevertheless still able to employ the symmetric Nash bargaining solution to model the outcome when rational players bargain under common knowledge of each other's preferences.

If Adam and Eve were to bargain face-to-face, there would be little more to say. Their bargaining problem would be the pair (X, ξ). Our prediction of the bargaining outcome would be the Nash bargaining solution ν illustrated in Figure 1.5(a). Since Adam and Eve would be bargaining in the absence of mutual trust, it would be necessary to follow Section 1.7 in finding a continuous path along which Adam and Eve can move from ξ to ν without either party having an incentive to interrupt the transition by demanding a renegotiation of the terms of the agreement. As illustrated in Figure 1.13(b), the straight-line path joining ξ and ν suffices for this purpose. For any point ξ' on this path, the symmetric Nash bargaining solution of the problem (X, ξ') is ν. Any attempt to renegotiate the terms of the deal will therefore simply reproduce the original agreement.

Section 3.7.2 explains why the same straight-line path will also suffice when Adam and Eve do not bargain face-to-face, but buffer their interaction through the device of the original position, as required by the Game of Morals. However, when their bargaining problem is (T, τ) rather than (X, ξ), we shall find plenty of other problems to keep us occupied.

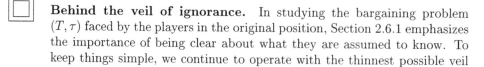

Behind the veil of ignorance. In studying the bargaining problem (T, τ) faced by the players in the original position, Section 2.6.1 emphasizes the importance of being clear about what they are assumed to know. To keep things simple, we continue to operate with the thinnest possible veil

---

[35]Alternatively, one can argue that a player in the original position believes that his discount rate is equally likely to be ρ_A or ρ_E. An analysis of the Alternating Offers Game with this assumption yields the symmetric Nash bargaining solution when the time interval between successive proposals becomes vanishingly small.

of ignorance. In the original position, Adam forgets his identity and is referred to instead as player I. Similarly, Eve forgets her identity and is referred to as player II. However, each player retains his own empathetic preferences. It follows that the players must also forget which empathetic preference derives from Adam and which from Eve. But everything else is assumed to be common knowledge between players I and II. They therefore know the rules of both the Game of Life and the Game of Morals. Each also knows Adam and Eve's personal preferences, and his own empathetic preferences together with those of his bargaining partner.[36]

Sections 2.6.1 and 2.6.3 already break much of the necessary ground in setting up the bargaining problem (T, τ) faced by the players in the original position. Recall that they must take account of two possible events, AE and EA. If AE occurs, player I turns out to be Adam and player II to be Eve. If EA occurs, player I turns out to be Eve and player II to be Adam. To evaluate such a contingent social contract, the players consult their empathetic preferences and come up with two payoff pairs y and z. The payoff pair y lies in the set X_{AE} of Figure 2.1(c) and represents their evaluation of the social contract that will be operated if the event AE occurs. The payoff pair z lies the set X_{EA} of Figure 2.1(d) and represents their evaluation of the social contract that will be operated if the event EA occurs. Since AE and EA are held to be equally likely, the players evaluate the contingent contract as a whole to be worth

$$t = \tfrac{1}{2}y + \tfrac{1}{2}z,\qquad(4.3)$$

as illustrated in Figure 2.2(a).

I do not follow Rawls [417] in rejecting Bayesian decision theory. Players I and II are therefore assumed to maximize expected utility rather than employing the maximin criterion or some other decision rule. As explained in Section 4.5 of Volume I, the hypotheses of the original position create a microcosm that might have been tailor-made for applications of Bayesian decision theory. In rejecting Bayesian decision theory in the circumstances of the original position, Rawls should therefore be taken to be rejecting the theory altogether. Section 4.6 of Volume I reviews the reasons he gives for taking such an iconoclastic line without finding anything at all convincing. I suspect the truth is that the younger Rawls[37] was so dismayed at the utilitarian conclusions to which he believed that Bayesian reasoning leads

---

[36]Section 2.6.1 notes that these strong informational assumptions can be weakened by appealing to Harsanyi's [232] theory of games of incomplete information, albeit at the expense of creating a more complex theory. But the prospects of an extension are less rosy in this chapter because no analogue of Binmore's [66] extension of the Nash Demand Game is available for Rubinstein's Alternating Offers Game.

[37]The elder Rawls [419] is very much more circumspect.

that he didn't take the possibility that such reasoning might be correct seriously. He therefore seized on the nearest argument to hand that supported his intuition that the set T of feasible contingent social contracts in the original position should be identified with $X_{AE} \cap X_{EA}$ as in Figure 2.2(b), rather than $\frac{1}{2}(X_{AE} + X_{EA})$ as in Figure 2.2(a).

But following Harsanyi [233] in using Bayesian decision theory to locate t according to equation (4.3) does not entail identifying the feasible set T with $\frac{1}{2}(X_{AE} + X_{EA})$. This would be right only in the presence of a philosopher-king with the power to enforce the judgment of the hypothetical coin that determines who will occupy what role on leaving the original position. Without such a philosopher-king, the rules of the Game of Morals allow a party disadvantaged by the fall of the coin to demand a return to the original position, where the coin will be tossed anew.

Both players therefore have no choice but to recognize that they have to make do with a feasible set that is substantially smaller than the set $T = \frac{1}{2}(X_{AE} + X_{EA})$ of Harsanyi's Figure 2.2(a). In fact, Section 4.6.4 argues that only the payoff pairs in the set $T = X_{AE} \cap X_{EA}$ of Rawls' Figure 2.2(b) are feasible. However, since this conclusion is reached without appealing to the maximin criterion, there is no reason to follow Figure 2.2(b) in using the maximin criterion to locate the *status quo* point τ. We thereby escape being led into the meaningless calculations of Section 2.6.3.

4.6.4   When is Justice Dispensed?

Section 4.6.3 leaves off its exploration of the consequences of abandoning the fiction of a philosopher-king at the point where the issue of the *timing* of appeals to the original position becomes significant. Rawls [417] has been criticized on this subject by numerous authors including Dasgupta [146], Hare [226], Calvo [113], Howe and Roemer [262], and Rodriguez [432], all of whom could all just as easily have included Harsanyi [233] in their strictures.

Following Rawls [417, p.167], Section 1.2.4 of Volume I offers the following speech with which a slaveholder might seek to justify the institution of slavery to a slave: "If you had been asked to agree to a structure for society without knowing who you would be in that society, then you would have agreed to a slaveholding society because the prospective benefits of being a slaveholder would have outweighed in your mind the prospective costs of being a slave. Finding yourself a slave, you therefore have no just cause for complaint." Rawls gives this argument to challenge the validity of its utilitarian reasoning. He denies that the slave and the slaveholder would follow Harsanyi [233] in employing Bayesian decision theory behind the veil of ignorance. Instead, he proposes that they would employ the maximin criterion—and hence that there would be no chance of their agreeing to a slaveholding society in the original position. But Rawls takes pains to argue that the slaveholder's argument would be valid if it were true that a slave-holding society would be agreed behind the veil of ignorance—explicitly rejecting the slave's objection that he sees no grounds for honoring a hypothetical contract to which he never actually assented and which postulates a lottery for distributing advantage that never actually took place.

As Section 1.2.4 of Volume I observes, the slave's objection raises a number of difficulties for both Harsanyi and Rawls—not least of which is the question of *when* appeals to justice in their theories are envisaged as taking place. They clearly do not intend to follow early contractarians like Grotius and Pufendorf in postulating some ancient folk gathering at which a consensus was reached that somehow binds all those living today. Hume's [274] devastating criticisms of such theories of the "original contract" are unanswerable. Instead, Harsanyi and Rawls locate their original positions in some Kantian limbo outside space and time. Just as it is always six o'clock at the mad-hatter's tea party, so this hypothetical limbo contains a hypothetical coin forever frozen in the act of falling heads or tails.

God made them high or lowly? The idea that our status in society is preordained by such a phantom coin is popular in conservative circles. As a child, I remember being made to sing a hymn containing the verse:

> The rich man in his castle,
> The poor man at his gate,
> God made them high or lowly,
> And ordered their estate.

This piece of propaganda is meant to torpedo the subversive argument that it isn't fair that peasants should labor in the fields while the gentry take their ease. It combines an appeal to God as the ultimate philosopher-king with the suggestion that any new appeal to justice will merely confirm slaves in their servitude—since the only right and proper model of the hypothetical coin toss that determines who will occupy what role in society is the original Act of God that condemned the slaves to serfdom in the first place. This is an attractive argument for the gentry, but it is hard to imagine that any peasant was ever convinced by such an appeal to his intuition on how fairness norms work. One might as well ask a gambler to play roulette knowing that the wheel is fixed to ensure that he loses!

Diamond [161] uses a similar argument against utilitarianism. Imagine that Adam and Eve are both in need of a heart transplant but that only one heart is available. If the lives of both are regarded as equally valuable, then a utilitarian will be indifferent between giving the heart to Adam or

Eve. But Eve would regard it as grossly unfair if the heart were given to Adam on the grounds that he is a man—or because he is white or rich. Nor would she be at all mollified if told that she had an equal chance of being a man when the egg from which she grew was fertilized in her mother's womb. If such a random event is to be used to determine who gets the heart, she will argue that a real coin should be tossed right now.

It seems to me that she would clearly be right about how fairness norms work in practice. Just as in primitive food-sharing arrangements, or when a coin is tossed to determine the initial choice of ends at a soccer match, there has to be some uncertainty about who will get what. As it says in the Book of Proverbs, it is the lot that causeth contentions to cease.

Diamond's [161] heart transplant story makes it transparent that the Game of Morals cannot set the hypothetical coin toss that determines who will be Adam and who will be Eve in the far past, or in some timeless limbo. To capture the way that fairness norms actually work in practice, each appeal to the device of the original position must be independent of any other appeal. After each new appeal, a new coin must be tossed.

If at first you don't succeed. When fairness norms are employed in practice to settle day-to-day coordination problems, it is seldom the case that no source of external enforcement is available. Imagine, for example, the weight of public disapproval that would follow if a captain were to refuse to honor the fall of the coin used to determine the initial choice of ends at a soccer match, and were to insist instead that it be tossed again until it fell in his favor! But it is precisely this unpoliced case that we have to contemplate, both when considering the circumstances under which the fairness norms we use in our daily lives evolved, and in evaluating their suitability for bringing about large-scale reforms in a modern society.

The absence of any human analogue of a philosopher-king in modern hunter-gatherer societies was discussed in Section 4.5.2. But public disapproval of breaches of fairness norms is as strong in modern foraging bands as in any soccer stadium of an industrialized society. However, Section 4.5.3 argues that this social mechanism cannot predate the evolution of the primitive fairness norms with which we are concerned here. In the circumstances under which the human fairness algorithm evolved, Adam and Eve would have had no help from their neighbors in enforcing their coordination agreements. Their plight must therefore be analyzed without postulating a philosopher-king of any kind. As for large-scale reform in a modern society, to dispense with a philosopher-king is simply to recognize that nobody stands outside the system. In particular, nobody should be treated as being invulnerable to corruption.

Without a philosopher-king, a contingent social contract that assigns substantive advantage on the basis of the fall of a coin is untenable, since nothing prevents the disadvantaged party from refusing to honor the deal and demanding another toss of the coin. Nor must we lose sight of the fact that the coin tossed behind the veil of ignorance is hypothetical. Even if a philosopher-king were available, how would he know whether a newly tossed *hypothetical* coin had fallen heads or tails?

The remedy to this difficulty is draconian. We have to accept that contingent social contracts which offer an advantage to one of the players at the expense of the other are not viable. In particular, a player in the original position will not agree on a contingent social contract unless he expects to obtain the *same* payoff whoever turns out to be Adam or Eve.

Recall that the players in the original position evaluate a contingent social contract in terms of a payoff pair y in X_{AE} and a payoff pair z in X_{EA}. As Bayesians, they evaluate the whole contingent social contract as being worth $t = \frac{1}{2}(y + z)$. But if $y = z$, then t is in both X_{AE} and X_{EA}. That is to say, the feasible set of payoff pairs available to players I and II in the original position is $T = X_{AE} \cap X_{EA}$, as illustrated in Figure 2.2(b).

The effect of Rawls' appeal to the maximin criterion in Section 2.6.3 is therefore obtained here by abandoning altogether Rawls' [417, p.115] claim that we have a "natural duty" to honor hypothetical deals reached in the original position. Players in the original position are not assumed to be irrationally averse to Bayesian decision theory or absurdly cautious. They confine their attention to the set T of payoff pairs in both X_{AE} and X_{EA} because payoffs outside this set are unavailable without a philosopher-king.

Although I differ from the younger Rawls [417, p.176] in denying the existence of a natural duty to honor a fair social contract, the line I take remains compatible with his more general concern over the difficulty of sticking by one's commitments. It can even be argued that I am merely taking to its logical extreme his view that Adam and Eve will not consent to a contract that contains clauses which will strain their capacity to honor their commitments too far, because they know that they will be unable to resist the temptation to cheat (Rawls [417, p.178]).

On this as on other issues, I therefore still feel entitled to describe my approach as Rawlsian in spirit—especially since I am told that Rawls himself briefly experimented with a defense of the maximin criterion based on the idea that no coin toss can be regarded as final. Admittedly, Rawls' [419] later work shows no trace of this concern. His current defense of the maximin principle simply reduces to the claim that an egalitarian baseline is intuitive in the circumstances of the original position, the maximin criterion then being offered as the inevitable result of bargaining from an egalitarian *status quo*. From my point of view, the major difficulty with

such an attempt to short-circuit a serious attempt to analyze the problems
faced by the players behind the veil of ignorance is that it allows no handle
on the problem of interpersonal comparison of utility. But one can only
treat an egalitarian distribution as unproblematic if one knows in advance
what it means for something to be shared equally.

Evaluating the status quo. In seeking to specify the bargaining problem (T, τ) faced by players in the original position, we have so far argued
that $T = X_{AE} \cap X_{EA}$, as in Rawls' Figure 2.2(b). However, Section 4.4.3
concludes by remarking that we are not thereby forced to follow Figure
2.2(b) in using the maximin criterion to locate τ. On the contrary, we have
no choice but to take $\tau = \frac{1}{2}\eta + \frac{1}{2}\zeta$, as in Harsanyi's Figure 2.2(a).

However, a problem arises if $\eta \neq \zeta$. For the reasons given in Section
3.7.2, the lack of trust between players I and II implies that they must
move from τ to the Nash bargaining solution σ of (T, τ) along a straight-
line path. However, what is important here is that the path be continuous.
To describe this continuous path, they must first get to τ' in Figure 4.6(a).
To implement τ', the coin that determines who is Adam and who is Eve
must be tossed. If the toss determines the event AE, then the players will
move to η'. If it determines EA, they will move to ζ'. But in each case,
either player I or player II would prefer that the coin be tossed again. The
device of the original position is therefore *unworkable* in the absence of a
philosopher-king unless $\eta = \zeta$.

One can interpret the requirement that $\eta = \zeta$ to mean that the players'
empathetic preferences must accept the terms of the current social contract
as fair—in the sense that both player I and player II are indifferent between
being Adam and Eve in the state of nature. A traditional moral philosopher
will protest that there is no point in talking about justice if current practice
is always deemed to be fair. However, I think that such protests confuse
intuitions derived from observing how social contracts are *sustained* with
how new social contracts are *selected* when a new source of surplus becomes
available for exploitation.

Distinguishing between the rules for sustaining an equilibrium and the
rules for selecting an equilibrium is easy when the players are entirely rational, and hence never deviate from equilibrium. But real people do not
find the strait and narrow path so easy to follow. When they transgress
and infringe the rights of others, evolution has taught us to call for social
sanctions to be imposed upon them by complaining that their behavior
"isn't fair". Section 3.3.8 explains how the rules for selecting an equilibrium can then serve double duty by determining what punishment is fair. It
is therefore inevitable that our experience of the workings of an imperfect

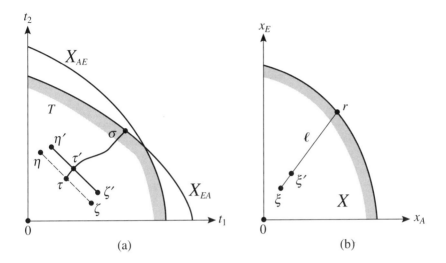

Figure 4.6: Getting the *status quo* right.

social contract should blur the distinction between the duties we accept in sustaining the current social contract and the social values to which we appeal when selecting a new social contract.

To cut through the confusion, observe that to behave unfairly is to break the rules of the Game of Morals by straying from the equilibrium path it specifies in the Game of Life. But what would it then mean to say that our current social contract is *itself* unfair? One must either accept that the relativism implicit in a naturalistic approach to ethics makes the question incoherent, or else abandon oneself to one or other of the traditional metaphysical notions of justice about which evolution cares not a whit.

Equations (2.5) and (2.6) allow the requirement $\eta = \zeta$ to be expressed in terms of the four constants U_1, U_2, V_1 and V_2 that characterize the players' empathetic utility functions:[38]

$$\eta_1 = U_1 \xi_A = 1 - V_1(1 - \xi_E) = \zeta_1 \, ; \qquad (4.4)$$

---

[38] Recall from Section 4.3.1 of Volume I that Adam and Eve's personal utility functions are normalized so that $u_A(\mathcal{L}) = u_E(\mathcal{L}) = 0$ and $u_A(\mathcal{W}) = u_E(\mathcal{W}) = 1$, where both players agree that the outcome $\mathcal{L}$ is worse and the outcome $\mathcal{W}$ better than any other relevant alternative. The worst outcome $\mathcal{L}$ is therefore being identified with the state of nature and the best outcome $\mathcal{W}$ with the utopian point of Section 1.3.2. Player i's empathetic utility function v_i is normalized so that $v_i(\mathcal{L}, A) = 0$ and $v_i(\mathcal{W}, E) = 1$. The constants U_i and V_i are then defined by $U_i = v_i(\mathcal{W}, A)$ and $1 - V_i = v_i(\mathcal{L}, E)$.

$$\zeta_2 = U_2\xi_A = 1 - V_2(1 - \xi_E) = \eta_2. \qquad (4.5)$$

Since Adam and Eve's personal utility scales have been normalized so that $\xi = 0$, it follows that $\eta = \zeta = 0$. Thus,[39]

$$V_1 = V_2 = 1. \qquad (4.6)$$

4.6.5 Rawls Vindicated!

Section 2.5.5 distinguishes between the short run, the medium run and the long run. In the short run, both personal and empathetic preferences are fixed. Section 2.6.1 continues by exploring the short-run implications of assuming that Adam and Eve have the same empathetic preferences, so that the game played in the original position is symmetric, with $U_1 = U_2 = U$ and $V_1 = V_2 = V$. In the presence of a philosopher-king to enforce commitments, we were led to endorse Harsanyi's [233] version of utilitarianism. As illustrated in Figure 2.5(a), the social contract that will be operated under such circumstances is represented by the point h in X at which the social welfare function

$$W_h(x) = Ux_A + Vx_E$$

is maximized. Section 2.6.3 examines the effect of repeating Harsanyi's analysis, using the maximin criterion à la Rawls in place of Bayesian decision theory. We were then led to the proportional bargaining solution r with weights U and V, as illustrated in Figure 2.5(b). But r is not the proportional bargaining solution of the bargaining problem $(X, 0)$, as would make sense with the state-of-nature point ξ normalized at 0. Instead, r is the proportional bargaining solution of the bargaining problem (X, α) with $\alpha = (0, 1 - 1/V)$. Not only is this a meaningless result in its own right, but it led us into nonexistence problems when an attempt was made to carry through a medium-run analysis.

Fortunately, the conclusions reached in Section 4.4.4 when considering the implications of operating the Game of Morals without a philosopher-king allow the analysis of Section 2.6.3 to be recycled meaningfully. Rawls' reliance on the maximin criterion formalized in equation (2.30) is abandoned. Instead of assuming that player i seeks to maximize $t_i = \min\{y_i, z_i\}$, we retain the Bayesian assumption that he seeks to maximize $t_i = \frac{1}{2}y_i + \frac{1}{2}z_i$. However, the new argument of Section 4.4.4 allows us to retain the conclusion that the feasible set T in the bargaining problem (T, τ) faced by the

---

[39]Alternatively, one may note that setting $\xi = 0$ is equivalent to identifying the state of nature with the outcome $\mathcal{L}$. The requirement that player i be indifferent between being Adam or Eve in the state of nature therefore reduces to $1 - V_i = v_i(\mathcal{L}, E) = v_i(\mathcal{L}, A) = 0$.

players in the original position consists of all payoff pairs that lie in both X_{AE} and X_{EA}. That is to say,

$$T = X_{AE} \cap X_{EA}.$$

The *status quo* point τ can also be located at 0, as in Section 2.6.3. But the new argument for placing τ at 0 depends, not on the maximin criterion, but on the fact that $\tau = \frac{1}{2}\eta + \frac{1}{2}\zeta$ and $\eta = \zeta = 0$. As recorded in (4.6), the requirement that $\eta = \zeta = 0$ also entails that $V_1 = V_2 = V = 1$. When we recycle the remaining part of the argument of Section 2.6.3, we are therefore led again to the social contract represented by the proportional bargaining solution r with weights U and V for the bargaining problem (X, α). But now $V = 1$ and so the bargaining problem (X, α) reduces to $(X, 0)$.

Figure 4.6(b) illustrates this conclusion for the general case in which the state-of-nature point ξ has been left unnormalized. Observe that r is the social contract in X at which the Rawlsian social welfare function

$$W_\rho(x) = \min \{U(x_A - \xi_A), V(x_E - \xi_E)\}$$

is maximized when X is strictly comprehensive (Section 4.3.2). As promised in Volume I, a maximin rabbit has therefore been extracted from a Bayesian hat. But since the arguments involved are mathematically trivial, it will be apparent that no legerdemain is involved.

Trustless renegotiation. The mention of commitment provides a reminder that we have still to tackle the problem of renegotiation in a trustless world. Section 1.6.2 argues that this difficulty entails finding a continuous path along which Adam and Eve can move from ξ to r without the need for either to trust the other except to an arbitrarily small degree. As this path is traversed, the Game of Morals allows a new appeal to be made to the device of the original position at any time. Since the players will predict the outcome of any such appeal, the result of the appeal can only be to renew the original agreement to move to r—otherwise an inconsistency would have been built into the system.

Fortunately, as explained in Section 3.7.2, the renegotiation problem is easily resolved by requiring that Adam and Eve move from ξ to r along the straight-line path ℓ joining them in Figure 4.6(b). Behind the veil of ignorance, this corresponds to moving from τ in Figure 2.6(b) to N along the straight-line path joining these two points. If we take a point ξ' on ℓ and apply the proportional bargaining solution to the problem (X, ξ'), we simply renew the agreement to move to r. Similarly, if we take a point τ' on the straight-line path from τ to N in Figure 2.6(b) and apply the symmetric Nash bargaining solution to the problem (T, τ'), we simply renew the agreement to move to N. No player therefore ever has any reason to revise his prediction of the final outcome of the coordinating agreement.

Inequitable but fair social contracts. In Figure 4.7(a), the effect of applying the proportional bargaining solution to a problem (X, ξ) in which the set X of feasible agreements is not comprehensive is badly Pareto-inefficient. One of the virtues that Rawlsians detect in the maximin criterion is that it avoids such inefficiencies, albeit at the expense of introducing inequities of division. When $\xi = 0$ and $U = V = 1$, Figure 4.7(a) shows that the maximin point R is not located at the point ρ on the boundary of X at which Adam and Eve each receive the same payoff. The maximin criterion therefore ranks efficiency above equality in such cases.

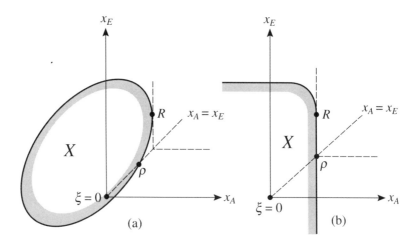

Figure 4.7: Unequal maximin outcomes.

However, too much significance should not be attached to this result, since X is always comprehensive in the relevant range—provided that each player is able to decrease his utility unilaterally if he so chooses, perhaps by burning some money (Section 3.3.4). As Figure 4.7(b) illustrates, the proportional bargaining solution ρ is then weakly Pareto-efficient and satisfies the maximin criterion. Nevertheless, we can still find an outcome R satisfying the maximin criterion that is strongly Pareto-efficient, in that neither player can be made strictly better off without his partner being made strictly worse off.

This example signals that the argument offered in defense of the proportional bargaining solution is not the last word in characterizing a fair social contract. The argument shows only that an appeal to the device of the original position will result in the implementation of the proportional bargaining solution. However, Section 4.4.1 does not define a fair social contract to be the outcome of a making an appeal to the original position. It defines a social contract to be fair if nobody playing the Game of Morals would have an incentive to appeal to the original position. But all social contracts satisfying the maximin criterion have this property. In particular, Adam has no incentive to upset the social contract R in Figure 4.7(b), because his payoff is not improved at the proportional bargaining solution.

Although this conclusion vindicates Rawls' use of the maximin criterion altogether by assigning it priority over the proportional bargaining solution, it is not the most exciting of results. Firstly, the two notions can differ only when X is weakly comprehensive. Secondly, the merest hint of irrational envy will be enough to destabilize a strongly Pareto-efficient social contract R that differs from the proportional bargaining solution ρ.

4.6.6   Interpersonal Comparison in the Medium Run

It is all very well for Section 4.6.5 to announce that Rawls has been vindicated, but its reworking of his ideas leaves two vital questions unanswered. Why do Adam and Eve have the same empathetic preferences, so that $U_1 = U_2 = U$? And what determines U? Given that $V = 1$, the value of U is of particular interest because it represents the number of Eve's utils that are equivalent to one of Adam's utils.

Neither Harsanyi nor Rawls has any answer to the second question. Rawls [417] plucks his index of primary goods out of a hat, while Harsanyi's [233] appeal to some rational algorithm that makes U a function of what we all have in common is unhelpful without some clue as to the nature of the algorithm (Section 4.3.2 of Volume I). As Section 2.3 documents, many other proposals have been made for comparing utils across individuals, any of which could be used to determine U. One might, for example, follow Gauthier [196] in his implicit appeal to the zero-one rule (Section 2.6.3). But I think it futile to impose such arbitrary standards for making interpersonal comparisons on the problem. The problem should impose its own standards on us.

Returning to the first and more fundamental question of why $U_1 = U_2 = U$, I agree with Harsanyi and Rawls that there is indeed a substantial level of consensus within a particular society on how utils should be compared across individuals. But to understand how such a consensus arises, it seems to me that we must look to the cultural history of the society under study.

Such an attitude is arguably not so distant from that of philosophers like Hare [228] and the teleological Harsanyi [233], who continue to nurture the ideal observer tradition of Adam Smith's [500] *Theory of Moral Sentiments*. One need not see their ideal observer as more than a metaphor—like the invisible hand of Adam Smith's [501] *Wealth of Nations* that equates supply and demand. The ideal observer is not a real entity, but a personification of a set of general cultural attitudes that we perhaps unconsciously feel no need to defend because we take for granted that they are shared by all those for whom we are writing. But, if we all share common cultural attitudes, it is because we have a common cultural history. So why not try to build the notion of a common cultural history directly into the analysis?

In following up this alternative to the ideal observer tradition, the surrogates for culture to be employed are the empathetic preferences that Adam and Eve bring with them into the original position. The question to be asked is how these empathetic preferences might be molded over time by the forces of social evolution.

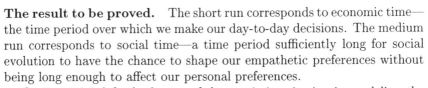

The result to be proved.   The short run corresponds to economic time—the time period over which we make our day-to-day decisions. The medium run corresponds to social time—a time period sufficiently long for social evolution to have the chance to shape our empathetic preferences without being long enough to affect our personal preferences.

Section 4.6.5 defends the use of the maximin criterion in modeling the way we employ fairness norms in the short run when external enforcement is unavailable. As explained in Section 4.6.4, this argument requires setting $V_1 = V_2 = 1$, but offers no guidance on how to determine the constants U_1 and U_2. Two questions therefore need to be addressed:

- Why is $U_1 = U_2 = U$?
- How is U determined?

The approach to be adopted requires attempting a medium-run study of what is likely to happen to the empathetic preferences of the citizens in a society who settle fairness questions using the Game of Morals. As in Section 2.5.4, the notion of an empathy equilibrium will be used to characterize the result of allowing social evolution to operate on our empathetic preferences in the medium run. Recall that an empathy equilibrium is defined by the requirement that no player in the original position would wish to misrepresent the empathetic preferences he finds himself holding if he had the opportunity to do so.

Section 2.6.2 explores the consequences for Harsanyi's nonteleological utilitarian theory of allowing social evolution to move our empathetic preferences until they settle into a symmetric empathy equilibrium. The result is a simple theory that determines how interpersonal comparisons of utility are determined by social evolution in the medium run. The conclusion that

$U_1 = U_2 = U$ and $V_1 = V_2 = V$ is bought cheaply by restricting attention to symmetric empathy equilibria. As explained in Section 2.7.2, the
constants U and V that determine how Adam's utils are weighed against
Eve's are found by first locating the symmetric Nash bargaining solution ν
of the problem (X, ξ). The ratio of the constants U and V is then chosen
so that the weighted utilitarian solution with weights U and V for (X, ξ) is
identical to ν, as illustrated in Figure 2.6(a).

This approach will now be adapted to my reworking of Rawls' egalitarian
theory. Although the technical details differ, the final result has a similar
structure to that obtained in Harsanyi's utilitarian theory. In the Rawlsian
case, only the constant U is available for shaping by the forces of social
evolution, since our study of the state of nature in Section 4.6.4 shows that
we must insist that $V = 1$. It is then unnecessary to impose a symmetry
requirement. At the only empathy equilibrium, $U_1 = U_2 = U$. The value
of U at the empathy equilibrium is found by following a two-step procedure
analogous to that of Section 2.7.2:

(1) Find the symmetric Nash bargaining solution ν of the bargaining
problem (X, ξ).

(2) Choose the weights U and $V = 1$ to make the proportional
bargaining solution for (X, ξ) coincide with ν.

This two-step procedure is illustrated in Figure 4.8(a). Those uninterested
in the technical details leading to the result should skip forward to Section
4.6.7.

One may ask why the value of U matters, since the Nash bargaining
solution of (X, ξ) can be computed without its aid. The answer is the same
as that offered in Section 2.7.2. The value of U is relevant in the *short
run*, after some change in the underlying environment expands the set of
feasible social contracts from X to Y. The fairness norm then operates
to shift society to a new Pareto-efficient social contract ω in the manner
illustrated in Figure 4.8(b). However, further discussion of such short-run
questions is left until Section 4.6.8.

Proving the result. Only the case when the set X of feasible social
contracts is *strictly* convex will be considered.

Four cases are illustrated in Figure 4.9. The diagrams on the left show
how matters appear to players I and II in the original position. The Nash
bargaining solution for the problem $(0, T)$ with $T = X_{AE} \cap X_{EA}$ is denoted
by σ. The diagrams on the right show how things appear when the players
emerge from behind the veil of ignorance to implement the contingent social

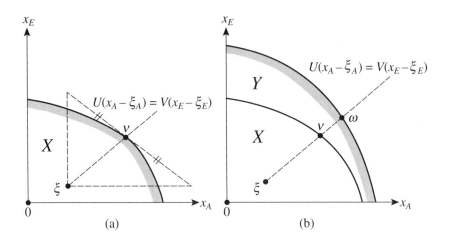

Figure 4.8: Morality in the medium and short run.

contract $(\mathcal{C}, \mathcal{D})$ corresponding to σ. If the event AE occurs, then the social contract $\mathcal{C}$ is implemented, with the result that Adam and Eve receive the personal payoff pair R. If the event EA occurs, then the social contract $\mathcal{D}$ is implemented, with the result that Adam and Eve receive the personal payoff pair S. To find the points R and S, recall that $V_1 = V_2 = 1$, and so $y = (U_1 R_A, R_E)$ and $z = (S_E, U_2 S_A)$. The requirement that $y = z$ therefore leads to the conclusion that $U_1 R_A = S_E$ and $R_E = U_2 S_A$. Since player I is actually Adam and player II is actually Eve, the personal payoffs the players receive in practice are R_A and R_E.

Case 1: The diagram on the left of Figure 4.9(a) reproduces Figure 2.5(b). It applies when the values of U_1 and U_2 are not too dissimilar. The Nash bargaining solution σ is then located at one of the points where the Pareto-frontiers of X_{AE} and X_{EA} cross. The diagram on the right is obtained by noting that the points (R_A, S_E) and (S_A, R_E) lie on rays through 0 with slopes U_1 and U_2 respectively. This fixes the location of R and S, since both these points must lie on the Pareto-frontier of X.

Case 2: Figure 4.9(b) reproduces Figure 4.9(a) in the symmetric case when $U_1 = U_2 = U$. We then have that $R = S = \rho$, where ρ is the proportional bargaining solution for $(X, 0)$ with weights U and $V = 1$. The fact that $R = S$ in this case follows from the fact that $\sigma_1 = \sigma_2$. This symmetric case can be implemented by a contingent social contract $(\mathcal{C}, \mathcal{D})$ in which

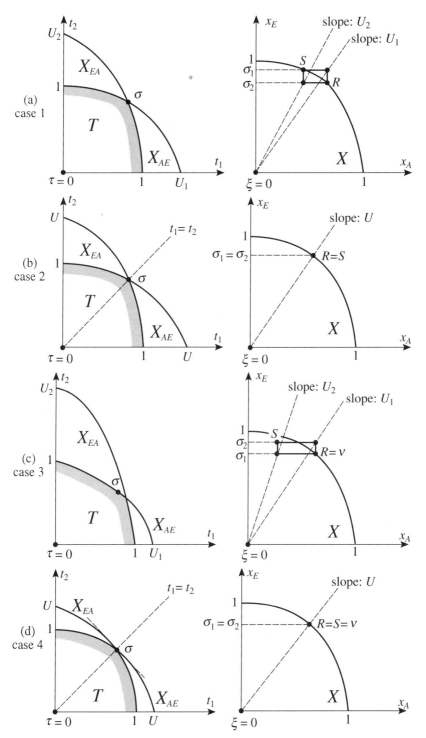

Figure 4.9: Expressing agreements in terms of personal payoffs.

$\mathcal{C} = \mathcal{D}$. That is to say, the only way matters differ in the social contract that operates when EA occurs as compared with AE is that Adam and Eve exchange roles.

Case 3: The diagram on the left of Figure 4.9(c) applies when the values of U_1 and U_2 are sufficiently dissimilar. The Nash bargaining solution σ for the problem $(T, 0)$ is then either identical with the Nash bargaining solution for $(X_{AE}, 0)$, or else with that for $(X_{EA}, 0)$. The case drawn represents the first of these two possibilities. The diagram on the right of Figure 4.9(c) is then determined by the observation that the point R must be identical with the Nash bargaining solution ν for $(X, 0)$, because of the axiom called Independence of Utility Calibration in Section 1.4.2.

Case 4: Figure 4.9(d) illustrates the case of most interest, since it will be shown to correspond to the unique empathy equilibrium. It occurs when the Pareto-frontiers of X_{AE} and X_{EA} are tangent where they cross. Cases 1 and 3 then both apply simultaneously. It follows that the situation must be symmetric, with $U_1 = U_2 = U$, and so Case 2 applies as well. The two-step procedure we are seeking to justify is therefore a method for finding the value of U that leads to Case 4.

Ruling out Case 1. The next step is to eliminate Case 1 configurations that do not belong in Case 4 as possible empathy equilibria.

Suppose that players I and II have empathetic preferences characterized by U_1 and U_2 that generate a Case 1 configuration as illustrated in Figure 4.9(a). To show that such a configuration is not in equilibrium, it is necessary to demonstrate that one of the players can gain relative to his true empathetic preferences by claiming to have a false set of empathetic preferences.

Suppose that player I falsely claims to have a set of empathetic preferences characterized by $U_1' < U_1$, but player II sticks by her true empathetic preferences, so that $U_2' = U_2$. Unless σ is the Nash bargaining solution for $(X_{AE}, 0)$, U_1' can be chosen sufficiently close to U_1 that Case 1 still applies. The Nash bargaining solution for the new problem $(T', 0)$ is therefore σ', as illustrated in Figure 4.10(a). Although $\sigma_1 < \sigma_1'$, it does not follow immediately that player I has improved his position. We need to note that $\sigma_1' = \frac{1}{2}U_1'R_A' + \frac{1}{2}S_E' < \frac{1}{2}U_1 R_A' + \frac{1}{2}S_E'$, where the final term is player I's evaluation of the *new* deal reached in the original position relative to his *old* empathetic preferences.[40]

Unless σ is the Nash bargaining solution for $(X_{EA}, 0)$, the same argument can be employed with player II replacing player I. It follows that the

---

[40]It is true that $U_1 R_A = S_E$ and $U_1' R_A' = S_E'$, but not that $U_1 R_A' = S_E'$.

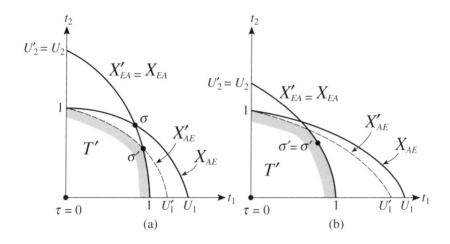

Figure 4.10: Ruling out Cases 1 and 3.

only possibility for an empathy equilibrium in Case 1 occurs when σ is the Nash bargaining solution for both $(X_{AE}, 0)$ and $(X_{EA}, 0)$ simultaneously. But this condition characterizes Case 4.

Ruling out Case 3.   The next step is to eliminate Case 3 configurations that do not belong in Case 4 as possible empathy equilibria.

Suppose that players I and II have empathetic preferences characterized by U_1 and U_2 that generate a Case 3 configuration as illustrated in Figure 4.9(c). Assume that the Nash bargaining solution σ for $(T, 0)$ coincides with that for $(X_{EA}, 0)$, but not with that of $(X_{AE}, 0)$. To show that such a configuration is not in equilibrium, it is necessary to demonstrate that one of the players can gain relative to his true empathetic preferences by claiming to have a false set of empathetic preferences.

Suppose that player I falsely claims to have a set of empathetic preferences characterized by $U_1' < U_1$, but player II sticks by her true empathetic preferences, so that $U_2' = U_2$. If U_1' is chosen sufficiently close to U_1, the Nash bargaining solution σ' for the new problem $(T', 0)$ remains σ, as illustrated in Figure 4.10(b). But then $\sigma_1 = \sigma_1' = \frac{1}{2}U_1'R_A' + \frac{1}{2}S_E' < \frac{1}{2}U_1 R_A' + \frac{1}{2}S_E'$, and so player I's position is improved.

If the Nash bargaining solution σ for $(T, 0)$ coincides with that for

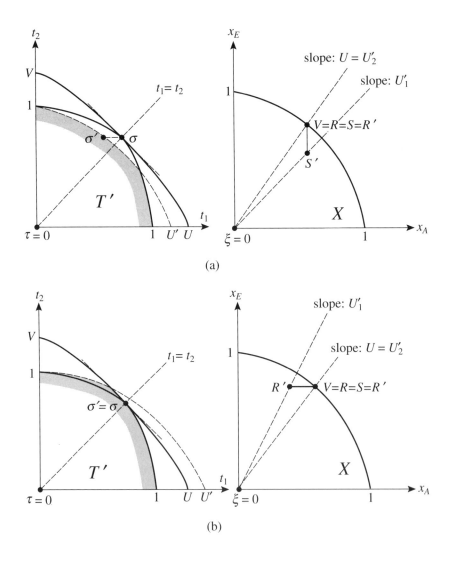

Figure 4.11: Ruling in Case 4.

$(X_{AE}, 0)$, but not with that for $(X_{EA}, 0)$, the same argument can be used with player II replacing player I. The only possibility for an empathy equilibrium in Case 3 therefore occurs when σ is the Nash bargaining solution for both $(X_{AE}, 0)$ and $(X_{EA}, 0)$ simultaneously. But this is Case 4.

Ruling in Case 4. The final step is to confirm that Case 4 is indeed an empathy equilibrium.

Suppose that players I and II have empathetic preferences characterized by $U_1 = U_2 = U$ that generate a Case 4 configuration as illustrated in Figure 4.9(d). To show that this configuration is in equilibrium, it is necessary to demonstrate that neither player can gain relative to his true empathetic preferences by claiming to have a false set of empathetic preferences. Because of the symmetry of the situation, it is enough to show that player I does not improve his position by falsely claiming to have empathetic preferences characterized either by $U_1' < U$ or by $U_1' > U$, on the assumption that player II continues to announce that $U_2' = U$.

Figure 4.11(a) illustrates the case $U_1' < U$. By the axiom of Independence of Utility Calibration (Section 1.4.2), the Nash bargaining solution σ' for $(X_{AE}, 0)$ assigns player II the same payoff as σ. Since σ' lies in the set $T' = X_{AE}' \cap X_{EA}'$, it must also be the Nash bargaining solution of $(T', 0)$. Case 3 therefore applies. As the diagram on the right of Figure 4.11(a) illustrates, $R_A = R_A'$ and $S_E' < S_E$. It follows that $\frac{1}{2}UR_A' + \frac{1}{2}S_E' < \frac{1}{2}UR_A + \frac{1}{2}S_E = \sigma_1$, and so player I is worse off.

Figure 4.11(b) illustrates the case $U_1' > U$. An appeal to the Independence of Irrelevant Alternatives (Section 1.4.2) shows that Case 3 applies again with $\sigma' = \sigma$. As the diagram on the right of Figure 4.11(b) illustrates, $S_E' = S_E$ and $R_A' < R_A$. It follows that $\frac{1}{2}UR_A' + \frac{1}{2}S_E' < \frac{1}{2}UR_A + \frac{1}{2}S_E = \sigma_1$, and so player I is worse off.

4.6.7 Consensus and Context

Rawls' [417] treatment of his index of primary goods reveals that he thinks achieving a consensus on how interpersonal comparisons should be made is relatively unproblematic compared with the other issues he discusses. Harsanyi [233, p.60] is equally sanguine about the prospects of consensus when he observes that: "In actuality, interpersonal utility comparisons between persons of similar cultural background, social status, and personality are likely to show a high degree of interobserver validity." I share Harsanyi and Rawls' impression that we do in fact see a substantial level of agreement on how to make interpersonal comparisons when a group of people are working smoothly together. But, for the reasons given in Section 2.7.2, my theory predicts that the consensus should *vary with the context*.

The view that fairness judgments are made in different ways in different circumstances has been defended by numerous authors. In recent years, Elster [173] and Young [573] have urged this point particularly strongly. However, it usually seems to be taken for granted that the standard for making interpersonal comparisons does not vary much between contexts, leaving differences in what is deemed to be fair to be explained by variations in the fairness algorithm in use. By contrast, my theory assumes that the fairness algorithm is always the device of the original position, no matter what the context may be. Differences in fairness judgments are therefore attributed to contextual variations in the standard for making interpersonal comparisons of utility.

My claim that differences in fairness judgments are not caused by changes in the fairness algorithm in use looks as though it needs to be qualified with the reservation that the device of the original position leads to different results depending on whether an external agency is available to enforce hypothetical deals reached behind the veil of ignorance. Without an external agency to act as a philosopher-king, this chapter argues that the use of the original position leads to a version of Rawls' egalitarian theory. With a philosopher-king, Chapter 2 argues that the use of the original position leads to a version of Harsanyi's utilitarian theory. Of course, when the original position is used in practice to settle small disputes in minisocieties, neither of these two polar extremes apply. There will be outside pressures that provide some measure of enforcement,[41] but such pressures will seldom be entirely decisive. If the impact of such partial commitment possibilities were modeled, presumably some mixture of utilitarian and egalitarian principles would emerge. One can therefore argue that the original position should not be thought of as a unique fairness algorithm, but as a collection of many fairness algorithms. Indeed, if the standard for making interpersonal comparisons of utility depended on the presence or absence of a philosopher-king, one would have no choice but to adopt the latter viewpoint.

However, it turns out to be false that the standard for comparing utils which emerges in the medium run depends on the availability of a philosopher-king. Whether or not Adam and Eve are committed to the deal reached behind the veil of ignorance, the value of U/V obtained at an empathy equilibrium is the same. To see this, recall that Figure 1.7(b) shows three different bargaining solutions, the symmetric Nash bargaining solution ν, the weighted utilitarian solution h, and the proportional bargaining solution ρ. The weights U and V in the latter cases are the same.

---

[41] Section 4.6.4 mentions the implications for a soccer captain who refuses to honor the result of tossing a coin to decide who chooses ends.

Yaari [570] has shown that, if U and V are chosen to make two of these solutions coincide, then the third solution lies at the same place as well.

4.6.8 Morality in the Short Run

As in Section 2.7.2, it is a mistake to take the fact that time will eventually erode all moral content from a fairness norm used regularly for a standard purpose as meaning that justice is an empty concept. It is true that Section 4.6.6 shows that the constant U is determined in the medium run by the two-step procedure illustrated in Figure 4.8(a). Once social evolution has taken the system to an empathy equilibrium, Adam and Eve will therefore get precisely the same personal payoffs when they play fair by using the device of the original position as they would if they were to bargain face-to-face with no holds barred.

In Plato's *Republic*, Socrates disputes the nature of justice with Glaucon and Thrasymachus. Glaucon gets short shrift for putting a view close to the social contract ideas defended in this book. Thrasymachus proposes that justice is simply what the powerful are able to impose. One might categorize the result obtained in Section 4.6.4 by saying that Thrasymachus turns out to be right *in the medium run*. However, decisions are not made in the medium run—they are made *in the short run*. Nor did the idea of justice evolve to settle medium-run coordination problems; it evolved to solve the equilibrium selection problem that arises in the short run when a new source of surplus becomes available (Section 2.7). Glaucon is therefore not be abandoned in favor of Thrasymachus as my advocate in arguing against the absurdly authoritarian state that Plato makes Socrates defend.

To study the moral implications of the theory, one must look at how it works in the short run when the feasible set X of social contracts expands to Y. Such an expansion of the feasible set is illustrated in Figure 4.8(b), with the *status quo* ξ no longer normalized at 0. To interpret this figure, one must imagine that the standard for making interpersonal comparisons of utility has adapted in the medium run to the problem (X, ξ). Players using the proportional bargaining solution with weights U and $V = 1$ therefore end up operating the Nash bargaining solution ν for (X, ξ). But an unanticipated shock expands the set of feasible social contracts from X to Y. Adam and Eve then use the device of the original position for the purpose for which it evolved—to shift from their current social contract ν to a social contract ω that is Pareto-efficient in Y. As explained when vindicating Rawls in Section 4.6.5, such a use of the original position when Y is strictly comprehensive is equivalent to adding a third step to the two-step procedure of Section 4.6.6:

(3) The new social contract ω is the proportional bargaining solution
for the bargaining problem (Y, ν), computed with the weights U and
$V = 1$ that evolved for the problem (X, ξ).

The state of nature for the third step is ν rather than ξ as in Section 2.7.2,
because the state of nature in this chapter is always the current social
contract. However, Figure 4.8(b) shows that ω would have been the same
even if the state of nature had been taken to be ξ. This reflects the fact that
the constant U that determines how interpersonal comparisons are made
is a function of *past history*. The standard for making such interpersonal
comparisons does not instantly adapt to the new feasible set Y. As in
Section 2.7.2, U would eventually adjust if the representative problem faced
by Adam and Eve were to remain (Y, ν) for long enough. Fairness would
then lose its bite. But U remains fixed in the short run.

To summarize the point made at greater length in Section 2.7.2, the suggestion is that our intuitions about justice derive from observing a fairness
norm in operation that is represented here in a stylized form by the device
of the original position. The standard of interpersonal comparison used
by Adam and Eve is modeled by a single parameter U, which is envisaged
as being part of a society's culture. It is therefore determined by history.
If the representative problem faced by Adam and Eve remains (X, ξ) for
long enough, U adapts until fair outcomes reflect nothing but the underlying balance of power. However, new problems faced in the short run are
resolved using the fairness norm with the historically established value of
U.

4.6.9 Egalitarianism versus Utilitarianism

Section 4.6.7 explains why an egalitarian and a utilitarian will not only end
up making the same interpersonal comparisons in the medium run—they
will also find themselves endorsing precisely the same social contract! This
conclusion is more than a little ironic, since debates between followers of
the Good and the Right typically reduce to bizarre philosophical duels, in
which each side shoots off absurdly contrived moral questions and begins
poking fun at the answers of their rivals without waiting to hear what these
answers might be. Even if they listened to each other, I cannot see how
either side would achieve anything useful by demonstrating that our moral
intuitions often generate very odd remedies when applied to sore thumb
problems. If a moral problem is sufficiently outlandish, it will have occurred
so infrequently in our evolutionary history that it would be amazing if our
moral intuitions were adapted to solving it in a way that made any sense.

The Alice in Wonderland flavor of such philosophical slanging matches can only be fully relished after noting that neither side thinks the standard of interpersonal comparison used in evaluating sore thumb moral problems is at all crucial. We are therefore entitled to choose U and V for them. As we have seen, a careful choice of these constants will ensure that both egalitarians and utilitarians endorse exactly the same split of a surplus. Tweedledum will then be unable to ridicule Tweedledee without ridiculing himself! But in making this point, I am at risk of being ridiculed myself. Am I seriously suggesting that there is no difference between egalitarians and utilitarians?

As so often, such questions have to be answered by denying the premise. To point out that egalitarianism and utilitarianism will converge on the same equilibrium *in the medium run* is very far from saying that the two doctrines always have the same moral implications. To accuse me of making such a silly mistake is to forget that morality has its bite *in the short run*. After social evolution has had a chance to work on U and V in the medium run, it is true that Adam and Eve will find themselves operating the same standard for comparing utils, whether the social contract that serves as their current state of nature is construed as egalitarian or utilitarian. But U and V influence the proportional bargaining solution and the weighted utilitarian solution in opposite ways.[42]

The example of Section 4.5.4 will help to illustrate the point. To study the short run, we need to expand Adam and Eve's feasible set from X to Y as described in 2.7.2. The simplest way to do this is to increase the size of the kill.[43] However, if anything new is to occur, it is necessary that the essential relationship between Adam and Eve also be altered at the same time. In the current example, the most interesting modification would seem to be to make Eve as needy as Adam by replacing her individual utility function by his. The symmetry of the set Y so obtained makes it particularly easy to predict how the kill will be divided in the medium run. Except for the intergenerational cases, the players' standards for making interpersonal comparisons will adjust until a fifty-fifty split is implemented.

However, U and V remain *fixed* in the short run. With the utilitarian social contract of a modern foraging band, the standards for making inter-

---

[42]The reason is already apparent in the very different ways that the standard of interpersonal comparison is operationalized in the two cases. One of Adam's utils is worth U/V of Eve's utils in a utilitarian context, because the value of the weighted utilitarian social welfare function is unchanged when one util is subtracted from Adam's payoff and U/V utils are added to Eve's payoff. But we say the same thing in an egalitarian context because a fair division of a surplus obtained by applying the proportional bargaining solution awards Adam an extra util if and only if it awards Eve an extra U/V utils.

[43]By replacing the old individual utility functions with new ones that satisfy $u_A(1) = u_E(1) = k$, where $k > 1$

personal comparisons listed in Table 4.1 then *always* lead to Eve's getting $\frac{5}{6}$ of the kill in the short run. Although both have the same utility function, *Eve's* need is now recognized to the exclusion of Adam's—even in the case when Eve is Adam's mother.

To compare like with like, it is necessary to use the proportional bargaining solution of egalitarianism with the *same* state of nature as in the utilitarian case. This is the equilibrium that was in place before X was expanded to Y. By the criterion of Section 4.6.1, we should therefore compute a fair social contract for each case in Table 4.1, using the listed pairs of shared payoffs as the appropriate state of nature.

The results reflect the fact that Adam wishes that U/V were *bigger* when the weighted utilitarian solution is applied, and *smaller* when the proportional bargaining solution is applied. If the size of the kill is sufficiently large, *Adam's* need is now recognized to the exclusion of Eve's—except in the case when Adam is Eve's father, when her needs again prevail over his.

To understand why such diametrically opposed results are obtained, it is helpful to think of an egalitarian and a utilitarian handing out morsels of the surplus one by one. An egalitarian takes each morsel and splits it between Adam and Eve in proportion to their respective worthiness coefficients, $1/U$ and $1/V$ (Section 4.4). A utilitarian gives the morsel to whoever is perceived to gain the most utility thereby. He therefore favors players that get a lot of utility from a small share of the surplus. But such players are precisely the folk who are deemed unworthy by an egalitarian. Since I believe that our evolutionary history leads real people to make fairness judgments on egalitarian or whiggish principles, utilitarians or socialists might therefore be characterized by saying that they are in favor of giving priority to the needs of the unworthy, while neglecting the fair claims of worthy folk like ourselves.

Section 3.4.5 observes that the leaders of neofeudal states put their authority at risk from the resentment of their followers if they use their coordinating powers to select an equilibrium perceived as blatantly unfair. But an authority governing on the socialist principles advocated by utopian levelers from John Ball to Karl Marx cannot avoid creating destabilizing resentment. Its principles require that it distribute goods in a manner that will *necessarily* be perceived as unfair in the short run. Socialists can emulate King Canute by ordering back the tide of evolution, and insisting that we *ought* to regard the unworthy as worthy, but such appeals can only be temporarily effective in large societies.[44] History shows that human

---

[44]Utilitarians may respond that the existence of the quasi-utilitarian social contracts of contemporary hunter-gatherers shows that at least some human beings have responded to the moral imperative that requires the unworthy to be reclassified as worthy. But

nature eventually reasserts itself. The pigs become indistinguishable from the farmers, worthiness reverts to its original sense, and robber barons reemerge to appropriate the product of the working classes in standard neofeudal style (Section 4.7.4).

4.6.10 Paradise Regained?

The discussion of the Game of Morals that began in Section 4.6 now needs to be rounded off with yet another reminder of the significance its analysis may have for whiggish programs of reform.

Although they lost their philosopher-king when expelled from the Garden of Eden, my theory still allows Adam and Eve to use fairness conventions for the purpose of coordinating on Pareto-efficient equilibria in new situations, without costly internal strife or the need for external enforcement. In eating of the fruit of the tree of knowledge, one might say that they swapped their honor for a set of street smarts that still allows them to get along together amicably. Although this swap is traditionally held to have been a bad bargain, the idealized world of Adam and Eve after the Fall is nevertheless a paradise of sorts when compared with our own.

Whigs believe that this admittedly second-best paradise is available to us as well. Access to the new Garden of Eden is gained by extending the use of the original position from the small-scale problems for which it evolved to large-scale coordinating problems.

In making this proposal, no claims are made that it is the Good or Right thing to do. Followers of the Seemly claim no special authority for the preferences they reveal when advocating one feasible reform rather than another. They simply invite others with similar prejudices about the society in which they would like their children to live to join them in trying to create such a society. However, as Section 2.7.2 insists, such an invitation would be futile if the feasibility constraint is relaxed. In particular, no matter how fancy the arguments offered in defense of some abstract principle of justice, people will not permanently accept a political compromise based on this principle as fair if they see those they regard as unworthy being rewarded.

---

they forget that such societies are sustained only at the cost of everybody taking a pathological interest in the accountancy of envy. If Adam becomes more productive than Eve, nobody says that his share of a surplus should be larger than hers because Adam is now more worthy. Instead, elaborate fictions are invented to explain why Adam isn't really more worthy than anyone else. In some societies, Adam's very enterprise is seen as a reason for treating him as a figure of fun. In others, he is accused of using witchcraft to steal his powers from others. Either way, our genes are deceived enough to allow the charade to continue.

4.7   Worthiness and Power

The Bible says that a laborer is worthy of his hire. The inference is that the laborer has fulfilled his side of a fair contract and that the time has now come for magnanimity in the matter of payment. Worthiness can therefore be determined by a bargain, and it is in this sense that Section 4.4 uses the term when identifying Adam and Eve's worthiness in psychological equity theory with the constants $1/U$ and $1/V$ derived from my own theory. Quite a large step is involved in proposing such an identification, since psychological experiments only suggest that context-dependent worthiness coefficients exist, whereas my theory offers some qualitative predictions of how these coefficients vary with the context.

The current section derives some predictions valid in particularly simple contexts, with a view to providing guidance for further experimentation or applied work. I also use these predictions to make sweeping generalizations about political systems, but it must be remembered that my suggestion that we use the fairness algorithm written into our genes to facilitate large scale sociopolitical reforms is at best only an aspiration. Currently, the fairness algorithm is used systematically only to solve small-scale coordination problems. The predictions are therefore only empirically relevant to the *local* social contracts that operate in *minisocieties*. Each such minisociety is guided by the rules of a different miniculture derived from a different history of experience, and is constrained by the rules of a different minigame of life. Different minisocieties will therefore recognize different standards of interpersonal comparison. A person's worthiness may therefore vary sharply as he moves between the many minisocieties, permanent or transient, to which we all belong. In brief, one must not forget that the worthiness coefficients of real life are likely to be very sensitive to the *context* in which fairness judgments are made (Section 4.6.7).

Returning to the grand scale, recall that both egalitarians and utilitarians make interpersonal comparisons in the same way in the medium run (Section 4.6.7). Since it is based on a medium-run analysis, the comparative statics carried out in this section therefore applies equally well to both egalitarianism and utilitarianism. However, the conclusions to which one is led have opposite implications when the two principles are applied in the short run. As we shall see, the worthiness coefficients $1/U$ and $1/V$ are determined by the underlying power structure of a society. One then has to face the fact that egalitarianism has nothing to offer those powerless folk who are unable to enlist the sympathy of others. On the other hand, utilitarianism singles out precisely such unfortunates for special treatment. But the helpless would be unwise to expect anything from utopians who use this fact to claim the moral high ground for socialism. Whether we like it or not, evolution made human nature egalitarian rather than utilitarian. Otherwise people would notice that it is hypocritical for utopians to trumpet their sympathy for the starving while dining in luxury restaurants.

4.7.1 Will to Power?

Talk of power is shocking to the modern ear. But some plain speaking is necessary when insisting on its importance in determining the value judgments that societies actually use in conducting their affairs—as opposed to those to which they render only lip service. The absurdly unrealistic propaganda on moral issues with which we mislead our children is so deeply embedded that a little *Sturm und Drang* seems forgivable if it serves to draw attention to the need for some rethinking on a subject about which we commonly feel very comfortable in our prejudices. In Section 0.2 and elsewhere, Machiavelli, von Clausewitz and the like are therefore quoted with approval for their refreshing frankness on the realities of state power. But the harmless delight to be found in shocking the bourgeoisie with such macho posturing needs to be tempered with a recognition that the wordplay of reactionaries attracted by the idea that might is right is just as dangerous as that of revolutionaries.

Consider, for example, Hobbes' [251] mischievous definition of the value of a man:

> The Value, or Worth of a man is, as in all other things, his Price; so much as would be given for the use of his Power.

To make such a definition is to invite misunderstanding. In particular, the right-wing conceit that the value of a man is determined only by his market power allowed Marx to be taken seriously when he claimed that the *only* value a capitalist society recognizes in a worker is the highest price at which he can sell his labor. Taking for granted that labor will always be in excess supply in a competitive market, he then followed David Ricardo in arguing that the iron law of supply and demand will necessarily reduce a worker's wages to subsistence levels.

But markets are only one of many vehicles for the exercise of power. Even in transactions normally governed by the rules of the market, political pull often trumps market power. Nor are markets immune from other imperfections and external influences. History shows that unskilled workers in excess supply are not totally powerless, since they cannot be prevented from unionizing in the medium run. Even when market rules apply, unskilled workers need not be paid starvation wages. As Henry Ford demonstrated, the profit motive often requires the use of apparently generous incentive schemes. Nor was Marx right to treat unskilled labor as the norm. Workers with relevant skills are always in short supply and so well able to look after their own interests.

In brief, political philosophers typically exaggerate the extent to which the fulcrum of power favors the high and mighty. Those of us who occupy less exalted rank have little individual power, but we exercise a great deal of power *collectively* because of our adherence to an implicit insurance contract which requires that we use our superiority in numbers to punish anyone who trespasses on the established rights and liberties of people sufficiently like ourselves. The downtrodden masses of popular tradition are therefore not as helpless as they might seem.[45] They are frequently victimized in the short run, but, just as water finds its own level, so society's underdogs gradually learn to exploit the weak points in the systems employed to oppress them. A regime operating out of equilibrium is thereby undermined and so becomes ripe for collapse when sufficiently stressed.

Modeling power. To say that the worthy are determined by the underlying power structure of a society is sometimes thought to imply the Thrasymachian doctrine that worthiness is to be equated with brute force. But it is necessary to remember that eunuchs have come to rule mighty empires. Nor has the hand that rocks the cradle often been without influence, even in the most patriarchial societies. How is the exercise of such different types of power to be modeled?

When studying equilibrium selection, my theory squeezes all the juice of our way of life into a formal bargaining problem (X, ξ) (Section 1.2). In applications on the grand scale, the set X describes all potential ways in which power can conceivably be exercised in a stable society. Its shape is therefore restricted only by the *rules* of the Game of Life. As Section 2.3.1 of Volume I explains, we must not risk the coherence of our analysis by thinking of the rules of this game as being under our control. We obey the rules of the Game of Morals most of the time because it is optimal to do so, but we obey the rules of the Game of Life for the same reason that we obey Newton's law of gravitation.

Information about who is big and strong or quick and nimble is therefore manifested in the shape of the set X. Along with the immutable realities

---

[45]Nor are they any less inclined than the mighty to kick sand in the face of weaklings who are genuinely powerless. Unless such weaklings have friends or relations who sympathize with their plight, they can expect little from a society whose moral values are shaped in the medium run by social evolution. Hence the misery and abuse endured by the aged who outlive their resources, by uncuddly animals, by evil-smelling and foulmouthed street people, by friendless orphans and the mentally handicapped, and by all the other unfortunates about whom we somehow manage to forget when congratulating ourselves on our benevolent outlook on the world.

of the physical world, such fundamental facts about what people can or cannot do determines which social contracts are feasible. Suppose, for example, that Adam and Eve's Game of Life specifies that they share an enclosure with a trough at one end and a lever to fill the trough at the other. If Adam is a small pig and Eve is a big pig, one can envisage two social contracts: one in which Adam pulls the lever and one in which Eve pulls the lever. But Eve's size makes the first of these infeasible. If Adam pulls the lever while Eve waits by the trough, then he will get nothing to eat, because her piggy nature insists that she exclude him from the trough by brute force. Experiment confirms that the only social contract compatible with the nature of pigs is for Eve to pull the lever while Adam gobbles as much as he can before she can gallop back to drive him off.

The Thrasymachian mistake is to think that power resides *only* in the rules of such a porcine Game of Life. But Glaucon understood that power also resides in the rules of the current morality game—which is manifested in the location of the state-of-nature point ξ. As Edmund Burke explained to utopian admirers of the French Revolution, history matters when social contracts are reformed. When would-be reformers select a social contract σ from the set X, it is therefore necessary that they specify a viable route from our current social contract ξ to the New Jerusalem represented by σ.

Consider, for example, the case of slavery. Before emancipation, the slaves were chained and imprisoned. But the rules of the Game of Life do not specify that it is part of the nature of some human beings that they be hung with chains and kept in close confinement. Their chains can be removed and their prison doors unlocked, provided that their masters can be persuaded or forced to do so. A slavemaster therefore has power over a slave only by virtue of the social contract that is currently in place. It is not the rules of the Game of Life that grant and defend his right to hold slaves, but the rules of the current morality game. Nor need such power over others involve physical trappings like chains and prison cells to be make itself felt. The gossamer threads of common belief can bind just as firmly as any physical chain. As Section 3.4.5 explains, a tyrant is obeyed because everybody believes that everybody else will obey him. To quote Hobbes [251, p.16]: "The power of the mighty hath no foundation but in the opinion and belief of the people." But to know that the root of a tyrant's power lies only in the beliefs people hold in common does not make the fact of his power any less real.

There will always be individuals or classes whom the current social contract makes too powerful to coerce. Finding some way of persuading them that their best interests are served by giving up some of their power in return for other benefits is crucial. If the new social contract to be operated after they have surrendered their power is too radical, they will be immune to persuasion. If reform is to be achieved by mutual consent, we must therefore restrict our attention to Pareto-improvements on the state of nature. However, a coordination problem is then created because many such Pareto-improvements commonly exist.

Fairness is Nature's way of solving such coordination problems, but not even the concept of justice is free from the taint of power. What counts as fair is determined by the worthiness of those to whom justice is dispensed. However, Adam or Eve's worthiness is determined by the historical accidents frozen into the power structure of the social contract currently being operated. As Hobbes [251] explains:[46]

> WORTHINESS is a thing different from the worth or value of a man, and also from his merit or desert, and consisteth in a particular power or ability ... usually named FITNESS, or *aptitude*.

Hobbes' concept of worthiness is narrower than mine. In the preceding passage, he is talking about what I call *social standing*.[47] But Hobbes' point is as relevant to my broader concept of worthiness as it is to his. Worthiness is certainly influenced by a person's native talent and his objective contributions to the public weal, but it must not be confused with either. Like beauty, worthiness lies in the eye of the beholder. It is not a universal property of human nature, but a way of describing the features of the power structure in the state of nature that are relevant to fairness transactions.

4.7.2 Comparative Statics

Having accepted that power is fundamental, we can ask how Adam and Eve's characteristics should be anticipated to affect their worthiness. In short, what makes one person more worthy than another? Economists regard such a question as raising a problem in comparative statics.

In the medium run, the dynamics of social evolution will settle on equilibrium values of U and V—the values that Section 4.6.6 deduces from the requirements for an empathy equilibrium. According to Section 4.4, Adam and Eve's worthiness coefficients can then be identified with $1/U$ and $1/V$, provided that one bears in mind that only the ratio of these quantities has any real significance. But what if social evolution had operated under the counterfactual hypothesis that the Game of Life or the state of nature were

---

[46]Showing that he understood perfectly well that there are values beyond those traded in markets.

[47]To maintain which it is necessary that a man show that he knows his place by paying proper deference to his social superiors, and by gracefully accepting the deference of his inferiors—even in the supposedly classless United States of America.

different? Then different values of U and V would have been obtained. In comparative statics, one studies how such alternative values of U and V vary with the underlying parameters of the problem.

What are the underlying parameters that matter for the comparative statics of worthiness? Although I am no admirer of Karl Marx, it has to be admitted that he put his finger on all the characteristics of a person that seem to be of interest in studying how a society identifies the worthy. Recall that, after the revolution, workers were to be rewarded according to their labor. The Marxist labor theory of value is certainly no jewel in the crown of economic thought, but it is nevertheless true that the relative levels of *effort* Adam and Eve need to exert in creating a source of surplus must be one of the major parameters requiring attention in determining their worthiness when it come to splitting the surplus they have jointly created. Of course, according to Marx, labor was to provide only a stopgap measure of worthiness. In the socialist utopia that would ensue after the state had eventually withered away, the rule was to be: from each according to his ability—to each according to his need. Human nature being what it is, such an incentive scheme seems designed to convert the able of a large society into the needy overnight. Nevertheless, both *ability* and *need* are parameters that must be taken into account when evaluating worthiness.

Finally, we need to add *social standing* or *political pull* to effort, ability and need. For Marx, social standing was a feudal survival to be swept away along with all of history's other failed experiments in human organization, but a realistic social contract theory cannot ignore the fact that the power structures of today evolved from the power structures of the past. A person's social standing, as measured by the role assigned to him in the social contract currently serving a society's *status quo*, is therefore highly relevant to how his worthiness is assessed by those around him.[48]

The background story for all the cases to be studied will be the problem faced by Adam and Eve in writing a marriage contract in the Garden of Eden. As in Section 2.6.1, the issues will be simplified so that the only question is how to share the dowry. In Chapter 2, the dowry was a once-and-for-all gift of one dollar, but here it needs to be reinterpreted as an income flow of one dollar per unit of time.

The worst outcome $\mathcal{L}$ in each story consists in the dowry being entirely wasted, so that Adam and Eve both receive nothing. The best outcome $\mathcal{W}$ is taken to be the counterfactual event in which all the dowry is assigned

---

[48]The hostility that often greets this suggestion tends to subside when it is discovered that individuals favored by the current social contract turn out to be *less* worthy than their country cousins. But is it reasonable to base one's reactions to a claim about how the world works on whether one likes or dislikes its consequences?

to each of Adam and Eve separately. These heaven and hell states always remain fixed and $u_A(\mathcal{L}) = u_E(\mathcal{L}) = 0$ and $u_A(\mathcal{W}) = u_E(\mathcal{W}) = 1$. Adam and Eve's personal utilities are therefore always measured in the same way, and so it is meaningful to compare how their utilities change as we move from case to case.

4.7.3   To Each According to His Need?

The language available for discussing moral issues is so suffused with the mistaken assumptions of folk psychology that it is a wonder we ever manage to say anything coherent at all. The idea of *need* is particularly confused. It is especially important that two of the strands wound together in the folk concept be unraveled. The sense I want to capture is that a person in need lacks something he regards as vital to his well-being. However, I don't mean to imply that he is in need because he is *unable* to get what he lacks—only that he doesn't have it at present.

Despair. In colloquial English, to say that Adam is needy always means that he is in want. But it may also carry the extra understanding that he is persistently in need. If so, then he is presumably powerless to alleviate his want, whether acting alone or collectively with others in the same boat. Nor does his plight raise sufficient sympathy in the breasts of others that those with the power to help are willing to exert themselves on his behalf.

This sense has to be stripped from the notion of need, because it confuses Adam's personal tastes with the social circumstances operating in his society. A powerless person is certainly likely to be in need, but a person in need may not be powerless. Where necessary, I shall say that a person who is unable to satisfy his needs is *in despair*. One can then comment, for example, that the Russian government is currently playing with fire by making its soldiers needy. The military is in despair right now, because a leader has yet to succeed in establishing himself as a coordinating focus for its latent collective power. But who knows what the future may bring?

As in Section 4.5.4, Adam's need is measured by the extent to which he is willing to take risks to get what he lacks. Far from rendering him powerless, such a definition makes Adam relatively powerful when he bargains on equal terms with a less needy Eve. As explained in Sections 1.4.2 and 4.8.2, the more risk averse Eve becomes, the smaller the share she will be awarded by the Nash bargaining solution. The intuition is very simple. Needy people get more because their desperation leads them to bargain harder. Being desperate is therefore very different from being in despair. A needy person will only despair if he is unable to interact with Eve except on her terms.

Empathizing with need. Having unraveled two of the strands that folk psychology tangles together in the concept of need, it is now necessary to address a third strand of meaning. To say that Adam is in need usually carries the connotation that we endorse Adam's perception of his need. Sometimes the endorsement is moral in character, but morality can play no part in the concept of need to be studied here, because all moral judgments are made in the original position. However, my notion of need necessarily requires that we be willing to endorse the rationality of Adam's beliefs about his plight.

For example, we would be reluctant to equate Eve's need for food to fill the bellies of her starving children with Adam's need for a red Porsche with fully reclining seats,[49] although both may be willing to take similar risks to get what they want. Postwelfarists will welcome this observation as an opportunity to respond to the diatribes of Sections 3.8.1 and 4.3.2 by pointing out that it is this kind of consideration that makes it necessary to measure well-being in objective terms. But they again confuse personal tastes with social values.

Eve's need fulfills the criterion required of a personal taste in Section 2.5.5—it remains constant whatever may happen in the medium run. But the same is not true of Adam's need for a Porsche. His biological programming urges his young heart to strive for status, so that young women will find him attractive. A Porsche is only a means to this end, rather than an end in itself. To put it crudely, Adam's ultimate aim is to get laid as often as possible, so that his genes can be spread far and wide. However, Adam's unconscious understanding that this aim is best achieved by acquiring a Porsche depends on his beliefs about his society's culture. The fact that he demands a Porsche rather than some other status symbol reveals his belief that a particular social value is embedded in the social contract of the minisociety in which he moves (Section 3.8.1). If Adam's social standing improves enough as a consequence of driving a Porsche, it will become necessary for him to find more subtle forms of sexual display if his status is to be enhanced further (Section 4.7.7). But it is a mistake to attribute such a medium-run change in his demand for status symbols to an alteration in his personal tastes. Personal tastes remain fixed in the medium run. The cause of the change in Adam's demand is that his altered social standing has changed his beliefs about how best to satisfy these fixed personal tastes.

After Adam's beliefs about his culture have been separated from his

---

[49]Recall that we are discussing payoff flows in a repeated game. An economist therefore dismally translates Adam's wet dream into an income flow sufficient to pay off the loan and to compensate for depreciation.

personal preferences, his need can be compared with Eve's on the same basis. Given that both are to respond optimally to their genetic imperatives, should Adam be willing to take the same sort of risks to get a Porsche as Eve is willing to take to feed her children? Obviously not. Children represent an enormous investment to their mother. If Eve's children die, she may never get the opportunity to make a similar investment in the future. But Adam is only denied one of many opportunities for the relatively costless activity of planting his wild oats in fertile soil.

But what if Adam irrationally persists in being willing to take risks that are comparable with Eve's? Then she will fail to empathize succcessfully with him. Having put herself in Adam's position with his personal preferences, she will predict that he will be much readier to compromise than she. In consequence, they will disagree about what is fair, and the opportunity to split the surplus will be lost. Given long enough, Adam will learn to be more realistic about the way the world works, although few adults recall the teenage traumas that accompany the process with any pleasure. While Adam remains immature, my theory does not apply, because his behavior does not reflect his underlying personal tastes.[50] But once he comes of age, the difficulties in using his risk behavior to assess his need disappear.

What are the needy worth? Much of what is to be said next about computing the worthiness of the needy is anticipated in Section 4.5.4. However, the purity of the example used here will perhaps clarify the principles involved.

Recall that we are using the marriage contract of Section 2.6.1 as our standard example in studying the comparative statics of worthiness. To examine the influence of need on worthiness, suppose that the dowry is provided by the families of the happy couple without Adam or Eve lifting a finger. Economists sometimes emphasize that such an assumption is being made by referring to the dowry as "manna from heaven". While discussing need, it will be further assumed that the dowry is an indivisible object D that can only be enjoyed by one person at a time. Attention is confined to sharing rules that only determine how much of the time Adam gets to enjoy D on average compared with Eve. With our standard simplifying assumptions about discounting over time (Section 3.3.3), a marriage contract can then be characterized in terms of the probability p that Adam will be in possession of the object at a randomly chosen time.

---

[50]Given the power to intervene, I would then follow the postwelfarists in finding in favor of Eve, as I did when ruling on disputes between my children when they were young. But I see no reason to over-rule the judgments of mature adults about their personal tastes. I certainly want nobody doing this to me.

What is the bargaining problem (X, ξ) that Adam and Eve face in negotiating a marriage contract? Their state of nature point ξ is located at 0. The Pareto-frontier of the set X of feasible social contracts is the straight line through the points $(a, 0)$ and $(0, e)$, where $u_A(\mathcal{D}) = a$ and $u_E(\mathcal{D}) = e$. Since both players regard possession of $\mathcal{D}$ as an outcome intermediate between $\mathcal{L}$ and $\mathcal{W}$, we need to assume that $0 \leq a \leq 1$ and $0 \leq e \leq 1$.

According to Section 4.6.6, the action of social evolution in the medium run will result in a fair marriage contract at the Nash bargaining solution σ for the problem $(X, 0)$. Since $\sigma = \frac{1}{2}(a, 0) + \frac{1}{2}(0, e)$, the probability p that Adam will then be in possession of the dowry at a randomly chosen time is always $\frac{1}{2}$, regardless of the values of a and e. Adam and Eve will therefore always share $\mathcal{D}$ equally. However, sharing the dowry equally does not necessarily assign the players equal payoffs relative to their personal utility scales—whose zero and unit we anchored arbitrarily[51] at $\mathcal{L}$ and $\mathcal{W}$. In fact, Adam receives a payoff of $\frac{1}{2}a$ and Eve a payoff of $\frac{1}{2}e$.

To determine how Adam and Eve's utils are compared in the medium run, it is necessary to choose U and V so that the proportional bargaining solution with these weights coincides with the Nash bargaining solution. Since the straight line joining $\xi = 0$ and $\sigma = (\frac{1}{2}a, \frac{1}{2}e)$ has slope e/a, we must therefore take $U/V = e/a$. It follows that we can identify Adam's worthiness $1/U$ with the parameter a and Eve's worthiness $1/V$ with the parameter e. A player's worthiness therefore increases with the value he or she places on the dowry.

Recall that need is to be measured in terms of the risks that people are willing to take to satisfy their lack of something important to them. Fortunately, this is precisely the information encoded in a Von Neumann and Morgenstern utility. In fact, as Section 4.2.2 of Volume I explains, to say that Eve assigns a utility of e to the possession of $\mathcal{D}$ means that she is just indifferent between $\mathcal{D}$ and a lottery in which she gets the prize $\mathcal{L}$ with probability $1 - e$ and the prize $\mathcal{W}$ with probability e. If e is almost one, it follows that she is so averse to the outcome $\mathcal{L}$ that the prospect of getting $\mathcal{W}$ instead of $\mathcal{D}$ is nowhere near attractive enough to persuade her to tolerate more than a tiny probability of ending up with $\mathcal{L}$. In brief, she is so desperate not to get the outcome $\mathcal{L}$ that she ceases to distinguish very much between the other possible outcomes. If it is sensible to say that a person with such preferences *needs* to avoid $\mathcal{L}$, then Eve's worthiness *increases* with her need, provided that Adam's need is held constant.

Other things being equal, an egalitarian using the proportional bargaining solution will therefore award higher gains to people who are more

---

[51] It is always necessary to be on guard against falling into the error of assuming that it is meaningful to make interpersonal comparisons using the zero-one rule (Section 2.3).

needy when a short-run opportunity to divide a surplus arises. A utilitarian will award the needy smaller gains. If this seems paradoxical, recall that my definition of need makes the needy desperate, and hence dangerous adversaries. But utilitarians discriminate against those able to look after themselves in order to reward the powerless. To merit favorable treatment from a utilitarian, Eve must not only be in need, she must have despaired of satisfying her need without a utilitarian white knight to ride to her rescue. But she will not then have been competing on equal terms with Adam.

 ### 4.7.4   Arbeit Macht Frei?

To study the impact of effort and ability on worthiness, we shall abandon both the assumption that the dowry is indivisible and the assumption that it is provided like manna from heaven. It will be assumed instead that the dowry is created as a consequence of Adam's effort. However, he can enjoy some fraction of the fruits of his labor only with the consent of Eve, who therefore fills the role of a traditional robber baron. But in spite of being a peasant, Adam is not entirely without power. If mutual agreement on a social contract cannot be achieved, he will produce nothing at all.

This is the ancient story of Capital versus Labor. In prehistoric times, the capital was the strength invested in male bodies by Nature. When a female gatherer brought home the fruits of her labor to feed her children, her mate would be waiting to commandeer her product in the manner typical of his lazy and brutal sex.[52] In the time of Adam Smith [501], the relevant capital consisted of the state-enforced property rights of rentiers, which allowed them to expropriate a sizable fraction of an agricultural surplus to whose production they had contributed nothing whatever. In modern times, capitalists commonly contrive to pay no taxes at all, while simultaneously enjoying the public goods provided from taxes subtracted directly from the pay checks of the poor. Performances of grand opera subsidized in this way are a particularly blatant example.

Figure 4.12(a) illustrates the new scenario. Adam's preferences over bundles (y_A, ℓ), in which y_A is the payoff he receives as his share of the dowry and ℓ is the labor he exerts in producing the dowry, are assumed to be quasilinear (Section 4.3 of Volume I). Each point in the set X of social contracts that would be available if the dowry were manna from heaven therefore needs to be displaced a distance ℓ to the left to obtain the set Y of social contracts that require effort from Adam to produce. How does such a displacement affect Adam's medium-run worthiness?

---

[52]Perhaps this is why the fruits and vegetables gathered largely by women are less universally shared than the meat hunted by men.

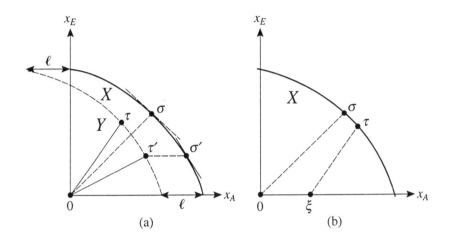

Figure 4.12: Computing worthiness.

In the manna-from-heaven case, U/V is the slope of the straight line joining 0 and the Nash bargaining solution σ for the problem $(X, 0)$. In the new scenario, U/V is the slope of the straight line joining 0 and the Nash bargaining solution τ for the problem $(Y, 0)$. As in Figure 1.5(c), the slope of the Pareto-frontier of X at σ is the same as the slope of the ray joining 0 and σ. It follows that τ' in Figure 4.12(a) cannot be the Nash bargaining solution of $(Y, 0)$ because the slope of the Pareto-frontier of Y at τ' is the same as that of X at σ', which is steeper than at σ, whereas the ray from 0 to τ' is less steep than the ray from 0 to σ. The ray joining 0 and the Nash bargaining solution τ for $(Y, 0)$ must therefore be steeper than that joining 0 and σ.

If we keep $V = 1$, the argument shows that U must become larger when the dowry ceases to be manna from heaven and is obtained instead from the sweat of Adam's brow. The need for Adam to exert effort to produce a surplus therefore *diminishes* his worthiness.[53]

---

[53] Veblen [535] is a delight on this subject. Why did men of his time assert their worthiness by wearing high starched collars? Because nobody can earn their living by the sweat of his brow dressed in such an instrument of torture. Why do women still wear long white gloves at ceremonial events? To show that their lifestyle doesn't even require that they touch anything.

If this conclusion seems counterintuitive, it is because we do not nowadays commonly encounter situations in which an *individual* has Eve's power to reduce all of Adam's effort to naught. Under more normal circumstances, Eve must perforce concede the lion's share of the surplus to Adam lest he opt out of their relationship. Even a prehistoric forager did not have to bring home all of her product. Her outside option was to feed herself and her children wherever the fruits of the field were gathered. Similarly, a modern worker has the outside option of taking another job or going on the dole. Eve will only perceive Adam as being unworthy because he earns their bread by the sweat of his brow, if he has no option but to submit to her power. He is then effectively a serf.

But although individuals in Western democracies are seldom able to exercise such power over their fellows, it does not follow that the feudal memes that made forced labor a source of unworthiness are dead. In the neofeudal social contracts operated by modern Western democracies, contemporary robber barons no longer hire troops of mercenaries to extract tribute from those powerless to resist. Instead, they use the apparatus of the state as an intermediary in funneling money from our pockets to theirs. We peasants do not have the option of avoiding taxation legally. We are therefore perceived as being unworthy *even by ourselves* when compared with the neo-aristocrats who have used their power to make their tax-evasions schemes legal. We even admire their displays of *noblesse oblige,* when they demonstrate their social superiority by giving a little of their leisure or money to good causes (Section 4.7.7).

4.7.5   From Each According to His Ability?

Ability is not so difficult to define as need, but its meaning still requires some discussion. It is commonly identified with the individual characteristics that make one person more productive than another. Under this heading come physical and mental traits like native intelligence, deftness of hand, strength of body, and charm. Such characteristics are products of our genetic and social inheritance over which we have little control. Education and technical skill are examples of other traits that belong under the same heading, but are acquired as the result of a conscious investment decision. Education and technical skill are the leading examples of such items of human capital. It is important that the effort required to acquire such items of human capital represent a *sunk* cost[54] when being counted amongst a

---

[54]Economists emphasize that *sunk* costs should not influence current decisions. Poker players who continue betting with a mediocre hand to "protect their investment" are a case in point. The money you have put into the pot already is a sunk cost. The fact that you will lose this money is irrelevant to whether you should now fold. You lost the

person's abilities. Any continuing effort Adam exerts to maintain his skills will diminish the worthiness his ability would otherwise buy.

I always seem to be meeting people who combine brilliant and cultivated minds with beautiful and athletic bodies. Such birthday gifts from Fairy Nature and Fairy Nurture make them better able to get by in the world than the rest of us. But some comfort is to be found in the reflection that nature and nurture are less significant in magnifying a person's productive power than the technology available to his society. A man with a combine harvester is vastly more productive than a team of peasants armed with scythes. Organizational advances are often equally important. For example, Adam Smith [501] famously estimated that the division of labor in a pin factory magnified the productivity of a worker by a factor of at least 240. In most cases, Eve's ability to produce therefore depends more on how much her boss invests in maintaining and improving the technology of his enterprise than on her own personal gifts.[55]

Various factors that influence ability have been mentioned. We now abstract away these considerations and simply define an able person to be someone who can produce the same output as a less able competitor without exerting as much effort. With this definition, the problem of studying the effect of changes in ability on worthiness becomes trivial. The more able we make Adam, the less effort he will need to exert in order to create a given level of surplus. Since effort decreases worthiness, ability therefore increases Adam's worthiness, provided that his ability is channeled toward maximizing the available surplus to be divided between Adam and Eve.

One might summarize the results of Sections 4.7.4 and 4.7.5 by saying that, other things being equal, an egalitarian using the the proportional bargaining solution will favor talented aristocrats at the expense of retarded serfs. But utilitarians should hesitate before congratulating themselves on reversing this bias. We no longer live in a stagnant feudal economy. Before any judgments are made, even utopians need to consider the impact of their

---

money as soon as you contributed it to the pot, and it now belongs to whoever turns out to be the winner. Since you will need to bet a lot more to be present at the showdown, the only rational question is whether the probability that you will turn out to be the winner is sufficiently large to justify paying so much *extra* money to stay in the game. Similarly, the costs spent acquiring an education do not equate with effort because they were sunk in the past.

[55]By virtue of his continuing investment, Eve's boss lowers his relative worthiness, because his costs equate with effort in the relevant calculation. However, although employing skilled labor and paying back the interest on money borrowed to invest in equipment and good organization decreases his relative worthiness, he does not lose out by offering a fair wage. He simply substitutes a smaller share of a big cake for a larger share of a cake that may disappear altogether if he falls too far behind in the race with his competitors.

system on the *dynamics* of incentives and innovation (Section 4.7.7).

4.7.6   The High and the Lowly

If Aristotle's [18] concept of nobility is identified with social standing, it becomes apparent why temperance is regarded as a noble virtue in fair societies (Section 3.8.1). Other things being equal, worthiness *declines* with social standing. In accordance with the principle of *noblesse oblige*, the high and mighty therefore take a *smaller* share of a surplus than their lowly brethren.[56]

To confirm that social standing reduces worthiness, it is necessary to turn the spotlight away from the Pareto-frontier of the feasible set X to the state-of-nature point ξ. Adam's social standing in the current social contract will be measured by the payoff he receives at ξ. If we increase his social standing ξ_A, then Adam's payoff at the Nash bargaining solution for the problem (X, ξ) will also increase. High social standing therefore makes Adam better off. But what happens to his worthiness?

This question is easily resolved by recycling the argument of Section 4.7.4, which was used to show that effort reduces worthiness. Moving the *status quo* point ξ a distance ℓ to the right is equivalent to first displacing X a distance ℓ to the left and then shifting the whole configuration a distance ℓ to the right. Since the last operation merely shifts the Nash bargaining solution σ for $(X, 0)$ a distance ℓ to the right, it follows that the result of increasing Adam's social standing in the current social contract is to *reduce* his worthiness, as illustrated in Figure 4.12(b).

The dynamics of social status.   How is social standing acquired? I follow Homans [260] on this subject. After proposing a close analogue of the proportional bargaining solution as an empirical description of the way people actually make fairness judgments, Homans suggests characterizing the gains each receives in terms of units of esteem. His concept of esteem is somewhat fuzzy, but since social status presumably correlates closely with the extent to which people are imitated by others, it could perhaps be operationalized in terms of social fitness—which I identify with whatever makes a person a locus for the replication of his memes to the heads of other people. When the power to get anything done depends primarily on a

---

[56]A lowly Adam therefore sometimes insists on paying more than his fair share in the hope that Eve will be misled into overestimating his social standing. Hence the tiresome coordination failures when status seekers try to grab the bill in restaurants. The fierceness with which such attempts at patronage are resisted seems out of all proportion to the offense. We seem to be programmed to resent being patronized almost as much as being pitied.

person's social standing, it may therefore not be unreasonable to reinterpret my utils, first as units of social fitness, and then as units of esteem.

Units of esteem are valuable to Adam because they increase his social standing. In an egalitarian society, where esteem is acquired largely through fairness transactions, he will therefore benefit from having a large worthiness coefficient. As he accumulates esteem, his increased social standing will shift the state of nature in his favor. But such a shift in the location of the state of nature will lower his worthiness. Eventually, an equilibrium will be established in which his flow of esteem from fairness transactions is just sufficient to sustain a degree of worthiness that maintains the flow of esteem at its current level. Egalitarians can therefore congratulate themselves on subscribing to a self-regulating system that automatically corrects any tendency for inequalities between social classes to widen.[57] My claim that prehistoric foraging bands were organized along egalitarian lines would certainly look very thin otherwise.

4.7.7 Socialism versus Capitalism

Social esteem is a tricky concept, and the dynamics proposed for social standing are correspondingly speculative, but the dynamic effects of the other factors that influence worthiness are much more straightforward. Their study merely serves to sharpen the consensus view that old-style socialism stifles enterprise and initiative.

A utilitarian society neglects the needy because their willingness to take risks makes them able to look after themselves. It handicaps the able because they don't have to work as hard as others to produce the same output. Entrepreneurs who are willing to take risks and investors who are willing to sink money into new enterprises are punished for the same reasons. Getting more education than others amounts to sinking an investment in oneself, and so scholarship is similarly disadvantaged. More generally, utilitarianism frowns on anything whatever that empowers Adam and Eve to get on in the world under their own steam. Self-help, thrift, ingenuity, and enterprise are discouraged by steering resources away from those who display such whiggish virtues to unthinking workhorses like Boxer in Orwell's *Animal Farm*.

Section 2.8 argues that utilitarianism would be the first-best society in a *static* society—provided that the pigs in the farmhouse could be prevented from being corrupted by power. But it offers all the wrong incentives to

---

[57] I don't think the same argument shows that the classes in a utilitarian society will be driven further and further apart, because it is hard to believe that the powerless would genuinely be esteemed, even in a utopian society that puts the relief of their despair at the top of its priority list.

a society whose cultural survival depends on how well it competes on economic terms with other societies. Rather than belaboring the fall of the Soviet Empire and the retreat from social democracy elsewhere, let me offer a slightly doctored extract from the doggerel poem in which de Mandeville [341] expresses the moral to be drawn from his fable of the *Grumbling Hive*, in which the bees destroy their society by actually adopting the utopian virtues they had previously honored only in the breach:

> So Vice is beneficial found,
> When it's by Justice lopt and bound;
> Nay, where the People would be great,
> As necessary to the State
> As Hunger is to clear a plate.

Modern utilitarians commonly accept de Mandeville's prophetic observation that socialism cannot compete with capitalism in the creation of economic wealth, but they see no reason to give any ground to egalitarians on this count. Why should the economic success of capitalism be any more relevant to the Right or the Good than the fact that money can be made by mugging tourists in back alleys? In this at least, postwelfarist egalitarians agree. After all, the Rawlsian program of maximizing some objective measure of the well-being of the least successful class of society is not much better at providing incentives for the creation of wealth than utilitarianism.

I also share the view that wealth creation is irrelevant to the battle between the Right and the Good, but for the very different reason that such debates over invented Moral Absolutes have nothing to do with the real world. Morality evolved in the human race to coordinate human behavior on Pareto-improving equilibria in the Game of Life. The versions of egalitarianism and utilitarianism being compared in this section respect this insight of David Hume, and hence have been reconstructed as alternative approaches to the coordination problem within a theory of the seemly. They remain relevant to much of the traditional debate between egalitarians and utilitarians because absolutist intuitions about the Right and the Good actually derive from seeing genuinely workable social contracts in action. But followers of the seemly have to part company with moral absolutists when they insist that Morality demands that opportunities to share a surplus should sometimes be neglected. At this point, absolutists lose contact with the intuitive understanding of the workings of real social contracts that keeps them on the right track much of the time. Fairness norms evolved as a way of coordinating among the Pareto-efficient equilibria of the Game of Life. When properly formulated as a theory of the Seemly, egalitarianism respects this function, but utilitarianism does not.

4.8   The Market and the Long Run

If Hobbes' *Leviathan* represents his idea of a just polity, in which individual citizens coordinate their efforts like the cells in a healthy body, then the anarchic history of the British civil war he recounts in *Behemoth* seems an appropriate metaphor for the operation of the free market.[58] No hand, invisible or otherwise, directs the traders on the floor of the Chicago wheat market as they scream and shout and throw their arms in the air. But the sum of their actions takes prices to their market-clearing values with amazing rapidity.

The manner in which order springs spontaneously from chaos in such circumstances has led to the market being used as a general metaphor for self-organizing social mechanisms that operate without the intervention of any central authority.[59] So compelling is the metaphor that it has led a generation of right-wing thinkers to overlook the fact that it is only a metaphor. The mistake is then made of seeing all self-organizing social phenomena as markets whose failings must necessarily be treated with the same medicine that one would apply to an ailing market. Coase [128] even proposes modeling the propogation of knowledge as a market in ideas!

The most dangerous version of this mistake occurs when the market is proposed as a model of the way an ideal society should work, with the role of government reduced to providing public goods and internalizing externalities. I agree that part of the role of a government can usefully play is to extend the range of available goods and to assist in the creation of new markets, but to see a government only in such terms is to wear blinders. Aside from other considerations, it seems obvious that the existence of

---

[58]What's the market? A place, according to Anachasis, wherein they cozen another, a trap, nay, what's the world itself? A vast chaos, a confusion of manners, as fickle as the air, a crazy house, a turbulent troop full of impurities, a mart of walking spirits, goblins, the theatre of hypocrisy, a shop of knavery, flattery, a nursery of villainy, the scene of babbling, the school of giddiness, the academy of vice; a warfare where, willing or unwilling, one must fight and either conquer or succumb, in which kill or be killed; wherein every man is for himself, his private ends, & stands upon his own guard—Burton's [110] *Anatomy of Melancholy.*

[59]Hayek [241, 242] is commonly credited by libertarian thinkers with having freed political philosophy from the social contract tradition by inventing the revolutionary concept of *spontaneous order*. But the notion goes back at least as far as Lucretius [335], and must surely have been familiar to Hayek from the works of Hume [267] and Darwin [145]. Nor does the fact that a political philosopher makes use of a contractarian metaphor imply that he believes that our societies were planned by ancient social architects. One might as well argue that Adam Smith's use of the metaphor of an invisible hand implies that he believed in the real existence of the fictional auctioneer of neoclassical economics! In my theory, the social contract is similarly a metaphor for the spontaneous order generated in a society by the action of biological and cultural evolution.

a well-developed social contract is a *precondition* for the emergence of a market. Even the notion of a private good would not be meaningful in the absence of some of the common understandings built into our culture that right-wing thinkers insist should be envisaged as public goods. The idea that law and order is something that can be measured adequately only in terms of the amount spent by government on its enforcement has proved particularly disastrous.

Nor does it seem particularly useful to assess social institutions in terms of how far they are forced to deviate from market ideals by transaction costs that would be zero in the case of perfect competition. Indeed, the Coasian vision of the world as a perfectly competitive arena, marred by occasional patches where the market model does not apply, because transaction costs become prohibitive seems to me like a photographic negative. Our arena is the Game of Life, which is played according to market rules only in a very restricted set of circumstances.

To deny the universality of the market model is not to overlook the fact that markets often provide a flexible and robust tool for the efficient distribution of resources. Nor is there any doubt that Coase was right to emphasize the importance of assigning property rights unambiguously when the market mechanism is applied in a new context. But setting up a market is only one of many ways we can plan to allocate resources.

Whigs like myself are at one with marketeers in our suspicion of command structures administered by armies of bureaucrats whose selfless devotion to the service of the community is a precondition of the system's successful operation. However, markets are not the only alternative to the type of command economy advocated by old-time socialists. Nor are capitalist economies at all closely modeled by the paradigm of perfect competition. Many different kinds of socioeconomic organization are in use, and new types are being experimented with all the time. Indeed, part of the reason for the success of game theory is that it provides a language that can be used to describe such structures as they evolve.

If we are sufficiently clever, we may even learn to use the freedom of thought offered by the language of game theory to escape the false dichotomy perceived by traditionalists on both the left and the right. We do not need to choose between the market and a command economy. It is not necessary for the left to deny that a stable society needs to allocate resources efficiently, and that decisions must therefore be decentralized to the level where the relevant information resides. Nor need the right pretend not to notice that a stable society must plan to allocate resources fairly lest those who find themselves unjustly treated seek violent or criminal redress. The subject of *mechanism design* suggests that it may be possible to have things both ways by using game-theoretic ideas on a grand scale. In such a vision of the future, the virtues of the market would be retained by leaving decisions to be made by the people on the spot, but with their behavior constrained by rules selected to provide incentives that make it optimal for decision makers to choose in accordance with an agreed plan.

However, we are a long way yet from knowing how such a scheme would work in detail—otherwise this book would not have been forced into the simplifying assumption that Adam and Eve are perfectly informed about everything that matters. The major problem is that the theory of mechanism design, as so far developed, relies on an external enforcement agency to ensure that the mechanism's rules are honored. But to be relevant to social contract theory, the rules would need to be self-policing.

Justice or market values?   Not only must a whig take issue with marketeers over the range of equilibria in the Game of Life available to a society, he also has to contend with the almost religious fervor with which it is commonly maintained that markets are necessarily good for us. The justification of this claim lies in the First and Second Welfare Theorems to be found in all textbooks on economic theory.

The First Welfare Theorem asserts that Walrasian equilibria are Pareto-efficient under certain conditions. But these conditions are much more restrictive than is commonly appreciated. For example, returns to scale in production need to be decreasing and markets in futures must be complete. No guarantee that most of the markets that operate in the real world will generate an efficient outcome therefore exists.

As for the Second Welfare Theorem, this asserts that any Pareto-efficient allocation can be achieved using the market mechanism by transferring goods between the agents before trading begins so as to relocate the endowment allocation suitably.[60]   For example, an egalitarian government might choose to equalize the endowments of all citizens and then allow them to trade until Pareto-efficiency is achieved (Section 4.3.1). In textbooks, the Second Welfare Theorem is therefore accompanied by a piece of philosophy which argues that the market mechanism need not be associated with any particular political position, since even Robin Hood might find it useful as a tool for attaining efficiency if propelled into a position of power. But this piece of political philosophy is not consistent with the practice of identifying Pareto-efficiency with social optimality (Section 4.3).

To say that an objective is socially optimal is to claim that society has no further grounds for intervention once the objective has been attained. To identify social optimality with Pareto-efficiency is therefore to assert

---

[60]The argument is not hard, since the endowment point can be relocated at the Pareto-efficient outcome in question.

that it is irrelevant to a society whether an allocation is unfair provided it is efficient.

Right-wingers may respond that such an approach to social optimality is no different in principle to the line I take on the evolution of fairness norms. Section 2.5.4 argues that fairness norms did not evolve because our ideas about fairness lead to goods being distributed in a manner that has some intrinsic survival value. Fairness norms evolved because groups need to get to Pareto-efficient equilibria in the games they play without costly internal conflict. If some other norm had evolved for this purpose—so that goods were distributed unfairly according to current ideas—we would be provided instead with equally strong intuitions about the essential rightness of this alternative norm.

In those special cases where the First Welfare Theorem applies, such a right-wing line of reasoning leads to the suggestion that evolution has already thrown the market into the ring to compete with fairness as a means of getting society to a Pareto-efficient outcome. Indeed, the historical spread of market institutions to wider and wider circles of economic activity suggests that the market is a very successful rival. As when the laws of supply and demand finally overthrew the medieval concept of a fair price, has the time not come to admit that our intuitions about a just society have been superseded by a new paradigm? Once the market meme has replaced the fairness meme altogether, so the story goes, it will become irrelevant whether the market is unfair in the traditional sense. Indeed, right-wing commentators like Sennet [485, p.43] are already only too willing to assist in the process by redefining a fair outcome to be whatever results when the invisible hand operates without constraint.

The purpose of this section is to make a case against such a dramatic scenario. I agree that evolution is eventually likely to generate social contracts that operate like markets in the restricted class of subsidiary games of the Game of Life that admit such a solution to their equilibrium selection problem—but not because the market mechanism embodies a set of striking new values that trump a bunch of outdated fairness norms, which survive only as fossil memories of social contracts operated in prehistoric times by hunter-gatherer communities. On the contrary, insofar as the market is fated to triumph when the circumstances are favorable, I shall argue that it is because the market is precisely what one should expect to see in the long run as a result of people's personal preferences adapting over time to the use of the device of the original position as a fairness norm.

We have already studied what happens when allocations are made using the device of the original position after *empathetic* preferences have been allowed to adjust in the medium run. As Section 4.6.6 documents, although Adam and Eve may continue to describe their social contract as fair, they will actually agree on the Nash bargaining solution to their bargaining problem. All moral content in the fairness procedure they employ will then have been eroded away, since the compromise they reach is the same as if they had negotiated face-to-face using whatever bargaining power lay at their disposal.

This medium-run argument takes for granted that the players' personal preferences remain fixed while their empathetic preferences vary. But the classification of time spans introduced in Section 2.5.5 maintains this assumption only in the short and medium run. In the long run, *personal* preferences may also vary in adapting to whatever allocation mechanism is in use. We then need to ask what personal preferences Adam and Eve would choose to reveal if both were seeking to maximize their fitness on the assumption that the preferences they reveal were to serve only as inputs to the Nash bargaining solution. If neither player would wish to misrepresent his or her personal preferences in a situation that can be modeled in market terms, then it turns out that the use of the Nash bargaining solution implements a Walrasian equilibrium.

The market is therefore the final step in a process that first leaches out the moral content of a culture and then erodes the autonomy of its citizens by shaping their personal preferences. Such a conclusion is at odds with the right-wing rhetoric that idealizes the market system as the embodiment of true freedom. Neither will it be welcome to those left-wingers who advocate a welfare state that provides a fair deal for all at the expense of stifling entrepreneurial activity, since it suggests that any such system is likely to degenerate over time into something resembling a market. Whigs may congratulate themselves on the fact that the market mechanism goes a long way toward meeting their concerns about corruption, but the reasons offered in my story are not particularly comforting. Just as a rock that has fallen into a pit cannot easily be induced to fall any further, so a market cannot easily be corrupted because its institutions have been corrupted already.

However, although nobody has any grounds for delighting in the fact that the moral values of a culture operating in a fixed Game of Life will be eroded away in the long run, neither is there cause for despair. One is certainly not entitled to the conclusion that morality has no role to play in a developed society. The short-run and medium-run considerations summarized in Figure 4.8 remain valid whatever may happen in the long run. A short-run modification of the Game of Life, perhaps caused by a technological or environmental change, will therefore provoke the same demands for a fair distribution of the surplus in an advanced economy as in a band of hunter-gatherers.

Thaler's [520] down-home example of the market for shovels after a snowfall captures the realities of the phenomenon very neatly. When the snow falls, the demand for shovels increases. If supply is to be equated with demand and shovels are in fixed supply, then the price of shovels must rise. But consumers then react by condemning the rise in price as unfair.[61] They therefore punish the retailers for cheating by refusing to buy.

Of course, such a refusal to buy is a short-run response. If it began to snow every day at the onset of a new ice age, people would cease to insist that fairness requires operating at the prices that cleared the market before the snowfall. In the long run, they would learn to regard the new market-clearing prices as fair.

Right-wing politicians like Mrs. Thatcher, who abstract away such out-of-equilibrium phenomena, put the social contracts entrusted to their care in peril.[62] The anger generated by envy is a powerful force, especially when rationalized as righteous indignation. To ignore the social mechanisms by which it is generated is to make the classic mistake of neoclassical economics—to assume that everything is always in equilibrium. But all social systems take time to find their way to an equilibrium. To find their way to a long-run equilibrium, they need a *long* time. However, as Section 2.7 explains, fairness evolved to provide *short-run* resolutions to the equilibrium selection problem.

4.8.1   The Walrasian Bargaining Solution

Section 4.3.1 is orthodox in emphasizing the role of prices in coordinating behavior on a Walrasian equilibrium. Once prices have been established, all commodity bundles can then be valued in money. But the banknotes we use as money are valuable only by convention. In societies operating different social contracts, different conventions have operated. In Hume's time, gold and silver were used as currency. In prisoner-of-war camps, cigarettes served the same purpose during the Second World War. But such conventions cannot have predated the appearance of established market rates at which goods were bartered. So how did markets get going before money was available to facilitate exchange? How did social evolution come up with the idea of a Walrasian equilibrium?

Core.   The standard textbook story is that of the core. If all members of a potential coalition can get more than society is currently allocating

---

[61]The retailer's worthiness is unchanged after the snowfall, because he bought the shovels from the wholesaler at pre-snowfall prices, while the consumers see no reason why their increased need should make them any less worthy.

[62]Recall that Mrs. Thatcher was ousted from power as a consequence of a widespread tax revolt created by the perception that her poll tax was unfair.

to them by redistributing their own endowments amongst themselves, then they have a reason for blocking the current allocation by jointly refusing to cooperate in its implementation. The set of allocations that cannot be blocked is called the core (Section 0.5.1). In a sufficiently large pure exchange economy, it turns out that the core coincides with the set of Walrasian equilibria (Section 4.3.1). If blocking coalitions can always be guaranteed to form whenever the conditions are ripe, the story is that only Walrasian equilibria can survive.

But who says there is some process by which allocations have to be unanimously agreed before they get implemented, rather than a piecemeal system in which the final allocation is determined by large numbers of independently bartered deals among small numbers of agents? Even if such a process were established, why should the mere fact that a coalition *can* block an allocation be thought to imply that it *will* block the allocation? Nothing could ever be agreed in zero-sum games if this assumption were valid, since the core of such games is always empty. The assumption overlooks the folly of participating in a blocking coalition without first asking yourself whether you will *eventually* get more or less than you are currently being offered, after *all* blocking activity is over.

Consider, for example, an economy based on the Edgeworth box of Figure 4.2(a). If Adam and Eve negotiate a deal that implements the Nash bargaining solution, then they will divide each of the two available goods equally. By contrast, if they settled on the Walrasian equilibrium, then the fig leaves would still be divided in roughly equal shares, but Eve would get nearly all the apples. Adam therefore prefers the Nash outcome to the Walrasian outcome. Now follow Edgeworth by replicating Adam and Eve a large number of times. We know from Section 4.3.1 that the result of blocking everything blockable is that the Walrasian equilibrium will be the final outcome. But if the Nash allocation is on the table, why should an Adam join a blocking coalition, since any advantage he may thereby gain can only be temporary? As Von Neumann and Morgenstern [538] realized when rejecting the core in favor of their concept of a stable set (Section 0.5.1), the blocking story only provides a defense of the core in cases when the players are quite unreasonably myopic.

In rejecting the core story as a viable explanation of how Walrasian equilibria first appeared on the scene, I am not denying that the core is an important concept in game theory. It arises repeatedly in situations to which the criticisms offered here do not apply. But like much else, the weaknesses of its foundations as a fundamental concept are ignored in the literature because of the seductive conclusions to which its adoption leads.

Before money. The beginnings of a much more plausible account of the origins of the market mechanism are to be found in the matching and bargaining literature.[63] In this literature, traders are modeled as meeting in pairs, who then bargain, taking into account the outside options awaiting them if their negotiation were to break down. These outside options include the possibility of finding another agent with whom to trade. A typical result is that the system converges on a Walrasian equilibrium when the underlying frictions or transaction costs become vanishingly small.

Such work on bartering systems serves to refocus attention away from the traditional stories with which the idea of a Walrasian equilibrium is commonly defended to its merit as bargaining solution. This may seem a perverse suggestion after Chapter 1 worked so hard and long to advertise the merits of the Nash bargaining solution. But Nash is not to be dethroned in favor of Walras. When information is perfect and the players are able to bargain without institutional constraints, the Nash bargaining solution will remain the foundation stone of our analysis. However, in other circumstances, the door remains open for alternative bargaining concepts.

 Bargaining in an economic environment. Nash's [380] theory of bargaining assumes that the players' preferences are common knowledge. To distinguish his approach to bargaining from the general case, Nash proposed using the word "haggling" to describe situations in which players negotiate without being fully informed about the personal preferences of their opponents.

Harsanyi's [232] theory of incomplete information provides a tool that allows haggling to be modeled in some circumstances, but a convincing analysis of the problems that arise when players seek to take strategic advantage of the ignorance of their opponents remains beyond our reach at this time. The reason is that realistic haggling models typically have large numbers of equilibria, and so one faces the equilibrium selection problem in a particularly aggravated form (Binmore *et al* [80]). My guess is that the state of the art in this area will remain unsatisfactory until substantial progress has been made on how the trial-and-error adjustment processes that take real people to equilibria in games actually work.

The haggling problem is mentioned only to deny that the material on bargaining presented in this section is intended as a contribution to the literature on how the ignorance of players can be exploited by their opponents. The much less ambitious questions to which the ideas apply will

---

[63]See Rubinstein and Wolinsky [450, 451], Binmore and Herrero [78, 77], Gale [191, 192, 193], and numerous others.

become apparent in Section 4.8.2. In the interim, the axioms from Binmore [66] given below can be treated as a contribution to what Roemer [433] calls the problem of bargaining in economic environments.

When Nash formulated an abstract bargaining problem as a pair (X, ξ) (Section 1.3), he took for granted that only the utility information encoded in this data is relevant to the bargaining behavior of rational players. This is a reasonable assumption when the bargainers are unhampered by informational or institutional constraints. But what happens when players are so constrained?

For example, in Section 4.3.1 Adam and Eve are only be able to communicate in terms of prices. In such cases, the economic environment on which their utility functions are defined becomes relevant. One can no longer follow Nash [380] in defining a bargaining solution as a function f that maps a pair (X, ξ) to a point of X. The class of bargaining functions must be expanded to take account of the possibility that the outcome on which rational bargainers agree when faced with the pair (X, ξ) might depend on the economic environment underlying their problem.

Only two simple economic environments will be considered. In the first case, the economic environment is based on an Edgeworth box $\mathcal{E}$ with the properties described in Sections 1.2.5 and 4.3.1. When bargaining in the Edgeworth box is discussed, it is usually taken for granted that the *status quo* should be placed at the endowment point e, but this assumption requires some justification.

Section 1.2.2 distinguishes between the breakdown point and the deadlock point in a negotiation. The former represents the most that the players can get if either abandons the negotiations unilaterally for an outside option, leaving the other no alternative but to do the same. The deadlock point is what the players will get if they bargain forever without reaching an agreement. As Section 1.2.2 explains, although the *status quo* in Nash's bargaining theory belongs at the deadlock point, it is commonplace in applied work to locate the *status quo* at the breakdown point instead. Fortunately, this issue need not concern us in this chapter since we are interpreting payoffs as *flows*. We must therefore think of Adam and Eve's endowments as constant commodity streams. Assuming further that the commodities are perishables that must be consumed in the period in which they are acquired, the result of a failure by Adam and Eve to agree in a particular period is that both will perforce consume the commodities with which they have been endowed in that period. As in Section 1.2.5, the endowment point e therefore serves as a deadlock point for their negotiations.[64]

---

[64] When Adam and Eve's endowments are *stocks* in a once-and-for-all negotiation, the endowment point needs to be treated as a breakdown point, since the players only

The second economic environment to be considered is simpler than the first. It derives from the divide-the-dollar problem introduced in Section 1.2.4. In this case, the economic environment can be identified with a triple consisting of the set $\mathcal{E} = [0, 1]$ of possible physical outcomes, the physical outcome $(0, 0)$ that will result if there is no agreement, and Adam and Eve's Von Neumann and Morgenstern's utility functions $u_A : \mathcal{E} \to \mathbb{R}$ and $u_E : \mathcal{E} \to \mathbb{R}$. Each x in $[0, 1]$ corresponds to a division of the dollar in which Adam's share is x and Eve's is $1 - x$.

Bargaining axioms that characterize Walrasian equilibria. The standard approach of cooperative game theory consists of listing axioms that characterize various bargaining solutions. Once we allow bargaining solutions to depend on the underlying economic environment, the opportunities for exercising the imagination in inventing lists of axioms that might conceivably be of interest become unlimited. But only axioms appropriate to the simple exchange economy of Section 1.2.5 are studied here. Thus $\mathcal{E}$ is an Edgeworth box.

In seeking to study bargaining solutions that are sensitive to the underlying economic environment, it is natural to begin by formulating a set of axioms that retain the spirit of Nash's [380] axioms (Section 1.4.2). Somewhat surprisingly, the axioms that emerge turn out to characterize the set of Walrasian equilibria (Binmore [66]). It therefore sometimes makes sense to refer to a Walrasian equilibrium as a Walrasian bargaining solution.

As always in this book, only a simplified version of the necessary argument will be described. In particular, attention will be confined to the case when there is a unique Walrasian equilibrium. Gevers [202] and Nagahisa [378] offer more sophisticated treatments.

The Walrasian bargaining solution will be characterized as a trade w in the Edgeworth box. At this trade, Adam and Eve will receive the payoff pair x. As with the other bargaining solutions studied, it is taken for granted that x belongs to the bargaining set of the underlying problem. This implies that w lies on the contract curve (Section 1.2.5) in the Edgeworth box. The remaining axioms characterizing the Walrasian bargaining solution are:

- Independence of Utility and Commodity Calibration.
- Symmetry.
- Independence of Irrelevant Alternatives.

---

consume their own endowments after the negotiation has failed. The location of the deadlock point is determined by factors unrelated to the Edgeworth box. This point eluded me in Binmore [66].

The first of the axioms that characterize the Nash bargaining solution in Section 1.4.2 requires that the solution be independent of the manner in which zeros and units are chosen on the players' utility scales. The first of the axioms characterizing the Walrasian bargaining solution differs only in the requirement that the solution also not depend on how the commodities are measured. The second axiom requires only the comment that a symmetric Edgeworth box is identical to its reflection in the diagonal on which the endowment point e lies.

As in the case of the Nash bargaining solution, the vital axiom is the Independence of Irrelevant Alternatives. Like the first axiom it needs to be split into two parts, one of which says something about utilities and the other about commodities. Taking the commodity part first, the axiom requires that deleting trades from the Edgeworth box so as to create a new Edgeworth box that still contains both e and w should leave the solution invariant. The utility part is more interesting. It requires that the solution remain invariant when each player's utility function is changed so that no possible trade increases in utility, and the utilities of e and w remain the same. If the players are willing to agree on w originally, the idea of the second part is that the utility change will offer no incentives to alter the agreement, since all that has happened is that some alternatives have been made even less attractive.

"Proof". The first step is to expand the Edgeworth box from $\mathcal{E}$ to $\mathcal{E}_0$ as illustrated in Figures 4.13(a) and 4.13(b). It is important that the secondary diagonal of $\mathcal{E}_0$ should be the straight line through e and w. The second step is to transform $\mathcal{E}_0$ into a square by changing the units in which commodities are measured. In the new commodity units, e and w appear in this diagram as e_1 and w_1. The third step is to change the utility scales so that Adam and Eve both assign a utility of 0 to e_1, and a utility of 1 to w_1. We have now reached the Edgeworth box $\mathcal{E}_1$ illustrated in Figure 4.13(c).

The fourth step is to reflect Adam and Eve's indifference curves in the secondary diagonal of $\mathcal{E}_1$. Figure 4.13(c) shows Eve's indifference curve $u_E(t) = c$ and the reflection $r \circ u_A(t) = c$ of Adam's indifference curve $u_A(t) = c$ in the secondary diagonal of $\mathcal{E}_1$. From these two indifference curves, construct a new indifference curve $v_E(t) = c$ as illustrated in Figure 4.13(c).[65] When v_A is constructed in a similar way, the result is the symmetric Edgeworth box $\mathcal{E}_2$ shown in Figure 4.13(d).

---

[65]The function v_E is the smallest quasiconcave function that dominates both u_E and $r \circ u_A$. One cannot carry through the proof using only concave utility functions without relaxing an assumption somewhere (Binmore [66]).

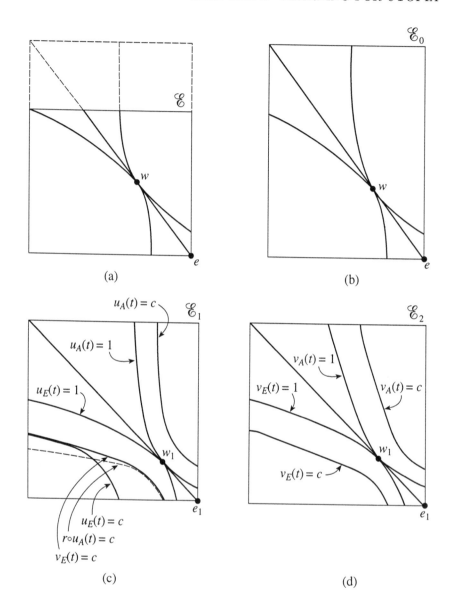

Figure 4.13: Characterizing the Walrasian bargaining solution.

The only symmetric trade on the contract curve in $\mathcal{E}_2$ is the Walrasian equilibrium w_1. To prove the result, it then only remains to reverse the steps used to construct $\mathcal{E}_2$ from $\mathcal{E}$, noting that the axioms ensure that the Walrasian equilibrium remains the bargaining solution at each stage. The second part of the Independence of Irrelevant Alternatives is required in reversing the fourth step, and the first part in reversing the first step.

4.8.2 Misrepresenting Personal Preferences

According to the Nash program of Section 0.6, the usefulness of a proposed cooperative bargaining solution should be assessed by constructing models of the type of bargaining process whose outcome the bargaining solution supposedly predicts. When such a model of the bargaining process is analyzed as a noncooperative game, its solution outcome should then coincide with the cooperative bargaining solution under study. The Nash bargaining solution was tested in this way in Chapter 1 with the aid of noncooperative bargaining games of complete information in which Adam and Eve's personal preferences are common knowledge. But how do preferences become common knowledge in real life?

Their belly prepareth deceit. As Section 2.3.5 of Volume I explains, something becomes common knowledge if and only if it is implied by a public event—an event that everybody knows cannot occur without everyone knowing it. So what public events can lead to Adam and Eve's personal preferences becoming common knowledge?

The answer is to be found in how they behave when others are watching. Indeed, according to revealed preference theory, it is tautological that Eve cannot *consistently* choose an apple over a fig leaf in Adam's presence without his deducing that she prefers to eat an apple rather than wear a fig leaf when he is about. But the question then arises as to whether it may not be in Eve's interests to mislead Adam about her preferences. If they are later to trade Eve's apples for Adam's fig leaves at an agreed rate, she stands to gain a substantial advantage if she can convince Adam that she does not care greatly for what he has on offer. However, misleading Adam in this way is costly for her because it requires her to eat large numbers of apples when she would prefer to preserve her modesty by wearing fig leaves instead. She must therefore weigh the costs of establishing a reputation for disliking fig leaves against the benefits that this reputation will secure for her when she bargains with Adam. Unless she thinks that Adam is unusually naive, her calculations about the benefits to be obtained from deceiving him must also take into account the fact that he is unlikely to take her simulated liking for apples at its face value.[66]

Even if it were possible to unravel the general strategic problems raised by the opportunities for mutual deceit in such haggling situations, I think it unlikely that the result would tell us much about the relevance of the Walrasian bargaining solution. To make progress in this direction, it seems to me that we have to look at two extremal cases. In the first case, deceit is too costly to contemplate. The players' personal preferences are therefore truthfully revealed and so the informational requirements necessary for applications of the Nash bargaining solution are satisfied.

In the second case, deceiving others about one's preferences outside the bargaining context is essentially costless. Any reputation that Eve may seek to build up by eating apples will therefore be treated as mere "cheap talk" by Adam (Section 3.2.3 of Volume I). Her only opportunity to establish a reputation for liking apples is therefore when bargaining in previous encounters. In accordance with this chapter's simplifying assumptions about Adam and Eve's continuing relationship, it will be assumed that the payoff each can obtain in such an encounter is negligible compared with the total payoff anticipated from all similar future encounters. It then makes sense to study the case in which whatever lie the players choose to live is sustained forever.

I model this situation as a problem in preference misrepresentation. If Adam and Eve could publicly commit themselves for bargaining purposes to act as though they had a set of personal preferences other than those they really have, what phony preferences would they announce? What outcome would result from their bargaining with such phony personal preferences?

There is an obvious parallel between such an approach to the revelation of personal preferences and that used in Section 2.5.4 when defining an empathy equilibrium. To make the parallel explicit, I define a *misrepresentation equilibrium* to be a pair of Von Neumann and Morgenstern utility functions announced by the players, with the property that both players answer *no* when asked:

> Suppose that you could deceive everybody into believing that your personal preferences are whatever you find it expedient to claim them to be when bargaining. Would such an act of deceit seem worthwhile to you relative to the personal preferences *you actually hold* if you had to bargain as though holding the personal preferences *you have announced?*

Of course, the answer to this question will depend on how the bargaining

---

[66]It is naught, it is naught, saith the buyer. But when he is gone his way, he boasteth— Proverbs 20:14. The title of this subsection quotes a similar piece of wisdom from the Book of Job (15:35.)

is conducted. I shall simply assume that the bargaining outcome is the symmetric Nash bargaining solution calculated as though the phony utility functions announced by the players were genuine.

Living a lie? As in the case of an empathy equilibrium, it would be a mistake to take the teleological language in which a misrepresentation equilibrium is expressed *literally* when considering its application to the social contract theory developed in this book. Adam and Eve are not to be envisaged as consciously committing themselves by some means or other to behaving as though they had the phony preferences they announce in the model. I certainly do not want to be thought to be reviving the Transparent Disposition Fallacy, which purports to rationalize cooperation in the one-shot Prisoners' Dilemma (Section 3.2 of Volume I). As always when game theory is applied in an evolutionary context, the outcome achieved is *as if* the players chose in the manner described in the model, but the mechanism by which the outcome is *actually* achieved is much more complex.

If we stick to the story of Section 2.5.5, which identifies the long run with biological time, then the players' true personal preferences should be identified with their biological fitnesses. Assuming that their bargaining behavior is the only important factor in determining their fitness, then evolutionary stability demands that they behave optimally when bargaining. If bargaining in their society is conducted using the Nash bargaining solution calculated on the basis of the personal preferences that Adam and Eve have revealed in the past, then they will be genetically programmed to reveal personal preferences that maximize the fitness they derive from the bargaining outcome. Otherwise they will eventually be displaced by mutants who do.

Even if we were to abandon the fiction that personal preferences are determined only by biological evolution, we need not picture Adam and Eve as consciously living a lie. My own Humean view about how and why we carry preference-belief models in our heads was briefly described in Section 2.5.4. If this is only halfway right, then we not only do what we like, but we also have a tendency to learn to like what we habitually do—or see others doing. In a model in which personal preferences adapt in the medium run like empathetic preferences, one might even make a case for restating the definition of a misrepresentation equilibrium so that the players do not evaluate the effect of changing the phony preferences they announce in terms of an exogenously given set of true preferences, but in terms of the phony preferences they actually do announce.

Such an approach would deny the very concept of a true personal preference, and therefore seems to open the door to the Kantian line of thought which argues that rationality in the one-shot Prisoners' Dilemma lies in the players' changing their preferences to make cooperation possible (Sen [480]). Let me therefore emphasize again that the players are not to be thought of as *choosing* their preferences, whether personal or empathetic. Their preferences are determined by evolutionary pressures that derive from considerations of biological and social fitness. The link between an expressed preference and fitness may be complicated—as in my story of how empathetic preferences evolve, or relatively simple—as in the case of personal preferences that are biologically hardwired. But one cannot escape the fact that fitness must ultimately be *exogenously* determined.[67]

In summary, to speak of misrepresentation is only a metaphor that allows complicated phenomena to be discussed in simple terms—albeit at the risk of providing ammunition for silly critics who think that genes or memes cannot act as though they were selfish without being equipped with free will "like human beings".

Modeling in the medium and long run. Recall from Section 2.5.5 that all preferences are assumed to be fixed in the short run. The short run therefore corresponds to economic time—the period in which decisions are actually taken. The medium run corresponds to social time, during which empathetic preferences adjust until an empathy equilibrium is reached, but personal preferences remain fixed. The long run corresponds to biological time, in which I assume that personal preferences adjust until a misrepresentation equilibrium is achieved.

Section 2.5.5 explains the modeling program borrowed from the economic theory of the firm, for which these time spans need to be distinguished. Section 4.6.6 brings this program to the stage where our study of the medium-run process culminated in the determination of Adam and Eve's empathetic preferences from the idea of an empathy equilibrium. When the equilibrium values of these medium-run variables are employed, Adam and Eve receive the same personal payoff whether they use the proportional bargaining solution or the symmetric Nash bargaining solution. The final step is to ask how Adam and Eve's personal payoffs are determined in the long run if the result of their being assigned personal preferences is that these are used to compute their payoffs using the symmetric Nash bargaining solution. It is for this purpose that the idea of a misrepresentation equilibrium has been formulated.

---

[67]It has become fashionable to revive some of the fallacies reviewed in Chapter 3 of Volume I by telling evolutionary stories in which organisms can choose how fit to be in certain circumstances.

Playing it cool. Before turning to misrepresentation equilibria in simple exchange economies, it may be helpful to study the same problem for the divide-the-dollar problem.

Section 1.7.1 explains that the personal characteristics that determine who gets how much when Adam and Eve bargain over the division of a dollar are their attitudes to risk taking, and their patience in the face of delays in reaching an agreement. If the personal preferences that guide a player's choice of bargaining strategy are to become common knowledge, it is therefore advantageous to him if he has low risk-aversion and a high discount factor. A reputation for playing it cool by never seeming hurried or anxious is therefore a valuable asset in a bargaining context—hence our admiration for the absurdly phlegmatic hero portrayed by Clint Eastwood in the spaghetti Western series.

Section 1.7.1 advocates modeling face-to-face bargaining using Rubinstein's [445] Alternating Offers Game. The bargaining outcome can then be predicted using a weighted Nash bargaining solution in which the bargaining powers are inversely proportional to the rates at which the players discount time. However, we are not concerned with face-to-face bargaining here. We are using the *symmetric* Nash bargaining solution, because it is our prediction of the personal payoff pair that Adam and Eve will receive as a result of employing the device of the original position when their empathetic preferences have achieved an empathy equilibrium in the medium run. A more sophisticated theory might still find a role for the patience of the players, but I leave this possibility aside and focus only on the incentives that Adam and Eve have for misrepresenting their aversion to taking risks when the symmetric Nash bargaining solution is employed.

In studying misrepresentation in the divide-the-dollar problem, the most interesting case assumes that the players are risk averse and cannot pretend to be risk loving (Section 4.6 of Volume I). In this case, Crawford and Varian [137] have shown that it is a dominant strategy in the misrepresentation game to announce that you are risk neutral. Such a player aims to maximize his expected monetary gain, thereby equating Von Neumann and Morgenstern utils with dollars. In short, a misrepresentation equilibrium requires that each player claim to be as indifferent to taking risks as is permitted. Since both players then announce the same phony preferences, the symmetry of the Nash bargaining solution then ensures that each player receives a payoff of fifty cents.

Figure 4.14 is a revamped version of Figure 1.3(a) that seeks to explain the result geometrically. Recall that Figure 1.3(a) shows how to map a divide-the-dollar problem into a Nash bargaining problem $(X, 0)$. The symmetric Nash bargaining solution for this problem is denoted by N. It is implemented by dividing the dollar as specified by the point T.

In Section 1.2.4, the Von Neumann and Morgenstern utility functions graphed in the southeast and northwest parts of the diagram represented Adam and Eve's true preferences. But now they represent a pair of phony preferences that Adam and Eve have announced. No matter what phony preferences Eve claims to have, it will be argued that Adam can get at least as much money by changing his announcement to the risk-neutral utility function whose graph is the straight-line segment joining P and Q. Since his true utility function is assumed to be increasing in money, pretending to be risk neutral is therefore a best reply to whatever phony preferences Eve may choose to announce.

Consider the effect of Adam announcing a new utility function obtained by replacing the part of the graph of his old phony utility function between the points P' and Q' by the straight-line chord joining these points. He will then be pretending to be less risk averse than before. The bargaining problem $(X, 0)$ will will be replaced by $(X', 0)$. To divide the dollar as specified by T, the players would now have to agree on the payoff pair M on the Pareto-frontier of the set X'.

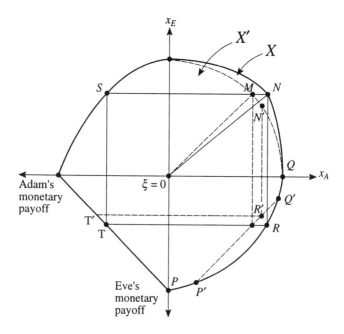

Figure 4.14: Pretending to be risk neutral in divide-the-dollar.

The characterization of the symmetric Nash bargaining solution to be

used is illustrated in Figure 1.5(c). It requires that the tangent to the Pareto-frontier of X at N make the same angle to the horizontal as the ray from O to N. To exploit this fact, the slope of the chord joining P' and Q' has been chosen to be no steeper than the tangent to Adam's old phony utility function at R. The slope to the Pareto-frontier of X' at M is therefore less steep than the slope to the Pareto-frontier of X at N. At the same time, the ray from O to N is less steep than the ray from O to M. Thus M cannot be the symmetric Nash bargaining solution for $(X', 0)$. This must be located at a point N' at which the slope to the Pareto-frontier of X' is steeper than at M, and the ray from O to N' is less steep than the ray from O to M. Such a payoff pair corresponds to a division T' of the dollar that Adam prefers to T.

Snipping off chords that are not too steep from the graph of a phony utility function therefore makes a player at least as well off as before, no matter what phony utility function his opponent may have announced. The biggest chord that can be snipped off in this way has $P' = P$, but there is no guarantee that $Q' = Q$. However, once R' is on such a chord, the slope of the chord can be steepened without changing the division T' of the dollar at all.[68] It follows that Adam can never lose by pretending that the graph of his Von Neumann and Morgenstern utility function is the straight line joining P and Q.

Misrepresenting preferences in exchange economies. At a misrepresentation equilibrium in the divide-the-dollar problem, the dollar is split fifty-fifty. However, something more interesting is obtained when we move to the case of a simple exchange economy, which provides an underlying economic environment that may be asymmetric. Under mild assumptions, a Pareto-efficient misrepresentation equilibrium results in a Walrasian allocation (Hurwicz [276], Schmeidler [466], Binmore [66]).

Room for confusion exists because the Walrasian allocation could be calculated either with the players' true preferences or their phony preferences. Pareto-efficiency is similarly ambiguous. Let me therefore insist that the result to be obtained here requires that both notions are to be calculated using the players' true preferences. Whatever phony preferences the players announce, the allocation that results will automatically be Pareto-efficient relative to the players' *phony* preferences, because we are assuming that

---

[68]The Nash bargaining solution will then be the same as if the graph of Adam's phony utility function were simply the prolongation of the chord PQ' to the horizontal axis. Steepening such a straight-line graph corresponds to recalibrating Adam's utils by multiplying them by a positive constant. But such a transformation leaves the outcome generated by the Nash bargaining solution invariant, by the Independence of Utility Calibration axiom.

the outcome of the bargaining that follows is determined by the symmetric Nash bargaining solution. It also turns out that a misrepresentation equilibrium always implements a Walrasian equilibrium relative to the players' *phony* preferences. Both of these results are important, but should not be allowed to distract attention from the fact that the result to be proved concerns the players' *true* preferences.

The environment $\mathcal{E}$ to be studied is the Edgeworth box, which was introduced in Section 1.2.5 and discussed more extensively in Section 4.3.1. The assumptions of these sections will continue to be made here. In particular, each player's preferences are assumed to be convex. This means that the set of trades a player likes at least as much as any particular trade is always a convex set. A player with a concave utility function has convex preferences, and so one can assure that the players have convex preferences by assuming them to be risk averse. However, it isn't true that a player with convex preferences necessarily has a concave utility function. Utility functions consistent with convex preferences are said to be *quasiconcave.*

Section 4.3.1 already reveals the essence of the argument that connects misrepresentation equilibria and Walrasian allocations, since we can reinterpret Figure 4.2(c) in terms of a misrepresentation equilibrium. Assume that Adam announces a phony set of preferences that are identical to his true preferences inside the set C of Figure 4.2(b), but make everything outside C worse than the endowment point. If the final allocation is to lie on the contract curve of the phony Edgeworth box,[69] then nothing outside C is now feasible for Eve. The best she can hope for is therefore the allocation in C that she genuinely prefers most. Similarly, if Eve announces a phony set of preferences that are identical with her true preferences inside the set D, but make everything outside D worse than the endowment point, then the best that Adam can hope for is the allocation in D that he genuinely prefers most. If C and D are chosen as in Figure 4.2(c), both players will e choosing a phony set of preferences that constitutes a best reply to the phony preferences chosen by their opponent. Figure 4.2(c) therefore provides an example of a Pareto-efficient misrepresentation equilibrium that implements the Walrasian allocation w. The example can easily be modified as in Figure 4.15(a) to show that a misrepresentation equilibrium which is not Pareto-efficient need not implement a Walrasian allocation.

Note that the phony preferences in the example do not satisfy the conditions with which we restricted the players' genuine preferences. One would have to examine the limit of a sequence of continuous, concave utility func-

---

[69] As explained in Section 1.2.5, the contract curve corresponds to the bargaining set of Section 1.2.3. The symmetric Nash bargaining solution, or any other bargaining solution, will therefore yield an allocation on the contract curve.

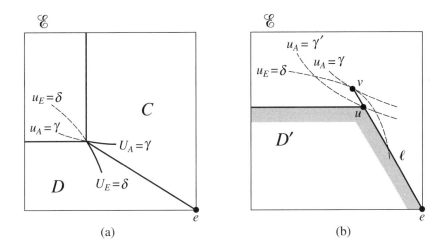

Figure 4.15: Misrepresentation equilibria in the Edgeworth box.

tions to come up with such an exotic set of indifference curves. However, it seems reasonable to add such possibilities to the set of strategies available to Adam and Eve for the same reason that one would wish to add the point 1 to the open interval $(0, 1)$ when seeking the maximum value of x on this interval. If one does not compactify the interval in this way, the maximum does not exist. All one can do is to give progressively better approximations from the interval $(0, 1)$ to the maximum that would exist if one had compactified the interval.

It remains to show that a pair of phony preferences that implements an allocation v that is Pareto-efficient but not Walrasian cannot be a misrepresentation equilibrium. A detailed proof is given in Binmore [66]. However, Figure 4.15(b) makes it easy to see why the result is true. Since v is not Walrasian, the straight-line segment ℓ joining e and v must contain an allocation u that one of the players prefers to v. In Figure 4.15(b), this player is Eve. If u is taken sufficiently close to v, then continuity assures that Adam will prefer u to any allocation on ℓ between e and u. Moreover, u lies in the set C because C is convex and contains both e and v. It follows that Eve would do better to switch from the phony preferences she has announced to a new set of phony preferences based on the set D' shown in Figure 4.15(b). Her old phony preference was therefore not a best reply to Adam's phony preference, and so the old pair of phony preferences was not a misrepresentation equilibrium.

Of course, people watching Adam and Eve bargain will be unaware of their true preferences. They will know only the phony preferences that Adam and Eve reveal. Indeed, the evolutionary story that is used to interpret misrepresentation equilibria within my theory makes it unlikely that the players themselves will perceive their fitnesses as representing their true preferences. Just as young men laugh when it is suggested that the female bodies they find most sexy are those that show signs of being good at bearing children, so Adam and Eve will insist that their phony preferences are entirely genuine. Nor are we entitled to insist that we know their true preferences better than they do themselves. After all, they have the theory of revealed preference on their side.

It is therefore not enough to characterize the bargaining outcome that results when Adam and Eve operate a misrepresentation equilibrium in terms of their "true" preferences—it also needs to be characterized in terms of their "phony" preferences. As noted above the answer is again that a Walrasian allocation will be observed, but now the proviso that the result be Pareto-efficient is unnecessary since the allocation is automatically Pareto-efficient with respect to the player's phony preferences. The reasons why this result holds will probably be evident from Figure 4.15(a)—although there is no reason why $C \cap D$ should have an empty interior as in this diagram; nor need the indifference curves of the phony utility functions U_A and U_E trace out the same paths as the indifference curves of the true utility functions u_A and u_E.

4.8.3   The Concept of a Fair Price

The phony preferences of the misrepresentation equilibrium illustrated in Figure 4.2(c) describe a familiar set of consumer attitudes. When Adam refuses to contemplate any trade outside the set C, he will rationalize this self-denying ordinance by telling himself that he is going to be rushed neither into buying more than he needs—nor into paying an *unfair* price.

Recall that Adam and Eve see the social contract operating at an empathy equilibrium as fair. The same therefore goes for the prices that clear the market at a misrepresentation equilibrium. The concept of a *fair price* in Adam's rationalization of the attitudes he reveals when operating a misrepresentation equilibrium are therefore entirely consistent with Rawls' characterization of fairness in terms of the original position. Presumably, the medieval concept of a fair price enforced by the Church evolved as a consequence of a long period of market stability caused by entrepreneurial and technological stagnation.[70]

Many caveats need to be observed in interpreting the idea of a fair price and the other simple results on Walrasian equilibria discussed in this section. The long-run story about the evolution of personal preferences is just one of many weak links in the chain. One must also remember that the preferences we reveal are not only determined by what we get when bargaining, they also depend on all kinds of other factors. One can gain an advantage in bargaining by convincing others of being willing to take large risks, but the costs of gaining such a reputation may be prohibitive if it can only be acquired by betting heavily in casinos. Nor will Adam always be bargaining with Eve. They will be matched with different partners holding different endowments at different times. The difficulties discussed in Section 2.7.2 over interpreting empathy equilibria therefore arise here also. Nor do players usually know the preferences to which their opponents are committed when they bargain. Nor yet do traders necessarily want the goods they acquire for their own purposes. Most trading is done by middlemen who make a living by selling to one person what they buy from another.

Although such difficulties with the simple model studied here could be multiplied indefinitely, one does not need a model to appreciate the pressures that exist in bargaining situations to pretend that one does not care much if the negotiation fails, or to exaggerate the distaste one feels for a deal that offers less than one might be able to get. With all its failings, the model not only succeeds in capturing such phenomena, but explains why Walrasian equilibria are more robust to the habitual use of such bargaining tricks than other bargaining solutions. My rhetoric on the subject of markets goes much further, but one must always remember that it applies only to the extent that the two-person microcosm in which our idealized Adam and Eve dwell is relevant to the world as it really is.

4.8.4 Time Corrupts All

People are often unaware of the power they have at their disposal, especially when they need to coordinate with others like themselves to use their power effectively. But while a potential source of power remains unused, the stability of a social system remains in question. If enough time is available, a

---

[70]The idea of a fair price must surely also be relevant to the sticky price phenomena that microeconomic textbooks notably fail to explain with such discredited notions as the "kinked demand curve" (Stigler [509]).

gene or meme will eventually appear whose fitness is enhanced by exploiting the existing power vacuum. People who fall victim to the new replicator may not be aware that they are being used as instruments in a power struggle, and so Lord Acton's famous dictum that people are corrupted by power is not necessarily entirely apt.[71] To bowdlerize the poet Horace, it is *time* that corrupts out-of-equilibrium societies.[72]

I have argued that the Game of Morals is an idealized version of an equilibrium selection device we use on a small-scale to get to Pareto-efficient equilibria without damaging internal conflict. It is then natural to suggest that we deliberately go on to redeploy the fairness norm that guides our behavior when using the Game of Morals to resolve small-scale problems of human coordination to a large-scale applications. But such a program can only be effective if we avoid telling ourselves fairy tales about how the Game of Morals actually works at present. As Section 2.7.2 observes, appeals to fairness that ignore the realities of power are doomed, because the underlying balance of power is what ultimately shapes the interpersonal comparisons necessary for fairness judgments to be meaningful. This section extends this warning to the market. Where a market has long been established in a stable environment, we have seen that there are grounds for thinking it futile to appeal to social justice in motivating a reform of the allocation mechanism. If time has had long enough to work on our perceptions, the market will seem fair already.[73]

To summarize my story of Adam and Eve in the Garden of Eden, the result of using of the device of the original position in the short run is that the players compromise on the proportional bargaining solution when no philosopher-king is available to enforce commitments. In the medium run, the standards for making interpersonal comparisons of utility that deter-

---

[71] Even when an out-of-equilibrium social system collapses overnight, it is usually a mistake to attribute the shift in regime to human agents. The normal pattern is that the common understandings that underpin a social contract are gradually eroded over a long time span as the environment changes, until the system becomes vulnerable to destabilizing influences that would be irrelevant to a society operating an equilibrium deep within its basin of attraction. Once it has been shifted from its traditional resting place, the system will eventually restabilize at a new social contract, whose character we like to attribute to the great names of history, but which is actually determined to a large extent by chance events during the transition period.

[72] The late Kenneth Tynan preferred to bowdlerize Lord Acton himself. I guess it is no surprise to find that theater critics think that "power is delightful, and absolute power is absolutely delightful."

[73] Marketeers should note that this sentence does not imply that markets ought to be applied everywhere because they will eventually come to be regarded as fair. Such an argument not only makes the mistake of arguing that morality lies in hurrying evolution along, it also misses the point that support for a reform is gathered by appealing to what people *currently* hold to be fair.

mine who is worthy of special treatment in operating this norm evolve until fairness reduces to assigning the players what they would have got if they had bargained face-to-face using the Nash bargaining solution. When their problem reduces to trading goods, their personal preferences then adjust in the long run until they see it as fair to operate a Walrasian equilibrium.

In brief, time will necessarily corrupt our frail souls until the forum for moral debate becomes the marketplace. But this truth invites misunderstanding. It applies only when the Game of Life is *fixed* in the long run. But, as Section 2.7 argues, fairness norms evolved to solve the *short-run* coordination problem created when a new source of surplus appears as a result of an unanticipated change in the Game of Life. In a market context, one should therefore expect to observe morality kicking in after an exogenous shock alters the fundamental structure of the economy. Just as in real economies, it will manifest itself as a reluctance to move away from the old market-clearing prices that Adam and Eve have come to regard as fair.

4.9 Unfinished Business

With the completion of my theory of the Seemly in the last section, the time has come to emulate Diogenes [149] when he saw the blank space coming up on Khrysippos's scroll at a public lecture. Turning to the audience, he said "Cheer up, guys. Land is in sight!" However, before signing off, I want to draw attention to some of the issues left untouched and to register some hopes for the future.

My strategy in writing this book was to describe the simplest possible models that capture enough of what goes on in the world to make them instructive. This approach has the advantage of making it clear where traditional reasoning needs to be replaced by a more rational form of analysis. The disadvantage is that specific plans for constitutional reform need to be based on something more solid than just-so stories. Three of the oversimplifications of which I am guilty will be assessed in this section as part of an agenda that future whigs may wish to follow.

4.9.1 Large Societies and Coalitions

The first deficiency of this book's basic model lies entirely on the surface. A society containing only Adam and Eve has too few citizens to provide an adequate representation of a large modern state. It would be easy to extend my results to societies with an arbitrarily large number of citizens if I were willing to follow Harsanyi [233] and Rawls [417] in assuming that the only coalitions that matter are the degenerate cases in which either each citizen acts alone, or else everybody joins in a grand coalition so that society acts as a single body. However, it seems to me obvious that authors like Harcourt and de Waal [223] are correct to identify the power that coalitions confer on their members as a major determinant in the development of our capacity for sociality. Even in chimpanzee and baboon societies, the evidence supporting the importance of coalition formation and social networking is overwhelming.

It is therefore necessary to envisage a society as a system of overlapping subsocieties, each of which has a culture of its own. To understand such a society is to appreciate how these different subcultures interact with each other.

As Section 0.5.1 documents, game theorists have proposed many models that are intended to explain coalition formation among rational players, but none can be said to have succeeded in capturing everything that matters. Until further advances have been made on this front, I therefore thought it wise to study only the case in which current theory is adequate—that of a two-person society containing only Adam and Eve. The opportunities for applying the theory directly are therefore limited to situations in which Adam and Eve can usefully be seen as representatives of monolithic blocs. However, one must be careful not to fall into the trap of treating such abstractions as Man and Woman, or Capital and Labor, like individuals motivated by a consistent set of preferences (Section 1.2.8 of Volume I). A more promising area of application lies in international relations, where philosopher-kings are especially notable by their absence.

4.9.2 Incomplete Information and Mechanism Design

Not only has game theory failed as yet to solve the problem of coalition formation, it still has no definitive resolution to the problem of bargaining with incomplete information. All the models in this book therefore assume that Adam and Eve's preferences are common knowledge among the players. Under this hypothesis, it is often reasonable to assume that they will bargain to a Pareto-efficient outcome (Section 1.2.3). But what if Adam and Eve's preferences are not common knowledge?

As a case study, imagine that Eve wants to sell Adam an apple. Assume that it is common knowledge that each player thinks the other's reservation price[74] is equally likely to be any number between zero and one dollar, but neither player knows the precise value of his or her opponent's reservation price. When Adam's reservation price exceeds Eve's, she is willing to sell

---

[74] Adam's reservation price is the most that he will pay for the apple; Eve's is the least that she will accept.

the apple at a price he is willing to pay. A Pareto-efficient and individually rational bargain will therefore consist in the apple being sold to Adam at some price intermediate between their reservation prices (Section 3.2.3 of Volume I).

The difficulty in achieving such a Pareto-efficient outcome lies in the fact that neither player will be willing to reveal the true value of their reservation prices. If Adam can persuade Eve that he doesn't want the apple much, she will lower the price she demands. Similarly, if Eve can persuade Adam that she is reluctant to part with the apple, he will raise the price he offers. How much information is revealed about Adam and Eve's reservation prices as the bargaining proceeds will depend on the rules of the bargaining game they play. But let us consider the most favorable case, when the rules are chosen to maximize the expected total surplus generated by their encounter. Myerson and Satterthwaite [376] then obtain the remarkable result that an exchange will occur only if Adam's reservation price exceeds Eve's by twenty-five cents or more. Nearly half the time that Adam and Eve can both benefit by doing a deal, the deal will therefore not take place.

The preceding story shows that the theory developed in this book is hopelessly inadequate for the case when Adam and Eve have incomplete information about their common plight. But the result of Myerson and Satterthwaite [376] signals that game theorists are not altogether helpless in such situations. Indeed, it is generally held that the branch of game theory likely to be most useful as a practical social tool in the immediate future is *mechanism design.*

The need for mechanism design arises when solutions are sought to the principal-agent problem. In the jargon of economics, a principal faces an implementation problem—like King Solomon in Section 0.7. She has the power to force one or more agents to honor the rules of a game of her choice. Her aim in choosing the rules is to provide incentives that lead the agents to select strategies which advance her aims. If information were complete, she would simply offer her agents the choice of doing whatever maximizes her payoff, or else suffering some dreadful penalty. Her problem is therefore only interesting when the agents know things that she doesn't know but which are relevant to her goals. For example, a philosopher-king in government may wish to redistribute wealth more fairly but not know for sure how much each citizen has hidden under his mattress. Or a firm may wish to increase its production without being able to monitor the effort that each worker is exerting. Or a houseowner may wish to choose a form of auction that maximizes her expected revenue from selling her house without knowing the reservation prices of the prospective buyers.

The big success story for game theory has been in tackling such problems

of mechanism design. In some cases, game theorists are able to specify
the optimal mechanism precisely. Even when current theory has not been
entirely adequate for the task, it can provide valuable insights—as in the
recent series of major auctions of federally owned assets in the United States
(McMillan [350]). In view of its runaway success, it is an enormous pity that
the theory of mechanism design cannot be applied to the social contract
problems studied in this chapter. But its reliance on the enforcement powers
of the principal means that it is relevant to the ideas pursued in this book
only when a philosopher-king is available. Progress must therefore await
the proving of folk theorems for mechanism design problems so that insight
can be gained into when the rules of a mechanism can be expected to be
self-policing.

4.9.3   A Changing Game of Life

The mainspring of human sociality is to be found in the reciprocity mechanism. Since this cannot be captured in one-shot games, this book turned
to the theory of indefinitely repeated games for the simplest possible model
of the Game of Life that allows the reciprocity mechanism to operate in a
reasonably realistic manner. But the decision to work with repeated games
makes it necessary to be unrealistic about another important matter.

As Section 2.5.4 explains, I believe that fairness norms evolved to make
it possible for communities to exploit new sources of surplus without destabilizing internal strife. But we cannot model the appearance of a new
source of surplus without stepping outside the framework of repeated game
theory.[75] Section 2.7 deals with this modeling difficulty by imagining that
the players never anticipate the possibility that a new source of surplus may
become available. They suddenly wake up one morning to find that their
feasible set X has expanded to Y, but never take into account the possibility that the same might happen again at some future time. Still less do
they contemplate the uncomfortable possibility that they might wake up
one morning to find that their feasible set has shrunk overnight, leaving
less to go round than there was before.

Such a crude modeling approach is excusable only as the first step toward a more sophisticated theory in which the Game of Life is modeled as
what game theorists misleadingly call a *stochastic game*. This terminology
makes it seem as though a stochastic game is merely a repeated game in
which a chance move determines which of a number of one-shot games is
to be played in the next period. Such stochastic games certainly are of

---

[75]In a repeated game, each new day presents the players with exactly the same problem
as the day before.

interest. For example, the Fashionwise Prisoners' Dilemma of Section 3.2.2 is a stochastic game introduced to explain how positive freedoms arise in a social contract framework. But when questions are asked about the extent to which the current generation is entitled to consume exhaustible resources or to pollute the environment at the expense of future generations, it becomes necessary to consider stochastic games in which the strategies chosen by today's players can influence not only their own payoffs, but also the game played by tomorrow's players.

Stochastic games in which players can make possibly irreversible strategic choices about the games to be played in the future are obviously much harder to analyze than repeated games. As things stand, we do not even know the class of stochastic games in which some recognizable analogue of the folk theorem holds. Probably no such analogue exists if the players are able to restrict the options available in future games sufficiently quickly. If this is the nature of the modern game of life, then the prospects of our being able to sustain a civilized social contract seem more than a little gloomy. But we may perhaps hope that technological advances will open up new options faster than than they they allow us to close others down.

4.10   A Perfect Commonwealth?

In his essay *On the Idea of a Perfect Commonwealth*, David Hume [272] indulged in a little *jeu d'esprit* that took the form of a practical program of whiggish reform for the society of his time to follow. In view of all the unfinished business reviewed in the previous section, it would be a major piece of hubris to seek to follow his example, but I cannot resist giving some voice to the aspirations of a modern whig.

4.10.1   What is Whiggery?

There is sometimes much to be learned from how an intellectual or political position is misrepresented by its enemies. For example, modern American politicians who resent tax dollars being spent on helping out the needy reveal a great deal about themselves when they choose the word *liberal* to insult their enemies. The following piece of doggerel by the poet Yeats is equally instructive about the attitudes of an earlier set of reactionaries:

What is Whiggery?
A levelling, rancorous, rational sort of mind
That never looked out of the eye of a saint
Or out of a drunkard's eye.

Section 1.1 of Volume I observes that Yeats is right to say that whigs respect rationality. Whigs believe that the way to a better society lies in appealing to the rational self-interest of all concerned. Yeats is also right that whigs are levelers, and if this seems rancorous to unreconstructed Tories like Yeats, it is because they do not see what is in their own long-term interests. Yeats also tells us that whigs are not saints. He is right about this also. Not only are whigs not saints, they do not think that people have the capacity to become saints, as the more naive thinkers of the left would have us suppose. People can temporarily be persuaded to put the interests of the community as a whole ahead of their own private concerns, but a community based on the assumption that its citizens can be relied upon to behave unselfishly toward those outside their own extended family most of the time simply will not work.

Finally, Yeats is right that whigs see no reason to behave like drunkards, lurching from crisis to crisis. Planning and reform need not be dirty words. They do not require the existence of some mythical "common good". We can plan instead to institutionalize "common understandings". Nobody need make great sacrifices in the process, once it is understood that it is not in the self-interest of the strong that they let the weak fall by the wayside. We can go from the old to the new *by mutual consent*. We do not need to set up stultifying and inefficient bureaucracies along the way. Nothing prevents our planning to use markets wherever markets are appropriate, but a society that relies *only* on markets is leaving much of its potential unfulfilled.

In less poetic terms, whigs first recognize that utopian aspirations should not be allowed to conceal the fact that stability is the prime need of a society. When contemplating reform, the feasible set of new social contracts should therefore be restricted to equilibria in the Game of Life. Moreover, the new social contract should be be reachable from the current social contract by a process that is not itself destabilizing. But how should new social contracts be chosen?

My pragmatic suggestion is that we adapt the fairness norms that are in daily use for settling small-scale coordinating problems to large-scale problems of social reform. I argue that the underlying fairness algorithm was probably biologically fixed when we lived in anarchic hunter-gatherer communities, at which time any workable fairness understandings would necessarily have been self-enforcing. The structure of our instinctive fairness intuitions is therefore well suited for application to the problem of constitutional reform, where external enforcement is absent by the very nature of the problem.

Although the basic fairness algorithm is probably instinctive, the standards of interpersonal comparison we use as inputs when making fairness judgments must surely be culturally determined. Like all our attitudes, these standards adapt over time to reflect the underlying power structure of a society. However, reformers need to remember that people will only be persuaded that a new social contract is fair if it rewards the worthy according to the standards of interpersonal comparison *currently* in place. Utopians who insist on pressing metaphysical arguments about what would be regarded as fair in some ideal world simply muddy the waters for practical reformers who actually have some hope of reaching peoples' hearts.

In responding to such utopian critics, the first step is first to challenge their claim to the moral high ground by pointing out that you share their dislike of the objective fact that only second-best societies are consistent with what evolution has made of human nature, but what we like or dislike is irrelevant to what is true. The next step is to press home the advantage by accusing your critics of actually obstructing the relief of human misery with pie-in-the-sky schemes that are doomed to fail. Finally, kick them while they are down by explaining that their kind of armchair reasoning has contributed not one whit to the scientific study of the structure of the societies in which real people actually live. And who but a fool would try to fix something of whose workings he is ignorant?

Future aspirations. I hope that the simple models studied in this book are adequate not only to show that whiggery is a coherent creed, but also to explain how the leveling aspect of whiggish political thought operates, and why attention to fairness is essential if reforms are to stand the test of time. However, there is much else that would yield to a whiggish approach if more sophisticated models were available.

If the analysis of this book could be extended to stochastic games, it would be possible to discuss the stability of social contracts in the face of an uncertain future. How much disruption can an institution based on reciprocity endure before falling apart? It would also be possible to discuss the planning of irreversible reforms. How many people should there be? (Section 2.5.2) What share of our current resources should we leave to our children? How should we ration health care and welfare benefits? To what extent should government subsidize education and research?

The informational problem is also serious. An equilibrium in a game of incomplete information is not a simple thing. One might think of it as a self-enforcing mechanism—a set of rules whose infringement is *observable*. The point here is that nothing is to be gained by inventing rules that specify punishment for those who do not pull their weight in the Game of Life, unless it is possible to detect such free riding when it occurs. Nor can one assign blame or desert on the basis of unobservable personal characteristics.

It should be noted that the rules of of such a self-enforcing mechanism—society's Game of Morals—will look much more like the real-life morality game that governs our daily lives than the simplified Game of Morals studied in this book. In particular, a Game of Morals that took informational constraints properly into account would leave enormous scope for the exercise of the positive freedoms of Section 3.2.4. In a society in which all information is available to everybody, it would be possible to operate a social contract that regulates the behavior of each citizen in minute detail. With no information at all, a social contract would necessarily leave each citizen free to choose any action at all. But the mechanism designed by a whiggish society to operate under realistic informational constraints will constitute a social contract that occupies an intermediary position between such authoritarian and anarchic extremes. As such it may gather support from each side. Those who worship order can take delight in the rules of the mechanism. Lovers of freedom can simultaneously take comfort in the fact that the rules are fairly chosen by the citizens themselves to provide socially acceptable incentives for agents who choose freely from a wide menu of possibilities.

Finally, there is the all-important issue of coalitions, which are clearly even more important in human societies than in the societies of other primates. It seems to me pointless to follow Harsanyi [233] and Rawls [417] in proceeding as though no power structures intervene between the level of the individual and the state. After all, the community of all humans on the planet operates a social contract of sorts, but our relations with citizens of other countries are mediated through our respective governments. Consider, for example, the conduct of international relations between France and Germany. The French and Germans each coordinate on appreciably different social contracts within their own national games of life. But France and Germany can themselves be seen as players in an international game of life in which yet another social contract operates.

Nor is the phenomenon of coalitions acting somewhat in the manner of individual players restricted to the international scene. On the contrary, the character of a modern democracy is largely defined by how its coalitional patterns shift in response to stresses and strains. Modern societies consist of a complex system of interlocking organizational hierarchies that coordinate behavior at many different levels. Any attempt to describe the current *status quo* that took no account of this reality would clearly be futile, let alone the social contracts to which a society might aspire. The subsocieties into which society as a whole is split are not just epiphenomena, they are essential to its efficient operation.

One can imagine the device of the original position being used to resolve coordination problems between several such subsocieties with fully briefed delegates serving as representatives of each subsociety. The brief of each such delegate would in turn be decided by a similar use of the original position in resolving coordination problems within each subsociety—which would itself typically be split into factions and splinter groups. However, it is inevitable that different standards of interpersonal comparison will arise at different levels and in different contexts. What counts as fair in society as a whole will therefore depend in a complicated manner on what counts as fair in all the subsocieties of which it is formed.

Although the current state of game theory excludes all but the most speculative of models, I am willing to hazard some guesses about the structure of the ideal whiggish state in the multiperson case. I think that its social contract will be pluralist in nature, with decision making decentralized to the level where the information is available to make choices most effectively. But the units of the plural or federal state I have in mind will not be distinguished by geography alone. All coalitions with some power will be institutionalized within the social contract. The units of the federal state will therefore be an overlapping system of organizations, with each citizen connected to government through many avenues. What form will these organizations take and what policies should they pursue? My guess is that the answer to the first question will largely be decided by the accidents of history that have determined the current *status quo*. As for the second question, only an ipsedixist would presume to make this choice on behalf of the people whose choice it is to make.

4.10.2 Where is Whiggery?

The anthropology literature on hunter-gatherer societies surveyed in Section 4.5 uses the word "egalitarian" in a broad sense to include both widespread sharing of resources and freedom from authority. But I guess nobody would maintain that there is any necessary linkage between the two notions. Indeed, my own theory suggests that one might usefully classify social contracts, both real and hypothetical, using the two-dimensional scheme of Figure 4.16(a), in which freedom from coercion and equality of resources appear on orthogonal axes. It is perhaps no accident that the psychologist Eysenck [177] found that the data obtained from attempting to match personality types with political attitudes fits much more comfortably into such a scheme than the classical one-dimensional political spectrum between left and right. Figure 4.16(b) attempts to place some of the great names of political philosophy within a similar scheme.

With neofeudalism appearing prominently in Figure 4.16(a), followers of Marx might reasonably expect to see variants of capitalism and socialism appearing also. It is not hard to place the idealized form of capitalism in

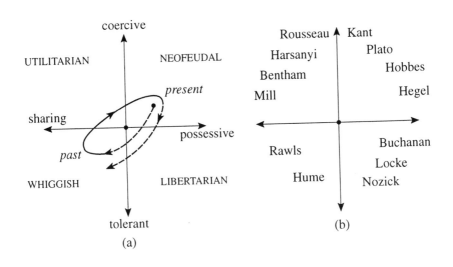

Figure 4.16: Classifying political attitudes.

which all social interaction is supposedly transacted through the market in the libertarian category. Nor is it difficult to place the idealized form of socialism in which state officials love the powerless as much as they love themselves in the utilitarian category. But what of communism as once practiced in the Soviet Union, or the mixed economies of the West?

Unlike Marx, I don't think we ever graduated from the hierarchical authority systems that typify feudal societies. We simply found new feudal forms to practice. The forms of socialism and capitalism that have been practiced in the world therefore all belong in the neofeudal category. After all, who were the officials of the Soviet Communist Party if not a self-appointed aristocracy? What more does a modern democracy offer than the periodic opportunity to replace one bunch of oligarchs with another? Who is the president of the United States if not an elected monarch?

Far from seeing the problems of political organization we face as a battle between the ideals of the left and right, I see utilitarianism and libertarianism as the Scylla and Charybdis between which reformers must steer a course if we are to escape our feudal past. Scylla provides no safe port of call, because nothing can prevent the bosses in an authoritarian society from becoming acquisitive. Charybdis is similarly unsafe, because possessions cannot be held securely in an anarchic society. Our choice is between neofeudalism and whiggery.

As Section 3.4.5 explains, the distinction between neofeudalism and whiggery is one of degree. Elite groups that blatantly ignore the standards of fairness currently operating in their societies merely destabilize their own regimes (Balandier [39]). On the other hand, no society can dispense with the need for leaders and entrepreneurs to handle decisions that need to be made quickly, and to seek out new opportunities to exploit. Even the most egalitarian of modern foraging societies take advice from their more successful hunters on how hunts should be organized, while the indigenous tribes of the Great Plains of North America understood the necessity of granting temporary authority to war chiefs.[76]

I believe the reason that romantic authors see such savage societies as noble has little to do with the reasons proposed by Rousseau for admiring the noble savage. The lifestyle of hunter-gatherers strikes a chord in our hearts because they do not need to suppress the instincts that make us resentful of the unfair exercise of personal authority. By contrast, as argued in Section 4.5, we occupy what Maryanski and Turner [343] call a social cage, constructed by our ancestors when population pressures forced them to adopt a farming lifestyle. The bars of this cage mark the front line of an ongoing war between two forms of social contract—those in which equilibria in the Game of Life are chosen using fairness as a coordinating mechanism, and those in which equilibrium selection is delegated to leaders.

Perhaps the distant future will see technological advances that free us altogether from our neofeudal social cage, but only the most utopian of anarchists would wish to argue that a large modern society can survive without putting power into the hands of its officials. We cannot dispense with the need for a human police force and a codified punishment system. While we have militant neighbors, an army is necessary to defend our freedoms. Taxes need to be raised and administered. Nor does the evidence suggest that we are capable of exploiting the returns to scale possible in large commercial or industrial enterprises without bosses to direct our efforts. Without leaders and entrepreneurs, the social contract in a large society cannot possibly come close to the Pareto-frontier of the feasible set.

There is nothing we can do to alter the fact that Pareto-efficient social contracts in large societies must be authoritarian to some extent. Nor can we rewrite the history of a society with a view to changing the standards of fairness it has inherited from the past. But, like the founding fathers of the American Republic, we can attempt to persuade our fellow citizens not to waste the opportunities for reforming our social contract as new

---

[76]Like Cincinnatus, they served only for the duration of the emergency.

opportunities for Pareto-improvements arise. The whig proposal is that we select whichever of the *Pareto-efficient* contracts in our current feasible set is *fairest* according to current thinking.

Whigs who yearn for utopia therefore propose steering a course away from neofeudalism, heeding the siren songs of neither the utilitarian left nor the libertarian right, toward the noble savagery of our foraging ancestors. On the way, we will give up the unnatural habits of authority worship and conspicuous consumption that currently keep us locked in our social cage. There will still be bosses in Ithaca, but they will be seen for what they then will be—people like ourselves, who are paid to help us coordinate on a fair and Pareto-efficient social contract.

4.11   Humean and Humane

As the founder of modern social contract theory, Thomas Hobbes was allowed the last word in first volume of *Game Theory and the Social Contract*. But David Hume is the true hero of this book. He is universally accepted as one of the great philosophers, but the full magnitude of his creative genius is obscured by the fact that he is also one of the great stylists of the English language. As though merely making conversation, he would today be explaining quantum physics to someone he met on a bus without either party feeling any sense of incongruity. But the tendency to assess the quality of a good by its price carries over into philosophy. Surely the hours spent struggling to make sense of Immanuel Kant must be a measure of the depth and originality of his thought? How could the insights bought at such a low price from David Hume be comparable in value?

I shared in this folly until I compared what Hume and Kant have to say about game theory. As all the fuss about the one-shot Prisoners' Dilemma demonstrates, people find this a deep and difficult subject. Kant goes hopelessly wrong, but buries his mistakes in muddled prose whose obscurity is matched only by Heidegger or Hegel. By contrast, Hume effortlessly formulates the idea of an equilibrium in such a transparent fashion that the depth of his insight passes almost unnoticed.

Hume is the original inventor of reciprocal altruism—the first person to recognize that the equilibrium ideas now studied in game theory are vital to an understanding of how human societies work. He understood that one must look to evolution for a solution of the equilibrium selection problem— a conclusion that modern game theorists have reached only after fifty years of fruitless attempts to tackle the difficulty by inventing high-flown definitions of rationality in the style of Kant. He even anticipated modern game theorists in seeing constitutional reform as a problem in mechanism design.

But people who write about Hume commonly fail to notice the political implications of his game-theoretic approach, even though his essays make his views entirely explicit. Almost incredibly, there are even those who deny that Hume can legitimately be seen as an icon of whiggish reform. By the same kind of Humpty-Dumpty reasoning that contrives to represent Immanuel Kant as a rational liberal, such modern authors as Miller [361], Livingston [329] and Whelan [551] follow the tradition of representing David Hume as a political conservative.

In his lifetime, David Hume was certainly held to be a Tory, but that was because Whigs and Tories were orginally separated by their views on the legitimacy of the Stuart kings, to whose line the whiggish Glorious Revolution of 1688 denied the succession. However, in his *History of England,* Hume [268] refused to toe the Whig party line, which denied any virtue whatsoever to the deposed Stuarts. John Stuart Mill [359] added to the confusion by arguing that Hume had to be a political conservative, content to continue with whatever social practices were current, because he was too skeptical a philosopher to be able to repose any trust in the power of reason to devise workable reforms! Presumably Mill was unaware of Hume's [272] essay, *Of the Idea of a Perfect Commonwealth,* and the numerous other essays and letters in which Hume put the rational case for various whiggish projects of social and political reform. When such evidence is examined, as in Stewart [508], it seems hard to understand how anyone can ever have thought that David Hume was anything other than a straightforward Scots Whig on political questions.[77]

As for the relation between Hume's philosophy and his politics, Hume [267] repeatedly denies that his skepticism leaves no role for reason. He is certainly no friend to the kind of *a priori* reasoning that philosophers traditionally employ when advocating social reforms, believing that the study of human societies is an empirical science like any other. In particular, he had no time for utopian schemes that ignore the facts of human nature. But it is hard to see how such realism could be interpreted as being supportive of authoritarian regimes. After all, it was Hume [267] who first clearly saw how reciprocity provides the means to sustain a society in equilibrium without the need for an autocrat from outside the system to enforce its decrees (Section 1.2.3 of Volume I). Nor was the folk theorem of repeated game theory the only important game-theoretic idea with which he was playing two hundred years before Von Neumann was born. In the following piece of whiggery, Hume [273] even anticipates the idea of mechanism design:

When there offers, therefore, to our censure and examination, any

---

[77]Although, of course, he had no time for their fanaticism in matters religious.

plan of government, real or imaginary, where the power is distributed among several courts, and several orders of men, we should always consider the separate interest of each court, and each order; and, if we find that, by the skilful division of power, the interest must necessarily, in its operation, concur with the public, we may pronounce that government to be wise and happy.

However, it is not Hume's political philosophy that I wish to celebrate in this final section, but his personal life, which provides a definitive refutation of the oft-repeated claim that the Hobbesian psychology and Humean morals embraced in this book should carry a public health warning. What more convincing demonstration could there be that there is no reason to feel gloomy because life has no ultimate purpose or because conventional conceptions of moral responsibility are built on foundations of sand? So what if our fine feelings and intellectual achievements are just the stretching and turning of so many springs or wheels, or our value systems are mirrored by those of chimpanzees and baboons. Our feelings are no less fine and our values no less precious because the stories we have traditionally told ourselves about why we hold them turn out to be fables.

In discarding the metaphysical baggage with which the human race bolstered its youthful sense of self-importance, Hume taught us that we throw away nothing but a set of intellectual chains. Far from being dehumanized, dispirited or depressed by his beliefs, Hume was the most civilized, companiable and contented of men, especially when compared with such oddities as Rousseau or Kant. Even on his deathbed, Hume retained his good humor—totally disarming James Boswell, who tactlessly quizzed him on how it felt to be at death's door without a belief in the afterlife.[78] As Boswell [96] reports, "Mr. Hume's pleasantry was such that there was no solemnity in the scene, and death for the time did not seem so dismal."

In a less ghoulish deathbed conversation, Hume [269, p.xliii] told Adam Smith that he had been reading Lucian's *Dialogues of the Dead*, in which various notables offer reasons to Charon why they should not be ferried across the Styx. When his own time came, he proposed to say, "Have a little patience, good Charon, I have been endeavouring to open the eyes of the Public. If I live a few years longer, I may have the satisfaction of seeing the downfall of some of the more prevailing systems of superstition." But then, says Hume, Charon would lose all patience, "You loitering rogue, that will not happen these many hundred years. Do you fancy I will grant you a lease for so long a term? Get into the boat, you lazy, loitering rogue."

---

[78] Although the world of fashion classified Hume alongside Hobbes and Spinoza as an atheist, Hume was a deist. He was prepared to accept God as the creator of the universe, but the available evidence led him to deny that He took any further interest in man.

David Hume was right to predict that superstition would survive for hundreds of years after his death, but how could he have anticipated that his own work would inspire Kant to invent a whole new package of superstitions? Or that the incoherent system of Marx would move vast populations to engineer their own ruin? Or that the infantile rantings of the author of *Mein Kampf* would be capable of bringing the whole world to war?

Perhaps we will one day succeed in immunizing our societies against such bouts of collective idiocy by establishing a whiggish social contract in which each child is systematically instructed in Humean skepticism. Such a new Emile would learn about the psychological weaknesses to which *homo sapiens* is prey, and so would understand the wisdom of treating all authorities—political leaders and social role-models, academics and teachers, philosophers and prophets, poets and pop stars—as so many potential rogues and knaves, each out to exploit the universal human hunger for social status. He would therefore appreciate the necessity of doing all of his own thinking for himself. He would understand why and when to trust his neighbors. Above all, he would waste no time in yearning for utopias that are incompatible with human nature.

Would Adam and Eve be happy in such a second-best Garden of Eden? On this subject at least, Hume's own experience is immensely reassuring. We don't need to tell ourselves lies in order to be content. We certainly don't need to believe that utopia can be achieved by some quick fix. It isn't even necessary to be optimistic that things will get better in the long run. We need only the freedom to create a microsociety within which we can enjoy the respect of those whose respect we are able to reciprocate. As Hume's example shows, even death need not then seem so terrible.

William Blake's portrait of Ezekiel captures the sincerity
with which the prophet details the bloodthirsty punishments God has in store for those who fail to worship Him
with adequate enthusiasm. Modern prophets also really
mean it when they assure us of God's infinite mercy.

Appendix A

Really Meaning It!

By God, this is impossible!

Thomas Hobbes, on the 47th Proposition of Euclid

A.1 Naturalism

This appendix reiterates those elements of my theory which critics find it hardest to believe can be intended to be taken literally. Sometimes, as with evolutionary approaches to ethics, they have been taught as undergraduates to laugh at such ideas.

Moral philosophers typically hold that the purpose of their discipline is to uncover universal principles that we all ought to follow when interacting with our fellows. A naturalist denies that such principles exist. He may well wish that society would adopt his own pet schemes for reform, but his alliegance to naturalism prevents his claimimg any special authority for his views, since he knows that he would be advocating different reforms if he had been brought up in a different culture. In the words of Xenophanes: "The gods of the Ethiopians are black and flat-nosed, and the gods of the Thracians are red-haired and blue-eyed."

A naturalist sees himself as a scientist, exploring the biological and social facts on which our moral intuitions are based. Such facts are contingent. They would have been otherwise if biological and social history had taken a different course. Moral behavior in chimpanzees and baboons differs from moral behavior in humans because their biological history differs from ours. Moral behavior in other human societies differs from moral behavior in our society because their social history differs from ours. Homophobes need

511

to recognize that that they would almost certainly have found adolescent boys sexually attractive if they had lived in classical Greece. Feminists must surely see that they would have been likely to defend their husband's right to beat them if they had lived in the time of the prophets. None of us would find it unseemly to hide food from hungry children if we lived in one of the more pathologically selfish societies described by anthropologists.

Such frank relativism is too much for many to swallow, because it denies that there are any absolute moral standards. As always, David Hume faced up squarely to the issue, but Adam Smith was the first of many to seek to have it both ways. However, those who wish to enter the pulpit to preach that one society is better than another are not entitled to appeal to naturalistic theories of ethics. Even the wishy-washy liberal doctrine that all societies are equally meritorious receives no support from naturalism. There is no culture-free Archimedean standpoint from which to apply a moral lever to the world. If we could liberate ourselves from all cultural prejudices, we would find that morality no longer had any meaning for us.

When advocating reform, the temptation for Adam to claim naturalistic authority for his goals when disputing with Eve is hard to resist. But naturalism offers authority only when means are discussed. Adam may criticize the feasibility of the reforms proposed by Eve, or point out that they will not achieve the objectives for which they are intended. He may suggest that the social tools with which evolution has equipped a society may be used to achieve other ends than those for which they are currently employed. But when it comes to the determination of ends, Eve has no reason to regard Adam's naturalistic expertise as any more relevant than her own experience of life. Nor has Adam any reason to respect Eve's claims for any noncontingent Moral Facts she may imagine she can demonstrate by metaphysical means. Still less need he be defensive if accused of being unable to justify any Moral Facts of his own. In brief, values do not predate the social contract—our choice of social contract creates our values.

A.1.1   Causal Reversals

The causal reversal of the last sentence is only one of many that this book has proposed. For example, an equitable compromise does not assign more to Eve than to Adam because she is more worthy. She is deemed to be more worthy because the concept of equity generated by social evolution in the medium run assigns her more than Adam. Societies do not become dysfunctional because the old virtues are abandoned. The old virtues cease to be honored because the social contract has shifted. We do not punish people because they are morally responsible for their actions. We say that they are morally responsible because our social contract requires that they be punished. We are not unpredictable because we have free will. We say that we have free will because we are not always predictable. A society does not choose a social contract because it promotes the common good. Our definition of the common good rationalizes our choice of social contract. Finally, we have the archetypal reversal. In modern utility theory, Adam does not prefer action a to action b because its utility is larger. We assign a larger utility to a than b because Adam never chooses b when a is available.

This last causal reversal causes immense problems for moral philosophers, which would be greatly eased if we could agree to distinguish between the prescriptive utility of Bentham [53] and Mill [360] on the one hand, and the descriptive utility of modern economic theory on the other. Since *utility* is universally agreed to be an unfortunate misnomer for the excess of pleasure over pain envisaged by our Victorian forbears, I suggest substituting the word *felicity*, in honor of Bentham's description of his approach as the "Felicific Calculus". Postwelfarists like Sen [477] could then justifiably denounce Harsanyi [233] for not being a felicitarian, while the rest of us could continue to congratulate him for being a utilitarian (Appendix B).

A.2 Modeling Man

Much dissatisfaction has been expressed in recent years with *homo economicus* as a model of man. It is true that too much has been claimed for him in the past, but the last thing we ought to do in seeking a more realistic model is to abandon the discipline he forces upon us in favor of one of the many varieties of *homo ethicus* that utopians of the left urge upon us. It would be particularly ironic if economists were to throw *homo economicus* overboard at the very moment when evolutionary biologists are reaping a rich harvest from exploiting the techniques economists have developed for his study.

As I see it, we need to be willing to study optimizing models at numerous different levels:

- The level of the gene;
- The level of the meme;
- The level of the individual;
- The level of the group.

In this book, genetic evolution is seen as determining the more primal of our personal preferences and the justice algorithm that allows us to operate as social animals. Social evolution is held responsible for the empathetic preferences that serve as inputs to the justice algorithm. Individual optimization is seen as the stabilizing force that maintains a social contract.

Intergroup dynamics determines the mechanism used to select among social contracts.

Reductionism. Only a fool would maintain that *all* phenomena are best described by optimizing models. It is certainly easy to give examples where optimizing models fail at some level. But when this happens, the research strategy of first resort is to look for an optimizing model at a lower level.

For example, why do laboratory subjects sometimes honor fairness norms when these conflict with individual maximization? A plausible answer is that we have placed them in a sore thumb situation to which their fairness norm is not adapted (Sections 3.5.1 and 4.2). But if we want to know why our subjects apply one particular fairness norm rather than another, we need to begin by looking at an optimizing model that studies the evolution of fairness norms in the environment that actually shaped them. The disastrous research strategy to which advocates of *homo ethicus* are prone reverses this reductionist approach. Instead of moving *down* a level when they find their current optimizing model fails to work well, they move *up* a level. But it is a fatal error to attempt to explain low-level phenomena in terms of high-level phenomena. Biologists have demonized Wynne-Edwards [569] for falling prey to this mistake. Economists and philosophers who try to explain individual behavior in terms of various notions of "group rationality" have so far escaped similar treatment, but I would be very happy to place them in at least the outer circle of hell among the noble pagans.

Group selection. The time to move up a level is when everything significant at the current level of analysis has been adequately modeled. Emergent properties at the next level can then be tackled. Unfortunately, evolutionary biologists seem to have been so traumatized by the demonization of Wynne-Edwards that they condemn all attempts to climb up a level as the Group Selection Fallacy. This book therefore speaks of equilibrium selection rather than group selection, but I still sometimes find myself being accused of denying the selfish gene paradigm. Such accusations overlook the fact that selection at the lower level has *already* been taken into account when selection at the higher level is restricted to choosing among the *equilibria* that represent the endproduct of selection at the lower level.

Individual optimization. Abandoning *homo economicus* seems to me like throwing away your toolbox when you notice that the house needs fixing. Nor am I convinced that the house that economists have built is in such bad repair. The problem is rather that its architects insist on claiming palatial dimensions for what is essentially a comfortable bourgeois residence (Section 2.2 of Volume I). In particular, it is silly to insist that models based on optimization at the level of the individual *always* predict human behavior. In my experimental work (Binmore [76, 79, 82]), I insist that we should only expect the optimizing models of neoclassical economics to predict in the laboratory when three conditions are satisfied:

- The problem faced by the subjects is not only simple but is presented to the subjects in a simple way. One can only hold the attention of laboratory subjects for a limited time in the laboratory, during which it is unreasonable to expect them to master complex tasks.

- The incentives provided are adequate. There is no point asking subjects what they *would do* if a large sum of money were hanging on their decision. What people say and what they actually do are demonstrably not the same.

- There must be sufficient time for trial-and-error learning. It is this consideration that I think is most important, but most experimental studies report the behavior only of subjects with little or no experience of the task they are asked to perform.

When these criteria fail to hold in the laboratory or elsewhere, we should be doubtful about using optimizing models at the level of the individual. For example, *homo economicus* is probably worthless as a model of how consumers behave in the rapidly changing environment of a supermarket. But it does not follow that he is a failure as a model in more favorable economic environments.

Nor is it sensible to ignore the flexibility afforded by the *homo economicus* framework by insisting that those of us who use him as a model are saying that people care only about money. Genes are admittedly selfish insofar as they have to be modeled as caring only about fitness, but to accept selfishness at the level of the gene does not imply that a similarly naive concept of self-interest is appropriate at the level of the individual.

In denying that *homo economicus* need be identified with Stigler's [510] Mr. Hyde, or that he is necessarily incompatible with Sen's [478] preference for Dr. Jekyll, Section 1.2.2 of Volume I quotes Hobbes [251] on the subject of the nature of man, which he tells us can be entirely summarized in terms of his strength of body, his reason, his passions, and his experience. These categories are considered systematically below. Much of the criticism directed against *homo economicus* arises from a failure to understand the extent to which the optimizing paradigm leaves one free to vary how each category is modeled.

A.2.1 Strength of Body

Strength of body translates into what is feasible for an individual. In a social context, we therefore have to specify what aspects of the Game of Life are relevant to the issue under study. Criticism of *homo economicus* would often be better directed against the naively simple games that are commonly used to model the human predicament.

For example, Chapter 3 of Volume I discusses at length the confusion that arises when the one-shot Prisoners' Dilemma is used to represent the Game of Life. A contradiction is then thought to arise because real people often cooperate with each other, whereas optimizing players in the one-shot Prisoners' Dilemma always behave like hawks. The conclusion is then drawn that real people are not optimizers. But this deduction ignores the alternative explanation that the Game of Life does not really resemble the Prisoners' Dilemma—or any other one-shot game like the Battle of the Sexes or Chicken. Nor is it a zero-sum game, as Hobbes' [249] analogy of running a race would suggest. Real people cooperate consistently when everybody stands to gain something. Hence my insistence in Section 2.2.6 of Volume I on the importance of using *repeated games* to model the Game of Life. Enough structure is then available for cooperation to evolve as a consequence of the reciprocity mechanism coming into play. If the human Game of Life were adequately modeled by the one-shot Prisoners' Dilemma, we would not have developed into social animals.

Modeling the Game of Life as a one-shot Prisoners' Dilemma and then insisting on the existence of communal solidarity in order to escape its Pareto-inefficient equilibrium is a left-wing mistake. The communitarianism that is currently fashionable somehow contrives to combine this Marxist error with views that one would otherwise classify as right-wing. However, the traditional right-wing mistake is to treat the Game of Life as a market, and then to hope that the conditions of the First Welfare Theorem will turn out to represent reality sufficiently well that the Walrasian equilibria of the market will generate outcomes that are approximately Pareto-efficient. But the Game of Life is not a market. Nor do the conditions of the First Welfare Theorem necessarily apply when subsidiary games of the Game of Life actually do have a market structure. Returns to scale are frequently increasing, transaction costs are seldom negligible, and externality problems are typically serious. Under such conditions, there is no reason why markets should produce Pareto-efficient outcomes

However, those who think of everything in market terms make a more serious error than misapplying the theorems of neoclassical economics. They confuse themselves by referring to such intangibles as law-and-order or freedom-of-speech as "public goods", as though their supply were just a matter of spending enough money. Similarly for the currently fashionable attempt to discuss community spirit in terms of "social capital". Far from all social contract issues being reducible to market problems, the existence of an orderly social contract is a precondition for the appearance of a market. But to study the social contract that a society operates, we must be prepared to examine games that have a much more complicated structure than the markets of neoclassical economics.

A.2.2 Reason

In accordance with Hume's [267] dictum that reason is the slave of the passions, *homo economicus* uses his reason when deciding which of his feasible actions will generate the result he most prefers.[1]

In reviewing the silly paradox that arises when one asks how *homo economicus* can be free to choose any action from his feasible set if his nature always compels him to choose the optimal action, Section 3.2.4 points out the central role of counterfactual reasoning in rational choice theory. *Homo economicus* will necessarily choose the optimal action, but the action he chooses is optimal only because he knows or believes he would get a smaller payoff if he were to take one of the other actions in his feasible set.[2]

Evaluating the counterfactuals that arise in game theory is particularly difficult, because each player must seek to assess the impact a deviation from the equilibrium path would have on the predictions the other players would then make about his future plans. His reasoning must therefore anticipate how others will reason. But their reasoning must then take into account his reasoning about their reasoning. And so on.[3] The problem of counterfactual reasoning in games has led to the invention of a whole variety of new fictional hominids that one might group together as varieties of the species *homo ludens*.[4] Each such hominid has different beliefs about how to model what would happen if a rational player were to behave irrationally

---

[1]In the small worlds considered in this book, it is always appropriate to model the optimizing process using Bayesian decision theory, but it does not follow that I believe that Bayesianism does more than nibble at the edges of the general problem of rational choice. Like Savage [461], it seems to me both "ridiculous" and "preposterous" to apply his theory to what he calls a large-world context (Binmore [69]).

[2]Aumann [26] suggests altering the requirement that a rational choice should generate a payoff that a player knows to be at least as high as any other available payoff to the requirement that a player should not know that any other available payoff generates a payoff higher than the rational choice (Binmore [72, 74]).

[3]Section 2.5.4 endorses the view that we have big brains because much computational power is required to sustain the complicated models of ourselves and other people that are necessary for this purpose.

[4]But not the *homo ludens* of the charming book of Huizinga [264], whose thought is so distant from mine that he maintains that play ceases where morality begins.

(Section 3.4.1 of Volume I). In consequence, societies made up of different varieties of *homo ludens* would find themselves playing different refinements of the Nash equilibria in their Game of Life.

Personally, I think the huge literature on refinements of Nash equilibria will turn out to have little or no relevance to the way real people actually play games. As Section 0.4.2 records, I even think it necessary to apologize for restricting attention to subgame-perfect equilibria in the context of this book. My reason for being dubious about the refinement literature is that I believe it operates at the wrong level in its theory of how people get to equilibrium. At the level of the individual, *homo economicus* may provide an adequate model of what people eventually do when they have adjusted to a new set of circumstances, but he is seldom likely to be a good model of how real players adjust their behavior over time. Real people are not the omniscient, lightning calculators of the traditional microeconomic textbook. They therefore do not think their way to an equilibrium; they mostly find their way to an equilibrium—insofar as they do—via some interactive learning process involving both imitation of others and trial-and-error experimentation. During this adjustment process, Section 0.4.2 argues that the players are better modeled as *homo sociologicus* than *homo economicus*. The optimizing paradigm then needs to be applied, not at the level of the individual, but at the level of the meme or gene.

The point of reiterating this familiar theme is to warn against those theorists who insist on modeling man as some variety of *homo ludens* without any regard to the historical processes that shaped the behavior which sustains the equilibrium he is assumed to operate. However tempting a theorist's metaphysical speculations about what should be assumed about a rational player who is behaving irrationally, the truth is that it is the players' *history* of experience that determines how the counterfactuals which need to be evaluated in sustaining an equilibrium actually get interpreted. One should be especially suspicious of claims that modeling the players as *homo economicus* requires that we focus our attention on one equilibrium rather than another in a particular game in those cases when a study of simple processes of interactive learning in the game shows that they are not robust to small perturbations.

The Ultimatum Game of Sections 0.4.1 and 3.5.2—which has become a focus for demands that economists replace *homo economicus* by *homo ethicus*—is an example of a game in which simple processes of interactive learning are both slow and easily disrupted (Binmore *et al* [76], Roth and Erev [440]). But it would be foolish to abandon *homo economicus* at the individual level in games where interactive processes of learning are fast and robust merely because it is necessary to move down to the level of the meme or gene in games for which interactive learning operates less reliably.

A.2.3 Passions

In modern terms, a person's passions are his preferences. When using *homo economicus* as a modeling tool, nothing constrains the preferences we attribute to him beyond the requirement that they be consistent. There is therefore no reason why *homo economicus* should be assumed to have the narrowly selfish preferences of a Mr. Hyde. On the contrary, one can stir in as much of Dr. Jekyll as seems realistic. However, Section 1.2.2 of Volume I warns against mixing too much of Dr. Jekyll into the preferences of *homo economicus* when studying interactions among strangers. As Section 2.5.2 argues, Nature has programmed us with a capacity for sympathy toward our kinfolk that seems to extend to those with whom we interact in small groups over extended periods, but it is a bad mistake to assume that the same goodwill can be relied upon when we move outside our immediate circle of family and friends.

Not only is it unrealistic to write overly altruistic motivations into the preferences of *homo economicus* when using him as a model of man, there is also a danger that allowing oneself too much freedom in this respect will remove any bite from the model. When critics of neoclassical economics are not taking economists to task for assuming that people care only about money, they accuse them instead of operating a model in which things are defined in such a way that the optimizing paradigm becomes tautological. As Section 2.2 of Volume I explains, a system of tautologies can be very useful in organizing a body of data, but it is certainly true that a tautology is without substantive content and hence cannot explain anything in itself. In particular, if one is seeking to explain why Eve chose a rather than b, it does not help to say that she prefers a to b when the definition of preferring a to b is that she didn't choose b when a was available.

More generally, it is too easy to explain a piece of human behavior by saying that the subject wanted to do it. As a minimum, such an explanation needs to be supported by some evidence that the preferences to which one is appealing are sufficiently persistent that they can be observed at other times and in other circumstances. For example, my explanation of the behavior of experimental subjects in the Ultimatum Game as a sore thumb phenomenon (Section 3.5.2) is commonly rejected in favor of the hypothesis that subjects have a taste for fairness. But the behavior of subjects demonstrably changes as they gain more experience of the game (Roth and Erev [440]). To accommodate such data, the taste-for-fairness explanation must therefore be modified so that the subjects' taste for fairness varies over time. It therefore ceases to have any content unless we can come up with some explanation of how people come to have a taste for fairness, and why this should change with time.

My own view, of course, is that it is a mistake to seek to explain fairness norms at the level of individual preferences. Fairness arises at the next level up, as a component in a convention for coordinating on equilibria in games. The fact that it needs to be modeled as a social value rather than a personal taste is signaled by the fact that it changes in the medium run, whereas a personal taste should remain fixed (Sections 3.8.1).

A.2.4   Experience

Experience translates as what a player knows or believes. It is on this subject that game theory has perhaps had its greatest influence on modern economics. Before Von Neumann and Morgenstern [538] formulated the notion of information sets and expected utility, economic theory had perforce to remain silent on issues where uncertainties about the future were central. But, although Von Neumann and Morgenstern's ideas on modeling knowledge and belief have revolutionized economic thinking, the significance of their ideas for social questions remains widely unappreciated. It is more than twenty years since Aumann [24] and Lewis [325] taught us to be sensitive to the difference between *individual* knowledge and *common* knowledge—thereby allowing us to build conceptually more sophisticated models of human sociality than traditional thinking allows. Surely the time has come to put these insights to work in moral philosophy.

I am sure that currently fashion in communitarian thinking is right to emphasize community spirit as a vital ingredient of human sociality. But, in studying human communities, we no longer need to be shackled to models that are more appropriate to other phyla. Human communities do not operate like ant hills. Outside the context of the family, to appeal to natural laws of interaction within communities, or to mythical ideals of collective rationality, or to one of the many conceptions of the common good that have been invented over the years, is to make what I believe to be a major modeling error. I think that Adam Smith taught us well in making *homo economicus* the mainspring of *The Wealth of Nations*. Nor, as Section 3.8.1 insists, was he wrong in believing that the more inchoate ideas of his *Theory of Moral Sentiments* can be brought into line with an optimizing model of man. But we must then not try to build his conception of propriety in a large society into a citizen's strength of body, his reason or his passions. Such ideas belong in a discussion of a society's *culture*, which Section 2.3.6 of Volume I identifies with its pool of common knowledge.

A Song of Liberty

1. The Eternal Female groand! it was heard over all the Earth:

2. Albions coast is sick silent; the American meadows faint!

3. Shadows of Prophecy shiver along by the lakes and the rivers and mutter acrofs the ocean? France rend down thy dungeon;

4. Golden Spain burst the barriers of old Rome;

5. Cast thy keys O Rome into the deep down falling, even to eternity down falling,

6. And weep and bow thy reverend locks!

7. In her trembling hands she took the new born terror howling;

8. On those infinite mountains of light now barr'd out by the atlantic sea, the new born fire stood before the starry king!

9. Flag'd with grey brow'd snows and thunderous visages the jealous wings wav'd over the deep.

10. The speary hand burned aloft, unbuckled was the shield, forth went the hand of jealousy among the flaming hair, and

The argument that William Blake uses here to reconcile his love of liberty with religious authority is more than a little obscure. I fear that the same is true of the arguments of John Stuart Mill and his followers when they similarly try to reconcile personal liberty with utilitarianism.

Appendix B

Harsanyi Scholarship

> The ground of obligation must be looked for, not in the nature of
> man ... but solely *a priori* in the concepts of pure reason.
>
> Immanuel Kant

B.1 Introduction

Throughout this book, Harsanyi's [233] work has provided a constant source
of inspiration and support, but I chose to ignore the continuing scholarly
debate over the viability of his contribution to utilitarianism in order to
avoid diluting the importance of the fact that my approach heretically replaces Harsanyi's metaphysical interpretation of his ideas by a naturalistic
interpretation. This appendix[1] seeks to redress the balance by gathering
together the references to Harsanyi's work scattered through the two long
volumes of *Game Theory and the Social Contract* with a view to defending
them from some of his critics. This exercise will simultaneously provide
an opportunity to reiterate some of my own differences with Harsanyi on
questions of interpretation.

My distinction between Harsanyi's teleological and nonteleological approaches confuses some orthodox social choice theorists. Harsanyi does
not recognize the distinction himself, but other authors sometimes separate Harsanyi's ideas into two similar categories by distinguishing between
his Aggregation Theorem and his Ideal Observer Theorem. I make things
difficult for such authors by using the metaphor of the ideal observer in discussing the Aggregation Theorem that serves as the cornerstone for what I

---

[1]A longer version appears as *Naturalizing Harsanyi and Rawls* (Binmore [75]).

call Harsanyi's teleological approach, whereas authors like Weymark [550] prefer to simplify the common interest game played behind Harsanyi's veil of ignorance in his nonteleological approach by treating each player as a replica of a single ideal observer. But my own adaptation of Harsanyi's ideas requires that each player retain a distinct identity. As for the current fashion of treating the words *teleological* and *consequential* as synonyms, this seems to me entirely retrograde.

B.2 Teleological Utilitarianism

Section 2.2.3 of Volume II proposes three questions that all utilitarians need to answer:

- What constitutes utility?
- Why should individual utilities be added?
- Why should I maximize the sum of utilities rather than my own?

In discussing Harsanyi's teleological approach to utilitarianism, these questions will be considered one by one.

What constitutes utility? In answering the first question, Bentham and Mill felt comfortable in talking about happiness, but modern authors have learned to be more circumspect. One school of thought led by Scanlon [462], Sen [479, 481], Dworkin [167], Cohen [129], and others argues that welfarism is an inadequate foundation on which to build a moral theory. Since Sen [481] defines welfarism to be the approach to social choice that pays attention only to the utilities that citizens assign to the available social alternatives, this postwelfarist movement challenges the claim of neoclassical economics that a person's well-being can be adequately assessed in terms of the extent to which his preferences are satisfied. As observed in Section 4.3.2, some authors even deny that subjective criteria like individual preferences have any place at all in a moral theory.

As an alternative to utility as normally understood in economics, postwelfarists give lists of supposedly objective criteria that need somehow to be weighed against each other in determining how well off someone is. The most famous example is Rawls' [417] index of primary goods. Attempts are then made to deny legitimacy to those who stick by orthodox utility theory. For example, Sen [477] and Roemer [435, 434] deny that Harsanyi [233] can properly be counted as a utilitarian because he interprets utility like Von Neumann and Morgenstern.

The ipsedixism of such postwelfarists is not attractive, but their chief crime in my book lies in their refusal to recognize that modern utility theory was invented *because* attempts to encompass human aspirations in terms of *a priori* objective criteria have been universally unsuccessful. In particular, it is incoherent to attack modern welfarism on the grounds that the utilities attributed to the citizens in a society do not adequately reflect their motivations. According to the modern view, if their utilities did not adequately reflect their motivations, they would be the wrong utilities.

Having argued that Harsanyi [233] should not be denied the utilitarian label for interpreting utility in the sense of Von Neumann and Morgenstern, I must now defend him for sticking by their definition of a utility function. To discuss the criticism that Weymark [550] and others direct at Harsanyi's approach to interpersonal comparison, it is necessary to begin by identifying a potentially confusing ambiguity in terminology. Recall that Von Neumann and Morgenstern's consistency postulates imply that a rational player makes risky decisions as though seeking to maximize the expected value of a utility function. More precisely, a function $\phi : \Omega \to \mathbb{R}$ exists such that his preference relation $\preceq$ over the set $\text{lott}(\Omega)$ of lotteries with prizes in Ω is described by $\mathcal{E}\phi : \text{lott}(\Omega) \to \mathbb{R}$, where $\mathcal{E}$ is the expectation operator.[2] The question at issue is whether ϕ or $\mathcal{E}\phi$ is said to be a Von Neumann and Morgenstern utility function. The choice matters because $\phi : \Omega \to \mathbb{R}$ is a *cardinal* utility function for the restriction of $\preceq$ to Ω, whereas $\mathcal{E}\phi : \text{lott}(\Omega) \to \mathbb{R}$ is only an *ordinal* utility function for the unrestricted preference relation $\preceq$ on $\text{lott}(\Omega)$. Since it can only make sense to compare utils on cardinal scales, someone who follows Harsanyi in comparing Von Neumann and Morgenstern utility functions must therefore necessarily have settled the nomenclature question in favor of ϕ.

A second difficulty with Harsanyi's [233] use of the Von Neumann and Morgenstern theory of utility arises from the commonly held view that a function ϕ which measures individual welfare should admit the interpretation that $\phi(b) - \phi(a) > \phi(d) - \phi(c)$ if and only if the citizen's preference for b over a is more intense than his preference for d over c. Otherwise

---

[2]To say that a utility function $\Phi : \text{lott}(\Omega) \to \mathbb{R}$ describes a preference relation $\preceq$ on the set $\text{lott}(\Omega)$ means that $\mathbf{L} \preceq \mathbf{M}$ if and only if $\Phi(\mathbf{L}) \leq \Phi(\mathbf{M})$. I follow Von Neumann and Morgenstern [538] in defining a Von Neumann and Morgenstern utility function $\phi : \Omega \to \mathbb{R}$ by the requirement that $\mathcal{E}\phi : \text{lott}(\Omega) \to \mathbb{R}$ be a utility function for $\preceq$ on $\text{lott}(\Omega)$. Note that the Von Neumann and Morgenstern utility function $\phi : \Omega \to \mathbb{R}$ does *not* describe $\preceq$ on $\text{lott}(\Omega)$: it describes the *restriction* of $\preceq$ to Ω. A function ϕ is a Von Neumann and Morgenstern utility function if and only of the same is true of $A\phi + B$, for all constants $A > 0$ and B. As with a temperature scale, one is therefore free to choose the zero and the unit on a Von Neumann and Morgenstern utility scale, but then everything else is determined. After zeros and units have been chosen on Adam and Eve's utility scales, it then makes sense to compare Adam and Eve's utils—just as it makes sense to compare the degrees on a centigrade thermometer with the degrees on a fahrenheit thermometer. In brief, it is meaningful to compare *cardinal* utility functions.

a util given to a person when his welfare is high might be worth more or less than when his welfare is low. Those who draw this conclusion merely from the fact that ϕ is a Von Neumann and Morgenstern utility function fall prey to one of the standard fallacies listed in Luce and Raiffa's [334] textbook. To obtain the conclusion legitimately, one has to make an *assumption* about what intensities of preference are to be taken to mean. Harsanyi's use of Von Neumann and Morgenstern's risk-based definition of intensity provides just such a legitimizing assumption, and hence provides one of many possible foundations for the concept of individual welfare.

Suppose that $a \prec b$ and $c \prec d$. Then Harsanyi follows Von Neumann and Morgenstern [538, p.18] in arguing that a citizen should be deemed to hold the first preference more intensely than the second if and only if he would be always be willing to swap a lottery $\mathbf{L}$ in which the prizes a and d each occur with probability $\frac{1}{2}$ for a lottery $\mathbf{M}$ in which the prizes b and c each occur with probability $\frac{1}{2}$. To see why Harsanyi proposes this definition, imagine that the citizen is in possession of a lottery ticket $\mathbf{N}$ that yields the prizes b and d with equal probabilities. Would he now rather exchange b for a in the lottery or d for c? Presumably, he should prefer the latter swap if and only if he thinks that b is a greater improvement on a than c is on d. But to say that he prefers the first of the two proposed exchanges to the second is to say that the citizen prefers $\mathbf{M}$ to $\mathbf{L}$.

In terms of the citizen's Von Neumann and Morgenstern utility function ϕ, the fact that $\mathbf{L} \prec \mathbf{M}$ reduces to the proposition that

$$\tfrac{1}{2}\phi(a) + \tfrac{1}{2}\phi(d) < \tfrac{1}{2}\phi(b) + \tfrac{1}{2}\phi(c) .$$

Thus the citizen holds the preference $a \prec b$ more intensely than the preference $c \prec d$ if and only if $\phi(b) - \phi(a) > \phi(d) - \phi(c)$.

Of course, other definitions of intensity of preference will lead to other conceptions of individual welfare. One might then be led like Weymark [550] to suggest replacing ϕ as a welfare measure by some strictly increasing transformation of ϕ. But Harsanyi's definition would seem the obvious first avenue of exploration.

Why add utilities? This question includes the problem of comparing utils across individuals. Unless utilities are suitably weighted before being added, you might as well pretend, as Bentham [53] observes, to add 20 apples to 20 pears. Postwelfarists have no answer to the problem of interpersonal comparison. The criteria for human well-being they compile are simply asserted to be universally applicable. Harsanyi has two answers. This section examines what I call his teleological answer.

A teleological moral theory postulates an *a priori* conception of the common good. Harsanyi [233] follows a well-trodden path by writing down a list of axioms that are said to characterize such a common good. The nature of the common good is then deduced from these axioms, rather than simply being asserted to be self-evident. The axioms used by utilitarians like Harsanyi take the personal preferences of individual citizens as given, and describe how these are to be related to the preferences of society as a whole. Society is therefore treated as a single person written large. Or, to follow Adam Smith [500], the interests of society are identified with those of an impartial spectator or an *ideal observer*, whose preferences are some kind of average or aggregate of the preferences of all the individuals in the community. This ideal observer's utility function is then said to be a welfare function for the society as a whole.

The mathematics with which Harsanyi [231, 233] defends his teleological approach have been criticized by Weymark [550], Roemer [435] and others. However, the trivial mathematics of Section 4.2.4 of Volume I do not seem vulnerable to similar criticism. Nor are the axiomatic approaches of Broome [102], Hammond [219, 221], Maskin [344], and numerous other authors.

Consider a society with two citizens, Adam and Eve. If Adam and Eve honor the Von Neumann and Morgenstern rationality conditions, their preferences over lotteries can be summarized by Von Neumann and Morgenstern utility functions defined on a finite set S of possible social states. To keep things simple, let us assume that Adam and Eve agree that there is a worst social state $\mathcal{L}$ and a best social state $\mathcal{W}$. We can then select the zeros and the units on Adam and Eve's utility scales so that $u_A(\mathcal{L}) = u_E(\mathcal{L}) = 0$ and $u_A(\mathcal{W}) = u_E(\mathcal{W}) = 1$. The Von Neumann and Morgenstern utility functions u_A and u_E will serve as inputs to our algorithm for determining the preferences of the ideal observer.

The next step is very much less innocent. Not only are Adam and Eve assumed to honor the Von Neumann and Morgenstern rationality conditions, but the same is assumed to be true of the ideal observer, who is taken to be no different in kind from Adam and Eve. In particular, he may be assigned a Von Neumann and Morgenstern utility function u_i that describes his preferences over lotteries in which the prizes are social states. As with Adam and Eve, the zero and the unit on the ideal observer's utility scale will be chosen so that $u_i(\mathcal{L}) = 0$ and $u_i(\mathcal{W}) = 1$.

The third step requires that, in assessing the consequences of choosing an action that leads to one lottery over social states rather than another, the ideal observer is assumed to take into account *only* Adam and Eve's personal preferences over the lotteries. One can translate this requirement into formal terms by requiring that, for each lottery $\mathbf{L}$,

$$\mathcal{E}u_i(\mathbf{L}) = v_i(\mathcal{E}u_A(\mathbf{L}), \mathcal{E}u_E(\mathbf{L})), \qquad (\text{B.1})$$

where the values $v_i(x_A, x_E)$ of the function v_i depend only the pair (x_A, x_E)

of utilities that Adam and Eve receive as a consequence of the ideal observer's decision. In brief, the ideal observer's expected utility for a lottery is assumed to depend only on Adam and Eve's expected utilities for the lottery.

Equation (B.1) implies that the function v_i is *linear* on its domain of definition D.[3] A standard result from elementary linear algebra then tells us that constants U and V must exist for which

$$u_i(x_A, x_E) = U x_A + V x_E.$$

The ideal observer will therefore make his choice from S by maximizing a suitably weighted sum of Adam and Eve's utilities. The common good he personifies is therefore utilitarian.

Why not maximize your own utility? When asked why a citizen should selflessly pursue the common good rather than his own interests, a Kantian like Harsanyi [233] simply asserts that the citizen has a moral obligation to do so.[4] But is this not simply to say that he ought to do it because he ought to do it?

As observed in Section 2.2.3, I think it a major error for utilitarians to fudge the issue of why citizens should pursue the aims of some ideal observer rather than furthering their own individual interests. The question that needs to be decided is whether utilitarianism is a moral system to be employed by *individuals* in regulating their interactions with others, or set of tenets to be followed by a *government* that has the power to enforce its decrees.

My own view is that utilitarians would be wise to settle for the public policy option, which Hardin [224] refers to as *institutional utilitarianism*. If those in power are inclined to personify the role of the government of which

---

[3]If x is in D, then there exists a lottery $\mathbf{M}$ such that $x_A = u_A(\mathbf{M})$ and $x_E = u_E(\mathbf{M})$. A corresponding lottery $\mathbf{N}$ can be associated with any y in D. Let $\mathbf{L}$ be the compound lottery that yields the prize $\mathbf{M}$ with probability p and $\mathbf{N}$ with probability $1 - p$. Then the left-hand side of (B.1) is $\mathcal{E}u_i(\mathbf{L}) = p\mathcal{E}u_i(\mathbf{M}) + (1 - p)\mathcal{E}u_i(\mathbf{N}) = pv_i(x) + (1 - p)v_i(y)$. The right-hand side of (B.1) is given by $v_i(\mathcal{E}u_A(\mathbf{L}), \mathcal{E}u_E(\mathbf{L})) = v_i(pu_A(\mathbf{M}) + (1-p)u_A(\mathbf{N}), pu_E(\mathbf{M}) + (1-p)u_E(\mathbf{N})) = v_i(px + (1-p)y)$. It follows that $v_i(px + (1-p)y) = pv_i(x) + (1-p)v_i(y)$. Mathematicians will recognize this equation as the requirement that v_i be affine. But $u_A(\mathcal{L}) = u_E(\mathcal{L}) = u_i(\mathcal{L}) = 0$. Hence $v_i(0) = 0$, and so v_i is linear.

[4]Kant [297, p.63] is good for a laugh on this subject. The arch-deontologist first asserts the teleological principle that each organ of the body is ideally suited to its purpose. He then observes that a suitably complex system of refined instincts would have better served the purpose of securing our survival and welfare than the reasoning abilities of our brains. Hence we have reason for another purpose: to figure out that we ought to obey the categorical imperative! (Section 2.4.2 of Volume I)

they form a part, then they may be open to the suggestion that its actions should be rational in the same sense that an individual is rational. In a society with liberal traditions, Bentham's [53] "everyone to count for one, nobody to count for more than one" will also be attractive. If the powerful are persuaded by such propaganda, then Harsanyi's [233] teleological argument will then require that the government act as though it had the preferences of a utilitarian ideal observer.

In my view, Harsanyi's [233] teleological defense of utilitarianism therefore applies best to the problem that welfare economics sets out to solve. In its idealized form, a benign but paternalistic government asks what behavior it should enforce on the citizens subject to its authority. Harsanyi's answer makes sense when the government regards itself as an individual written large who is immune to personal prejudice.

B.3 Nonteleological Utilitarianism

Evolutionary biologists shrink in horror from teleological explanations of natural phenomena. The idea that evolution is designed to fulfill some *a priori* purpose is nothing less than heretical. Authors like myself, who offer naturalistic explanations of moral phenomena in human societies, feel much the same about teleological explanations of social phenomena. However, it is commonplace for critics to fail to understand a nonteleological approach to moral theory altogether, since they see no point in discussing a common good function G unless the fact that $G(a) < G(b)$ provides a reason why a society should prefer state b to a. But a nonteleological theory follows Rawls [417] in taking the *procedure* by means of which a society makes its decisions as fundamental. A common good function that arises in such a theory has no causal role. One simply asserts that to use the procedure under study is to behave *as though* maximizing a particular common good function. Rawls [417] refers to the process of constructing a common good function from a fair procedural system as deducing the Good from the Right.

In Chapter 2, I develop a utilitarian theory based on what I claim to be the current consensus on political legitimacy in western democracies. The people use a fair process to deliver a mandate to the government, which then acts to enforce the laws that the people have made for themselves. In seeking to model this consensus, I invent an impartial philosopher-king, who first receives a mandate from the people to pursue certain ends, and then coerces each person into taking whatever action is necessary to achieve these ends.

In modeling the fair procedure used to determine his mandate, I follow

Harsanyi [233] and Rawls [417] in employing the device of the original position. However, I differ sharply from both Harsanyi and Rawls in seeing the original position as nothing more than a coordinating device that evolved along with the human species. I therefore find it necessary to restrict attention to social contracts that are equilibria in the Game of Life and hence police themselves. By contrast, the metaphysical approach adopted by Harsanyi and Rawls allows them to argue that the social contract agreed behind the veil of ignorance is binding on the citizens of a society because they are subject to some metaphysical prior commitment to behave morally.

In expressing Harsanyi's nonteleological approach in such terms, I risk confusing Harsanyi scholars because, like Hume [274] before him, Harsanyi [234] denies that contractarianism makes any sense (Section 1.2.3 of Volume I). I agree with both that the respective forms of contractarianism they attack are indefensible, but I don't find it helpful to insist that one can only be a contractarian if one holds indefensible views. My own form of contractarianism is not vulnerable to the criticisms of either Hume of Harsanyi, since I do not argue that a social contract can be treated as though it were a legal document.

Harsanyi's original position. When presenting Harsanyi's own nonteleological defense of utilitarianism, the analysis of Section 2.6.1 of Volume II can be greatly simplified. Recall from equations (2.9) and (2.10) that the utilities that players I and II respectively expect from an agreement to implement $\mathcal{C}$ if AE occurs and $\mathcal{D}$ if EA occurs are:

$$w_1(\mathcal{C}, \mathcal{D}) \;=\; \tfrac{1}{2} U_1 u_A(\mathcal{C}) + \tfrac{1}{2}\{1 - V_1(1 - u_E(\mathcal{D}))\}\,, \qquad \text{(B.2)}$$

$$w_2(\mathcal{C}, \mathcal{D}) \;=\; \tfrac{1}{2}\{1 - V_2(1 - u_E(\mathcal{C}))\} + \tfrac{1}{2} U_2 u_A(\mathcal{D})\,. \qquad \text{(B.3)}$$

To reduce these equations to something recognizable in Harsanyi's work, begin by suppressing the irrelevant constants $\tfrac{1}{2}(1 - V_1)$ and $\tfrac{1}{2}(1 - V_2)$. Observe next that Harsanyi does not admit contingent social contracts, and so (B.2) and (B.3) can be simplified by writing $\mathcal{C} = \mathcal{D}$, as in Sections 4.3 and 4.4 of Volume I. Finally, it is necessary to appeal to Harsanyi's assumption that players I and II will necessarily have the *same* empathetic preferences behind the veil of ignorance, so that $U_1 = U_2 = U$ and $V_1 = V_2 = V$.

When these simplifications are incorporated into (B.2) and (B.3), the bargaining problem faced by players I and II in the original position becomes trivial, since each player then wants to maximize the *same* utility function. They will therefore simply agree on the social contract $\mathcal{C}$ that maximizes the utilitarian social welfare function:

$$w(\mathcal{C}) = \tfrac{1}{2} U u_A(\mathcal{C}) + \tfrac{1}{2} V u_E(\mathcal{C})\,.$$

Metaphysical pole-vaulting. As documented in Sections 4.3 and 4.4 of Volume I, my difficulty with Harsanyi's nonteleological defense of utilitarianism lies in the argument he deploys to ensure that $U_1 = U_2 = U$ and $V_1 = V_2 = V$. In order to guarantee that players I and II both agree in the original position that V of Adam's utils are worth U of Eve's, Harsanyi proposes that the veil of ignorance should be assumed so thick that Adam and Eve forget their empathetic preferences along with their personal preferences. Each must therefore formulate new empathetic preferences in the original position. An appeal to the Harsanyi doctrine—which says that rational folk in identical circumstances will be led to identical conclusions—then guarantees that the empathetic preferences they construct are identical. However, Harsanyi offers no insight into the considerations to which players I and II would appeal in determining the weights U and V that characterize their new empathetic preferences.

My own approach is based on how it seems to me that the device of the original position is actually used in real life. As argued in Section 4.7 of Volume I, if practice is to be our guide, then we must not follow Harsanyi and Rawls in postulating a thick veil of ignorance. On the contrary, the veil of ignorance needs to be as thin as possible. Adam and Eve must certainly still forget their personal preferences along with their identities, but it is essential that they do not forget the *empathetic* preferences with which their culture has equipped them. To isolate Adam and Eve in a Kantian void from the cultural data summarized in their empathetic preferences, and then to ask them to make interpersonal comparisons seems to me like inviting someone to participate in a pole-vaulting competition without a pole.

As explained in Section 2.5, my guess is that we have empathetic preferences at all only because we need them as inputs when employing fairness norms for which the device of the original position serves as a stylized representative. Section 2.5.4 then argues that social evolution is likely to equip Adam and Eve with the *same* empathetic preferences, so that a version of Harsanyi's argument can then go through.

From a practical point of view, the advantage of such a naturalistic approach is that it tells us in principle how interpersonal comparisons of utility are made. With Harsanyi's metaphysical approach, one can only find out by asking a rational superbeing how he would feel if placed in a Kantian limbo. However, my guess is that, even if we could find such a rational superbeing, he or she would find the question meaningless.

William Blake's engraving shows some of Plato's students exercising the mathematical skills without which they were denied entry to his Academy. Modern writers of footnotes to Plato might usefully go and do likewise.

Appendix C

Bargaining Theory

And what is this? Theorems.

Epictetus, *Moral Discourses*

C.1 Introduction

The purpose of this appendix is twofold.[1] The primary aim is to provide a simple proof of a version of Rubinstein's [445] bargaining theorem in a setting sufficiently general to cover the situations that commonly arise in applications where information is complete. In particular, the feasible set is not assumed to be convex, and a reasonably general view is taken of how disagreements may arise.

C.2 Alternating Offers Game

Rather than setting the story directly in utility space, it will be told in terms of the classic problem of dividing the dollar (Section 1.2.4). A philanthropist donates a dollar to Adam and Eve on condition that they are able to agree how to share it. Disagreement may arise in various ways. A player may abandon the negotiations in favor of his best outside option, leaving the other no choice but to do the same. Or the philanthropist may lose patience if agreement is delayed too long, and withdraw the money. If either of these eventualities occurs, the negotiations will be said to have

---

[1]Appendix C is reprinted from Binmore [71] with the kind permission of the editor of *Investigaciones Economicas*.

broken down. Agreement may also fail to be reached because the players bargain forever without a proposal being accepted or the negotiations breaking down. When the players use strategies that would lead to such a perpetual disagreement unless interrupted by the philanthropist withdrawing his money, we say that they are *deadlocked.*

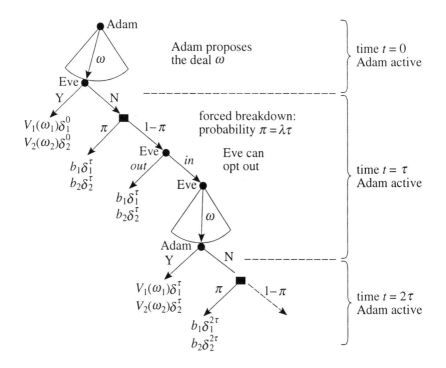

Figure C.1: Rules of the bargaining game.

The result of bargaining under precisely determined rules will be studied. All action takes place at times $n\tau(n = 0, 1, 2, \ldots)$, where $\tau > 0$. Adam is active when n is even. Eve is active when n is odd. If the game has not already ended at time $n\tau > 0$, the philanthropist withdraws his dollar with probability $\pi = \lambda\tau < 1$. Otherwise the game continues with probability $\bar{\pi} = 1 - \pi$. The active player then decides whether to opt in or out. If he opts in, then he continues by making a proposal on how to divide the dollar. The passive player than accepts or refuses. Only after a refusal does the clock advance by τ. The passive player then becomes active, and the above sequence of events is repeated.

The game begins at time $n = 0$ but the steps in which the negotiations may break down are omitted in the first period. The very first move therefore consists of Adam's making a proposal. If Eve refuses, the sequence of events described in the previous paragraph commences with $n = 1$. Figure C.1 illustrates the order of moves.[2]

C.3 Preferences

Adam is taken to be player I and Eve to be player II. The set of possible deals is identified with

$$\Omega = \{\omega : \omega_1 + \omega_2 \leq 1\},$$

where the result of implementing the deal $\omega = (\omega_1, \omega_2)$ is that player i receives a payment of $\$\omega_i$. Notice that free disposal is taken for granted (Section 1.2.1).[3] A point $\beta \in \Omega$ represents the pair of payoffs that the players receive if the negotiations break down.[4]

Player i's utility for the outcome $\omega \in \Omega$ at time t is taken to be

$$u_i(\omega_i)\delta_i^t, \tag{C.1}$$

where the discount factor δ_i satisfies $0 < \delta_i < 1$. (The corresponding discount rate ρ_i is given by $\delta_i = e^{-\rho_i}$.) To economize on notation, we write $\Delta_i = \delta_i^\tau$. Recall that $\pi = \lambda\tau$. Thus Δ_i and π are always functions of τ. The function $v_i : \mathbb{R} \to \mathbb{R}$ is assumed to be continuous and strictly increasing, but not necessarily concave. Its range is an open interval R_i. We take $b_i = v_i(\beta_i)$. The breakdown point then lies in $R_1 \times R_2$.

The Pareto-frontier of the set X of utility pairs available at time 0 is the graph of the function f defined by

$$f(x_1) = v_2(1 - v_1^{-1}(x_1)).$$

Both $f : R_1 \to R_2$ and $f^{-1} : R_2 \to R_1$ are continuous and strictly decreasing. Note that $b \in X$, so that $b_2 \leq f(b_1)$ and $b_1 \leq f^{-1}(b_2)$.

---

[2]Most of the sequencing in this specification is unimportant to the results. The specification chosen is for mathematical convenience. However, it is important that the active player's opting-out decision does not occur immediately *after* the passive player has refused an offer (Shaked [487]). Section 1.7.2 argues that the model of the text is more realistic.

[3]Osborne and Rubinstein [396] assume otherwise, and hence find extra equilibria. The assumption in the text seems more natural.

[4]Binmore *et al* [80] review the case when different payoffs may be received depending on how the breakdown occurs. The only new difficulties are combinatorial.

It remains to discuss how the players assess the consequences of the perpetual disagreement outcome D. This is assigned utility $u_i(D) = 0$ so as to be consistent with taking the limit $t \to \infty$ in (C.1). In addition, it is assumed that $0 \in \overline{R}_1 \times \overline{R}_2$ and $b \geq 0$.

A deadlock arises when players use strategies that rule out the possibility of agreement, but the result will be perpetual disagreement only if $\pi = 0$. In either case, the expected utility of a deadlock is

$$\pi b_i \Delta_i + \pi b_i (\overline{\pi} \Delta_i) + \pi b_i (\overline{\pi} \Delta_i)^2 + \cdots = \frac{\pi b_i \Delta_i}{1 - \overline{\pi} \Delta_i}. \tag{C.2}$$

If v_1 and v_2 are concave, then so is f. It follows that X is convex in this case. If X fails to be convex, we specifically do *not* replace X by its convex hull. This standard practice is usually justified by assuming that players can resolve issues by agreeing to implement a lottery with prizes in X. But such an assumption is frequently unrealistic. How can a union boss report to his members that their wage settlement was decided by tossing a coin? Where lotteries *are* feasible, Ω should be replaced by the set of all lotteries with prizes in Ω. A simpler theory is then obtained.

 Notes.

1. Fishburn and Rubinstein [182] show that relatively mild assumptions on preference relations guarantee a utility representation of the form (C.1). In particular, (C.1) is a substitute for assumptions (A1–A5) of Osborne and Rubinstein [396]. Their condition (A6) is not required.

2. Expressing the problem in terms of dividing the dollar clarifies the interpretation, but may obscure the generality of the approach. All that really matters is the shape of the set X of feasible utility pairs and the values of δ_1, δ_2, π and b. Geometric characterizations of the equilibrium outcomes in such a setting are easily obtained (Binmore [67]). Algebraic characterizations require more labor.

3. If $\pi > 0$, the utility functions must be understood in the sense of Von Neumann and Morgenstern. If $\pi = 0$, then expected utility calculations are not necessary, and so any utility representation will suffice. This observation permits the study of various cases not obviously included in the scope of the analysis. In particular, Rubinstein's [445] case of fixed costs of disagreement is accessible. Player i's utility for the deal ω at time t is then $U_i(\omega, t) = \omega_i - \gamma_i t$, where $\gamma_i > 0$. Rubinstein's version has to be supplemented here by requiring that, if someone opts out at time t, then player i gets $-\gamma_i t$.

One takes $u_i(\omega, t) = \exp U_i(\omega, t)$ and $\delta_i = e^{-\gamma_i}$. Then $v_i(x) = e^x$ is far from concave. Notice that $f : (0, \infty) \to (0, \infty)$ is given by $f(x) = e/x$ and $b = (1, 1)$. The example is useful because it exhibits various pathologies.

C.4 Stationary Subgame-Perfect Equilibria

In this section, η and ζ are Pareto-efficient agreements in Ω. The utilities assigned to these deals at time 0 by the two players are $y_i = v_i(\eta)$ and $z_i = v_i(\zeta)$. Notice that $y_2 = f(y_1)$ and $z_2 = f(z_1)$.

Adam will be assumed to use a stationary pure strategy s which requires that he follow two instructions:

a. Propose η whenever called upon to make a proposal;
b. Accept ζ or better and refuse anything worse, whenever called upon to make a response.

Eve will be assumed to use the stationary pure strategy t which has the same properties as s, but with η and ζ exchanged.

When is a pair (s, t) of strategies with these properties a subgame-perfect equilibrium? The current section explores this question in a series of lemmas.

Define $m_i : R_i \to \mathbb{R}$ by

$$m_i(x) = \Delta_i(\pi b_i + \overline{\pi} \max\{b_i, x\})$$

and restrict attention to those η and ζ for which

$$\left. \begin{array}{rcl} y_2 & = & m_2(z_2) \\ z_1 & = & m_1(y_1) \end{array} \right\} \tag{C.3}$$

These equations express the fact that the passive player will always be indifferent between accepting and refusing what has been proposed.

Lemma C.1 *If* (C.3) *holds, then* $y_1 \geq b_1$ *or* $z_2 \geq b_2$. *Also* $z_1 \leq y_1$ *and* $y_2 \leq z_2$.

Proof. If $y_1 < b_1$, then $z_1 = \Delta_1 \leq b_1$. Since z is Pareto-efficient, it therefore cannot be that $z_2 < b_2$. If $y_1 \geq b_1$, then $z_1 = \Delta_1(\overline{\pi} y_1 + \pi b_1) \leq \Delta_1 y_1 \leq y_1$. Thus, $z_2 = f(z_1) \geq f(y_1) = y_2$. A similar argument applies if $z_2 \geq b_2$.

Lemma C.2 *If* y *and* z *satisfy* (C.3), *then there exists a corresponding subgame-perfect equilibrium pair* (s, t).

Proof. The properties given for s and t do not specify whether or not a player should opt out when the opportunity arises. In this lemma, Adam and Eve's instructions are completed by requiring that Adam should opt in if $y_1 \geq b_1$ and opt out if $y_1 < b_1$. Eve should opt in if $z_2 \geq b_2$ and opt out if $z_2 < b_2$.

The proposing actions specified by s and t are always optimal because the active player cannot demand more without being refused. But an active player never wants to be refused. With one exception, the various cases that need to be checked in verifying this claim are all made immediate by Lemma C.1. In the exceptional case, the active player opts out in equilibrium.

Suppose that $y_1 \geq b_1$ and $z_2 < b_2$. Then $y_2 = \Delta_2 b_2$, and so

$$z_2 \geq= \Delta_2 b_2 \geq \Delta_2(\overline{\pi}\Delta_2 b_2 + \pi b_2) = \Delta_2(\overline{\pi} y_2 + \pi b_2) \,.$$

It follows that Eve prefers z to be accepted than to have her proposal refused. A similar argument applies when $y_1 < b_1$ and $z_2 \geq b_2$.

Lemma C.3 *The pair (y, z) satisfies (C.3), and hence characterizes a a stationary subgame-perfect equilibrium, if and only if y_1 is a zero of the function $g : R_1 \to \mathbb{R}$ defined by*

$$g(x) = f(x) - (m_2 \circ f \circ m_1)(x) \,,$$

and $y_2 = f(y_1)$, $z_1 = m_1(y_1)$, $z_2 = f(z_1)$.

It will be necessary to investigate the properties of the function g in some detail. Note that

$$g(x) = \begin{cases} f(x) - \Delta_2[\pi b_2 + \overline{\pi} f(\Delta_1 b_1)] \,, & \text{if } x \leq b_1 \,; \\ f(x) - \Delta_2[\pi b_2 + \overline{\pi} f(\Delta_1(\pi b_1 + \overline{\pi} x))] \,, & \text{if } b_1 \leq x \leq c_1 \,; \\ f(x) - \Delta_2 b_2 \,, & \text{if } x \geq c_1 \,; \end{cases}$$

where $\overline{\pi} c_1 = \Delta_1^{-1}(b_2) - \pi b_1$. The of g on τ is not made explicit. When the dependence matters, it turns out to be more convenient to study the function $G : R_1 \times (0, \infty) \to \mathbb{R}$ defined by

$$G(x, \tau) = \tau^{-1} g(x) \,. \tag{C.4}$$

Lemma C.4 *The function $g : R_1 \to \mathbb{R}$ has the following properties:*

a.	$x < \Delta_1 b_1$	$\Rightarrow \quad g(x) > 0 \,;$
b.	$x < f^{-1}(\Delta_2 b_2)$	$\Rightarrow \quad g(x) < 0 \,.$

Proof.   **a.** If $x < \Delta_1 b_1$, then

$$
\begin{aligned}
g(x) &> f(\Delta_1 b_1) - \Delta_2[\pi b_2 + \overline{\pi}f(\Delta_1 b_1)] \\
&= (1 - \Delta_2\overline{\pi})f(\Delta_1 b_1) - \Delta_2\pi b_2 \\
&\geq (1 - \Delta_2\overline{\pi})b_2 - \Delta_2\pi b_2 \\
&= b_2(1 - \Delta_2) \geq 0 \,.
\end{aligned}
$$

b. If $x > f^{-1}(\Delta_2 b_2)$, then $f(x) < \Delta_2 b_2$. If it is also true that $g(x) \geq 0$, then

$$
\begin{aligned}
\Delta_2 b_2 - \Delta_2[\pi b_2 + \overline{\pi}f(\Delta_1(\pi b_1 + \overline{\pi}x))] &> 0 \,. \\
b_2 &> f(\Delta_1(\pi b_1 + \overline{\pi}x)) \,. \\
f^{-1}(b_2) &< \Delta_1(\pi b_1 + \overline{\pi}x) \,. \\
x &> c_1 \,.
\end{aligned}
$$

Now suppose that $f^{-1}(\Delta_2 b_2) \leq c_1$. Then $g(x) < 0$ for $f^{-1}(\Delta_2 b_2) < x \leq c_1$. Hence, $g(x) < 0$ for $x > f^{-1}(\Delta_2 b_2)$, because g is continuous and decreases on (c_1, ∞). On the other hand, if $f^{-1}(\Delta_2 b_2) > c_1$, then $x > f^{-1}(\Delta_2 b_2)$ implies that $g(x) = f(x) - \Delta_2 b_2$.

Lemma C.5 *The function $g : R_1 \to \mathbb{R}$ has a zero in $[\Delta_1 b_1, f^{-1}(\Delta_2 b_2)]$, and hence a stationary subgame-perfect equilibrium always exists.*

Proof. This follows from the previous lemma because g is continuous.

C.5 Nonstationary Equilibria

This section proves a generalized version of a theorem of Rubinstein [445], following Binmore [67], Shaked and Sutton [488], and Binmore *et al* [90].

Let S be the set of all subgame-perfect equilibrium outcomes. The first result demonstrates that S is necessarily a large set when more than one stationary subgame-perfect equilibrium exists. In particular, Pareto-inefficient outcomes then lie in S.

Multiple stationary subgame-perfect equilibria exist when $g : R_1 \to \mathbb{R}$ has more than one zero. Let $(\underline{s}, \underline{t})$ be the strategy pair that corresponds to the smallest zero $\underline{y}_1$ of g. Let $(\overline{s}, \overline{t})$ be the strategy pair corresponding to the largest zero $\overline{y}_1$.

Let T be the set of all feasible payoff pairs that satisfy $x \geq (\underline{y}_1, \overline{y}_1)$.

Lemma C.6 *If $x \in T$, then there exists a subgame-perfect equilibrium (s, t) in which Adam proposes a deal ξ at time 0 worth x, and Eve accepts. Thus $T \subseteq S$.*

Proof. Three "states of mind", UP, DOWN and MIDDLE are distinguished. Players begin in the MIDDLE state. In this state, the subgame-perfect equilibrium (s, t) to be constructed requires that Adam propose ξ when called upon to make a proposal. Eve accepts ξ or better, and refuses anything else.

In the UP state, (s, t) requires that Adam and Eve play according to $(\overline{s}, \overline{t})$ in the remainder of the game. In the DOWN state, (s, t) requires playing according to $(\underline{s}, \underline{t})$ in the remainder of the game.

Once in the UP state the players remain there. The same goes for the DOWN state. Transitions from the MIDDLE state are made as follows. If Adam proposes ξ, then a refusal by Eve shifts both players to the UP state. If Eve refuses any other proposal, both players shift to the DOWN state.

Why is the schedule for proposal and response in the MIDDLE state optimal? Eve should accept ξ because $x_2 \geq \overline{y}_2 = m_2(\overline{z}_2)$. Her response to other proposals is optimal because she gets $\underline{y}_2 = m_2(\underline{z}_2)$ from refusing. Adam should propose ξ because he gets at most $\underline{y}_1$ from deviating, and $x_1 \geq \underline{y}_1$. (If Adam deviates to a proposal that is refused, either Eve opts out because $\underline{z}_2 < b_2$, or she opts in because $\underline{z}_2 \geq b_2$. In the latter case, Adam gets $\Delta_1(\pi b_1 + \overline{\pi}\underline{z}_1) \leq \Delta_1 \underline{z}_1 \leq \underline{z}_1 \leq \underline{y}_1 \leq x_1$. In the former case, he gets $\Delta_1 b_1 \leq b_1 \leq \underline{y}_1 \leq x_1$, by Lemma C.1.)

Next it will be shown that $S \subseteq T$. A preliminary lemma is needed.

Lemma C.7 *The set Y of all subgame-perfect equilibrium payoffs to Adam is $[\underline{y}_1, \overline{y}_1]$.*

Proof. Let $a = \inf Y$ and $A = \sup Y$. Let Z be the set of all subgame-perfect equilibrium payoffs to Eve in the companion game in which it is Eve who makes the first proposal at time 0. Write $e = \inf Z$ and $E = \sup Z$.

1. It is open to Eve to refuse whatever Adam proposes at time 0. If equilibrium strategies are used in the continuation of the game, then Eve will get an expected payoff of at least $m_2(e)$, because the companion game will be played after a time delay of τ, unless the dollar is withdrawn or Eve opts out in the interim. If equilibrium strategies are used, it follows that Eve gets at least $m_2(e)$, and so Adam gets at most $f^{-1}(m_2(e))$. Thus $A \leq f^{-1}(m_2(e))$, and so

$$f(a) \geq m_2(e). \tag{C.5}$$

On applying a similar argument in the companion game,

$$E \leq f(m_1(a)). \tag{C.6}$$

2. It is optimal for Eve to accept any proposal from Adam that assigns her a payoff $w_2 > m_2(E)$, provided that equilibrium strategies are used after a refusal. Thus Adam must get at least $f^{-1}(w_2)$. Hence $f^{-1}(w_2)$ is a lower bound for Y whenever $w_2 > m_2(E)$. It follows that $a \geq f^{-1}(m_2(E))$ and so

$$f(a) \leq m_2(E). \tag{C.7}$$

On applying a similar argument in the companion game,

$$e \geq f(m_1(A)). \tag{C.8}$$

3. From (C.5) and (C.8),

$$f(A) \geq m_2(e) \geq (m_2 \circ f \circ m_1)(A),$$

and hence $g(A) \geq 0$. But $g(x) < 0$ for $x > \bar{y}_1$. Thus $A \leq \bar{y}_1$. But $\bar{y}_1 \in Y$, and so $A = \bar{y}_1$.

From (C.6) and (C.7),

$$f(a) \leq m_2(E) \leq (m_2 \circ f \circ m_1)(a),$$

and hence $g(a) \leq 0$. But $g(x) > 0$ for $x < \underline{y}_1$. Thus $a \geq \underline{y}_1$. But $\underline{y}_1 \in Y$, and so $a = \underline{y}_1$.

4. It remains to confirm that $\underline{y}_1 \leq y_1 \leq \bar{y}_1$ implies $y_1 \in Y$. This follows immediately from Lemma C.6.

Theorem C.1 $\qquad S = T$.

Proof. By the preceding lemma, Adam's equilibrium payoffs lie in the set $[\underline{y}_1, \bar{y}_1]$. Similarly, Eve's equilibrium payoffs in the companion game lie in the set $[\underline{z}_1, \bar{z}_1]$. It follows that her equilibrium payoffs in the original game lie in $[\bar{y}_2, \underline{y}_2]$, since $\bar{y}_2 = m_2(\bar{z}_2)$ and $\underline{y}_2 = m_2(\underline{z}_2)$. Thus $S \subseteq T$. On the other hand, Lemma C.6 shows that $T \subseteq S$.

Notes. The second and third notes concern two cases of special interest considered by Rubinstein [445]. The conclusions differ from his because the bargaining models considered are not identical.

1. When multiple equilibria exist, there may be subgame-perfect equilibria in which agreement is not reached immediately (Section 3.10.1 of Osborne and Rubinstein [396]).

2. In the notation of Section C.3, take $v_1(x) = v_2(x) = x$ and $\pi = 0$. Then $R_1 = R_2 = \mathbb{R}$ and $f(x) = 1 - x$. The zeros of the function g are the stationary subgame-perfect equilibrium outcomes for Adam. In this case, g is strictly decreasing. It follows that there is a unique

equilibrium outcome. If $u = (1 - \Delta_2)/(1 - \Delta_1\Delta_2)$ satisfies $b_1 \leq u < \Delta_1^{-1}(1 - b_2)$, then Adam gets u. If $u < b_1$, then Adam will be planning to take his outside option should the opportunity arise. He then gets $1 - \Delta_2(1 - \Delta_1 b_1)$ when proposing. If $u > \Delta_1^{-1}(1 - b_2)$, Eve is planning to take her outside option and Adam gets $1 - \Delta_2 b_2$.

3. We now analyze the case described in Note 3 of Section C.3. When $\Delta_1 > \Delta_2$, g has a unique zero at Δ_2^{-1}. In terms of money payoffs, this means that, when $\gamma_1 < \gamma_2$, Adam gets $1 + \gamma_2\tau$ in equilibrium. (Eve plans to take her outside option.) When $\Delta_1 < \Delta_2$, g has a unique zero at $\Delta_1\Delta_2^{-1}$. In terms of money payoffs, this means that, when $\gamma_1 < \gamma_2$, Adam gets $(\gamma_2 - \gamma_1)\tau$ in equilibrium. (Adam plans to take his outside option.) When $\Delta_1 = \Delta_2 = \Delta$, g is zero on $[1, e\Delta^{-1}]$. With the notation of Section C.4, it follows that $\underline{y}_1 = 1$ and $\bar{y}_1 = e\Delta^{-1}$. Hence $\bar{y}_2 = \Delta$. Any feasible pair $x \geq (1, \Delta)$ is therefore an equilibrium outcome. In terms of money payoffs, this means that, when $\gamma_1 = \gamma_2 = \gamma$, the set of equilibrium outcomes is $\{\omega : \omega_1 + \omega_2 \leq 1, \omega_1 \geq 0, \omega_2 \geq -\gamma\tau\}$.

C.6   Generalized Nash Bargaining Solutions

The bargaining model described in the previous sections is of interest largely because the equilibrium outcomes in the limiting case as $\tau \to 0$ can be described in terms of a generalized Nash bargaining solution. The case when $\tau \to 0$ is of special interest for at least two reasons. The first is that there will often be nothing that constrains the players to keep to the timetable specified in the model. After refusing a proposal, they will then wish to register a counterproposal at the earliest possible opportunity. The second reason is that Adam's first-mover advantage disappears in the limit as $\tau \to 0$.

As in Section 1.2.2, a generalized Nash bargaining problem is identified with a triple (X, b, d), in which X is interpreted as the set of feasible payoff pairs, b is a breakdown point whose coordinates are the players' outside options, and d is a deadlock point. The sets R_1 and R_2 are open intervals as in Section C.2, $f : R_1 \to R_2$ is a strictly decreasing surjection, and $X = \{(x_1, x_2) \in R_1 \times R_2 : x_2 \leq f(x_1)\}$. Also $0 \in \bar{R}_1 \times \bar{R}_2$ and $0 \leq d \leq b \in X$. For the remainder of the appendix, it will also be assumed that f is twice differentiable on R_1.

A generalized Nash product with bargaining powers α and β is defined to be an expression of the form

$$P(x_1, x_2) = (x_1 - d_1)^\alpha (x_2 - d_2)^\beta . \tag{C.9}$$

When X is convex and $b = d$, the symmetric Nash bargaining solution introduced by Nash [381] identifies the solution of (X, b, d) with the point n at which the Nash product P with $\alpha = \beta$ is maximized, subject to the constraints $x \in X$ and $x \geq d$.

When X is not convex, a more elaborate definition is necessary. Let $p : R_1 \to \mathbb{R}$ be given by

$$p(x) = P(x, f(x)). \tag{C.10}$$

Attention will be restricted to the case in which $p'(x)$ is zero neither at an end point of the interval $[b_1, f^{-1}(b_2)]$, nor at an interior point where $p''(x) = 0$. Only pathological cases are excluded by this restriction. The function $H : R_1 \to \mathbb{R}$ is defined by

$$H(x) = \begin{cases} +\infty, & \text{if } x < b_1 \\ f'(x)\left(\frac{x - d_1}{\alpha}\right) + \left(\frac{f(x) - d_2}{\beta}\right) & \text{if } b_1 \leq x \leq f^{-1}(b_2) \\ -\infty, & \text{if } x > f^{-1}(b_2) \end{cases} \tag{C.11}$$

This has the same sign as p' on the interval $[b_1, f^{-1}(b_2)]$. A 'zero' of H will be understood to be any $z \in R_1$ which has the property that all of its neighborhoods contain both positive and negative values of H. In view of our restrictions on p, a 'zero' of H is either an interior point z of the interval $[b_1, f^{-1}(b_2)]$ at which $H(z) = 0$, or else an endpoint of the interval.

A Nash bargaining point n can now be defined as a Pareto-efficient point of X for which n_1 is a 'zero' of H. Of all Nash bargaining points, let $\bar{n}$ be that which assigns Adam the greatest payoff (and Eve the least payoff). Let $\underline{n}$ be that which assigns Adam the smallest payoff (and Eve the greatest payoff). The *generalized Nash bargaining solution* corresponding to the bargaining powers α and β for the bargaining problem (X, b, d) is then the set N of all feasible payoff pairs $x \geq (\underline{n}_1, \bar{n}_2)$. The definition is illustrated in Figure C.2.

Proposition C.1 *The point $\bar{n}$ is the local maximum of the Nash product (C.9), subject to $x \in X$ and $x \geq b$, that assigns Adam the greatest payoff. The point $\underline{n}$ is the local maximum that assigns Eve the greatest payoff.*

Proposition C.2 *A sufficient condition that N consist of a single point is that $(x - d_1)f'(x)$ be concave on $[b_1, f^{-1}(b_2)]$. In particular, it is sufficient that f be concave and so X is convex.*

Proof. The condition implies that H is strictly decreasing on $[b_1, f^{-1}(b_2)]$.

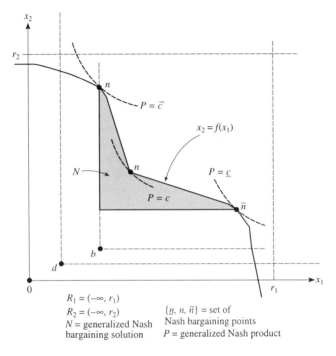

Figure C.2: A generalized Nash bargaining solution.

C.7 Nash Program

In accordance with the Nash program of Section 0.6, it is now necessary to show that the generalized Nash bargaining solution coincides with the subgame-perfect equilibrium outcomes of the Alternating Offers bargaining model.

Theorem C.2 *As* $\tau \to 0$, *the set of subgame-perfect equilibrium outcomes in the Alternating Offers model converges to the generalized Nash bargaining solution* N *with bargaining powers* $\alpha = 1/(\lambda + \rho_1)$ *and* $\beta = 1/(\lambda + \rho_2)$ *for the bargaining problem* (X, b, d) *in which the deadlock payoffs* $d_i = \lambda b_i / (\lambda + \rho_i)$ *are the limiting values of* (C.2) *as* $\tau \to 0$.

Proof. It will be shown that the set of values x for which the function G defined by (C.4) is zero converges to the set of 'zeros' of the function H defined by (C.11). The first step is to show that, for each $x \in R_1$,

$$G(x, \tau) \to H(x) \text{ as } \tau \to 0 .$$

On the interval $[b_1, f^{-1}(b_2)]$, this follows from L'Hôpital's rule, because $c_1 > f^{-1}(b_2)$. Outside the interval, one may appeal to Lemma C.4 to determine the sign of G.

1. Every neighborhood of a 'zero' of H contains a zero of G provided that τ is sufficiently small. With the restrictions introduced in Section C.6, each such neighborhood contains a point at which H is positive and a point at which H is negative. The same is therefore true of G if τ is sufficiently small. Since G is continuous, it follows that G has a zero in the neighborhood.

2. Every neighborhood that contains zeros of G for all sufficiently small τ also contains a 'zero' of H. The interval $[b_1, f^{-1}(b_2)]$ is compact, and hence G converges uniformly to H on this interval. This observation takes care of neighborhoods centered at points in the interior of the interval. A trivial argument then extends the conclusion to the endpoints as well.

In summary, when defending the use of the Nash bargaining theory with an Alternating Offers model, the *status quo* should correspond to the consequences of a deadlock (during which the players remain at the negotiation table but never reach an agreement). The outside options that they may obtain by abandoning the negotiations serve only as constraints on the range of validity of the Nash bargaining solution. Often it is convenient to apply these principles to *flows*. In a wage negotiation, for example, the deadlock flows may be the income per period during a strike.

The final result of this section is offered without a proof. It provides a criterion for the uniqueness of an equilibrium in the Alternating Offers model that does not depend on τ being small.

Theorem C.3 *A necessary condition that the Alternating Offers model have a unique subgame-perfect equilibrium is that N consist of a single point.*

Notes. Some special cases of Theorem C.3 deserve special mention. In each case X is assumed to be convex so that N consists of a single point.

1. $\rho_1 = \rho_2 = 0$. In this case, the equilibrium outcome converges to the symmetric Nash bargaining solution for the problem (X, b, d). Here the breakdown and deadlock points are the same, and there is no difficulty in deciding on an appropriate *status quo*. This case arises when it is not impatience that motivates an early agreement, but fear that the opportunity to reach an agreement may be lost if an agreement is delayed.

2. $\rho_1 = \rho_2 = \rho > 0$. In this case, the equilibrium converges to the symmetric Nash bargaining solution for the problem (X, b, d) in which

$d = \lambda b/(\lambda + \rho)$. Note the displacement of the *status quo* from b. In *symmetric* bargaining problems, this leaves the location of the Nash bargaining solution unchanged. Models that mistakenly place the *status quo* at b will therefore nevertheless lead to correct conclusions.

3. $\lambda = 0$. In this case, the equilibrium outcome converges to an *asymmetric* Nash bargaining solution with bargaining powers $\alpha = 1/\rho_1$ and $\beta = 1/\rho_2$ for the problem $(X, b, 0)$. Recall that the payoff pair 0 corresponds to the perpetual disagreement point D, which therefore serves as the appropriate *status quo* in these circumstances. This case arises when the players are unconcerned about the risk of losing the opportunity to reach an agreement, and are motivated simply by their impatience with delays.

Bibliography

[1] D. Abreu. On the theory of infinitely repeated games with discounting. *Econometrica*, 56:383–396, 1988.

[2] D. Abreu and H. Matsushima. Virtual implementation in iteratively undominated strategies: Complete information. *Econometrica*, 60:993–1008, 1992.

[3] D. Abreu and D. Pearce. A perspective on renegotiation in repeated games. In R. Selten, editor, *Game Equilibrium Models II.* Springer-Verlag, Berlin, 1991.

[4] D. Abreu, D. Pearce, and E. Stachetti. Renegotiation and symmetry in repeated games. *Journal of Economic Theory*, 60:217–240, 1993.

[5] D. Abreu and A. Rubinstein. The structure of Nash equilibrium in repeated games with finite automata. *Econometrica*, 56:1259–1282, 1988.

[6] D. Abreu and P. Sen. Virtual implementation in Nash equilibria. *Econometrica*, ?:993–1008, 1991.

[7] J. Adams. Towards an understanding of inequity. *Journal of Abnormal and Social Psychology*, 67:422–436, 1963.

[8] J. Adams. Inequity in social exchange. In L. Berkowitz, editor, *Advances in Experimental Social Science, Volume II.* Academic Press, New York, 1965.

[9] J. Adams and S. Freedman. Equity theory revisited: Comments and annotated bibiliography. In L. Berkowitz, editor, *Advances in Experminental Social Science, Volume IX.* Academic Press, New York, 1976.

[10] R. Alexander. The evolution of social behavior. *Annual Review of Ecology and Systematics*, 5:325–383, 1974.

[11] R. Alexander. *The Biology of Moral Systems.* Aldine de Gruyter, Hawthorne, New York, 1987.

[12] R. Alexander, K. Noonan, and B. Crespi. The evolution of eusociality. In P. Sherman *et al*, editor, *The Biology of the Naked Mole Rat.* Princeton University Press, Princeton, 1991.

[13] M. Allais. *Economie et Interet.* Imprimie Nationale, Paris, 1947.

[14] L. Anderlini and H. Sabourian. Cooperation and effective computability. *Econometrica*, 63:1337–1369, 1995.

[15] E. Andrew. *Shylock's Rights: A Grammar of Lockean Claims*. Cambridge University Press, Cambridge, 1988.

[16] Aristotle. Politics. In J. Barnes, editor, *The Complete Works of Aristotle, Volume II*. Princeton University Press, Princeton, 1984.

[17] Aristotle. Rhetoric. In J. Barnes, editor, *The Complete Works of Aristotle, Volume II*. Princeton University Press, Princeton, 1984.

[18] Aristotle. *Nicomachean Ethics*. Hackett, Indianapolis, 1985. (Translated by T. Irwin).

[19] K. Arrow. *Social Choice and Individual Values*. Yale University Press, New Haven, 1963.

[20] K. Arrow. Extended sympathy and the problem of social choice. *Philosophia*, 7:233–237, 1978.

[21] G. Asheim. Extending renegotiation-proofness to infinite horizon games. *Journal of Games and Economic Behavior*, 3:278–294, 1991.

[22] J. Aubrey. *Brief Lives*. Penguin, Harmondsworth, UK, 1962. (Edited by O. Lawson).

[23] R. Aumann. Values of markets with a continuum of traders. *Econometrica*, 43:611–646, 1975.

[24] R. Aumann. Agreeing to disagree. *Annals of Statistics*, 4:1236–1239, 1976.

[25] R. Aumann. Correlated equilibrium as an expression of Bayesian rationality. *Econometrica*, 55:1–18, 1987.

[26] R. Aumann. Backward induction and common knowledge of rationality. *Games and Economic Behavior*, 8:6–19, 1995.

[27] R. Aumann and M. Maschler. The bargaining set for cooperative games. In L. Shapley M. Dresher and A. Tucker, editors, *Advances in Game Theory*. Princeton University Press, Princeton, 1964.

[28] R. Aumann and M. Maschler. *Repeated Games with Incomplete Information*. MIT Press, Cambridge, MA, 1995.

[29] W. Austin and E. Hatfield. Equity theory, power and social justice. In G. Mikula, editor, *Justice and Social Interaction*. Springer-Verlag, New York, 1980.

[30] W. Austin and E. Walster. Reactions to confirmations and disconfirmations of expectancies of equity and inequity. *Journal of Personality and Social Psychology*, 30:208–216, 1974.

[31] R. Axelrod. *The Evolution of Cooperation*. Basic Books, New York, 1984.

[32] R. Axelrod. The evolution of strategies in the iterated Prisoners' Dilemma. In L. Davis, editor, *Genetic Algorithms and Simulated Annealing*. Morgan Kaufmann, Los Altos, CA, 1987.

[33] R. Axelrod. *The Complexity of Cooperation.* Princeton University Press, Princeton, 1997.

[34] R. Axelrod and W. Hamilton. The evolution of cooperation. *Science,* 211:1390–1396, 1981.

[35] A. Ayer. *Language, Truth and Logic.* Gollancz, London, 1936.

[36] K. Baier. *The Moral Point of View.* Cornell University Press, Ithaca, 1958.

[37] R. Bailey. The behavioral ecology of Efe pygmy men in the Atari Forest, Zaire. Technical Report Anthropological Paper 86, University of Michigan Museum of Anthropology, 1991.

[38] R. Baker and M. Bellis. Human sperm competition: Infidelity, the female orgasm and kamikaze sperm. Lecture delivered to the fourth annual meeting of the Human Behavior and Evolution Society in Albuquerque, New Mexico, 1992.

[39] G. Balandier. *Political Anthropology.* Random House, New York, 1970. (Translated by A. Sheridan Smith).

[40] A. Banerjee. A simple model of herd behavior. *Quarterly Journal of Economics,* 110:797–817, 1992.

[41] J. Banks and R. Sundaram. Repeated games, finite automata and complexity. *Games and Economic Behavior,* 2:97–117, 1990.

[42] D. Barash. *Sociobiology and Behavior.* Elsevier, New York, 1982.

[43] J. Barkow, L. Cosmides, and J. Tooby. *The Adapted Mind: Evolutionary Psychology and the Generation of Culture.* Oxford University Press, Oxford, 1992.

[44] J. Baron. Heuristics and biases in equity judgments: A utilitarian approach. In B. Mellors and J. Baron, editors, *Psychological Perspectives on Justice: Theory and Applications.* Cambridge University Press, Cambridge, 1993.

[45] R. Barro. Are government bonds net wealth? *Journal of Political Economy,* 82:1095–1117, 1974.

[46] R. Barro. The neoclassical approach to fiscal policy. In *Modern Business Cycle Theory.* Harvard University Press, Cambridge, MA, 1989.

[47] R. Barro. The Ricardian approach to budget deficits. *Journal of Economic Perspectives,* 3:37–54, 1989.

[48] B. Barry. *Theories of Justice.* University of California Press, Berkeley and Los Angeles, 1989.

[49] W. Baumol. *Superfairness.* MIT Press, Cambridge, MA, 1986.

[50] E. Ben-Porath. Repeated games with finite automata. *Journal of Economic Theory,* 59:17–32, 1993.

[51] J. Bentham. Pannonial fragments. In *Works of Jeremy Bentham.* ?, Edinburgh, 1863. (Edited by J. Bowring).

[52] J. Bentham. *Bentham MSS, University College London Catalogue of the Manuscripts of Jeremy Bentham*. London, 1962. (Edited by A. Milne. Second edition).

[53] J. Bentham. An introduction to the principles of morals and legislation. In *Utilitarianism and Other Essays*. Penguin, Harmondsworth, UK, 1987. (Introduction by A. Ryan. Essay first published 1789).

[54] J. Bergin and B. McCloud. Efficiency and renegotiation in repeated games. Working Paper, Queen's University, 1989.

[55] T. Bergstrom. On the evolution of altruistic ethical rules for siblings. *American Economic Review*, 85, 1995.

[56] I. Berlin. *Four Essays on Liberty*. Oxford University Press, Oxford, 1969.

[57] B. Bernheim. Is everything neutral? *A Neoclassical Perspective on Budget Deficits*, 3:55–73, 1989.

[58] B Bernheim and K. Bagwell. Is everything neutral? *Journal of Political Economy*, 96:308–338, 1988.

[59] E. Berscheid, D. Boye, and E. Walster. Retaliation as a means of restoring equity. *Journal of Personality and Social Psychology*, 10:370–376, 1968.

[60] C. Bicchieri. *Rationality and Coordination*. Cambridge University Press, New York, 1993.

[61] S. Bikhchandru, J. Hirshleifer, and I. Welch. A theory of fads, fashion, custom and cultural change as informational cascades. *Journal of Political Economy*, 100:992–1026, 1991.

[62] K. Binmore. Bargaining conventions. *International Journal of Game Theory*, 13:193–200, 1984.

[63] K. Binmore. Modeling rational players, I and II. *Economics and Philosophy*, 3 and 4:179–214 and 9–55, 1987.

[64] K. Binmore. Nash bargaining and incomplete information. In K. Binmore and P. Dasgupta, editors, *Economics of Bargaining*. Cambridge University Press, Cambridge, 1987.

[65] K. Binmore. Nash bargaining theory I. In K. Binmore and P. Dasgupta, editors, *Economics of Bargaining*. Cambridge University Press, Cambridge, 1987.

[66] K. Binmore. Nash bargaining theory, parts I, II and III. In K. Binmore and P. Dasgupta, editors, *Economics of Bargaining*. Cambridge University Press, Cambridge, 1987.

[67] K. Binmore. Perfect equilibria in bargaining models. In K. Binmore and P. Dasgupta, editors, *Economics of Bargaining*. Cambridge University Press, Cambridge, 1987.

[68] K. Binmore. Social contract III: Evolution and utilitarianism. *Constitutional Political Economy*, 1:1–26, 1990.

[69] K. Binmore. Debayesing game theory. In B. Skyrms, editor, *Studies in Logic and the Foundations of Game Theory: Proceedings of the Ninth International Congress of Logic, Methodology and the Philosophy of Science*. Kluwer, Dordrecht, 1992.

[70] K. Binmore. *Fun and Games*. D. C. Heath, Lexington, MA, 1992.

[71] K. Binmore. Bargaining theory without tears. *Investigaciones Economicas*, 18:403–419, 1994.

[72] K. Binmore. A note on backward induction. *Games and Economic Behavior*, 17:135–137, 1996.

[73] K. Binmore. Right or seemly? *Analyse & Kritik*, 18:67–80, 1996.

[74] K. Binmore. Rationality and backward induction. *Journal of Economic Methodology*, 4:23–41, 1997.

[75] K. Binmore. Naturalizing Harsanyi and Rawls. In M. Salles and J. Weymark, editors, *Justice, Political Liberalism, and Utilitarianism: Proceedings of the Caen Conference in Honor of John Harsanyi and John Rawls*. Cambridge University Press, Cambridge, 1998.

[76] K. Binmore, J. Gale, and L. Samuelson. Learning to be imperfect: The Ultimatum Game. *Games and Economic Behavior*, 8:56–90, 1995.

[77] K. Binmore and M. Herrero. Matching and bargaining in dynamic markets. *Review of Economic Studies*, 55:17–32, 1988.

[78] K. Binmore and M. Herrero. Security equilibrium. *Review of Economic Studies*, 55:33–48, 1988.

[79] K. Binmore, P. Morgan, A. Shaked, and J. Sutton. Do people exploit their bargaining power? An experimental study. *Games and Economic Behavior*, 3:295–322, 1991.

[80] K. Binmore, M. Osborne, and A. Rubinstein. Noncooperative models of bargaining. In R. Aumann and S. Hart, editors, *Handbook of Game Theory I*. North Holland, Amsterdam, 1992.

[81] K. Binmore, M. Piccione, and L. Samuelson. Evolutionary stability in alternating offers bargaining models. ELSE discussion paper, University College London, 1996.

[82] K. Binmore, C. Proulx, L. Samuelson, and J. Swierzbinski. Hard bargains and lost opportunities. ELSE discussion paper, University College London, 1995.

[83] K. Binmore, A. Rubinstein, and A. Wolinsky. The Nash bargaining solution in economic modelling. *Rand Journal of Economics*, 17:176–188, 1982.

[84] K. Binmore and L. Samuelson. Evolutionary stability in repeated games played by finite automata. *Journal of Economic Theory*, 57:278–305, 1992.

[85] K. Binmore and L. Samuelson. Muddling through: Noisy equilibrium selection. (To appear in *Journal of Economic Theory*), 1996.

[86] K. Binmore and L. Samuelson. Evolutionary drift and equilibrium selection. (ELSE discussion paper, University College London, 1997.

[87] K. Binmore, L. Samuelson, and R. Vaughan. Musical chairs: Modeling noisy evolution. *Games and Economic Behavior*, 11:1–35, 1995.

[88] K. Binmore, A. Shaked, and J. Sutton. Testing noncooperative game theory: A preliminary study. *American Economic Review*, 75:1178–1180, 1985.

[89] K. Binmore, A. Shaked, and J. Sutton. Noncooperative bargaining theory: a further test: Reply. *American Economic Review*, 78:837–839, 1988.

[90] K. Binmore, A. Shaked, and J. Sutton. An outside option experiment. *Quarterly Journal of Economics*, 104:753–770, 1989.

[91] K. Binmore, J. Swierzsbinski, S. Hsu, and C. Proulx. Focal points and bargaining. *International Journal of Game Theory*, 22:381–409, 1993.

[92] C. Blackorby, W. Bossert, and D. Donaldson. Intertemporal population ethics: A welfarist approach. Working Paper, University of British Columbia, 1993.

[93] C. Blackorby and D. Donaldson. Social criteria for evaluating population change. *Journal of Public Economics*, 25:13–33, 1984.

[94] C. Boehm. Impact of the human egalitarian syndrome on Darwinian selection dynamics. *The American Naturalist*, 150:S100–S121, 1997.

[95] G. Bolton. A comparative model of bargaining: Theory and evidence. *American Economic Review*, 81:1096–1136, 1991.

[96] J. Boswell. *The Journals of James Boswell, 1762-1795*. Heinemann, London, 1991.

[97] T. Bowley. Edgeworth's conjecture. *Econometrica*, 43:425–454, 1973.

[98] R. Boyd and P. Richerson. *Culture and the Evolutionary Process*. University of Chicago Press, Chicago, 1985.

[99] R. Boyd and P. Richerson. Group selection among alternative evolutionary stable strategies. *Journal of Theoretical Biology*, 145:331–342, 1990.

[100] S. Brams and A. Taylor. *Fair Division: From Cake-Cutting to Dispute Resolution*. Cambridge University Press, Cambridge, 1996.

[101] L. Bredvold and R. Ross. *The Philosophy of Edmund Burke*. University of Michigan Press, Ann Arbor, 1967.

[102] J. Broome. *Weighing Goods*. Blackwell, Oxford, 1991.

[103] J. Buchanan. *The Limits of Liberty*. University of Chicago Press, Chicago, 1975.

[104] J. Buchanan. A Hobbsian interpretation of the Rawlsian difference principle. *Kyklos*, 29:5–25, 1976.

[105] J. Buchanan. Towards the simple economics of natural liberty. *Kyklos*, 40, 1987.

[106] J. Buchanan. *The Economics and the Ethics of Constitutional Order.* University of Michigan Press, Ann Arbor, 1991.

[107] J. Buchanan and G. Tullock. *The Calculus of Consent: Logical Foundations of Consitutional Democracy.* University of Michigan Press, Ann Arbor, 1962.

[108] J. Burgess. Social spiders. *Scientific American*, 243, 1976.

[109] E. Burke. An appeal from the new to the old Whigs. Pamphlet, 1791.

[110] R. Burton. *The Anatomy of Melancholy.* Tudor, New York, 1927.

[111] J. Butler. *Fifteen Sermons Preached at the Rolls Chapel.* 1726.

[112] R. Byrne and A. Whiten. *Machiavellian Intelligence.* Clarendon Press, Oxford, 1988.

[113] G. Calvo. Some notes on time inconsistency and Rawls' maximin principle. *Review of Economic Studies*, 45:97–102, 1978.

[114] H. Carlsson and E. Van Damme. Global games and equilibrium selection. CenTER discussion paper 9052, Tilburg University, 1990.

[115] H. Carmichael and B. McCleod. Gift giving and the evolution of cooperation. Queen's University Discussion Paper, 1995.

[116] E. Cashden. Coping with risk: Reciprocity among the Basarwa of Northern Botswana. *Man*, 20:454–474, 1985.

[117] E. Cashden. *Risk and Uncertainty in Tribal and Peasant Communities.* Westview Press, Boulder, CO, 1990.

[118] L. Cavelli-Sforza and M. Feldman. *Cultural Transmission and Evolution.* Princeton University Press, Princeton, 1981.

[119] N. Chagnon. *Yanomamo: The Free People.* Harper Colophon, New York, 1966.

[120] N. Chomsky. A review of B. F. Skinner's "Verbal Behavior". *Language*, 35:26–58, 1959.

[121] N. Chomsky. *Aspects of the Theory of Syntax.* MIT Press, Cambridge, MA, 1965.

[122] N. Chomsky. *Rules and Representations.* Columbia University Press, New York, 1980.

[123] P. Churchland. *Matter and Consciousness.* MIT Press, Cambridge, MA, 1988.

[124] P. Churchland. *The Engine of Reason: The Seat of the Soul.* MIT Press, Cambridge, MA, 1995.

[125] C. von Clausewitz. *On War.* Princeton University Press, Princeton, 1976. (First published 1832).

[126] R. Coase. The problem of social costs. *Journal of Law and Economics*, 3:1–44, 1960.

[127] R. Coase. *The Problem of Social Costs*. W. A. Benjamin, Amsterdam, 1966.

[128] R. Coase. *Essays on Economics and Economists*. University of Chicago Press, Chicago, 1991.

[129] G. Cohen. Equality of what? On welfare, goods and capabilities. In M. Nussbaum and A. Sen, editors, *The Quality of Life*. Clarendon Press, Oxford, 1989.

[130] M. Cohen. *The Food Crisis in Prehistory: Overpopulation and the Origins of Agriculture*. Yale University Press, New Haven, 1977.

[131] R. Cohen and J. Greenberg. The justice concept in social psychology. In R. Cohen and J. Greenberg, editors, *Equity and Justice in Social Behavior*. Academic Press, New York, 1982.

[132] B. Cooper. Copying fidelity in the evolution of finite automata that play the repeated Prisoners' Dilemma. Working paper, Wolfson College, Oxford, 1996.

[133] D. Cooper. Supergames played by finite automata with finite costs of compexity in an evolutionary setting. *Journal of Economic Theory*, 68:266–275, 1996.

[134] L. Cosmides. The logic of social exchange: Has natural selection shaped how humans reason? *Cognition*, 31:187–276, 1989.

[135] L. Cosmides and J. Tooby. Evolutionary psychology and the generation of culture ii: A computational theory of social exchange. *Ethology and Sociobiology*, 10:51–97, 1989.

[136] L. Cosmides and J. Tooby. Better than rational: Evolutionary psychology and the invisible hand. *Americal Economic Review (Papers and Proceedings)*, 84:327–332, 1994.

[137] V. Crawford and H. Varian. Distortion of preferences and the Nash theory of bargaining. *Economics Letters*, 3:203–206, 1979.

[138] H. Cronin. *The Ant and the Peacock*. Cambridge University Press, Cambridge, 1991.

[139] J. Cross. *The Economics of Bargaining*. Basic Books, New York, 1969.

[140] W. Cutler. On the optimal strategy for pot-limit Poker. *American Mathematical Monthly*, 82:368–376, 1975.

[141] D. Damas. The Copper Eskimo. In M. Bicchieri, editor, *Hunters and Gatherers Today*. Holt, Rinehart and Winston, New York, 1972.

[142] S. Darwell, A. Gibbard, and P. Railton. *Moral Discourse and Practice: Some Philosophical Approaches*. Oxford University Press, Oxford, 1997.

[143] C. Darwin. *The Descent of Man and Selection in Relation to Sex*. Murray, London, 1871.

[144] C. Darwin. *The Expression of the Emotions in Man and Animals.* University of Chicago Press, Chicago, 1965.

[145] C. Darwin. *The Origin of Species by Means of Natural Selection.* Penguin, Harmondsworth, UK, 1985.

[146] P. Dasgupta. Of some alternative criteria for justice between generations. *Journal of Public Economics*, 3:405–423, 1974.

[147] P. Dasgupta. Population and savings: Ethical issues. Working paper, University of Cambridge, 1993.

[148] P. Dasgupta. Savings and fertility: Some ethical issues. *Philosophy and Public Affairs*, 23:99–127, 1994.

[149] G. Davenport. *Herakleitos and Diogenes.* Grey Fox Press, San Francisco, 1976.

[150] M. Davis. *Empathy: A Social Psychological Approach.* Westview Press, Boulder, CO, 1996.

[151] R. Dawkins. *The Selfish Gene.* Oxford University Press, Oxford, 1976.

[152] T. de Laguna. Stages of the discussion of evolutionary ethics. *Philosophical Review*, 15:583–598, 1965.

[153] G. Debreu and H. Scarf. A limit theorem on the core of an economy. *International Review of Economics*, 48:235–246, 1963.

[154] D. Dennett. Mechanism and responsibility. In G. Watson, editor, *Free Will.* Oxford University Press, Oxford, 1982.

[155] D. Dennett. *Elbow Room: The Varieties of Free Will Worth Wanting.* Oxford University Press, Oxford, 1984.

[156] D. Dennett. *The Intentional Stance.* MIT Press, Cambridge, MA, 1987.

[157] D. Dennett. *Consciousness Explained.* Allen Lane: Penguin Press, London, 1991.

[158] D. Dennett. *Darwin's Dangerous Idea.* Allen Lane: Penguin Press, London, 1995.

[159] M. Deutsch. *Distributive Justice: A Social Psychological Perspective.* Yale University Press, Newhaven, 1985.

[160] A. Dhillon and J-F. Mertens. Relative utilitarianism. Technical report, CORE, 1994.

[161] P. Diamond. Cardinal welfare, individualistic ethics and interpersonal comparison of utility: Comment. *Journal of Political Economy*, 75:765–766, 1967.

[162] F. Dostoyevsky. *House of the Dead.* Penguin, London, 1985.

[163] L. Dugnatov. *Cooperation among Animals: An Economic Perspective.* Oxford University Press, Oxford, 1997.

[164] R. Dunbar. Neocortex size as a constraint on group size in primates. *Journal of Human Evolution*, 20:469–493, 1992.

[165] R. Dunbar. Coevolution of neocortex size, group size and language in humans. *Behavioural Brain Science*, 16, 1993.

[166] R. Dworkin. *Taking Rights Seriously*. Duckworth, London, 1977.

[167] R. Dworkin. What is equality? I and II. *Philosophy and Public Affairs*, 10:185–345, 1981.

[168] M. Edel and A. Edel. *Anthropology and Ethics*. Thomas, Springfield, Ill, 1959.

[169] F. Edgeworth. *Mathematical Psychics*. Kegan Paul, London, 1881.

[170] G. Ellison. Cooperation in the Prisoners' Dilemma with anonymous matching. *Review of Economic Studies*, 61:567–588, 1994.

[171] J. Elster. Sour grapes–Utilitarianism and the genesis of wants. In A. Sen and B. Williams, editors, *Utilitarianism and Beyond*. Cambridge University Press, Cambridge, 1982.

[172] J. Elster. *The Cement of Society: A Study of Social Order*. Cambridge University Press, Cambridge, 1989.

[173] J. Elster. *Local Justice: How Institutions Allocate Scarce Goods and Necessary Burdens*. Russell Sage Foundation, New York, 1992.

[174] Epicurus. *Letters, Principal Doctrines and Vatican Sayings*. Bobbs-Merrill, Indianapolis, 1964. (Translated by R. Greer).

[175] D. Erdal and A. Whiten. Egalitarianism and Machiavellian intelligence in human evolution. In P. Mellars and K. Gibson, editors, *Modelling the Early Human Mind*. Oxbow Books, Oxford, 1996.

[176] E. Evans-Pritchard. *The Nuer*. Clarendon Press, Oxford, 1940.

[177] H. Eysenck. *Sense and Nonsense in Psychology*. Pelican Books, Harmondsworth, UK, 1957.

[178] P. Farber. *The Temptations of Evolutionary Ethics*. University of California Press, Berkeley, 1994.

[179] J. Farr. So vile and miserable an estate: The problem of slavery in Locke's political thought. *Political Theory*, 14:269–283, 1986.

[180] J. Farrell and E. Maskin. Renegotiation in repeated games. *Games and Economic Behavior*, 1:327–360, 1989.

[181] R. Fernandez and J. Glazer. Striking for a bargain between two completely informed agents. *American Economic Review*, 81:240–252, 1991.

[182] P. Fishburn and A. Rubinstein. Time preference. *International Economic Review*, 23:677–694, 1982.

[183] D. Foley. Resource allocation and the public sector. *Yale Economic Essays*, 7:45–98, 1967.

[184] P. Foot. *Vices and Virtues*. University of California Press, Los Angeles, 1978.

[185] R. Frank. *Passions within Reason*. Norton, New York, 1988.

[186] D. Fudenberg and E. Maskin. Evolution and cooperation in noisy repeated games. *American Economic Review*, 80:274–279, 1990.

[187] D. Fudenberg and J. Tirole. *Game Theory*. MIT Press, Cambridge, MA, 1991.

[188] L. Furby. Psychology and justice. In R. Cohen, editor, *Justice: Views from the Social Sciences*. Harvard University Press, Cambridge, MA, 1986.

[189] C. Furer-Haimendorf. *Morals and Merit*. Weidenfeld and Nicolson, London, 1967.

[190] W. Gaertner, P. Pattanaik, and K. Suzumura. Individual rights revisited. *Economica*, 59:161–178, 1992.

[191] D. Gale. Bargaining and competition I: Characterization. *Econometrica*, 54:785–806, 1986.

[192] D. Gale. Bargaining and competition II: Existence. *Econometrica*, 54:807–818, 1986.

[193] D. Gale. Limit theorems for markets with sequential bargaining. *Journal of Economic Theory*, 43:20–54, 1987.

[194] P. Gardenfors. Fairness without interpersonal comparisons. *Theoria*, 54:57–74, 1978.

[195] P. Gardner. The Paliyans. In M. Bicchieri, editor, *Hunters and Gatherers Today*. Holt, Rinehart and Winston, New York, 1972.

[196] D. Gauthier. *Morals by Agreement*. Clarendon Press, Oxford, 1986.

[197] D. Gauthier. Uniting separate persons. In D. Gauthier and R. Sugden, editors, *Rationality, Justice and the Social Contract*. Harvester Wheatsheaf, Hemel Hempstead, UK, 1993.

[198] D. Gauthier. Assure and threaten. *Ethics*, 104:690–721, 1994.

[199] D. Gauthier and R. Sugden. *Rationality, Justice and the Social Contract*. Harvester Wheatsheaf, Hemel Hempstead, U. K., 1993.

[200] E. Gellner. *Plough, Sword and Book*. Paladin, Grafton Books, London, 1988.

[201] E. Gellner. *Reason and Culture*. Blackwell, Oxford, 1992.

[202] L. Gevers. Walrasian social choice: Some simple axiomatic approaches to social choice and public decision making. In R. Heller *et al*, editor, *Essays in Honor of K. J. Arrow*. Cambridge University Press, New York, 1988.

[203] A. Gibbard. Manipulation of voting schemes: A general result. *Econometrica*, 41:587–601, 1973.

[204] A. Gibbard. A Pareto-consistent libertarian claim. *Journal of Economic Theory*, 7:388–410, 1974.

[205] A. Gibbard. *Wise Choices and Apt Feelings: A Theory of Normative Judgment*. Clarendon Press, Oxford, 1990.

[206] D. Glass. Changes in liking as a means of reducing cognitive discrepancies between self-esteem and aggression. *Journal of Personality*, 32:520–549, 1964.

[207] J. Glazer and A. Ma. Efficient allocation of a prize—King Solomon's dilemma. *Games and Economic Behavior*, 1:222–223, 1989.

[208] W. Godwin. *Enquiry Concerning Political Justice*. Toronto University Press, Toronto, 1946. (Edited by F. Priestley. First published 1798).

[209] R. Goodin. *Utilitarianism as a Public Philosophy*. Cambridge University Press, Cambridge, 1995.

[210] J. W. Gough. *The Social Contract*. Clarendon Press, Oxford, 1938.

[211] A. Grafen. The hawk-dove game played between relatives. *Animal Behavior*, 27:905–907, 1979.

[212] A. Grafen. Natural selection, kin selection and group selection. In J. Krebs and N. Davies, editors, *Behavioural Ecology (Second Edition)*. Blackwell, Oxford, 1984.

[213] J. Griffin. *Well-Being: Its Meaning, Measurement and Moral Importance*. Clarendon Press, Oxford, 1986.

[214] W. Guth, R. Schmittberger, and B. Schwarze. An experimental analysis of ultimatum bargaining. *Journal of Behavior and Organization*, 3:367–388, 1982.

[215] A. Hamilton, J. Jay, and J. Madison. *The Federalist*. Everyman, London, 1992. (Edited by W. Brock. First published 1787–1788).

[216] W. Hamilton. The evolution of altruistic behavior. *American Naturalist*, 97:354–356, 1963.

[217] W. Hamilton. The genetic evolution of social behavior, parts I and II. *Journal of Theoretical Biology*, 7:1–52, 1964.

[218] P. Hammond. Charity: Altruism or cooperative egoism. In E. Phelps, editor, *Altruism, Morality and Economic Theory*. Russell Sage, New York, 1975.

[219] P. Hammond. Consequentialist foundations for expected utility. *Theory and Decision*, 25:25–78, 1988.

[220] P. Hammond. Interpersonal comparisons of utility: Why and how they are and should be made. In J. Elster and J. Roemer, editors, *Interpersonal Comparisons of Well-Being*. Cambridge University Press, London, 1991.

[221] P. Hammond. Harsanyi's utilitarian theorem: A simpler proof and some ethical connotations. In R. Selten, editor, *Rational Interaction: Essays in Honor of John Harsanyi*. Springer-Verlag, Berlin, 1992.

[222] P. Hammond. Social choice of individual and group rights. In M. Salles W. Barnett, H. Moulin and N. Schofield, editors, *Social Choice, Welfare and Ethics*. Cambridge University Press, Cambridge, 1995.

[223] A. Harcourt and F. de Waal. *Coalitions and Alliances in Humans and Other Animals*. Oxford University Press, Oxford, 1992.

[224] R. Hardin. *Morality within the Limits of Reason*. University of Chicago Press, Chicago, 1988.

[225] R. Hare. *Freedom and Reason*. Oxford University Press, Oxford, 1965.

[226] R. Hare. Rawls' theory of justice. In B. Daniels, editor, *Reading Rawls*. Blackwell, Oxford, 1975.

[227] R. Hare. *Moral Thinking: Its Levels, Method and Point*. Clarendon Press, Cambridge, 1981.

[228] R. Hare. Ethical theory and utilitarianism. In A. Sen and B. Williams, editors, *Utilitarianism and Beyond*. Cambridge University Press, Cambridge, 1982.

[229] R. Hare. Rights, utility and universalization. In R. G. Frey, editor, *Utility and Rights*. University of Minnesota Press, Minneapolis, 1984.

[230] R. Hare. Could Kant have been a utilitarian? *Utilitas*, 5:1–16, 1993.

[231] J. Harsanyi. Cardinal welfare, individualistic ethics, and the interpersonal comparison of utility. *Journal of Political Economy*, 63:309–321, 1955.

[232] J. Harsanyi. Games with incomplete information played by "Bayesian" players, I–III. *Management Science*, 14:159–182, 1967.

[233] J. Harsanyi. *Rational Behavior and Bargaining Equilibrium in Games and Social Situations*. Cambridge University Press, Cambridge, 1977.

[234] J. Harsanyi. Review of Gauthier's "Morals by Agreement". *Economics and Philosophy*, 3:339–343, 1987.

[235] J. Harsanyi. A case for a utilitarian ethic. In H. Siebert, editor, *Ethical Foundations of the Market Economy*. Mohr, Tubingen, 1993.

[236] J. Harsanyi and R. Selten. A generalized Nash solution for two-person bargaining games with incomplete information. *Management Science*, 18:80–106, 1972.

[237] H. Hart. *The Concept of Law*. Clarendon Press, Oxford, 1961.

[238] D. Hausman. The impossibility of interpersonal utility comparisons. Technical Report Working Paper DP1/94, LSE Centre for Philosophy of Natural and Social Sciences, 1994.

[239] K. Hawkes, J. O'Connell, and N. Burton-Jones. Hunting income patterns among the Hadza. In A. Whitten and E. Widdowson, editors, *Foraging Strategies and Natural Diet of Monkeys, Apes and Humans*. Clarendon Press, Oxford, 1993.

[240] F. Hayek. *The Constitution of Liberty*. University of Chicago Press, Chicago, 1960.

[241] F. Hayek. *Studies in Philosophy, Politics and Economics*. University of Chicago Press, Chicago, 1967.

[242] F. Hayek. *The Road to Serfdom*. Routledge and Kegan Paul, London, 1976.

[243] B. Heine. The mountain people: Some notes on the Ik of north-eastern Uganda. *Africa*, 55:3–16, 1985.

[244] J. Helm. The Dogrib Indians. In M. Bicchieri, editor, *Hunters and Gatherers Today*. Holt, Rinehart and Winston, New York, 1972.

[245] Herodotus. *The History of Herodotus*. Dutton, New York, 1910.

[246] Hesiod. *Hesiod: The Homeric Hymns and Homerica*. Heinemann, London, 1929. (Edited by H. Evelyn-Waugh).

[247] W. Hildenbrand. *Core and Equilibria of Large Economies*. Princeton University Press, Princeton, 1974.

[248] W. Hines and J. Maynard Smith. Games between relatives. *Journal of Theoretical Biology*, 79:19–30, 1979.

[249] T. Hobbes. *De Cive*. In W. Molesworth, editor, *The Englsih Works of Thomas Hobbes II*. John Bohn, London, 1840. (First published 1642).

[250] T. Hobbes. *Elements of Law*. Frank Cass, London, 1969. (Edited by F. Tönnies).

[251] T. Hobbes. *Leviathan*. Penguin Classics, London, 1986. (Edited by C. B. Macpherson. First published 1651).

[252] L. Hobhouse. *The Rational Good*. Allen and Unwin, London, 1921.

[253] L. Hobhouse. *Elements of Social Justice*. Allen and Unwin, London, 1922.

[254] L. Hobhouse. *Morals in Evolution*. Chapman and Hall, London, 1951. (Seventh edition. First published 1906).

[255] M. Hoffman. Interaction of affect and cognition in empathy. In J. Kagan C. Izand and R. Zajonc, editors, *Emotions, Cognition and Behavior*. Cambridge University Press, Cambridge, 1984.

[256] M. Hoffman. The contribution of empathy to justice and moral judgment. In N. Eisenberg and J. Strayer, editors, *Empathy and its Development*. Cambridge University Press, Cambridge, 1987.

[257] J. Holland. *Adaption in Natural and Artificial Systems*. Uniiversity of Michigan Press, Ann Arbor, 1992. (Second edition, first published 1975).

[258] J. Holland. Genetic algorithms. *Scientific American*, 267:66–72, 1992.

[259] A. Holmberg. Nomads of the long bow. Technical Report Publications of the Institute of Social Anthropology 10, Smithsonian Institute, 1950.

[260] G. Homans. *Social Behavior: Its Elementary Forms*. Harcourt, Brace and World, New York, 1961.

[261] J. Howard. Cooperation in the Prisoners' Dilemma. *Theory and Decision*, 24:203–213, 1988.

[262] R. Howe and J. Roemer. Rawlsian justice as the core of a game. *American Economic Review*, 71:880–895, 1981.

[263] A. Hughes. *Evolution and Human Kinship*. Oxford University Press, Oxford, 1988.

[264] J. Huizinga. *Homo Ludens: A Study of the Play Element in Culture*. Beacon Press, Boston, 1950.

[265] D. Hume. The natural history of religion. In *Hume on Religion*. Fontana, London, 1963. (Edited by R. Wollheim. Essay first published 1757).

[266] D. Hume. *Enquiries Concerning Human Understanding and Concerning the Principles of Morals*. 3rd edition. Clarendon Press, Oxford, 1975. (Edited by L. A. Selby-Bigge. Revised by P. Nidditch. First published 1777).

[267] D. Hume. *A Treatise of Human Nature (Second Edition)*. Clarendon Press, Oxford, 1978. (Edited by L. A. Selby-Bigge. Revised by P. Nidditch. First published 1739).

[268] D. Hume. *History of England*. Liberty Classics, Indianapolis, 1983. (First published 1777).

[269] D. Hume. Introduction. In *Essays Moral, Political and Literary, Part I*. Liberty Classics, Indianapolis, 1985. (Edited by E. Miller).

[270] D. Hume. Of suicide. In *Essays Moral, Political and Literary*. Liberty Classics, Indianapolis, 1985.

[271] D. Hume. Of the first principles of government. In *Essays Moral, Political and Literary, Part I*. Liberty Classics, Indianapolis, 1985. (Edited by E. Miller. Essay first published 1758).

[272] D. Hume. Of the idea of a perfect commonwealth. In *Essays Moral, Political and Literary, Part I*. Liberty Classics, Indianapolis, 1985. (Edited by E. Miller. Essay first published 1742).

[273] D. Hume. Of the independency of Parliament. In *Essays Moral, Political and Literary, Part I*. Liberty Classics, Indianapolis, 1985. (Edited by E. Miller. Essay first published 1742).

[274] D. Hume. Of the original contract. In *Essays Moral, Political and Literary*. Liberty Classics, Indianapolis, 1985. (Edited by E. Miller. Essay first published 1748).

[275] N. Humphrey. The social function of intellect. In P. Bateson and R. Hinde, editors, *Growing Points in Ethology*. Cambridge University Press, London, 1976.

[276] L. Hurwicz. On allocations attainable through nash equilibria. *Journal of Economic Theory*, 21:140–165, 1979.

[277] F. Hutcheson. *System of Moral Philosophy*. London, 1755.

[278] T. Huxley. *Evolution and Ethics: With New Essays on its Victorian and Sociobiological Context*. Princeton University Press, Princeton, 1989. (Edited by J. Paradis and G. Williams. First published 1893).

[279] A. Hylland. Subjective interpersonal comparisons. In J. Elster and J. Roemer, editors, *Interpersonal Comparisons of Well-Being*. Cambridge University Press, Cambridge, 1991.

[280] G. Isaac. The food-sharing behavior of protohuman hominids. *Scientific American*, 238:90–108, 1978.

[281] J. Isbell. A modification of Harsanyi's bargaining model. *Bulletin of the American Mathematical Society*, 66:70–73, 1960.

[282] Y. Ito. *Behaviour and Evolution of Wasps: The Communal Aggregation Hypothesis*. Oxford University Press, Oxford, 1993.

[283] F. Jacob. Evolution and tinkering. *Science*, 96:1161–1166, 1977.

[284] J. Kagel, R. Battalio, and L. Green. *Economic Choice Theory: An Experimental Analysis of Animal Behavior*. Cambridge University Press, Cambridge, 1995.

[285] D. Kahneman, J. Knetsch, and R. Thaler. Fairness and the assumptions of economics. *Journal of Business*, 59:258–300, 1986.

[286] D. Kahneman and A. Tversky. Rational choice and the framing of decisions. In *Decision Making*. Cambridge University Press, Cambridge, 1988.

[287] E. Kalai. Nonsymmetric Nash solutions and replications of two-person bargaining. *International Journal of Game Theory*, 6:129–133, 1977.

[288] E. Kalai. Solutions to bargaining situations: Interpersonal utility comparisons. *Econometrica*, 45:1623–1630, 1977.

[289] E. Kalai and M. Smorodinsky. Other solutions to Nash's bargaining problem. *Econometrica*, 45:1623–1630, 1975.

[290] M. Kandori. Repeated games played by overlapping generations of players. *Review of Economic Studies*, 59:81–92, 1992.

[291] M. Kandori. Social norms and community enforcement. *Review of Economic Studies*, 59:63–80, 1992.

[292] M. Kandori, G. Mailath, and R. Rob. Learning, mutation, and long run equilibria in games. *Econometrica*, 61:29–56, 1993.

[293] I. Kant. Critique of judgment. In *The Philosophy of Kant*. Random House, New York, 1949. (Edited by C. Friedrich. First published 1788).

[294] I. Kant. Idea of a universal history. In *The Philosophy of Kant*. Random House, New York, 1949. (Edited by C. Friedrich. First published 1788).

[295] I. Kant. *The Philosophy of Kant*. Random House, New York, 1949. (Edited by C. Friedrich).

[296] I. Kant. Theory and practice. In *The Philosophy of Kant*. Random House, New York, 1949. (Edited by C. Friedrich. First published 1793).

[297] I. Kant. *Groundwork of the Metaphysic of Morals.* Harper Torchbooks, New York, 1964. (Translated and analyzed by H. Paton. First published 1785).

[298] I. Kant. *Perpetual Peace and Other Essays.* Hackett, Indianapolis, 1983. (Edited by T. Humphrey. First published 1784-95).

[299] I. Kant. *Critique of Practical Reason.* Macmillan, New York, 1989. (Translated by L. Beck. First published 1788).

[300] I. Kant. *Metaphysics of Morals.* Cambridge University Press, Cambridge, 1991. (Edited by M. Gregor. First published 1797).

[301] H. Kaplan and K. Hill. Food sharing among Ache foragers: Tests of explanatory hypotheses. *Current Anthropology*, 26:223–245, 1985.

[302] G. Kavka. Hobbes' war of all against all. *Ethics*, 93:291–310, 1983.

[303] E. Kayser, T. Schwinger, and R. Cohen. Layperson's conceptions of social relationships: A test of contract theory. *Journal of Social and Personal Relationships*, 1:433–548, 1984.

[304] E. Khilstrom, A. Roth, and D. Schmeidler. Risk aversion and solutions to Nash's bargaining problem. In O. Moeschlin and D. Pallaschke, editors, *Game Theory and Mathematical Economics*. North Holland, Dordrecht, 1981.

[305] R. Kirk. *Edmund Burke: A Genius Reconsidered.* Arlington House, New York, 1967.

[306] M. Klein. *Envy and Gratitude.* Virago Press, London, 1988.

[307] H. Kliemt. Papers on Buchanan and related subjects. *Studies in Economics and Social Sciences*, 1, 1990.

[308] B. Knauft. Violence and sociality in human evolution. *Current Anthropology*, 32:223–245, 1991.

[309] S. Kolm. *Theories of Justice.* MIT Press, Cambridge, MA, 1997.

[310] J. Krebs and N. Davies. *Behavioural Ecology (Second Edition).* Blackwell, Oxford, 1984.

[311] P. Kropotkin. *Mutual Aid.* Allen Lane, London, 1972. (Edited by P. Avrich).

[312] J. Ladd. *The Structure of a Moral Code.* Harvard University Press, Cambridge, MA, 1957.

[313] J. Ledyard. Public goods: A survey of experimental research. In J. Kagel and A. Roth, editors, *Handbook of Experimental Game Theory*. Princeton University Press, Princeton, 1995.

[314] I. Lee. On the convergence of informational cascades. *Journal of Economic Theory*, ?:395–411, 1993.

[315] R. Lee. *The !Kung San: Men, Women and Work in a Foraging Society.* Cambridge University Press, Cambridge, 1979.

[316] P. Legant and D. Mettee. Turning the other cheek versus getting even: Vengeance, equity and attraction. *Journal of Personality and Social Psychology*, 25:243–253, 1973.

[317] M. Lerner. Evaluation of reward as a function of performer's reward and attractiveness. *Journal of Personality and Social Psychology*, 1:355–360, 1965.

[318] M. Lerner. Observer's evaluation of a victim: Justice, guilt and veridical perception. *Journal of Personality and Social Psychology*, 20:127–135, 1971.

[319] M. Lerner. *The Belief in a Just World*. Plenum, New York, 1980.

[320] M. Lerner. The justice motive in human relations: Some thoughts about what we need to know about justice. In M. Lerner and S. Lerner, editors, *The Justice Motive In Social Behavior*. Plenum, New York, 1981.

[321] M. Lerner. Integrating societal and psychological rules of entitlement: The basic task of each social actor and a fundamental problem for the social sciences. In R. Vermunt and H. Steensa, editors, *Social Justice in Human Relations I: Societal and Psychological Origins of Justice*. Plenum, New York, 1991.

[322] M. Lerner and G. Matthews. Reactions to suffering of others under conditions of indirect responsibility. *Journal of Personality and Social Psychology*, 5:319–325, 1967.

[323] M. Lerner and C. Simmons. Observer's reaction to the innocent victim: Compassion or rejection. *Journal of Personality and Social Psychology*, 4:203–210, 1966.

[324] D. Levine. Modeling altruism and spite in experiments. UCLA Working Paper, 1995.

[325] D. Lewis. *Conventions: A Philosophical Study*. Harvard University Press, Cambridge, MA, 1969.

[326] N. Lin and C. Michener. Evolution of eusociality in insects. *Quarterly Review of Biology*, 47:131–159, 1972.

[327] B. Linster. *Essays on Cooperation and Competition*. PhD thesis, University of Michigan, 1990.

[328] B. Linster. Evolutionary stability in the repeated Prisoners' Dilemma played by two-state Moore machines. *Southern Economic Journal*, pages 880–903, 1992.

[329] D. Livingston. *Hume's Philosophy of Common Life*. Chicago University Press, Chicago, 1984.

[330] J. Locke. *Two Treatises of Government*. Cambridge University Press, Cambridge, 1963. (First published 1690).

[331] C. Lovejoy. The origin of man. *Science*, 211:341–350, 1981.

[332] W. Lucas. The proof that a game may not have a solution. *Transactions of the American Mathematical Society*, 137:219–229, 1969.

[333] R. Luce. Semiorders and a theory of utility discrimination. *Econometrica*, 24:178–191, 1956.

[334] R. Luce and H. Raiffa. *Games and Decisions*. Wiley, New York, 1957.

[335] Lucretius. *On the Nature of Things*. Johns Hopkins Press, Baltimore, 1945.

[336] C. Lumsden and E. Wilson. *Genes, Mind and Culture*. Harvard University Press, Cambridge, MA, 1981.

[337] A. Macintyre. *After Virtue: A Study in Moral Theory*. Duckworth, London, 1982.

[338] J. Mackie. *Ethics, Inventing Right and Wrong*. Penguin, London, 1977.

[339] R. MacMullen. *Corruption and the Decline of Rome*. Yale University Press, New Haven, 1988.

[340] T. Malthus. *An Essay on the Principle of Population*. Norton, New York, 1976. (Edited by P. Appleman. First Published 1798).

[341] B. de Mandeville. *The Fable of the Bees—or Private Vices, Publick Benefits*. Liberty Classics, Indianapolis, 1988. (Edited by F. Kaye. First published 1714).

[342] A. Marshall. *Principles of Economics*. MacMillan, London, 1961.

[343] A. Maryanski and J. Turner. *The Social Cage: Human Nature and the Evolution of Society*. Stanford University Press, Stanford, 1992.

[344] E. Maskin. A theorem on utilitarianism. *Review of Economic Studies*, 11:319–337, 1978.

[345] E. Maskin. The theory of implementation in Nash equilibrium: A survey. In D. Schmeidler L. Hurwicz and H. Sonnenschein, editors, *Social Goals and Social Organization*. Cambridge University Press, Cambridge, 1985.

[346] K. May. A set of independent, necessary and sufficient conditions for simple majority decision. *Econometrica*, 20, 1952.

[347] J. Maynard Smith. *Evolution and the Theory of Games*. Cambridge University Press, Cambridge, 1982.

[348] J. Maynard Smith and G. Price. The logic of animal conflict. *Nature*, 246:15–18, 1972.

[349] R. McKelvey and T. Palfrey. An experimental study of the Centipede Game. *Econometrica*, 60:803–836, 1992.

[350] J. McMillan. Selling spectrum rights. *Journal of Economic Perspectives*, 8:145–162, 1994.

[351] S. Medema. *The Legacy of Ronald Coase in Economic Analysis*. Edward Elgar, London, 1995.

[352] T. Megarry. *Society in Prehistory: The Origins of Human Culture*. MacMillan Press, London, 1995.

[353] M. Meggitt. *Desert People: A Study of the Walbiri Aborigines of Central Australia.* University of Chicago Press, Chicago, 1962.

[354] B. Mellers. Equity judgment: A revision of Aristotelian views. *Journal of Experimental Biology,* 111:242–270, 1982.

[355] B. Mellers and J. Baron. *Psychological Perspectives on Justice: Theory and Applications.* Cambridge University Press, Cambridge, 1993.

[356] D. Messick and K. Cook. *Equity Theory: Psychological and Sociological Perspectives.* Praeger, New York, 1983.

[357] S. Milgram. *Obedience to Authority.* Harper Colophon, New York, 1975.

[358] J. S. Mill. On liberty. In *Utilitarianism.* Collins, London, 1962. (Edited by M. Warnock. Essay first published 1859).

[359] J. S. Mill. Bentham. In *Utilitarianism and Other Essays.* Penguin, Harmondsworth, UK, 1987. (Introduction by A. Ryan. Essay first published 1863).

[360] J. S. Mill. Utilitarianism. In *Utilitarianism and Other Essays.* Penguin, Harmondsworth, UK, 1987. (Introduction by A. Ryan. Essay first published 1863).

[361] D. Miller. *Philosophy and Ideology in Hume's Thought.* Clarendon Press, Oxford, 1981.

[362] J. Monod. *Chance and Necessity.* Knopf, New York, 1971.

[363] G. E. Moore. *Ethics.* Oxford University Press, Oxford, 1966. (First published 1911).

[364] G. E. Moore. *Principia Ethica.* Prometheus Books, Buffalo, N.Y., 1988. (First published 1902).

[365] J. Moore. Implementation, contracts and renegotiation in environments with complete information. In J-J. Laffont, editor, *Advances in Economic Theory I.* Cambridge University Press, Cambridge, 1979.

[366] H. Moulin. Implementing the Kalai-Smorodinsky bargaining solution. *Journal of Economic Theory,* 33:32–45, 1984.

[367] H. Moulin. Interpreting common ownership. *Recherches Economiques de Louvain,* 56:303–326, 1990.

[368] A. Muthoo. Bargaining without commitment. *Games and Economic Behavior,* 2:291–297, 1990.

[369] A. Muthoo. A note on bargaining over a finite number of agreements. *Economic Theory,* 1:290–292, 1991.

[370] A. Muthoo. Revocable commitment and sequential bargaining. *Economic Journal,* 102:378–387, 1991.

[371] A. Muthoo. A bargaining model with players' perceptions on the retractibility of offers. *Theory and Decision,* 38:85–98, 1995.

[372] A. Muthoo. A bargaining model based on the commitment tactic. *Journal; of Economic Theory*, 69:134–152, 1996.

[373] R. Myerson. Two-person bargaining and comparable utility. *Econometrica*, 45:1631–1637, 1977.

[374] R. Myerson. *Game Theory: Analysis of Conflict*. Harvard University Press, Cambridge, MA, 1991.

[375] R. Myerson, G. Pollock, and J. Swinkels. Viscous population equilibria. *Games and Economic Behavior*, 3:101–109, 1991.

[376] R. Myerson and M. Satterthwaite. Efficient mechanisms for bilateral trading. *Journal of Economic Theory*, 29:265–281, 1983.

[377] J. Nachbar. Evolution in the finitely repeated Prisoners' Dilemma. *Intrernational Journal of Game Theory*, 19:307–326, 1990.

[378] R-I. Nagahisa. A necessary and sufficient condition for Walrasian social choice. *Journal of Economic Theory*, 62:186–208, 1994.

[379] T. Nagel. Moral luck. In *Mortal Questions*. Cambridge University Press, Cambridge, 1979.

[380] J. Nash. The bargaining problem. *Econometrica*, 18:155–162, 1950.

[381] J. Nash. Non-cooperative games. *Annals of Mathematics*, 54:286–295, 1951.

[382] J. Nash. Two-person cooperative games. *Econometrica*, 21:128–140, 1953.

[383] A. Neyman. Bounded complexity justifies cooperation in the finitely repeated Prisoners' Dilemma. *Economic Letters*, 19:227–229, 1986.

[384] F. Nietzsche. *The Will to Power, Volume II*. Foulis, Edinburgh, 1910.

[385] F. Nietzsche. *Thus Spake Zarathustra*. Tudor, New York, 1938.

[386] F. Nietzsche. *The Birth of Tragedy and the Geneology of Morals*. Doubleday, New York, 1956.

[387] M. Nitecki and D. Nitecki. *Evolutionary Ethics*. State University of New York Press, Albany, NY, 1993.

[388] M. Nowak and K. Sigmund. The evolution of stochastic strategies in the Prisoners' Dilemma. *Acta Applicandae Mathematicae*, 20:247–265, 1990.

[389] M. Nowak and K. Sigmund. Tit for tat in heterogeneous populations. *Nature*, 355:250–253, 1992.

[390] M. Nowak and K. Sigmund. A strategy of win-shift, lose-stay that outperforms tit-for-tat in the Prisoners' Dilemma game. *Nature*, 364:56–57, 1993.

[391] M. Nowak, K. Sigmund, and E. El-Sedy. Automata, repeated games and noise. Technical report, Department of Zoology, Oxford University, 1993.

[392] R. Nozick. *Anarchy, State, and Utopia*. Basic Books, New York, 1974.

[393] L. Nunney. Group selection, altruism and structured-deme models. *American Naturalist*, 126:212–230, 1985.

[394] M. Nussbaum. *Upheavals of Thought: A Theory of the Emotions.* Cambridge University Press, Cambridge, 1997.

[395] B. O'Neill. *Honour, Symbols, and War.* Michigan University Press, Ann Arbor, 1998.

[396] M. Osborne and A. Rubinstein. *Bargaining and Markets.* Academic Press, San Diego, 1990.

[397] M. Osborne and A. Rubinstein. *A Course in Game Theory.* MIT Press, Cambridge, MA, 1994.

[398] G. Owen. *Game Theory (Second Edition).* Academic Press, New York, 1982.

[399] D. Parfit. *Reasons and Persons.* Oxford University Press, Oxford, 1982.

[400] P. Pattanaik and K. Suzumura. Professor Sen on minimal liberty. Discussion Paper A231, Institute of Economic Research, Hitotsubashi University, Tokyo, 1990.

[401] R. Penrose. *The Emperor's New Mind.* Oxford University Press, Oxford, 1989.

[402] R. Penrose. *Shadows of the Mind.* Oxford University Press, Oxford, 1994.

[403] H. Peters. *Bargaining Game Theory.* PhD thesis, Proefschritt Universitat Nijmegen, 1986.

[404] M. Piccione. Finite automata equilibria with discounting. *Journal of Economic Theory*, 56:1029–1050, 1992.

[405] S. Pinker. *The Language Instinct: The New Science of Language and Mind.* Penguin, London, 1994.

[406] G. Pollock. Suspending disbelief: Of Wynne-Edwards and his reception. *Journal of Evolutionary Biology*, 2:205–221, 1989.

[407] G. Pollock and D. Probst. Mutation paths and evolutionary stability in the iterated Prisoners' Dilemma. (Forthcoming in *Sociological Methods and Research*), 1998.

[408] K. Popper. *The Open Society and its Enemies.* Routledge, London, 1945.

[409] M. Power. *The Egalitarians: Human and Chimpanzee.* Cambridge University Press, Cambridge, 1991.

[410] R. Pritchard. Equity theory; A review and critique. *Organizational Behavior and Human Performance*, 4:176–211, 1969.

[411] D. Probst. *On Evolution and Learning in Games.* PhD thesis, University of Bonn, 1996.

[412] R. Radner. Can bounded rationality resolve the Prisoners' Dilemma. In W. Hildenbrand and A. Mas-Collel, editors, *Contributions to Mathematical Economics.* North Holland, Amsterdam, 1986.

[413] H. Raiffa.   Arbitration schemes for generalized two-person games.   In
       H. Kuhn and A. Tucker, editors, *Contributions to the Theory of Games
       II*. Princeton University Press, Princeton, 1953.

[414] D. Raphael. *Adam Smith*. Oxford University Press, Oxford, 1985.

[415] A. Rapoport and A. Chammah. *Prisoner's Dilemma*. University of Michi-
       gan Press, Ann Arbor, 1965.

[416] J. Rawls. Justice as fairness. *Philosophical Review*, 57:185–187, 1958.

[417] J. Rawls. *A Theory of Justice*. Oxford University Press, Oxford, 1972.

[418] J. Rawls. Social unity and primary goods. In A. Sen and B. Williams, edi-
       tors, *Utilitarianism and Beyond*. Cambridge University Press, Cambridge,
       1982.

[419] J. Rawls. *Political Liberalism*. Columbia University Press, New York, 1993.

[420] D. Regan. *Utilitarianism and Cooperation*. Oxford University Press, Ox-
       ford, 1980.

[421] H. Reichenbach. *The Rise of Scientific Philosophy*. University of California
       Press, Berkeley, 1959.

[422] H. Reis. The mutidimensionality of justice. In R. Folger, editor, *The Sense
       of Injustice: Social Psychological Perspectives*. Plenum, New York, 1984.

[423] P. Reny and M. Perry. A noncooperative bargaining model with strategi-
       cally timed offers. *Econometrica*, 62:795–817, 1994.

[424] D. Riches. Hunting, herding and potlatching: Toward a sociological account
       of prestige. *Man*, 19:234–251, 1982.

[425] M. Ridley. *The Red Queen: Sex and the Evolution of Human Nature*.
       Viking: Penguin, London, 1993.

[426] M. Ridley. *The Origins of Virtue*. Penguin, Harmondsworth, 1997.

[427] J. Riley. Rights to liberty in purely private matters I. *Economics and
       Philosophy*, 5:121–166, 1989.

[428] J. Riley. Rights to liberty in purely private matters II. *Economics and
       Philosophy*, 6:26–64, 1989.

[429] L. Robbins. Inter-personal comparisons of utility. *Economic Journal*,
       48:635–641, 1938.

[430] A. Robson. Efficiency in evolutionary games: Darwin, Nash, and the secret
       handshake. *Journal of Theoretical Biology*, 144:379–396, 1990.

[431] A. Robson. A biological basis for expected and non-expected utility. *Journal
       of Economic Theory*, 68:397–424, 1996.

[432] A. Rodriguez. Rawls' maximin criterion and time consistency: A generali-
       sation. *Review of Economic Studies*, 48:599–605, 1981.

[433] J. Roemer. *Free to Lose*. Radius, London, 1988.

[434] J. Roemer. *Egalitarian Perspectives: Essays in Philosophical Economics.* Cambridge University Press, Cambridge, 1996.

[435] J. Roemer. *Theories of Distributive Justice.* Harvard University Press, Cambridge, MA, 1996.

[436] E. Rogers. The Mistassini Cree. In M. Bicchieri, editor, *Hunters and Gatherers Today.* Holt, Rinehart and Winston, New York, 1972.

[437] R. Rosenthal. Games of perfect information, predatory pricing, and chain-store paradox. *Journal of Economic Theory,* 25:92–100, 1981.

[438] A. Roth. *Axiomatic Models of Bargaining.* Springer-Verlag, Berlin, 1979.

[439] A. Roth. A note on risk aversion in a perfect equilibrium model of bargaining. *Econometrica,* 53:207–211, 1985.

[440] A. Roth and I. Erev. Learning in extensive-form games: Experimental data and simple dynamic models in the medium term. *Games and Economic Behavior,* 8:164–212, 1995.

[441] J.-J. Rousseau. *Emile.* J. M. Dent, London, 1908. (First published 1762.

[442] J.-J. Rousseau. A discourse on political economy. In G. Cole, editor, *Rousseau's Social Contract and Discourses,* pages 249–287. J. M. Dent, London, 1913. (First published 1755).

[443] J.-J. Rousseau. The inequality of man. In G. Cole, editor, *Rousseau's Social Contract and Discourses,* pages 157–246. J. M. Dent, London, 1913. (First published 1755).

[444] J.-J. Rousseau. The social contract. In G. Cole, editor, *Rousseau's Social Contract and Discourses,* pages 5–123. J. M. Dent, London, 1913. (First published 1762).

[445] A. Rubinstein. Perfect equilibrium in a bargaining model. *Econometrica,* 50:97–109, 1982.

[446] A. Rubinstein. Finite automata play the repeated prisoners' dilemma. *Journal of Economic Theory,* 39:83–96, 1986.

[447] A. Rubinstein. Similarity and decision-making under risk. *Journal of Economic Theory,* 46:145–153, 1988.

[448] A. Rubinstein. Comments on the interpretation of game theory. *Econometrica,* 59:909–924, 1991.

[449] A. Rubinstein, Z. Srafa, and W. Thomson. On the interpretation of the Nash bargaining solution. *Econometrica,* page ?, 1990.

[450] A. Rubinstein and A. Wolinsky. Equilibrium in a market with sequential bargaining. *Econometrica,* 53:1133–1150, 1985.

[451] A. Rubinstein and A. Wolinsky. Middlemen. *Quarterly Journal of Economics,* 102:581–593, 1987.

[452] A. Rubinstein and A. Wolinsky. Renegotiation-proof implementation and time preference. *American Economic Review,* 82:600–614, 1992.

[453] B. Russell. Mathematics and the metaphysicians. In *Mysticism and Logic*. Doubleday, New York, 1917.

[454] O. Sachs. *An Anthropologist on Mars*. Knopf, New York, 1994.

[455] M. Sahlins. *Stone Age Economics*. Tavistock, London, 1974.

[456] E. Sampson. On justice as equality. *Journal of Social Issues*, 31:54–64, 1975.

[457] L. Samuelson. Does evolution eliminate dominated strategies? In A. Kirman K. Binmore and P. Tani, editors, *The Frontiers of Game Theory*. MIT Press, Cambridge, MA, 1994.

[458] L. Samuelson. *Evolutionary Games and Equilibrium Selection*. MIT Press, Cambridge, MA, 1997.

[459] P. Samuelson. An exact consumption-loan model with or without the social contrivance of money. *Journal of Political Economy*, 66:467–482, 1958.

[460] M. Satterthwaite. Strategy-proofness and Arrow's condition: Existence and correspondence theorems for voting problems and social welfare functions. *Journal of Economic Theory*, 52:187–217, 1975.

[461] L. Savage. *The Foundations of Statistics*. Wiley, New York, 1951.

[462] T. Scanlon. Preferences and urgency. *Journal of Philosophy*, 73:655–670, 1975.

[463] D. Scharfstein and J. Stein. Behavior and investment. *Quarterly Journal of Economics*, 80:797–817, 1990.

[464] T. Schelling. *The Strategy of Conflict*. Harvard University Press, Cambridge, MA, 1960.

[465] K. Schlag. Dynamic stability in the repeated Prisoners' Dilemma played by finite automata. Technical Report SFB 303 Discussion Paper B-243, University of Bonn, 1993.

[466] D. Schmeidler. Walrasian analysis via strategic outcome functions. *Econometrica*, 48:1585–93, 1980.

[467] A. Schotter. *The Economic Theory of Social Institutions*. Cambridge University Press, Cambridge, 1981.

[468] H. Schroek. *Envy: A Theory of Human Behavior*. Liberty Press, Indianopolis, 1987.

[469] S. Schwartz. The justice of need and the activation of humanitarian norms. *Journal of Social Issues*, 31:11–136, 1975.

[470] W. Scott. *Francis Hutcheson*. Cambridge University Press, Cambridge, 1966. (First published 1900).

[471] R. Selten. Reexamination of the perfectness concept for equilibrium points in extensive-games. *International Journal of Game Theory*, 4:25–55, 1975.

[472] R. Selten. The chain-store paradox. *Theory and Decision*, 9:127–159, 1978.

[473] R. Selten. Evolutionary stability in extensive 2-person games. *Mathematical Social Sciences*, 5:269–363, 1983.

[474] R. Selten and U. Leopold. Subjunctive conditionals in decision theory and game theory. In Stegmuller/Balzer/Spohn, editor, *Studies in Economics*. Springer-Verlag, Berlin, 1982. *Philosophy of Economics*, Vol. 2.

[475] A. Sen. *Collective Choice and Social Welfare*. Holden Day, San Francisco, 1970.

[476] A. Sen. The impossibility of a Paretian liberal. *Journal of Political Economy*, 78:152–157, 1970.

[477] A. Sen. Welfare inequalities and Rawlsian axiomatics. *Theory and Decision*, 7:243–262, 1976.

[478] A. Sen. Rational fools: A critique of the behavioral foundations of economic theory. *Philosophy and Public Affairs*, 6:317–344, 1977.

[479] A. Sen. Equality of what? In S. McMurrin, editor, *Tanner Lectures on Human Values I*. University of Utah Press, Salt Lake City, 1980.

[480] A. Sen. *On Ethics and Economics*. Blackwell, Oxford, 1987.

[481] A. Sen. Utilitarianism and welfarism. *Journal of Philosophy*, 76:463–489, 1988.

[482] A. Sen. Minimal liberty. *Economica*, 59:139–160, 1992.

[483] A. Sen. Internal consistency of choice. *Econometrica*, 61:495–522, 1993.

[484] A. Sen and B. Williams. *Utilitarianism and Beyond*. Cambridge University Press, Cambridge, 1982.

[485] R. Sennet. *Authority*. Faber and Faber, Boston, 1980.

[486] R. Shackleton. The greatest happiness of the greatest number: The history of Bentham's phrase. *Studies on Voltaire and the Eighteenth Century*, 90:1461–1482, 1972.

[487] A. Shaked. Opting out: Bazaars versus hi-tech markets. *Investigaciones Economicas*, 18:420–432, 1994.

[488] A. Shaked and J. Sutton. Involuntary unemployment as a perfect equilibrium in a bargaining model. *Econometrica*, 52:1351–1364, 1984.

[489] L. Shapley. Utility comparison and the theory of games. In A. Roth, editor, *The Shapley Value: Essays in Honor of Lloyd S. Shapley*. Cambridge University Press, Cambridge, 1988.

[490] M. Shubik. Edgeworth market games. In A. Tucker and R. Luce, editors, *Contributions to the Theory of Games IV (Annals of Mathematics Studies 40)*. Princeton University Press, Princeton, 1959.

[491] M. Shubik. *Game Theory in the Social Sciences*. MIT Press, Cambridge, MA, 1982.

[492] H. Sidgwick. *The Methods of Ethics*. Hackett, Indianapolis, 1981. (Seventh edition 1907. First published 1874).

[493] K. Sigmund. *Games of Life: Explorations in Ecology, Evolution and Behaviour.* Penguin, Harmondsworth, UK, 1993.

[494] M. Singer. Moral skepticism. In C. Carter, editor, *Skepticism and Moral Principles.* New University Press, Evanston, IL, 1973.

[495] P. Singer. *The Expanding Circle: Ethics and Sociobiology.* Farrar, Strauss and Giroux, New York, 1980.

[496] P. Singer. *How Are We to Live: Ethics in an Age of Self-Interest.* Opus: Oxford University Press, Oxford, 1997.

[497] B. Skinner. *Walden II.* Macmillan, New York, 1948.

[498] B. Skinner. *Beyond Freedom and Dignity.* Knopf, New York, 1972.

[499] B. Skyrms. *Evolution of the Social Contract.* Cambridge University Press, Cambridge, 1996.

[500] A. Smith. *The Theory of Moral Sentiments.* Clarendon Press, Oxford, 1975. (Edited by D. Raphael and A. Macfie. First published 1759).

[501] A. Smith. *The Wealth of Nations.* Liberty Classics, Indianopolis, 1976. (First published 1776).

[502] E. Smith. Risk and uncertainty in the original "affluent society". In T. Ingold *et al*, editor, *Hunters and Gatherers.* Berg, London, 1988.

[503] F. Snare. *The Nature of Moral Thinking.* Routledge, London, 1992.

[504] B. de Spinoza. Tractatus politicus. In *Chief Works of Benedict de Spinoza,* Volume I. Bell, London, 1909. (Translated by R. Elwes. First published 1674).

[505] B. de Spinoza. Theologico-political treatise. In *Chief Works of Benedict de Spinoza,* Volume I. Dover, New York, 1951. (Translated by R. Elwes. First published 1670).

[506] B. de Spinoza. Ethics. In *Collected Works of Spinoza, Volume I.* Princeton University Press, Princeton, 1985. (Edited by E. Curley. First published 1677).

[507] I. Stahl. *Bargaining Theory.* Economics Research Institute, Stockholm, 1972.

[508] J. Stewart. *Opinion and Reform in Hume's Political Philosophy.* Princeton University Press, Princeton, 1992.

[509] G. Stigler. *The Economist as Preacher.* Blackwell, Oxford, 1982.

[510] G. J. Stigler. Economics or ethics? In S. McMurrin, editor, *Tanner Lectures on Human Values.* Cambridge University Press, Cambridge, 1981.

[511] R. Sugden. Liberty, preference and choice. *Economics and Philosophy,* 1:213–219, 1985.

[512] R. Sugden. *The Economics of Rights, Cooperation and Welfare.* Blackwell, Oxford, 1986.

[513] P. Suppes. Some formal models of grading principles. *Synthèse*, 6:284–306, 1966.

[514] P. Suppes. The transcendental character of determinism. In P. French *et al*, editor, *Midwest Studies in Philosophy*. University of Notre Dame Press, 1993.

[515] P. Suppes. Voluntary motion, biological computation and free will. In P. French *et al*, editor, *Midwest Studies in Philosophy*. University of Notre Dame Press, 1994.

[516] K. Suzumura. On the consistency of libertarian claims. *Review of Economic Studies*, 45:329–342, 1978.

[517] J. Tanaka. *The San Hunter-Gatherers of the Kalahari Desert: A Study of Ecological Anthropology*. Tokyo University Press, Tokyo, 1980.

[518] R. Teichgraeber. *Free Trade and Moral Philosophy: Rethinking the Sources of Adam Smith's* Wealth of Nations. Duke University Press, Durham, NC, 1986.

[519] R. Thaler. Anomalies: The Ultimatum Game. *Journal of Economic Perspectives*, 2:195–206, 1988.

[520] R. Thaler. *The Winner's Curse: Paradoxes and Anomalies of Economic Life*. Free Press, New York, 1991.

[521] W. Thomson. A study of choice correspondences in economics with a variable number of agents. *Journal of Economic Theory*, 46:237–254, 1988.

[522] J. Tooby. The emergence of evolutionary psychology. In D. Pines, editor, *Emerging Syntheses is Science*. Santa Fe Institute, Santa Fe, 1987.

[523] J. Tooby and L. Cosmides. Friendship, cooperation and the banker's paradox. In J. Maynard Smith, editor, *Evolution of Social Behavior Patterns in Primates and Man*. Proceedings of the British Academy, London, 1996.

[524] R. Trivers. The evolution of reciprocal altruism. *Quarterly Review of Biology*, 46:35–56, 1971.

[525] R. Trivers. *Social Evolution*. Benjamin Cummings, Menlo Park, CA, 1985.

[526] C. Turnbull. *Wayward Servants*. Eyre and Spottiswoode, London, 1965.

[527] C. Turnbull. *The Mountain People*. Touchstone, New York, 1972.

[528] E. van Damme. Renegotiation-proof equilibria in repeated Prisoners' Dilemma. *Journal of Economic Theory*, 46:206–217, 1989.

[529] E. van Damme, R. Selten, and E. Winter. Alternating bid bargaining with a smallest money unit. *Games and Economic Behavior*, 2:188–201, 1990.

[530] V. Vanberg and R. Congleton. Rationality, morality and exit. *American Political Science Review*, 86:418–431, 1992.

[531] H. Varian. Equity, envy and efficiency. *Journal of Economic Theory*, 9:63–91, 1974.

[532] H. Varian. Distributive justice, welfare economics and the theory of fairness. *Philosophy and Public Affairs*, 13:319–333, 1975.

[533] H. Varian. Two problems in the theory of fairness. *Journal of Public Economics*, 5:249–260, 1976.

[534] H. Varian. *Microeconomic Analysis (Third Edition)*. Norton, New York, 1992.

[535] T. Veblen. The theory of the leisure class. In M. Lerner, editor, *The Portable Veblen*. Viking Press, New York, 1958. (First published 1899).

[536] G. Vlastos. Justice and inequality. In J. Waldren, editor, *Theory of Rights*. Oxford University Press, Oxford, 1992.

[537] J. Von Neumann. Zur Theorie der Gesellschaftsspiele. *Mathematische Annalen*, 100:295–320, 1928.

[538] J. Von Neumann and O. Morgenstern. *The Theory of Games and Economic Behavior*. Princeton University Press, Princeton, 1944.

[539] F. de Waal. *Good Natured: The Origins of Right and Wrong in Humans and Other Animals*. Harvard University Press, Cambridge, MA, 1996.

[540] M. Wade. Soft selection, hard selection, kin selection and group selection. *American Naturalist*, 125:61–73, 1985.

[541] G. Wagstaff. Equity, equality and need: Three principles of justice or one? *Current Psychology: Research and Reviews*, 13:138–152, 1994.

[542] G. Wagstaff. Making sense of justice. Draft book, Psychology Department, University of Liverpool, 1997.

[543] G. Wagstaff, J. Huggins, and T. Perfect. Equal ratio equity, general linear equity and framing effects in judgments of allocation divisions. *European Journal of Social Psychology*, 26:29–41, 1996.

[544] G. Wagstaff and T. Perfect. On the definition of perfect equity and the prediction of inequity. *British Journal of Social Psychology*, 31:69–77, 1992.

[545] L. Walras. *Elements d'Economie Politique Pure*. Pichon et Durand-Auzias, Paris, 1926.

[546] E. Walster, E. Berscheid, and G. Walster. New directions in equity research. *Journal of Personality and Social Psychology*, 25:151–176, 1973.

[547] E. Walster and G. Walster. Equity and social justice. *Journal of Social Issues*, 31:21–43, 1975.

[548] E. Walster, G. Walster, and E. Berscheid. *Equity: Theory and Research*. Allyn and Bacon, London, 1978.

[549] E. Westermarck. *The Origin and Development of Moral Ideas*. Macmillan, London, 1906.

[550] J. Weymark. A reconsideration of the Harsanyi-Sen debate on utilitarianism. In J. Elster and J. Roemer, editors, *Interpersonal Comparisons of Well-Being*. Cambridge University Press, Cambridge, 1991.

[551] F. Whelan. *Order and Artifice in Hume's Political Philosophy.* Princeton University Press, Princeton, 1985.

[552] P. Wiesner. Risk, reciprocity and social influences on !Kung San economics. In E. Leacock and R. Lee, editors, *Politics and History in Band Cultures.* Cambridge University Press, Cambridge, 1982.

[553] G. Wilkinson. Reciprocal food-sharing in the vampire bat. *Nature*, 308:181–184, 1984.

[554] B. Williams. *Moral Luck.* Cambridge University Press, Cambridge, 1981.

[555] G. Williams. *Adaption and Natural Selection.* Princeton University Press, Princeton, 1974.

[556] D. Wilson. The group selection controversy: History and current status. *Annual Review of Ecology and Systematics*, 14:159–187, 1983.

[557] D. Wilson and E. Sober. Reviving the superorganism. *Journal of Theoretical Biology*, 136:337–356, 1988.

[558] E. Wilson. *The Insect Societies.* Harvard University Press, Cambridge, MA, 1971.

[559] E. Wilson. *Sociobiology: The New Synthesis.* MIT Press, Cambridge, MA, 1975.

[560] E. Wilson. *On Human Nature.* MIT Press, Cambridge, MA, 1978.

[561] J. Wilson. *The Moral Sense.* Free Press, New York, 1993.

[562] L. Wispé. The distinction between sympathy and empathy: To call forth a concept a word is needed. *Journal of Personality and Social Psychology*, 50:314–321, 1986.

[563] L. Wispé. *The Psychology of Sympathy.* Plenum, New York, 1991.

[564] L. Wittgenstein. *Tractatus Logico-Philosophicus.* Routledge and Kegan Paul, London, 1992. (Translated by C. Ogden. First published 1922).

[565] R. Wolff. A refutation of Rawls' theorem on justice. *Journal of Philosophy*, 63:170–190, 1966.

[566] R. Wolff. *Understanding Rawls.* Princeton University Press, Princeton, 1977.

[567] R. Wright. *The Moral Animal.* Random House, New York, 1994.

[568] J. Wu and R. Axelrod. How to cope with noise in the iterated Prisoner's Dilemma. *Journal of Conflict Resolution*, 39:183–189, 1995.

[569] V. Wynne-Edwards. *Animal Dispersion in Relation to Social Behavior.* Oliver and Boyd, Edinburgh, 1962.

[570] M. Yaari. Rawls, Edgeworth, Shapley, Nash: Theories of distributive justice re-examined. *Journal of Economic Theory*, 24:1–39, 1981.

[571] P. Young. The evolution of conventions. *Econometrica*, 61:57–84, 1993.

[572] P. Young. An evolutionary model of bargaining. *Journal of Economic Theory*, 59:145–168, 1993.

[573] P. Young. *Equity*. Princeton University Press, Princeton, 1994.

[574] C. Zahn-Waxler, B. Hollenbeck, and M. Rootke-Yarrow. The origins of empathy and altruism. In M. Fox and L. Mickley, editors, *Advances in Animal Welfare Sciences*. Humane Society of the United States, Washington, DC, 1984.

[575] C. Zeeman. Dynamics of the evolution of animal conflicts. *Journal of Theoretical Biology*, 89:249–270, 1981.

[576] F. Zeuthen. *Problems of Monopoly and Economic Warfare*. Routledge and Kegan Paul, London, 1930.

Index

DATE DUE

12-1-14		
MAY 03 2016		